Universitext

Springer
Berlin
Heidelberg
New York
Barcelona
Budapest
Hong Kong
London
Milan
Paris
Santa Clara
Singapore
Tokyo

Vladimir Boltyanski
Horst Martini
Petru S. Soltan

Excursions into Combinatorial Geometry

With 264 Figures

Springer

Vladimir Boltyanski
Steklov Mathematical Institute
Vavilov Street 42
117966 Moscow, Russia

Horst Martini
Faculty of Mathematics
TU Chemnitz-Zwickau
PSF 964
09009 Chemnitz, Germany

Petru S. Soltan
Faculty of Mathematics
Moldavian State University
Street Mateevici 60
Kishinev, Moldova

The cover picture shows a Borsuk partition of the known polyhedral
cover constructed by Branko Grünbaum

Cataloging-in-Publication applied for

Die Deutsche Bibliothek - CIP-Einheitsaufnahme

Boltjanskij, Vladimir G.:
Excursion into combinatorial geometry / Vladimir Boltyanski ;
Horst Martini : Petru S. Soltan. - Berlin ; Heidelberg ; New
York ; Barcelona ; Budapest ; Hong Kong ; London ; Milan ;
Paris ; Santa Clara ; Singapore ; Tokyo : Springer, 1997
 (Universitext)
 ISBN 3-540-61341-2
NE: Martini, Horst:; Soltan, Petru S.:

Mathematics Subject Classification (1991) 52-01, 52-02, 52A30, 52A35, 52A37, 52C17

ISBN 3-540-61341-2 Springer-Verlag Berlin Heidelberg New York

© Springer-Verlag Berlin Heidelberg 1997
Printed in Germany

Typesetting: Editing and reformatting of the authors' input files by Springer-Verlag
SPIN: 10521765   41/3143-543210 – Printed on acid-free paper.

To our wives

Lilia, Angelika and Luba

with love and thanks

Preface

Geometry undoubtedly plays a central role in modern mathematics. And it is not only a physiological fact that 80 % of the information obtained by a human is absorbed through the eyes. It is easier to grasp mathematical concepts and ideas visually than merely to read written symbols and formulae. Without a clear geometric perception of an analytical mathematical problem our intuitive understanding is restricted, while a geometric interpretation points us towards ways of investigation.

Minkowski's convexity theory (including support functions, mixed volumes, finite-dimensional normed spaces etc.) was considered by several mathematicians to be an excellent and elegant, but useless mathematical device. Nearly a century later, geometric convexity became one of the major tools of modern applied mathematics. Researchers in functional analysis, mathematical economics, optimization, game theory and many other branches of our field try to gain a clear geometric idea, before they start to work with formulae, integrals, inequalities and so on. For examples in this direction, we refer to [Mal] and [B-M 2].

Combinatorial geometry emerged this century. Its major lines of investigation, results and methods were developed in the last decades, based on seminal contributions by O. Helly, K. Borsuk, P. Erdös, H. Hadwiger, L. Fejes Tóth, V. Klee, B. Grünbaum and many other excellent mathematicians. Many of the open questions in this field can be explained in simple terms, even to non-mathematicians. But the solutions to these questions are often so complicated that every problem resolved is a major event.

It is impossible to give a complete list of references to this area, but the following books and surveys cover the major developments in combinatorial geometry: [D-G-K], [Grü 3,7,8], [H-D-K], [B-G 1,2], [B-SP 3], [Ec], [Rea], [Si], [SV 4], [B-So], chapters D and E in [C-F-G], [Be 3], [E], [Schm], [G-G-L], and [P-A]. We view the present book as a natural continuation of [B-G 1] and [B-SP 3]. The wider field of geometric convexity is covered for the most part in [Bl 2], [Eg 2,3], [Kl 2], [Ha 5], [Grü 6], [T-W], [G-W 1,2], [Schn], [B-M-S-W], [Zi], [Gar], [Tho] and several other publications.

Our book contains a collection of characteristic problems, tied together by several concepts: d-convexity, H-convexity, generalizations of the classical Helly theorem and its relatives, the Helly dimension of convex bodies, different covering and illumination problems etc. In addition, we introduce a new class of convex bodies, namely the family of belt bodies. The notion of a "belt body" is a natural generalization of the notion "zonoid". The belt bodies are dense in the set of all compact, convex bodies, whereas the zonoids are not even dense in the set of centrally symmetric, compact, convex bodies. It is shown that some difficult questions from combinatorial geometry can be completely resolved for belt bodies.

Our book can be used for different purposes. First, it is suitable for advanced undergraduate and graduate classes. With this audience in mind, we have amply illustrated the material with figures and provided numerous exercises. We hope readers will find this book stimulating to their geometric insight. Second, the book may serve as a reference to the current state of the art in some fields belonging to combinatorial geometry. Finally, specialists in these fields will find a collection of open problems in the last chapter.

A diagram at the beginning of the book illustrates the interdependence of the sections, and in addition some special "excursions" are described by section numbers. Note furthermore that some sections contain passages of survey character, that may be of less relevance, in first reading, to students. This applies in particular to sections 31 and 34.

The authors are grateful to I. Gohberg, B. Grünbaum, P. M. Gruber, J. Kincses, V. Klee, E. Makai Jr., and B. Weissbach for many stimulating discussions. They are also indebted to Mrs. Diana Lange for typing the manuscript in LATEX.

In addition, the authors are indebted to "Deutsche Forschungsgemeinschaft" for having supported longer visits of the first named author and the third named author at the Technical University Chemnitz-Zwickau.

V. Boltyanski, H. Martini, and P.S. Soltan

Table of Contents

I.   **Convexity**... 1
     §1   Convex sets 1
     §2   Faces and supporting hyperplanes....................... 6
     §3   Polarity 12
     §4   Direct sum decompositions.............................. 15
     §5   The lower semicontinuity of the operator "exp" 20
     §6   Convex cones 27
     §7   The Farkas Lemma and its generalization 36
     §8   Separable systems of convex cones 41

II. **d-Convexity in normed spaces**.............................. 49
     §9   The definition of d-convex sets 49
     §10 Support properties of d-convex sets 58
     §11 Properties of d-convex flats 65
     §12 The join of normed spaces 74
     §13 Separability of d-convex sets 81
     §14 The Helly dimension of a set family 91
     §15 d-Star-shaped sets 99

III. **H-convexity** 109
     §16 The functional md for vector systems 109
     §17 The ε-displacement Theorem 115
     §18 Lower semicontinuity of the functional md 121
     §19 The definition of H-convex sets....................... 125
     §20 Upper semicontinuity of the H-convex hull............. 127
     §21 Supporting cones of H-convex bodies................... 135
     §22 The Helly Theorem for H-convex sets 143
     §23 Some applications of H-convexity 149
     §24 Some remarks on connection between d-convexity
          and H-convexity....................................... 154

IV. **The Szökefalvi-Nagy Problem**............................. 163
     §25 The Theorem of Szökefalvi-Nagy and its generalization 163

§26 Description of vector systems with md $H = 2$
that are not one-sided 173
§27 The 2-systems without particular vectors 177
§28 The 2-system with particular vectors 184
§29 The compact, convex bodies with md $M = 2$ 188
§30 Centrally symmetric bodies 198

V.   **Borsuk's partition problem** 209
§31 Formulation of the problem and a survey of results 209
§32 Bodies of constant width in Euclidean and normed spaces .. 227
§33 Borsuk's problem in normed spaces 239

VI.   **Homothetic covering and illumination** 255
§34 The main problem and a survey of results 255
§35 The hypothesis of Gohberg-Markus-Hadwiger 275
§36 The infinite values of the functionals b, b', c, c' 288
§37 Inner illumination of convex bodies 301
§38 Estimates for the value of the functional $p(K)$ 309

VII.   **Combinatorial geometry of belt bodies** 319
§39 The integral respresentation of zonoids 319
§40 Belt vectors of a compact, convex body 327
§41 Definition of belt bodies 333
§42 Solution of the illumination problem for belt bodies 339
§43 Solution of the Szökefalvi-Nagy problem for belt bodies.... 346
§44 Minimal fixing systems 352

VIII. **Some research problems** 365

Bibliography 393

Author Index 411

Subject Index 415

List of Symbols 419

Dependence of sections

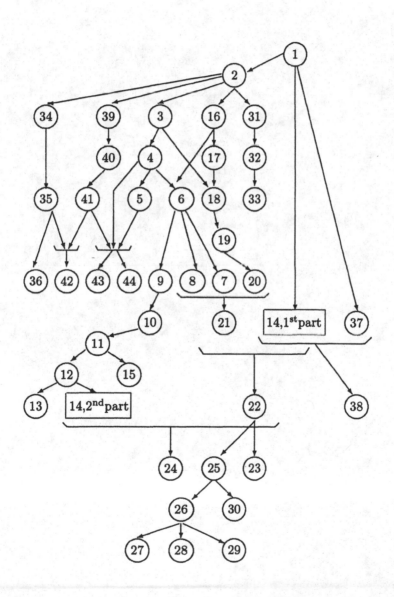

Some excursions

d-convexity:

   Sections   1, 2, 3, 4, 6, 9, 10, 11, 12, 13, 14, 15
                    (and the section 24 after studying *H*-convexity).

H-convexity:

   Sections   1, 2, 3, 4, 5, 6, 7, 8, 14 (1st part), 16, 17, 18, 19, 20, 21, 22, 23
                    (and the section 24 after studying *d*-convexity).

Szökefalvi-Nagy problem:

   Sections   1, 2, 3, 4, 5, 6, 7, 8, 14 (1st part), 16, 17, 18, 19, 20, 21, 22, 25,
                    26, 27, 28, 29, 30 (and, in addition, sections 39, 40, 41, 43).

Borsuk problem:

   Sections   1, 2, 31, 32, 33.

Gohberg-Markus-Hadwiger problem:

   Sections   1, 2, 34, 35, 36
                    (and, in addition, sections 39, 40, 41, 42).

Inner illumination problem:

   Sections   1, 2, 14 (1st part), 37, 38.

Minimal fixing systems:

   Sections   1, 2, 3, 4, 16, 39, 40, 41, 44.

I. Convexity

§1 Convex sets

In this monograph, we will be concerned with n-dimensional vector space $\mathbf{R}^n$. Elements of $\mathbf{R}^n$ will be called either *points* (when considerations have a geometrical character) or *vectors* (when algebraic operations are applied). If necessary, the scalar product $\langle \cdot, \cdot \rangle$ of Euclidean geometry is used.

For each pair of distinct points $a, b \in \mathbf{R}^n$ we denote by $[a, b]$ the *segment* with the endpoints a and b, i.e., the set of all points $\alpha a + \beta b$ with $\alpha \geq 0, \beta \geq 0, \alpha + \beta = 1$.

If $L \subset \mathbf{R}^n$ is an r-dimensional subspace and $a \in \mathbf{R}^n$, then $a + L$ (i.e., the set of all points $a + x$ with $x \in L$) is named a *flat* of dimension r or, for brevity, an *r-flat*. A flat of dimension $n - 1$ is called a *hyperplane*, a one-dimensional flat is a *line*. For every two points $a, b \in \mathbf{R}^n$ the r-flats $a + L$ and $b + L$ are said to be *parallel*.

Let M be an arbitrary point set in $\mathbf{R}^n$. The set M is said to be *convex* if for arbitrary points $x, y \in M$ the segment $[x, y]$ is completely contained in M. (If other types of convexity are introduced, we will speak about "linear convexity".) For example, each k-flat (of dimension $k \leq n$) is a convex set of $\mathbf{R}^n$. Another example is given by an arbitrary *closed half-space*, i.e. $\{x : \langle a, x \rangle \leq \lambda\}$ with $a \in \mathbf{R}^n$ different from the nullvector o and $\lambda \in \mathbf{R}$ (i.e., a real number). *Open half-spaces* $\{x : \langle a, x \rangle < \lambda\}$ and closed (open) *balls* with center $x_o \in \mathbf{R}^n$ and radius $r > 0$ are convex sets, too. Note, furthermore, that for any convex set $M \subset \mathbf{R}^n$ its ε-neighbourhood

$$U_\varepsilon(M) = \{x \in \mathbf{R}^n : d(x, M) < \varepsilon\}$$

is also a convex set. Here $\varepsilon > 0$, and by $d(x, M)$ the distance from x to M is denoted, i.e., $d(x, M) = \inf_{a \in M} \| x - a \|$.

Theorem 1.1: The intersection of an arbitrary family of convex sets is a (possibly empty) convex set. □

Since each closed half-space is a convex set, this theorem implies that the intersection of an arbitrary family of closed half-spaces is a closed, convex set. Vice versa, every closed, convex set $M \subset \mathbf{R}^n$ can be represented as the intersection of a family of closed half-spaces.

The intersection of finitely many closed half-spaces is called a *convex polyhedral set*. A set $M \subset \mathbf{R}^n$ is *bounded* if M is contained in a ball. In particular, one can refer to bounded, convex, polyhedral sets. They are called *convex polytopes*. A simple example of a convex polytope in $\mathbf{R}^n$ is an n-dimensional *parallelotope*, i.e., a set described in a system of Cartesian coordinates $x_1, x_2, \cdots, x_n$ by the inequalities $a_i \leq x_i \leq b_i; i = 1, 2, \cdots, n$, where $a_i < b_i$ for all i. Each of the inequalities $a_i \leq x_i, x_i \leq b_i$ gives a closed half-space, and therefore any n-dimensional parallelotope is the intersection of $2n$ half-spaces.

The smallest convex set containing $F \subset \mathbf{R}^n$, i.e., the intersection of all convex sets each of which contains F, is called the *convex hull* of F and denoted by conv F.

Theorem 1.2: A point $c \in \mathbf{R}^n$ belongs to the convex hull of a set $F \subset \mathbf{R}^n$ if and only if there exist (not necessarily distinct) points $a_o, a_1, \cdots, a_n \in F$ and nonnegative real numbers $\lambda_o, \lambda_1, \cdots, \lambda_n$, satisfying

$$\lambda_o + \lambda_1 + \cdots + \lambda_n = 1; \quad c = \lambda_o a_o + \lambda_1 a_1 + \cdots + \lambda_n a_n. \quad \square$$

For example, if arbitrary points $a_o, a_1, \cdots, a_k \in \mathbf{R}^n$ with $k \in \{1, \cdots, n\}$ are not contained in a $(k-1)$-plane, then conv $\{a_o, a_1, \cdots, a_k\}$ is said to be a k-dimensional *simplex*. By definition, such a simplex is a convex set. A one-dimensional simplex is a segment and a two-dimensional simplex is a nondegenerate triangle. By Theorem 1.2, the point $c \in \mathbf{R}^n$ belongs to the simplex conv $\{a_o, a_1, \cdots, a_k\}$ if and only if $c = \lambda_o a_o + \lambda_1 a_1 + \cdots + \lambda_k a_k$ with $\sum \lambda_i = 1$ and $\lambda_i \geq 0, i = 0, 1, \cdots, k$.

The set of all points given by $x = (1 - \lambda)a + \lambda b, 0 < \lambda < 1$, i.e., the *open interval* with endpoints $a \neq b$, is denoted by $]a, b[$. The *half-open interval* obtained by adding a to $]a, b[$ is denoted by $[a, b[$. Furthermore, $[a, b)$ denotes the *ray* with starting point a through $b \neq a$, i.e., the set of all points given by $x = (1 - \lambda)a + \lambda b, \lambda \geq 0$. The *line* (i.e., the one-dimensional flat) through the points $a \neq b$ is denoted by (a, b).

A point $x \in \mathbf{R}^n$ is said to be *interior point* of a set $M \subset \mathbf{R}^n$ if there is an $\varepsilon > 0$ such that the ball of radius ε, centered at x, is contained in M. The set of all interior points of the set M is named the *interior* of M and denoted by int M. A set $G \subset \mathbf{R}^n$ is *open* if it coincides with its interior. A point $x_o \in \mathbf{R}^n$ is said to be a *limit point* of the set $M \subset \mathbf{R}^n$ if each ball centered at x_o contains at least one point (and consequently infinitely many points) of M distinct from x_o.

The union of M and the set of all its limit points is called the *closure* of M and denoted by cl M. A set $F \in \mathbf{R}^n$ is *closed* if it coincides with its closure. In other words, a set $F \in \mathbf{R}^n$ is closed if and only if its complement $\mathbf{R}^n \setminus F$

is open. The difference cl $M \setminus$ int M between the closure and the interior of the set M is said to be the *boundary* of M and denoted by bd M. Each point $a \in$ bd M is named a *boundary point* of M. A bounded, closed set is said to be *compact*.

A convex set $M \subset \mathbf{R}^n$ is called a *convex body* if it is closed and has interior points (i.e., int M is a non-empty set). For instance, every n-dimensional simplex is a convex body in $\mathbf{R}^n$.

Theorem 1.3: Each convex set $M \subset \mathbf{R}^n$ with empty interior is contained in a k-dimensional flat, $k < n$. $\square$

The flat of the smallest dimension containing a convex set $M \subset \mathbf{R}^n$ is said to be the *carrying flat* or *affine hull* of M and denoted by aff M. The set M has a nonempty interior with respect to aff M.

A convex set is k-dimensional if its carrying flat is k-dimensional. The interior of M with respect to aff M is called the *relative interior* of M and denoted by ri M.

Theorem 1.4: If a set $M \subset \mathbf{R}^n$ is convex, then the sets cl M and ri M are also convex. $\square$

The set cl $M \setminus$ ri M, i.e., the set of boundary points of the convex set M with respect to its carrying flat aff M, is denoted by rbd M and named the *relative boundary* of M. For dim $M = n$ (i.e., M has a nonempty interior), the set rbd $M =$ cl $M \setminus$ ri $M =$ cl $M \setminus$ int M coincides with the boundary of $M \subset \mathbf{R}^n$. If dim $M < n$, then bd M and cl M coincide, since int $M = \emptyset$.

Theorem 1.5: Let $M \subset \mathbf{R}^n$ be an arbitrary convex set. For $a \in$ ri M and $b \in$ cl M, all the points of $[a, b[$ are contained in ri M. $\square$

In particular, this theorem implies that if $a, b \in$ rbd M, then the open interval $]a, b[$ is completely contained either in ri M or in rbd M. Furthermore, if $a \in$ ri M and $l \subset$ aff M is a ray with starting point a, then either l is completely contained in M or the intersection $l \cap$ rbd M is a point b with $l \cap$ ri $M = [a, b[$.

Theorem 1.6: If convex sets $M_1, M_2, \cdots, M_s$ satisfy ri $M_1 \cap$ ri $M_2 \cap \cdots \cap$ ri $M_s \neq \emptyset$, then the equalities

$$\text{ri } (M_1 \cap M_2 \cap \cdots \cap M_s) = \text{ri } M_1 \cap \text{ri } M_2 \cap \cdots \cap \text{ri } M_s,$$
$$\text{cl } (M_1 \cap M_2 \cap \cdots \cap M_s) = \text{cl } M_1 \cap \text{cl } M_2 \cap \cdots \cap \text{cl } M_s,$$
$$\text{aff} (M_1 \cap M_2 \cap \cdots \cap M_s) = \text{aff } M_1 \cap \text{aff } M_2 \cap \cdots \cap \text{aff } M_s$$

hold. $\square$

Let $M \subset \mathbf{R}^n$ be an unbounded, closed, convex set. A hyperplane L is said to be an *asymptotic hyperplane* of M [SP 2] if the following conditions are satisfied: *(i)* $d(M, L) = 0$ (i.e., for every $\varepsilon > 0$ there are points $x \in L$, $y \in M$ such that the Euclidean distance $||x - y|| = \sqrt{\langle x - y, x - y \rangle}$ is less than ε), *(ii)* M is contained in one of the closed half-spaces with respect to L, and

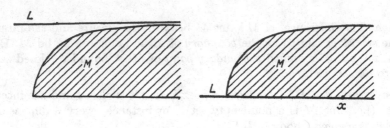

Fig. 1   Fig. 2

(iii) every hyperplane parallel to L has either the empty set or an unbounded set as its intersection with M. An asymptotic hyperplane L of M either has no common point with M (Fig. 1) or has with M an unbounded intersection (Fig. 2). For each asymptotic hyperplane L of M, we denote by Π_L the closed half-space with respect to L which contains M. The set $\tilde{M} = \bigcap_L \Pi_L$
(taking the intersection over all asymptotic hyperplanes of M) is said to be the *asymptotic scope* of M [SP 2]. (If an unbounded convex set $M \subset \mathbf{R}^n$ has no asymptotic hyperplane, then we set $\tilde{M} = \mathbf{R}^n$.) Obviously, $M \subset \tilde{M}$. We remark that the inclusion $M_1 \subset M_2$ does not imply $\tilde{M}_1 \subset \tilde{M}_2$. For instance, in the space $\mathbf{R}^2$ with coordinates x_1, x_2 one can consider the set M_1 defined by the inequality $x_2 \geq x_1^2$ and the set M_2 defined by the inequality $x_2 \geq 0$ (Fig. 3). Then $M_1 \subset M_2$, but $\tilde{M}_1 = \mathbf{R}^2$ and $\tilde{M}_2 = M_2 \neq \mathbf{R}^2$.

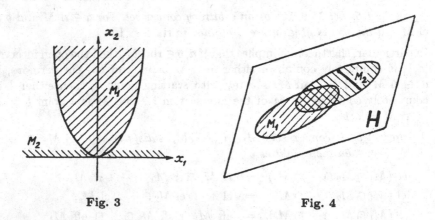

Fig. 3   Fig. 4

We say that convex sets $M_1, M_2 \subset \mathbf{R}^n$ are *separable* in $\mathbf{R}^n$ if there is a hyperplane $H \subset \mathbf{R}^n$ such that M_1 lies in one of the closed half-spaces regarding H and M_2 lies in the other one. Here H is called a *separating hyperplane*. In other words, $M_1, M_2 \subset \mathbf{R}^n$ are separable if there exists a vector $b \neq o$ and a number $\lambda \in \mathbf{R}$ such that $M_1 \subset \{x : \langle b, x \rangle \geq \lambda\}, M_2 \subset \{x : \langle b, x \rangle \leq \lambda\}$. If M_1, M_2 are separable and there is a separating hyperplane H such that at least one of the sets $M_1 \setminus H, M_2 \setminus H$ is nonempty (i.e., at least one of the sets M_1, M_2 is not contained in H), then M_1, M_2 are *strictly separated* by H. For example, if H is a hyperplane in $\mathbf{R}^n$ and M_1, M_2 are convex sets contained in H and satisfying ri $M_1 \cap$ ri $M_2 \neq \emptyset$ (Fig. 4), then M_1, M_2 are separable in $\mathbf{R}^n$, but not strictly separable.

Furthermore, convex sets $M_1, M_2 \subset \mathbf{R}^n$ are *strongly separable* if they have a separating hyperplane H for which the intersections $H \cap M_1, H \cap M_2$ are empty, i.e., M_1, M_2 are contained in the corresponding open half-spaces.

Theorem 1.7: If M_1, M_2 are convex sets in $\mathbf{R}^n$ with ri $M_1 \cap$ ri $M_2 = \emptyset$, then M_1, M_2 are strictly separable. $\square$

Theorem 1.8: If $M_1, M_2 \subset \mathbf{R}^n$ are closed, convex sets with $M_1 \cap M_2 = \emptyset$ and at least one of them is compact, then M_1, M_2 are strongly separable. $\square$

Exercises

1. Let a line l pass through an interior point of a convex set M. Prove that $l \cap \mathrm{bd}\, M$ consists of no more than two points.

2. Let $M \subset \mathbf{R}^n$ be a convex set distinct from a point. Prove that aff M is the union of all lines each of which pass through two points of M.

3. Let I denote the operation by which all segments with endpoints in M are added to a set M:

$$I(M) = \bigcup_{x,y \in M} [x,y].$$

Prove that for every (nonconvex) set $M \subset \mathbf{R}^n$ the n-fold application of the operation I gives the convex hull of M:

$$I^n(M) = \mathrm{conv}\, M.$$

Moreover, for obtaining conv M from $M \subset \mathbf{R}^n$ even the k-fold application is sufficient, where k denotes the smallest integer satisfying $2^k \geq n + 1$.

4. Let a k-dimensional flat L be contained in a closed, convex set $M \subset \mathbf{R}^n$. Prove that for each point $a \in M$ the translate of L through a is contained in M.

5. Let $M_1, M_2 \subset \mathbf{R}^n$ be convex sets. Prove that their vector sum $M_1 + M_2$ (i.e., the set of all points $x_1 + x_2$ with $x_1 \in M_1, x_2 \in M_2$) is a convex set.

6. Prove that if M_1, M_2 are non-parallel segments, then $M_1 + M_2$ is a parallelogram. What is the vector sum ot two parallel segments?

7. Prove that if $M_1, \cdots, M_s$ are nonparallel segments in the plane $\mathbf{R}^2$, then $M_1 + \cdots + M_s$ is a centrally symmetric, convex polygon. Prove that every centrally symmetric, convex polygon can be represented in such a form.

8. Prove that if M_1', M_2' are translates of convex sets M_1, M_2, then $M_1' + M_2'$ is a translate of $M_1 + M_2$.

9. Let $M_1, M_2 \subset \mathbf{R}^2$ be convex polygons and p_1, p_2 be their perimeters (i.e., the lengths of the broken lines bd M_1 and bd M_2), respectively. Prove that $M_1 + M_2$ is a convex polygon whose perimeter is equal to $p_1 + p_2$. Generalize this for any convex sets $M_1, M_2 \subset \mathbf{R}^2$.

10. Let a be an interior point of a convex body M and L be a flat through a. Prove that

$$\mathrm{ri}\ (M \cap L) = L \cap \mathrm{int}\ M.$$

11. A convex body $M \subset \mathbf{R}^n$ is said to be a *cylinder with k-dimensional generator* if it contains a k-flat but not any $(k+1)$-flat. Prove that each cylinder with k-dimensional generator can be represented in the form $N + L$, where L is a k-flat and N is a convex set not containing any line.

12. Let $M \subset \mathbf{R}^n$ be an unbounded, convex body. Prove that its asymptotic scope contains a line (i.e., is a cylinder).

13. Prove that the strict separability of convex sets $M_1, M_2 \subset \mathbf{R}^n$ implies ri $M_1 \cap$ ri $M_2 = \emptyset$. (This is the converse of Theorem 1.7.)

14. Let unbounded convex bodies M_1, M_2 with $M_1 \cap M_2 = \emptyset$ be separated by a hyperplane L. Prove that if the distance

$$d(M_1, M_2) = \inf_{x \in M_1, y \in M_2} \|x - y\|$$

is equal to 0, then L is an asymptotic hyperplane of M_1 as well as of M_2.

15. Is the condition "both the sets M_1, M_2 are closed" in Theorem 1.8 essential? Is the condition "at least one of the sets is compact" essential?

§2 Faces and supporting hyperplanes

Let $M \subset \mathbf{R}^n$ be a closed, convex set. For $x \in M$, denote by F_x the set containing x and the points y for which the line (x, y) contains an open interval $I =]a, b[$ with $x \in I \subset M$. Then F_x is said to be the *face* of the point $x \in M$.

For $x \in$ ri M the sets F_x and M coincide; in this case $M = F_x$ is the *improper face*. On the other hand, for $x \in$ rbd M also F_x is contained in rbd M; then we have a *proper face*.

For proper faces, there are two extreme cases which play an important role: if F_x does not contain any other face, then F_x is called a *minimal face*. Furthermore, a proper face is said to be *maximal*, if it is not contained in any other proper face. These two cases do not exclude each other, i.e., a proper face can be maximal and minimal simultaneously. For example, let $M \subset \mathbf{R}^3$ be a circular cylinder described in a coordinate system (x_1, x_2, x_3) by the inequality $x_1^2 + x_2^2 \leq 1$. The proper faces, simultaneously maximal and minimal, are the generators of M.

If the face F_x of $x \in$ rbd M does not contain any points, except for x, it is obviously minimal; a point x with this property is named an *extreme point* of the convex set M. In other words, a point x of a closed, convex set M is an extreme point of M if and only if there exists no segment $I \subset M$ with $x \in$ ri I. The set of all extreme points of the closed, convex set M is denoted by ext M.

If all maximal faces of a convex body $M \subset \mathbf{R}^n$ are points, M is *strictly convex*. In other words, a convex body $M \subset \mathbf{R}^n$ is strictly convex if and only if for each point $x \in$ rbd M the equality $F_x = \{x\}$ holds, i.e., all boundary points of M are extreme. A simple example of a strictly convex body is a ball in Euclidean n-space.

Theorem 2.1: The face F_x of $x \in M$ is the largest convex set $Q \subset M$ satisfying $x \in$ ri Q. If the intersection of two faces of M is nonempty, then it is also a face. The conditions $F_x = F_y$ and $y \in$ ri F_x are equivalent. On the other hand, for $y \in$ rbd F_x the set F_y is the face of y in F_x. $\square$

In particular, this theorem yields the following: If F is a face of a convex set M and F' is a face of the convex set F, then also F' is a face of M. Since the transition from a convex set M to one of its proper faces decreases the dimension for at least 1, the following statement is obvious: *Every closed convex set $M \subset \mathbf{R}^n$ with nonempty boundary (i.e., M is not a flat) has at least one minimal face.* This also implies that a face F of a convex set M is minimal if and only if F is a flat (possibly 0-dimensional), i.e., $F =$ aff F.

Applying these notions to convex polyhedral sets, we get the following assertions. The relative boundary rbd M of a k-dimensional convex polyhedral set M can be represented as the union of finitely many $(k-1)$-dimensional convex polyhedral sets, which have pairwise distinct carrying flats. These $(k-1)$-dimensional sets are said to be the *facets* of M. The facets of M (and only the facets) are the maximal faces of M. An arbitrary proper face of M is the intersection of several facets of M. Hence every convex polyhedral set M has a finite number of faces. The extreme points of M (i.e., the 0-dimensional faces of M) are called the *vertices* of M. We remark that there are convex polyhedral sets without vertices. An example is given by the set $M \subset \mathbf{R}^3$ defined in a coordinate system (x_1, x_2, x_3) by the inequalities $x_1 \geq 0, x_2 \geq 0$, and $x_1 + x_2 \geq 1$. This set has two minimal faces (both are one-dimensional), but no vertices (Fig. 5).

Theorem 2.2: An arbitrary compact, convex set M coincides with the convex hull of its extreme points, i.e., $M =$ conv ext M. $\square$

We show by an example that, in general, for a compact, convex body M, the set ext M is not closed.

EXAMPLE 2.3: Let (x_1, x_1, x_3) be an orthogonal cartesian coordinate system in $\mathbf{R}^3$. Denote by I the segment with the endpoints $(0,0,1)$, $(0,0,-1)$ and by K the circle in the (x_1, x_2)-plane centered at $(1,0,0)$ with radius 1:

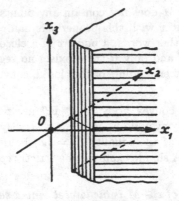

Fig. 5

$$K = \{(x_1, x_2, x_3) : (x_1 - 1)^2 + x_2^2 \leq 1, x_3 = 0\}.$$

Then $M = \operatorname{conv}(I \cup K)$ is a compact, convex body in $\mathbf{R}^3$ (Fig. 6). The points from rbd K, except for the origin, are extreme points of M. By $o \in \operatorname{ri} I$, the origin is not an extreme point of M. Thus the set ext M is not closed. □

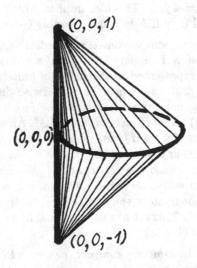

Fig. 6

In particular, Theorem 2.2 says that any convex polytope is the convex hull of its vertices. We note that for convex polytopes a statement holds which, in a sense, is a converse to Theorem 2.2: a set M is a convex polytope if and only if it is the convex hull of finitely many points.

For example, let $a_o, a_1, \cdots, a_k$ be points not contained in any $(k-1)$-flat. Then the simplex conv $\{a_o, a_1, \cdots, a_k\}$ is a convex polytope (of dimension k). Its vertices are the points $a_o, a_1, \cdots, a_k$, and its r-dimensional faces are the simplices determined by subsets of $\{a_o, a_1, \cdots, a_k\}$ with the cardinality $r+1$, in each case. In particular, any k-dimensional simplex has $k+1$ faces of dimension $k-1$ (its facets or maximal faces), and each k of them have a nonempty intersection. On the other hand, there is no point belonging to all the facets.

A hyperplane $\Gamma \subset \mathbf{R}^n$ is said to be a *supporting hyperplane* of a convex set $M \subset \mathbf{R}^n$ at a point $x \in$ bd M if $x \in \Gamma$ and the set M is contained in one of the closed half-spaces with respect to Γ. We remark that, for unbounded convex sets, supporting hyperplanes can be asymptotic ones (Fig. 2). On the other hand, an asymptotic hyperplane is not necessarily a supporting one (Fig. 1) and vice versa (Fig. 7).

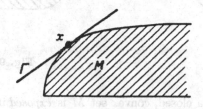

Fig. 7

Theorem 2.4: Let $M \subset \mathbf{R}^n$ be an arbitrary convex set. Each point $x \in$ bd M is contained in a supporting hyperplane of M. $\square$

If M is not a body (i.e., dim $M < n$), this is trivial, because a hyperplane containing aff M is a supporting hyperplane of M at each point $x \in M$. But for a body M this theorem describes a nontrivial, especially important fact. In addition we remark that for any convex set $M \subset \mathbf{R}^n$ and any hyperplane $\Gamma \subset \mathbf{R}^n$ there are no more than two supporting hyperplanes of M parallel to Γ. If M is a compact, convex body, there are exactly two such supporting hyperplanes.

Let $M \subset \mathbf{R}^n$ be a convex body and $x \in$ bd M. Furthermore, let Γ be a supporting hyperplane of M through x and Π be the closed half-space with bounding hyperplane Γ that contains M. This closed half-space is named a *supporting half-space* of M at x.

Theorem 2.5: Let $M \subset \mathbf{R}^n$ be a convex body. A supporting hyperplane Γ of M through $x \in$ bd M contains the face F_x of x. If the face F_x is even maximal, then $F_x = \Gamma \cap M$. $\square$

If Γ is a supporting hyperplane of a closed, convex set M, then the intersection $\Gamma \cap M$ is a face of M. Each face that can be represented in such a form is an *exposed face* of M. In other words, a face F of a closed, convex set M is exposed if and only if there is a supporting hyperplane Γ of M with $F = \Gamma \cap M$. Theorem 2.5 shows that any maximal face F of a closed, convex set M is exposed. The converse assertion is false. Indeed, the point x in Fig. 8 is extreme, i.e., $F_x = \{x\}$, and this face is exposed, since $F_x = \Gamma \cap M$. But this face is not maximal. We note that not each face is exposed. For example, in Fig. 9 the point x is extreme, i.e., $F_x = \{x\}$, but this face is not exposed. Indeed, the only supporting hyperplane Γ through x does not satisfy $\Gamma \cap M = F_x$.

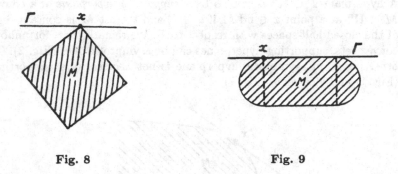

Fig. 8   Fig. 9

A boundary point x of a closed, convex set M is *exposed* if $F_x = \{x\}$ (i.e., this point is extreme) and, moreover, this face is exposed. In other words, a boundary point x is exposed if and only if there is a supporting hyperplane Γ of M with $\Gamma \cap M = \{x\}$. The set of all exposed points of a closed, convex body M is denoted by $\exp M$. Since each exposed point of M is extreme, the inclusion $\exp M \subset \operatorname{ext} M$ holds.

Theorem 2.6: Each extreme point of a compact, convex set M is the limit of a convergent sequence of exposed points, i.e., $\exp M \subset \operatorname{ext} M \subset \operatorname{cl} \exp M$. Consequently $M = \operatorname{conv} \operatorname{ext} M = \operatorname{conv} \operatorname{cl} \exp M = \operatorname{cl} \operatorname{conv} \exp M$.   □

Furthermore, Theorem 2.5 implies that an arbitrary supporting hyperplane of a strictly convex body M has only one point in common with M. The converse statement is also true: if each supporting hyperplane of a convex body M has only one point in common with M, then M is strictly convex. In other words, a convex body M is strictly convex if and only if $\operatorname{bd} M = \exp M$.

Let $M \subset \mathbf{R}^n$ be a convex body. A point $x \in \operatorname{bd} M$ is *regular* if there exists only one supporting hyperplane of M through x. Then there is only one supporting half-space of M at x. The *outward normal* of this half-space is said to be the outward normal of the body M at x.

A supporting hyperplane Γ of a convex body is said to be *regular* if it contains a regular boundary point. In other words, Γ is regular if and only if there exists a point $x \in \Gamma \cap M$ such that Γ is the only supporting hyperplane through x.

Theorem 2.7: The set P of all regular points $x \in$ bd M is dense in the boundary of the convex body M, i.e., cl $P =$ bd M.   □

Exercises

1. Let $M \subset \mathbf{R}^n$ be a closed, convex set. Prove that for every point $x \in M$ its face F_x is a closed, convex set.

2. Let $M \subset \mathbf{R}^n$ be a closed, convex set and F be a face of M. Prove that each minimal face of F is, at the same time, a minimal face of M.

3. Prove that a convex body is a cylinder if and only if it has no 0-dimensional face.

4. How many $(n-1)$-dimensional faces has each of the following n-dimensional polytopes?
   a) simplex; b) parallelotope; c) cross-polytope, i.e., the convex hull of the set

   $$\{e_1, \cdots, e_n, -e_1, \cdots, -e_n\},$$

   where $e_1, \cdots, e_n$ is a basis in $\mathbf{R}^n$.

5. Let $M \subset \mathbf{R}^n$ be an n-dimensional polytope, $x \in$ int M, and L be a k-flat through x. Prove that $L \cap M$ is a k-dimensional polytope.

6. Let $X \subset \mathbf{R}^n$ be a compact set which is not contained in a hyperplane and $M = \text{conv } X$ be its convex hull. Prove that no point $x \in (\text{bd } M) \setminus X$ is an extreme point of M. In other words, ext $M \subset (\text{bd } M) \cap X$.

7. Let $M \subset \mathbf{R}^n$ be a convex body and $x \in$ bd M. Denote by L the flat that is the intersection of all supporting hyperplanes of M through x. Prove that $L \cap M$ is a face of M. Is it possible that $L \cap M$ does not coincide with the face F_x of the point x?

8. Prove that for every face F of a convex, polyhedral set M there exists a supporting hyperplane Γ of M such that $F = \Gamma \cap M$.

9. Let $M \subset \mathbf{R}^n$ be a closed, convex body and F be one of its proper faces. Prove that $M \cap$ aff $F = F$. In other words, the flat $L =$ aff F satisfies the conditions dim $L =$ dim $F, L \cap$ int $M = \emptyset, F = M \cap L$.

10. Show that for every k, n, where $0 \le k < n$, there exists a convex body $M \subset \mathbf{R}^n$ and a k-dimensional face F of M such that there is no flat $L' \subset \mathbf{R}^n$ with dim $L' >$ dim $F, L' \cap$ int $M = \emptyset$, and $F = M \cap L'$.

11. Let $M_1, M_2 \subset \mathbf{R}^n$ be convex bodies and Π_1, Π_2 be their supporting half-spaces with the same outward normal p. Prove that $\Pi_1 + \Pi_2$ is the supporting half-space of the convex body $M_1 + M_2$ with the same outward normal p.

12. Prove that if $M_1, M_2 \subset \mathbf{R}^n$ are convex polytopes, then $M_1 + M_2$ is a convex polytope, too. Can you describe faces of $M_1 + M_2$?

13. Prove the equality conv $(M_1 + M_2) =$ conv $M_1 +$ conv M_2 for arbitrary $M_1, M_2 \subset \mathbf{R}^n$.

14. Let $M \subset \mathbf{R}^n$ be a convex body. Prove that the intersection of all supporting half-spaces of M coincides with M.

15. Let $M \subset \mathbf{R}^n$ be a convex body and $a \notin M$. Prove that there exists a regular boundary point x of M such that the supporting half-space of the body M at x does not contain a.

16. Let $M \subset \mathbf{R}^n$ be a convex body. Prove that the intersection of all supporting half-spaces of the body M at its regular boundary points coincides with M.

17. Prove that each proper face of a convex polyhedral set M is an exposed face of M.

18. Prove that a supporting hyperplane Γ of a convex polyhedral set M is regular if and only if Γ is the affine hull of a facet of M.

§3 Polarity

In this section, we denote by E the n-dimensional vector space: $E = \mathbf{R}^n$. A mapping $f : E \to \mathbf{R}$ is said to be a linear functional on E if

$$f(\alpha a + \beta b) = \alpha f(a) + \beta f(b)$$

for every $\alpha, \beta \in \mathbf{R}$ and $a, b \in E$. The set of all linear functionals defined on E is denoted by E^*. With the usual operations

$$(f + g)(a) = f(a) + g(a), \quad (\lambda f)(a) = \lambda f(a)$$

the set E^* becomes an n-dimensional vector space (consequently E^* is isomorphic to E). This vector space E^* is the *dual space* of E. For the sake of symmetry, we denote the value $f(x)$ as "scalar product" $\langle x, f \rangle$ of two vectors $x \in E, f \in E^*$.

We remark that under the convention $\langle x, f \rangle = x(f)$ the space E is identified with the dual space of E^*, i.e., $E^{**} = E$. In other words, each vector $x \in E$ can be considered as a linear functional on E^*, i.e., $x(f) = \langle x, f \rangle = f(x)$.

If $E = \mathbf{R}^n$ is a Euclidean space (i.e., a scalar product $\langle x, y \rangle$ is defined on E), then each vector $y \in E$ can be considered as a linear functional on E, i.e., $y(x) = \langle x, y \rangle$. In this case the dual space E^* *coincides* with E, i.e., E becomes a *selfadjoint space*.

The aim of this section is to give a "translation" of several notions referring to convex bodies to the language of the dual space. The obtained correspondence between the notions in E and in E^* is called the *polarity* (or *duality*).

Let $f \in E^*$ be a nonzero linear functional defined on E. By f^* we denote the set $\{x \in E : \langle x, f \rangle = 1\}$. The set f^* is a hyperplane in E not containing the point $o \in E$. Conversely, for every hyperplane $L \subset E$ not passing through

o, there exists a unique point $f \in E^*$ such that $f^* = L$. Thus we obtain a $1 - 1$-correspondence between the set of all hyperplanes in E, not passing through o, and the set of all points of E^* except for o.

On the other hand, let x be a point of E distinct from o. We denote by x^* the set $\{f \in E^* : \langle x, f \rangle = 1\}$. This set x^* is a hyperplane in E^* not containing the point $o \in E^*$. Again, we obtain a $1 - 1$-correspondence between the set of all points of E distinct from o and the set of all hyperplanes in E^* not passing through o. Moreover, for nonzero vectors $x \in E, f \in E^*$, the relations $x \in f^*, f \in x^*$ are equivalent, since each of them means that $\langle x, f \rangle = 1$.

Let now $M \subset E$ be a compact, convex body with the origin o in its interior. We say that a point $f \in E^*$ (i.e., a linear functional defined on E) is *compatible* with M if $\langle x, f \rangle \leq 1$ for every $x \in M$. The set M^* of all f compatible with M, i.e., the set

$$M^* = \{f \in E^* : \langle x, f \rangle \leq 1 \text{ for all } x \in M\},$$

is said to be the *polar body* of M (or, shortly, the *polar* of M). It is easily shown that M^* is a compact, convex body in the space E^*, also with the origin $o \in E^*$ in its interior. On the other hand, M is the set of all points $x \in E$ which are compatible with M^*, i.e.,

$$M = \{x \in E : \langle x, f \rangle \leq 1 \text{ for all } f \in M^*\}.$$

In other words, $M^{**} = M$. For this reason, we usually will formulate the polar versions only for notions and objects in the space E (for the space E^* they are analogous).

Theorem 3.1: For any nonzero $f \in E^*$, the hyperplane f^* is a supporting one for a body M if and only if f is a boundary point of M^*. □

PROOF. Admit that f is a boundary point of M^*. Then $f \in M^*$, but $\frac{1}{\lambda}f \notin M^*$ for any positive real number $\lambda < 1$, i.e., f is compatible with M, but $\frac{1}{\lambda}f$ is not compatible with M. In other words, $M \subset \{x : \langle f, x \rangle \leq 1\}$, but for any positive $\lambda < 1$ the body M is not contained in the half-space

$$\{x : \langle \frac{1}{\lambda}f, x \rangle \leq 1\} = \{x : \langle f, x \rangle \leq \lambda\}.$$

Hence $\{x : \langle f, x \rangle \leq 1\}$ is a supporting half-space of M, and consequently $\{x : \langle f, x \rangle = 1\} = f^*$ is a supporting hyperplane of M. Thus, if f is a boundary point of M^*, then f^* is a supporting hyperplane of M.

Conducting this reasoning in the opposite direction, we obtain the converse assertion: if f^* is a supporting hyperplane of M, then f is a boundary point of M^*. ∎

For the first time, we indicate the dual statement: For any nonzero $x \in E$ the hyperplane $x^* \subset E^*$ is a supporting one for a body M^* if and only if x is a boundary point of M.

Theorem 3.2: For any nonzero $f \in E^*$, the hyperplane f^* is a regular supporting hyperplane of M if and only if f is an exposed boundary point of M^*. □

PROOF. Admit that f is an exposed boundary point of M^*. Then there exists a supporting hyperplane x^* of M^* through f containing no boundary point of M^* except for f. In other words, no boundary point of the body M^*, distinct from f, is contained in the hyperplane x^*. In the language of the space E this means that no supporting hyperplane of the body M, distinct from f^*, contains the point x. In other words, f^* is the only supporting hyperplane through x. Consequently f^* is a regular supporting hyperplane of M. Thus, if f is an exposed boundary point of M^*, then f^* is a regular supporting hyperplane of M.

Conducting this reasoning in the opposite direction, we obtain the converse assertion: if f^* is a regular supporting hyperplane of M, then f is an exposed boundary point of M^*. ■

Exercises

1. Let e_1, e_2 be an orthonormal basis in $\mathbf{R}^2$. Find the polar of the convex set $M \subset \mathbf{R}^2$ in the following cases: a) M is the square with the vertices $\pm e_1, \pm e_2$; b) M is the triangle with the vertices $-e_2 \pm e_1, e_2$; c) $M = \operatorname{conv} K$, where K is the parabola defined (in the coordinate system with the basis e_1, e_2) by the equation $2x_2 = x_1^2 - 1$.

2. Let $M \subset \mathbf{R}^3$ be the cube defined in an orthogonal coordinate system x_1, x_2, x_3 by the inequalities $|x_i| \leq 1$, $i = 1, 2, 3$. Find the polar M^*.

3. Let $M \subset \mathbf{R}^n$ be a polytope with vertices $a_1, \cdots, a_q$ such that $o \in \operatorname{int} M$. Prove that $M^* = P_1 \cap \cdots \cap P_q$, where P_i is the half-space $\{x : \langle x, a_i \rangle \leq 1\}$, $i = 1, \cdots, q$.

4. Let $P \subset E$ be a k-dimensional flat. Prove that there exists an $(n - k - 1)$-dimensional flat P^* in E^* such that a point $f \in E^*$ belongs to P^* if an only if the polar hyperplane f^* in E contains P. The flat P^* is said to be the _polar_ of P. Thus we obtain a $1 - 1$-correspondence between the family of all k-dimensional flats in E, not containing o, and the family of all $(n - k - 1)$-dimensional flats in E^*, also not containing o.

5. Two flats $P, Q \subset E$ of different dimensions are said to be _incident_ if their intersection coincides with one of them, i.e., either $P \subset Q$ or $Q \subset P$. Prove that flats $P, Q \subset E$, not containing o, are incident if and only if their polars P^*, Q^* are also incident. More detailed, if $P \subset Q$, then $P^* \supset Q^*$ (and if $P \supset Q$, then $P^* \subset Q^*$).

6. Prove that flats $P, Q \subset E$, not containing o, have a nonempty intersection if and only if their polars P^*, Q^* are contained in a hyperplane L which does not contain o. In which case (described in terms of polars) do the flats P, Q have a k-dimensional intersection?

7. Let hyperplanes $P, Q \subset E$, which do not contain the origin, be parallel. What does this mean in polar terms? More detailed, if k-dimensional flats $P, Q \subset E$ (not passing through o) are parallel, what does this mean in polar terms?

8. Let $E = E^* = \mathbf{R}^n$, i.e., E is an Euclidean (selfadjoint) space, and $e_1, \cdots, e_n$ be an orthonormal basis in E. Denote by P the k-flat through λe_1 with the basis $e_2, \cdots, e_{k+1}$ and by Q the $(n - k - 1)$-flat through $\frac{1}{\lambda}e_1$ with the basis $e_{k+2}, \cdots, e_n$. Prove that $P^* = Q$.

9. Let $M \subset E$ be a compact, convex body with $o \in \operatorname{int} M$ and F be an exposed face of M. Denote by F^* the set of all points $f \in E^* \setminus \{o\}$, for which $F \subset f^*$. Prove that F^* is an exposed face of the body M^*. We say that F^* is the *polar face* for F. Prove that the polar face for the exposed face $F^* \subset M^*$ coincides with F, i.e., $F^{**} = F$. Thus we obtain the polar $1 - 1$-correspondence between the family of all exposed faces of M and the analogous family for M^*.

10. We say that an exposed face F of a compact, convex body $M \subset E$ with $o \in \operatorname{int} M$ is *tame* if

$$\dim F + \dim F^* + 1 = \dim E.$$

Prove that for any polytope M every face of M is exposed and, moreover, every face is tame.

11. Let $M_\alpha, \alpha \in A$, be a family of compact, convex bodies in E such that $M = \bigcap_\alpha M_\alpha$ contains the origin in its interior. Prove that

$$M^* = \operatorname{cl} \operatorname{conv} \left(\bigcup_\alpha M_\alpha^* \right).$$

12. Prove that if a compact, convex body has only a finite number of faces, then it is a polytope and consequently all its faces are tame. Show by an example that there exists a compact, convex body which is not a polytope and has only tame faces.

§4 Direct sum decompositions

Let E, F be finite-dimensional vector spaces and $g : E \to F$ be a homomorphism (i.e., a linear mapping). Let, furthermore, $f : F \to \mathbf{R}$ be a linear functional on F, i.e., $f \in F^*$. Then $f \circ g : E \to \mathbf{R}$ is a linear functional on E, i.e., $f \circ g \in E^*$. As usual, we denote this functional by $g^*(f)$, i.e., $g^*(f) = f \circ g$. Thus we obtain a mapping (in fact, a homomorphism) $g^* : F^* \to E^*$. It is named the *dual homomorphism* for g:

$$E \xrightarrow{g} F,$$
$$E^* \xleftarrow{g^*} F^*.$$

Obviously, for every $x \in E$ and $f \in F^*$, the equality

$$\langle x, g^*(f) \rangle = \langle g(x), f \rangle \tag{1}$$

holds.

An automorphism π of a finite-dimensional vector space E is said to be a *projection* if $\pi \circ \pi = \pi$. It is clear that $\pi(x) = x$ for every element x of the subspace Im $\pi = \pi(E)$ of the space E. A projection $\pi : E \to E$ is *nontrivial* if Im $\pi \neq \{o\}$. It is evident that if $\pi : E \to E$ is a projection, then its dual automorphism $\pi^* : E^* \to E^*$ is a projection, too (i.e., $\pi^* \circ \pi^* = \pi^*$).

A *direct sum decomposition* of E is defined if nontrivial projections $\pi_1, \cdots, \pi_s :$ $E \to E$ with $s \geq 2$ are given such that

$$\pi_p \circ \pi_q = o \text{ for } p \neq q, \text{ and } \pi_1 + \cdots + \pi_s = 1 \tag{2}$$

(the identical automorphism of E). In this case, the subspaces $E_1 = $ Im $\pi_1, \cdots, E_s = $ Im π_s are said to be *direct summands* of this decomposition and we write

$$E = E_1 \oplus \cdots \oplus E_s. \tag{3}$$

It is easily shown that if $x \in E_q$, then $\pi_q(x) = x$ and $\pi_p(x) = o$ for $p \neq q$ (consequently $E_p \cap E_q = \{o\}$ for $p \neq q$). Now (2) means that for every $x \in E$ the representation

$$x = x_1 + \cdots + x_s; \ x_1 \in E_1, \cdots, x_s \in E_s \tag{4}$$

holds (namely, $x_q = \pi_q(x), q = 1, \cdots, s$) and each $x \in E$ is *uniquely* representable in the form (4).

Consider now the dual automorphisms $\pi_1^*, \cdots, \pi_s^* : E^* \to E^*$. It follows from (2) that

$$\pi_p^* \circ \pi_q^* = o \text{ for } p \neq q, \text{ and } \pi_1^* + \cdots + \pi_s^* = 1^* \tag{5}$$

(the identical automorphism of E^*). Consequently we have a direct sum decomposition

$$E^* = E_1' \oplus \cdots \oplus E_s', \tag{6}$$

where $E_q' = $ Im π_q^*; $q = 1, \cdots, s$. This direct decomposition is said to be *dual* for (3). By analogy with (4), every $f \in E^*$ is uniquely representable in the form

$$f = f_1 + \cdots + f_s; \ f_1 \in E_1', \cdots, f_s \in E_s', \tag{7}$$

namely $f_q = \pi_q^*(f)$, $q = 1, \cdots, s$. It is easily shown that E_q' consists of the elements $f \in E^*$ which satisfy the condition $\langle x, f \rangle = o$ for every $p \neq q$ and $x \in E_p$.

In addition, we remark that the subspace $E_q' \subset E^*$ is naturally isomorphic to E_q^*. More precisely, for each $f_q \in E_q'$ denote by $j_q(f_q)$ the restriction of the functional f_q to E_q, i.e., $j_q(f_q) : E_q \to \mathbf{R}$ is defined by the equality $j_q(f_q(x)) = f_q(x)$ for every $x \in E_q$. Then $j_q : E_q' \to E_q^*$ is an isomorphism, and we agree to identify E'_q and E^*_q by this isomorphism. By virtue of this agreement, $E^*_q \subset E^*$ for every $q = 1, \cdots, s$.

Finally, for every $x \in E$ and $f \in E^*$ we have (cf. (4) and (7))

$$\langle x, f \rangle = \langle x_1 + \cdots + x_s, f_1 + \cdots + f_s \rangle = \langle x_1, f_1 \rangle + \cdots + \langle x_s, f_s \rangle. \tag{8}$$

Now we pass to the theory of convex sets. Let $E = E_1 \oplus \cdots \oplus E_s$, i.e., automorphisms $\pi_1, \cdots, \pi_s$ are given which satisfy (2). Suppose that for every $q = 1, \cdots, s$ a convex set $M_q \subset E_q$ is given. The set of all the points $x = x_1 + \cdots + x_s$, where $x_1 \in M_1, \cdots, x_s \in M_s$, is said to be the *direct sum* of the sets $M_1, \cdots, M_s$ and is denoted by $M_1 \oplus \cdots \oplus M_s$. It is easily shown that $M_1 \oplus \cdots \oplus M_s$ is a convex set in E. If each M_q is closed (compact), so $M_1 \oplus \cdots \oplus M_s$ is. If each M_q is a convex body in E_q (with the origin in its interior), then $M_1 \oplus \cdots \oplus M_s$ is a convex body in E (with the origin as interior point). A convex body $M \subset E$ is *decomposable* if there exists a direct sum decomposition $E = E_1 \oplus \cdots \oplus E_s$ with $s \geq 2$ and $\dim E_q \geq 1$ for each $q \in \{1, \cdots, s\}$ such that $M = M_1 \oplus \cdots \oplus M_s$, where M_q is a convex body in E_q, $q = 1, \cdots, s$.

Consider compact, convex bodies $M_q \subset E_q, q = 1, \cdots, s$, each of which contains the origin in its interior. The convex body conv $(M_1 \cup \cdots \cup M_s)$ is said to be the *join* of $M_1, \cdots, M_s$ and denoted by $M_1 \vee \cdots \vee M_s$. This join is again a compact, convex body in E with the origin in its interior. A compact, convex body $M \subset E$ is said to be *splittable* if there exists a direct decomposition $E = E_1 \oplus \cdots \oplus E_s$ with $s \geq 2$ and $\dim E_q \geq 1$ for all q such that $M = M_1 \vee \cdots \vee M_s$. (Again, M_q is a compact, convex body in E_q with the origin in its interior for each $q = 1, \cdots, s$.)

Theorem 4.1: Let $E = E_1 \oplus \cdots \oplus E_s$, and M_q be a convex body in the space E_q with the origin in its interior, $q = 1, \cdots, s$. A point $x \in E$ belongs to the body $M_1 \vee \cdots \vee M_s$ if and only if there are points $x_1 \in M_1, \cdots, x_s \in M_s$ and nonnegative real numbers $\lambda_1, \cdots \lambda_s$ such that

$$x = \lambda_1 x_1 + \cdots + \lambda_s x_s, \quad \lambda_1 + \cdots + \lambda_s = 1. \quad \square$$

This statement easily follows from Theorem 1.2, since $M_1 \vee \cdots \vee M_s = $ conv $(M_1 \cup \cdots \cup M_s)$.   ∎

Theorem 4.2: Let $E = E_1 \oplus \cdots \oplus E_s$, and M_q be a compact, convex body in the space E_q with the origin in its interior. Then

$$(M_1 \oplus \cdots \oplus M_s)^* = M_1^* \vee \cdots \vee M_s^*, \tag{9}$$

$$(M_1 \vee \cdots \vee M_s)^* = M_1^* \oplus \cdots \oplus M_s^*, \tag{10}$$

where M_q^* is the polar body of M_q in the space $E_q, q = 1, \cdots, s$. $\square$

PROOF. By virtue of duality, it suffices to prove only one of the equalities. We prove the first one.

Let $f \in M_1^* \vee \cdots \vee M_s^*$. More precisely (by virtue of our suggestion to identify E_q' with E_q^* by means of j_q), $f \in \mathrm{conv}\, (M_1' \vee \cdots \vee M_s')$, where $M_q' \subset E_q'$ and $j_q(M_q') = M_q^*$. Then $f = \lambda_1 f_1 + \cdots \lambda_s f_s$, where $\lambda_1, \cdots, \lambda_s$ are nonnegative numbers with $\lambda_1 + \cdots + \lambda_s = 1$ and $f_q \in M_q'$. Since $j_q(f_q) \in M_q^*$, the functional $j_q(f_q) : E_q \to \mathbf{R}$ is compatible with the convex body $M_q \subset E_q$, i.e., $j_q(f_q)(x_q) \leq 1$ for every $x_q \in M_q$ or, what is the same, $f_q(x_q) \leq 1$ for all $x_q \in M_q$. Furthermore, since $f_q \in M_q' \subset E_q'$, we have $f_q(x) = f_q(\pi_q(x))$ for each $x \in E$. Finally, for every $x \in (M_1 \oplus \cdots \oplus M_s)$ the equality (4) holds with $x_q \in M_q, q = 1, \cdots, s$, and consequently,

$$
\begin{aligned}
f(x) &= \langle \lambda_1 f_1 + \cdots + \lambda_s f_s, x_1 + \cdots + x_s \rangle = \\
     &= \lambda_1 \langle f_1, x_1 \rangle + \cdots + \lambda_s \langle f_s, x_s \rangle = \\
     &\leq \lambda_1 \cdot 1 + \cdots + \lambda_s \cdot 1 = 1.
\end{aligned}
$$

This means that f is compatible with $M_1 \oplus \cdots \oplus M_s$, i.e., $f \in (M_1 \oplus \cdots \oplus M_s)^*$. This proves the inclusion

$$
M_1^* \vee \cdots \vee M_s^* \subset (M_1 \oplus \cdots \oplus M_s)^*
$$

Conversely, suppose $f \in (M_1 \oplus \cdots \oplus M_s)^*$. We consider the expansion (7) and denote by λ_q' the maximal value of f_q on M_q. Thus $f_q(x_q) \leq \lambda_q'$ for all $x_q \in M_q$. This means that $f_q = \lambda_q' f_q'$, where $f_q' \in M_q'$. Furthermore, for $x \in (M_1 \oplus \cdots \oplus M_q)$ we have $f(x) \leq 1$, i.e., for every $x_1 \in M_1, \cdots, x_s \in M_s$ the inequality

$$
\begin{aligned}
\langle x, f \rangle &= \langle x_1 + \cdots + x_s, f_1 + \cdots + f_s \rangle = \\
                     &= \langle x_1, f_1 \rangle + \cdots + \langle x_s, f_s \rangle = \\
                     &= f_1(x_1) + \cdots + f_s(x_s) \leq 1
\end{aligned}
$$

holds. Conversely, $\lambda_1' + \cdots + \lambda_s' \leq 1$. Thus,

$$
f = f_1 + \cdots + f_s = \lambda_1' f_1' + \cdots + \lambda_s' f_s' \in \mathrm{conv}\, (M_1' \vee \cdots \vee M_s').
$$

By virtue of the agreement to identify M_q' and M_q^* (with the help of j_q), we obtain

$$
f \in \mathrm{conv}\, (M_1^* \cup \cdots \cup M_s^*) = M_1^* \vee \cdots \vee M_s^*.
$$

Thus we have shown the converse inclusion

$$
(M_1 \oplus \cdots \oplus M_s)^* \subset M_1^* \vee \cdots \vee M_s^*,
$$

and this completes the proof. ∎

Corollary 4.3: Let $M \subset E$ be a compact, convex body with the origin in its interior. The body M is decomposable if and only if M^* is splittable. □

Now we give some illustrations of Theorem 4.2.

<u>EXAMPLE 4.4:</u> Let $E = E^* = \mathbf{R}^n$ denote the n-dimensional Euclidean (self-adjoint) space, i.e., $E = E_1 \oplus \cdots \oplus E_n$, where each subspace $E_q = E_q^* = \mathbf{R}$ is one-dimensional. Furthermore, let M_q be the segment $[-1,1]$ contained in $E_q = E_q^* = \mathbf{R}$. Then $M_1 \oplus \cdots \oplus M_n$ is the unit cube in E centered at the origin, and $M^* = M_1 \vee \cdots \vee M_s$ is the cross-polytope, i.e., $M^* = \text{conv}\ (M_1 \cup \cdots \cup M_s)$. □

<u>EXAMPLE 4.5:</u> Let $E = E_1 \oplus E_2$, where $\dim E_1 = n-1$, $\dim E_2 = 1$. Furthermore, let $M_1 \subset E_1$ be an $(n-1)$-dimensional compact, convex body with the origin in its interior and $M_2 \subset E_2$ be a segment with $o \in$ ri M_2. Then $M = M_1 \oplus M_2$ is a *prism* in E with the basis M_1 and the generator M_2. Its polar $M^* = M_1^* \vee M_2^*$ is a *bicone* over M_1^*, i.e., $M^* = \text{conv}\ (M_1^* \cup \{a,b\})$, where a,b are the endpoints of the segment M_2^*. □

<u>EXAMPLE 4.6:</u> Let $M = M_1 \oplus M_2 \oplus M_3$ be the unit cube in $\mathbf{R}^3$ and the cross-polytope $M^* = M_1 \vee M_2 \vee M_3$ (the octahedron) be its polar (cf. Example 4.4). We take the edges of M and denote by Q the set of their midpoints. Similarly, the set of midpoints of the edges of M^* we denote by Q'. Then the *cuboctahedra* conv Q and conv Q' are homothetic to each other with the common center o and ratio 2. □

<u>Exercises</u>

1. Let $E = E_1 \oplus \cdots \oplus E_s$, and M_q be a compact, convex body in the subspace $E_q, q = 1, \cdots, s$. Prove that a point $x = x_1 + \cdots + x_s$ is a boundary point of $M_1 \oplus \cdots \oplus M_s$ if and only if x_q is a boundary point of M_q for at least one $q \in \{1, \cdots, s\}$.

2. Let $E = E_1 \oplus \cdots \oplus E_s$, and M_q be a compact, convex body in the subspace $E_q, q = 1, \cdots, s$. Prove that $x = x_1 + \cdots + x_s$ is a regular boundary point of $M_1 \oplus \cdots \oplus M_s$ if and only if x_q is a regular boundary point of M_q for an index q and $x_p \in$ ri M_p for all $p \neq q$.

3. Let $E = E_1 \oplus \cdots \oplus E_s$, and M_q be a compact, convex body in the subspace E_q with the origin in its interior, $q = 1, \cdots, s$. Prove the equalities
$$\text{ext}\ (M_1 \vee \cdots \vee M_s) = (\text{ext}\ M_1) \cup \cdots \cup (\text{ext}\ M_s),$$
$$\text{exp}\ (M_1 \vee \cdots \vee M_s) = (\text{exp}\ M_1) \cup \cdots \cup (\text{exp}\ M_s).$$

4. Let $E = E_1 \oplus \cdots \oplus E_s$, and M_q be a convex set in $E_q, q = 1, \cdots, s$. Furthermore, let F_q be a nonempty face of M_q (possibly improper, i.e., the equality $F_q = M_q$ is not excluded). Prove that $F_1 \oplus \cdots \oplus F_s$ is a face (possibly improper) of $M_1 \oplus \cdots \oplus M_q$. Prove in addition that each face of $M_1 \oplus \cdots \oplus M_s$ is representable in such a form.

5. Let $E = E_1 \oplus \cdots \oplus E_s$, and M_q be a compact, convex body in the subspace E_q with the origin in its interior, $q = 1, \cdots, s$. Furthermore, let F_q be a face of M_q (including ∅, as a face, but excluding the improper face M_q). Prove that conv $(F_1 \cup \cdots \cup F_s)$ is a face of the body $M_1 \vee \cdots \vee M_s$. In addition, prove that every face of $M_1 \vee \cdots \vee M_s$ (excluding the improper face $M_1 \vee \cdots \vee M_s$) can be represented in such a form.

6. In which case a face $F_1 \oplus \cdots \oplus F_s$ of the body $M_1 \oplus \cdots \oplus M_s$ (see Exercise 4) is an exposed one? In what case is a face $\mathrm{conv}\,(F_1 \cup \cdots \cup F_s)$ of the body $M_1 \vee \cdots \vee M_s$ (see Exercise 5) an exposed one?

7. Let $E = E_1 \oplus \cdots \oplus E_s$, and M_q be a compact, convex body in the space E_q with the origin in its interior, $q = 1, \cdots, s$. Prove that $M_1 \oplus \cdots \oplus M_s$ is symmetric with respect to the origin if and only if each body $M_q \subset E_q$ is centrally symmetric with respect to the origin. Prove the analogous assertion for the body $M_1 \vee \cdots \vee M_s$.

8. Let $E = E_1 \oplus \cdots \oplus E_n$, where each subspace E_q is one-dimensional, $q = 1, \cdots, n$. Furthermore, let $M_q \subset E_q$ be a segment with the midpoint at the origin. A polytope obtained from $M_1, \cdots, M_n$ by the operations "direct sum" and "convex hull" (in any order) is said to be a *Hanner polytope*, see [Han], [H-L], and [Bt 15]. For example,

$$((M_1 \oplus M_2) \vee (M_3 \oplus M_4) \vee M_5) \oplus M_6$$

is a Hanner polytope in $\mathbf{R}^6$. Prove that if $M \subset E$ is a Hanner polytope, then M^* is a Hanner polytope, too.

9. Let $E = E_1 \oplus \cdots \oplus E_s$, and $M_q \subset E_q$ be a polytope with the origin in its interior. Prove that if M_q has m_q nonempty faces (including M_q as its improper face), then $M_1 \oplus \cdots \oplus M_s$ has $m_1 m_2 \cdots m_s$ nonempty faces. Prove that $M_1 \vee \cdots \vee M_s$ has $m_1 m_2 \cdots m_s$ faces, too (including $\emptyset$, but excluding $M_1 \vee \cdots \vee M_s$ itself).

10. What is the number of faces of an n-dimensional Hanner polytope?

11. Let e_1, e_2, e_3 be an orthonormal basis in $\mathbf{R}^3$ and $E_1 = \lin f_1$, $E_2 = \lin f_2$, $E_3 = \lin (e_1, e_2)$, where $f_1 = \frac{\sqrt{3}}{2}e_1 + \frac{1}{2}e_2$, $f_2 = -\frac{\sqrt{3}}{2}e_1 + \frac{1}{2}e_3$. Prove that it is possible to define projections $\pi_i : \mathbf{R}^3 \to E_i$, $i = 1, 2, 3$, such that $\pi_1 + \pi_2 + \pi_3 = 1$ (but $\mathbf{R}^3 \neq E_1 \oplus E_2 \oplus E_3$). This example shows that the condition $\pi_p \circ \pi_q = o$ for $p \neq q$ in the definition of direct decomposition is essential, cf. (2) and (3).

12. Prove that if $\pi_1, \cdots, \pi_s$ are projections of $\mathbf{R}^n$ which are pairwise commutative ($\pi_p \circ \pi_q = \pi_q \circ \pi_p$ for $p \neq q$) and satisfy $\pi_1 + \cdots + \pi_s = 1$, then $\pi_p \circ \pi_q = o$ for $p \neq q$ (i.e., $\mathbf{R}^n = \mathrm{Im}\,\pi_1 \oplus \cdots \oplus \mathrm{Im}\,\pi_s$).

§5 The lower semicontinuity of the operator "exp"

First of all, we introduce the *Hausdorff distance* in the family of all compact sets in Euclidean space $E = \mathbf{R}^n$. Let $M \subset E$ be a set and ε be a positive real number. Denote by B_ε the open ball of radius ε centered at the origin. The vector sum $M + B_\varepsilon$ (i.e., the set of the points $y + b$, where $y \in M, b \in B_\varepsilon$) is said to be the ε-neighbourhood of M and denoted by $U_\varepsilon(M)$. In other words, $U_\varepsilon(M)$ is the set of all points $x \in E$ at a distance $< \varepsilon$ from M (Fig. 10), i.e., there exists a point $y \in M$ such that $\|x - y\| < \varepsilon$.

Let now $M, N \subset E$ be compact sets. The *Hausdorff distance* $\varrho(M, N)$ is defined by the equality

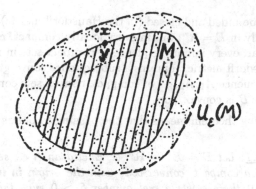

Fig. 10

$$\varrho(M, N) = \inf\{\varepsilon > 0 : M \subset U_\varepsilon(N) \text{ and } N \subset U_\varepsilon(M)\}.$$

We remark that, in general, the numbers

$$\varrho_1 = \inf\{\varepsilon > 0 : M \subset U_\varepsilon(N)\} \text{ and}$$
$$\varrho_2 = \inf\{\varepsilon > 0 : N \subset U_\varepsilon(M)\}$$

do not coincide; for example, $\varrho_1 \neq \varrho_2$ if M, N are the relative boundaries of two homothetic regular triangles with common center (Fig. 11). The largest of the numbers ϱ_1, ϱ_2 is equal to $\varrho(M, N)$.

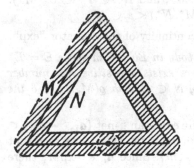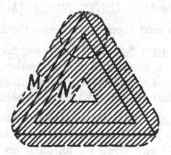

Fig. 11

The family S of all compact subsets of the space E is a metric space with respect to the Hausdorff distance. (In fact, this is true not only in $E = \mathbf{R}^n$, but in every metric space as well.) A sequence $A_1, A_2, \cdots$ is said to be *convergent* to a set B (where $A_i \in S$ for all i and $B \in S$) if $\varrho(A_i, B)$ tends to 0 as $i \to \infty$.

Furthermore, the metric space S of all compact subsets of E is complete and possesses the local compactness property. This means that every subspace of

S which is bounded and closed (in the Hausdorff metric) is compact. (This is true not only in $E = \mathbf{R}^n$, but in every locally compact, metric space as well.) In particular, every infinite sequence of compact sets in E, which is bounded in the Hausdorff metric, contains a subsequence converging to a compact set. And if a sequence $A_1, A_2, \cdots$ of compact, convex sets converges to a compact set B, then B is convex, too.

The following theorem means that the passing to polars is a continuous operator.

Theorem 5.1: Let $E = E^* = \mathbf{R}^n$ be Euclidean (i.e., selfadjoint) space and $M \subset E$ be a compact, convex body with the origin in its interior. Then for every $\varepsilon > 0$ there exists a real number $\delta > 0$ such that for each compact, convex body N, satisfying $\varrho(M, N) < \delta$, the inequality $\varrho(M^*, N^*) < \varepsilon$ holds. $\square$

PROOF. Let $Q^* \subset E^*$ be a polytope with the origin in its interior such that $Q^* \subset \operatorname{int} M^*$ and $U_\varepsilon(Q^*) \supset M^*$. Then, by duality, $Q \subset E$ is a polytope satisfying $\operatorname{int} Q \supset M$. Furthermore, let $P^* \subset E^*$ be a polytope with $\operatorname{int} P^* \supset M^*$ and $P^* \subset U_\varepsilon(M^*)$. Then, by duality, $P \subset E$ is a polytope such that $P \subset \operatorname{int} M$. We choose a positive real number $\delta > 0$ with $Q \supset U_\varepsilon(M)$ and $U_\delta(P) \subset M$. This δ is desired.

Indeed, let a convex body $N \subset E$ contain the origin in its interior and satisfy $\varrho(M, N) < \delta$. Then $N \subset U_\delta(M), M \subset U_\delta(N)$ and consequently $N \subset Q, U_\delta(N) \supset U_\delta(P)$, i.e., $N \supset P$. Passing to polar bodies, we obtain $N^* \supset Q^*, N^* \subset P^*$. Hence $U_\varepsilon(N^*) \supset U_\varepsilon(Q^*) \supset M^*$, and $N^* \subset P^* \subset U_\varepsilon(M^*)$. Thus $M^* \subset U_\varepsilon(N^*), N^* \subset U_\varepsilon(M^*)$, i.e., $\varrho(M^*, N^*) < \varepsilon$. $\blacksquare$

The next theorem expresses the lower semicontinuity of the operator "exp".

Theorem 5.2: Let M be a compact, convex body in Euclidean space $E = E^*$ and ε be a positive real number. Then there exists a positive real number δ such that for every compact, convex body $N \subset E$ with $\varrho(M, N) < \delta$ the inclusion $U_\varepsilon(\exp N) \supset \exp M$ holds. $\square$

PROOF. Since the set $\exp M$ is bounded, there exists an $\frac{\varepsilon}{2}$-net $\{a_1, \cdots, a_m\} \subset \exp M$, i.e., for every $x \in \exp M$ there is an index $i \in \{1, \cdots, m\}$ such that $\|x - a_i\| < \frac{\varepsilon}{2}$. We fix an index $i \in \{1, \cdots, m\}$. Since $a_i \in \exp M$, there exists a supporting hyperplane $\Gamma^{(i)}$ of M with $M \cap \Gamma^{(i)} = \{a_i\}$. Denote by $\Pi^{(i)}$ the half-space with the boundary $\Gamma^{(i)}$ containing M. Furthermore, for every $\alpha > 0$ we denote by $\Pi_\alpha^{(i)}$ the half-space contained in $\Pi^{(i)}$ such that the hyperplane bd $\Pi_\alpha^{(i)}$ is parallel to $\Gamma^{(i)}$ and has the distance α with respect to $\Gamma^{(i)}$ (Fig. 12). If α is small enough, then $M \setminus \operatorname{int} \Pi_\alpha^{(i)} \subset U_{\varepsilon/2}(a_i)$. Since $M \setminus \operatorname{int} \Pi_\alpha^{(i)}$ is compact, there exists a real number $\beta > 0$ such that $U_\beta(M) \setminus \operatorname{int} \Pi_\alpha^{(i)} \subset U_{\varepsilon/2}(a_i)$, and we denote by γ the smallest of the numbers α, β. Then $U_\gamma(M) \setminus \operatorname{int} \Pi_\gamma^{(i)} \subset U_{\varepsilon/2}(a_i)$.

Let now $\varrho(M, N) < \gamma$. Since

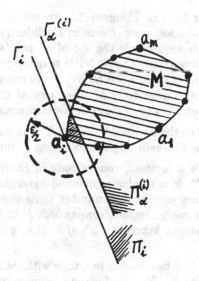

Fig. 12

$$N = \text{conv ext } N = \text{conv cl exp } N,$$

there exists a point $y \in \exp N$ which is not contained in $\Pi_\gamma^{(i)}$ (otherwise $N \subset \Pi_\gamma^{(i)}$ and consequently $a_i \notin U_\gamma(N)$, contradicting $\varrho(M,N) < \gamma$). Thus $y \in \exp N, y \notin \Pi_\gamma^{(i)}$, i.e.,

$$y \in \exp N \setminus \Pi_\gamma^{(i)} \subset N \setminus \text{int } \Pi_\gamma^{(i)} \subset U_\gamma(M) \setminus \text{int } \Pi_\gamma^{(i)} \subset U_{\varepsilon/2}(a_i),$$

and hence $\|y - a_i\| < \frac{\varepsilon}{2}$.

We see that for each $i = 1, \cdots, m$ there exists a positive number $\gamma = \gamma(i)$ with the following property: if $\varrho(M,N) < \gamma(i)$, then there exists a point $y \in \exp N$ with $\|y - a_i\| < \frac{\varepsilon}{2}$. We now denote by δ the least of the numbers $\gamma(1), \cdots, \gamma(m)$. Now, if $\varrho(M,N) < \delta$, then for every $i \in \{1, \cdots, m\}$ there is a point $y_i \in \exp N$ satisfying $\|y_i - a_i\| < \frac{\varepsilon}{2}$. Consequently, since $\{a_i, \cdots, a_m\}$ is an $\frac{\varepsilon}{2}$-net in $\exp M$, for each point $x \in \exp M$ there exists an index i such that $\|x - a_i\| < \frac{\varepsilon}{2}$, i.e.,

$$\|x - y_i\| \le \|x - a_i\| + \|y_i - a_i\| < \frac{\varepsilon}{2} + \frac{\varepsilon}{2} = \varepsilon.$$

In other words, $\exp M \subset U_\varepsilon(\exp N)$. ∎

It is possible to give another form of Theorem 5.2. Namely, let M be a compact, convex body in Euclidean space $E = E^*$; furthermore, let x be an exposed point of M and ε be a positive number. Then there exists a positive number δ such that for every compact, convex body $N \subset E$ with $\varrho(M,N) < \delta$ there is an exposed point y of N satisfying $\|y - x\| < \varepsilon$.

We now give a further form of Theorem 5.2, by using the notion of *lower limit*. Let $N_1, N_2, \cdots$ be a sequence of sets in Euclidean space E. The lower limit $\underline{\lim}_{k \to \infty} N_k$ of this sequence is the set of all points $x \in E$ such that, for every $\varepsilon > 0$, the intersection $U_\varepsilon(x) \cap N_k$ is nonempty for k large enough. Theorem 5.2 affirms that if a sequence $N_1, N_2, \cdots$ of compact, convex bodies converges to a compact, convex body M, then $\exp M \subset \underline{\lim}_{k \to \infty} \exp (N_k)$. In other words, the operator exp is *lower semicontinuous*.

Applying Theorem 5.2 to the bodies $M^*, N^* \subset E^*$ and passing to the polars, we obtain (by virtue of Theorem 5.1) the following result.

Theorem 5.3: Let M be a compact, convex body in Euclidean space $E = E^*$ with $o \in int\ M$ and f^* be a regular supporting hyperplane of M (i.e., $f \in \exp M^*$). Then for any positive real number ε there exists a positive real number δ such that for every compact, convex body $N \subset E$ with $\varrho(M, N) < \delta$ there is a regular supporting hyperplane g^* of N (i.e., $g \in \exp N^*$) such that $\|g - f\| < \varepsilon$. $\square$

EXAMPLE 5.4: Let $M \subset E$ be a convex polytope with the origin in its interior. There exists a sequence $N_1, N_2, \cdots$ of strictly convex bodies convergent to M. Thus $\exp M$ is a finite set (consisting of the vertices of M), whereas $\exp N_k = bd\ N_k$ and consequently $\lim_{k \to \infty} \exp N_k = \lim_{k \to \infty} bd\ N_k = bd\ M$. We see that the inclusion $\exp M \supset \underline{\lim}_{k \to \infty} \exp N_k$ does not hold. This shows that the operator exp does not possess the upper semicontinuity property. $\square$

Theorem 5.5: Let $M \subset E$ be a compact, convex body with the origin in its interior. If M is not splittable, then there exists a positive real number δ such that every compact, convex body $N \subset E$, satisfying $\varrho(M, N) < \delta$, is also not splittable. $\square$

PROOF: Since $M = conv\ cl\ \exp M$, the set $\exp M$ is not lying in a hyperplane, i.e., there exist linearly independent vectors $e_1, \cdots, e_n \in \exp M$. Let $\omega = (I^{(1)}, I^{(2)})$ be a nonempty disjoint partition of the index set $\{1, \cdots, n\}$ such that $1 \in I^{(1)}$. In other words, $I^{(1)} \cup I^{(2)} = \{1, \cdots, n\}, I^{(1)} \cap I^{(2)} = \emptyset$, and $1 \in I^{(1)}$. Denote by $L_\omega^{(1)}$ the subspace spanned by the vectors e_i with $i \in I^{(1)}$ and by $L_\omega^{(2)}$ the subspace spanned by all e_i with $i \in I^{(2)}$. Then $E = L_\omega^{(1)} \oplus L_\omega^{(2)}$. Since the body M is not splittable, the vector system ext $M \subset cl\ \exp M$ is not contained in $L_\omega^{(1)} \cup L_\omega^{(2)}$, i.e., there exists a vector in $\exp M$ which is not contained in $L_\omega^{(1)} \cup L_\omega^{(2)}$. For each partition ω, we fix one vector $f_\omega \in \exp M$ such that $f_\omega \notin L_\omega^{(1)} \cup L_\omega^{(2)}$. Finally, denote by F the vector system which consists of the vectors $e_1, \cdots, e_n$ and the vectors f_ω (for any partition ω). Thus $F \subset \exp M$.

It is easily shown that the finite vector system F is not splittable, i.e., there exist no nontrivial subspaces $E^{(1)}, E^{(2)}$ satisfying $E = E^{(1)} \oplus E^{(2)}$ and $F \subset E^{(1)} \cup E^{(2)}$. Indeed, admit that subspaces $E^{(1)}, E^{(2)}$ with the above properties exist. We may suppose (by changing the roles of the subspaces $E^{(1)}, E^{(2)}$ if necessary) that $e_1 \in E^{(1)}$. Now denote by $I^{(1)}$ the set of indices $i \in \{1, \cdots, n\}$ for which $e_i \in E^{(1)}$ and by $I^{(2)}$ the set of indices $i \in \{1, \cdots, n\}$ for which $e_i \in E^{(2)}$. Then $\omega = (I^{(1)}, I^{(2)})$ is a partition with $1 \in I^{(1)}$. Since $e_i \in E^{(1)}$ for each $i \in I^{(1)}$, the subspace $L^{(1)}_\omega$ is contained in E_1. Similarly, $L^{(2)}_\omega \subset E_2$. Consequently (by virtue of $E = L^{(1)}_\omega \oplus L^{(2)}_\omega, E = E^{(1)} \oplus E^{(2)}$) we obtain $L^{(1)}_\omega = E^{(1)}, L^{(2)}_\omega = E^{(2)}$. Now we conclude $f_\omega \notin L^{(1)}_\omega \cup L^{(2)}_\omega = E^{(1)} \cup E^{(2)}$, contradicting the assumption $F \subset (E^{(1)} \cup E^{(2)})$. This contradiction shows that the vector system F is not splittable.

A similar reasoning shows that there exists a positive number ε with the following property. Choose vectors $e'_1, \cdots, e'_n$ and a vector f'_ω for each partition ω such that $||e'_i - e_i|| < \varepsilon$ for all $i = 1, \cdots, n$ and $||f'_\omega - f_\omega|| < \varepsilon$ for each partition ω. Then the obtained vector system F' (that consists of e'_i and f'_ω) is not splittable. We fix a number ε with this property.

Finally, let δ be the number as in Theorem 5.2. If $\varrho(M, N) < \delta$, then $U_\varepsilon(\exp N) \supset \exp M$ and consequently $U_\varepsilon(\exp N) \supset F$. This means that there are some vectors e'_i, f'_ω in $\exp N$ such that $||e'_i - e_i|| < \varepsilon, ||f'_\omega - f_\omega|| < \varepsilon$ for all i and ω. Consequently, by virtue of the choice of ε, the system F' consisting of vectors e'_i, f'_ω is not splittable, and the system $\exp N \supset F'$ is not splittable a fortiori, i.e., the body N is not splittable. ∎

Applying Corollary 4.3 and Theorem 5.1, we obtain the following dual version of Theorem 5.5.

Theorem 5.6: Let $M \subset E$ be a compact, convex body. If M is indecomposable, then there exists a positive real number γ such that every compact, convex body $N \subset E$, satisfying the condition $\varrho(M, N) < \gamma$, is also indecomposable. □

We say that a polytope $C \subset E$ is a generalized cross-polytope if (up to a translation) there exist a basis $e_1, \cdots, e_n$ in E and numbers $\lambda_1, \cdots, \lambda_n$ with

$$C = \text{conv}\{\lambda_1 e_1, \cdots, \lambda_n e_n, -\lambda_1 e_1, \cdots, -\lambda_n e_n\}.$$

Theorem 5.7: If a compact, convex body $M \subset E$ is not a generalized cross-polytope, then there exists a positive real number δ such that every compact, convex body $N \subset E$, satisfying $\varrho(M, N) < \delta$, is also not a generalized cross-polytope. □

PROOF: As in the previous proof, it is possible to choose vectors $e_1, \cdots, e_n \in$ ext M which form a basis in E. Since M is not a generalized cross-polytope, there exists a vector $f \in \exp M$ such that $f = \lambda_1 e_1 + \cdots + \lambda_n e_n$ with at least two nonzero coefficients. We fix such a vector $f \in \exp M$ and choose an $\varepsilon > 0$ with the following property: if vectors $e'_1, \cdots, e'_n, f'$ satisfy the conditions $||e'_1 - e_1|| < \varepsilon, \cdots, ||e'_n - e_n|| < \varepsilon, ||f' - f|| < \varepsilon$, then $e'_1, \cdots, e'_n$ is a linearly independent vector system and $f' = \lambda'_1 e'_1 + \cdots + \lambda'_n e'_n$ with at least two nonzero coefficients. Let now δ be the number as in Theorem 5.2. If $\varrho(M, N) < \delta$, then $U_\varepsilon(\exp N) \supset \exp M$, i.e., $e_1, \cdots, e_n, f \in U_\varepsilon(\exp N)$. Consequently there are vectors $e'_1, \cdots, e'_n, f' \in \exp N$ differing correspondingly from $e_1, \cdots, e_n, f$ less than ε. Hence $f' = \lambda'_1 e'_1 + \cdots + \lambda'_n e'_n$ with at least two nonzero coefficients; this means that N is not a generalized cross-polytope. ∎

We remark that a compact, convex body $M \subset E$ (containing the origin in its interior) is a parallelotope if and only its polar M^* is a generalized cross-polytope. Thus we obtain the following *dual version* of Theorem 5.7.

Theorem 5.8: If a compact, convex body $M \subset E$ is not a parallelotope, then there exists a positive real number γ such that every compact, convex body $N \subset E$ with $\varrho(M, N) < \gamma$ is also not a parallelotope.   □

Exercises

1. Let $M \subset E = \mathbf{R}^n$ be a compact, convex body and ε be a positive number. Prove that there exists a polytope $P \subset$ int M satisfying $U_\varepsilon(P) \supset M$.

2. Let $M \subset E = \mathbf{R}^n$ be a compact, convex body and ε be a positive number. Prove that there exists a polytope P satisfying int $P \supset M$ and $P \subset U_\varepsilon(M)$.

3. Prove the triangle inequality $\varrho(M, N) + \varrho(N, P) \geq \varrho(M, P)$ for any compact, convex sets in E.

4. Prove that for every compact, convex set M, the body cl $U_\varepsilon(M)$ is compact and smooth, i.e., every its boundary point is regular.

5. Prove that for every compact, convex body M and positive number ε there exists a compact, strictly convex body N satisfying $\varrho(M, N) < \varepsilon$.

6. Prove that if M, N are disjoint compact, convex bodies, then there exists a number $\varepsilon > 0$ such that $U_\varepsilon(M) \cap U_\varepsilon(N) = \emptyset$. Does this remain true if only M is compact? ($\cdots M$ is compact and N is closed? $\cdots M, N$ are closed and unbounded?)

7. Prove that each of the operators "ext", "cl exp", "cl ext" is lower semicontinuous.

8. Prove that if a compact, convex body $M \subset E$ is not a simplex, then there exists a number $\varepsilon > 0$ such that every compact, convex body $N \subset E$ satisfying $\varrho(M, N) < \varepsilon$ is also not a simplex.

9. We say that a convex body $M \subset E$ is *k-decomposable* if there exists a decomposition $M = M_1 \oplus \cdots \oplus M_s$ with $s \geq k$ and dim $M_i \geq 1$, $i = 1, \cdots, s$. Furthermore, M is *k-splittable* if there is a representation $M = M_1 \vee \cdots \vee M_s$ with $s \geq k$ and dim $M_i \geq 1$, $i = 1, \cdots, s$. Prove that a compact, convex body $M \subset E$ with $o \in$ int M is k-splittable if and only if M^* is k-decomposable.

10. Prove that if a compact, convex body $M \subset E$ is not k-decomposable, then there exists an $\varepsilon > 0$ such that every compact, convex body N with $\varrho(M, N) < \varepsilon$ is not k-decomposable, too. Formulate the dual assertion.

11. Prove that if a compact, convex body $M \subset E$ is not a prism (see Example 4.5), then there exists a number $\varepsilon > 0$ such that each compact, convex body $N \subset E$, satisfying $\varrho(M, N) < \varepsilon$, is also not a prism.

12. a) Prove that the perimeter of a convex figure $M \subset \mathbf{R}^2$ is a continuous function on M (in the Hausdorff metric). Generalize this for the space $\mathbf{R}^n$.

   b) Prove that the area of a convex figure $M \subset \mathbf{R}^2$ is a continuous function on M (in the Hausdorff metric). Generalize this for the n-dimensional space.

13. Let $M \subset \mathbf{R}^2$ be a convex polygon with perimeter $p > 0$ and area $s > 0$. Furthermore, let B be a ball of radius r. Prove that the figure $M + B$ has perimeter $p + 2\pi r$ and area $s + pr + \pi r^2$. Generalize this for an arbitrary compact, convex figure $M \subset \mathbf{R}^2$. Can you generalize the statements for $\mathbf{R}^3$ or even for $\mathbf{R}^n$?

14. Let $\mathfrak{F} = \{M_\alpha\}$ be a bounded, infinite family of compact sets in $\mathbf{R}^n$. Dissect the space $\mathbf{R}^n$ into cubes by the system of hyperplanes $x_i = k$, where k is an integer and $i = 1, \cdots, n$. The union of the cubes, which have nonempty intersection with M_α, is said to be the *support* of M_α. Prove that there is an infinite subfamily of $\mathfrak{F}$ such that all the sets M_α, contained in this subfamily, have the same support. Deduce that the metric space of all compact subsets of $\mathbf{R}^n$ possesses the local compactness property.

15. Let $\mathfrak{F}$ be a family of convex bodies in $\mathbf{R}^n$, ordered by inclusion. Furthermore, suppose that the family $\mathfrak{F}$ is *chain-closed*, i.e., for every chain $\mathfrak{F}' \subset \mathfrak{F}$ the set cl $(\bigcup_{M \in \mathfrak{F}'} M)$ belongs to $\mathfrak{F}$. Prove (without the Zermelo axiom or an equivalent tool) that in this case the Zorn lemma holds.

16. Using the previous exercise, prove that there exists a convex body $B \subset \mathbf{R}^n$ which has maximal n-dimensional volume among all the bodies with the same surface area s (it is well known that B is the *ball* with the surface area s). Indicate some other examples for which the assertion of the previous exercise can be applied.

§6 Convex cones

A set $K \subset \mathbf{R}^n$ is a *cone with apex a* if for each point $b \in K$ with $a \neq b$ the ray $[a,b)$ is contained in K. In particular, one can consider *convex cones* (i.e., cones which are convex sets). For example, a closed half-space is a convex cone, where each point of the bounding hyperplane can be taken as its apex.

Let K be a convex cone with apex a in $\mathbf{R}^n$. It is easy to show that an arbitrary face F of K contains the point a. Since two distinct minimal faces cannot have common points (otherwise their intersection would be a minimal face), each convex cone has precisely one minimal face. The points belonging to this minimal face (and only these ones) can be chosen as apices of the convex cone K.

Theorem 6.1: Let K be a convex cone with apex a and $b_1, \cdots, b_k$ be points from K. Then for any nonnegative real numbers $\lambda_1, \cdots, \lambda_k$, the point $b = a + \lambda_1(b_1 - a) + \cdots + \lambda_k(b_k - a)$ belongs to K. Moreover, if b_1 is an interior point of K and $\lambda_1 > 0$, then also b belongs to the interior of K. □

Theorem 6.2: Let $Q_1, \cdots, Q_m$ be convex cones with apex o. Then $Q = \mathrm{conv}(Q_1 \cup \cdots \cup Q_m)$ is also a convex cone with apex o. The point b belongs to Q if and only if $b = b_1 + \cdots + b_m$ with $b_1 \in Q, \cdots, b_m \in Q_m$. In other words, $Q = Q_1 + \cdots + Q_m$. $\square$

If a convex cone K does not coincide with the whole space, then its apex a is a boundary point, and hence a is contained in a supporting hyperplane of the cone K.

Let now M be an arbitrary convex set in $\mathbf{R}^n$ and a be a boundary point of M. Consider the supporting half-spaces of M at a (i.e., the closed half-spaces Π with $\Pi \supset M$ and $a \in \mathrm{bd}\, \Pi$). The intersection of all these half-spaces is a closed, convex cone with the apex a, which is called the *supporting cone* of the convex set M at the point a and is denoted by sup cone$_a M$. If $a \in \mathrm{ri}\, M$, then the supporting cone of M at a is given by aff M.

There are some further possibilities for defining supporting cones. Let a belong to a convex set $M \subset \mathbf{R}^n$. Consider the rays $[a, b)$ with $b \in M, b \neq a$. The union of all these rays is a cone with apex a, which is also convex (but possibly not closed). Then the closure of K coincides with the supporting cone of M at a (Fig. 13). Finally one can formulate that the supporting cone of the convex set M at $a \in M$ is the *smallest* closed convex cone with apex a containing M.

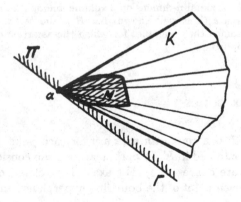

Fig. 13

The definition of the supporting cone yields the following assertion: Let Π be any supporting half-space of the convex set M at the point a. Then the supporting cone of M at a is contained in Π. Furthermore, each supporting hyperplane of the supporting cone K of the convex set M at $a \in M$ is also a supporting hyperplane of M at a and vice versa.

Let M be an unbounded, closed, convex set and $a \in M$. Consider the rays with starting point a, which are contained in M. The union of all these rays is a convex cone with apex a. It is called the cone inscribed to M at the point a or, shortly, the *inscribed cone* of M at a (Fig. 14). Up to translations, the inscribed cone does not depend on the choice of $a \in M$. In other words: let K, K' be cones inscribed to a closed, unbounded, convex set M at points $a \in M$ and $a' \in M$, respectively. Then K' is obtained from K by the translation with $a' - a$ as translation vector.

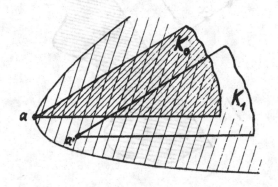

Fig. 14

A convex set $M \subset \mathbf{R}^n$ is said to be *half-bounded* if it does not contain any line. (In particular, each bounded convex set is half-bounded.) A half-bounded convex cone (i.e., the cone which does not contain any line) is called a *pointed cone*. If K is a pointed cone with an apex a, then for arbitrary points $b_1, \cdots, b_m \in K$ (different from a, in each case) and positive numbers $\lambda_1, \cdots, \lambda_m$ with $m \geq 1$, the point $b = a + \lambda_1(b_1 - a) + \cdots + \lambda_m(b_m - a) \in K$ is distinct from a. For non-pointed cones this implication is not correct.

Theorem 6.3: An arbitrary unbounded, closed, convex set $M \subset \mathbf{R}^n$ is representable as sum $M = L + M'$, where L denotes a k-flat $(0 \leq k \leq n)$ and M' denotes a half-bounded convex set contained in the complement of L. An inscribed cone of the set M is the sum of the flat L and the inscribed cone of M'. □

We remark that one can take here an arbitrary minimal face of the convex set M instead of L.

We now introduce the notion of *polar cone*. For this aim, consider an n-dimensional vector space E and its dual space E^*. Let $l \subset E$ be a ray emanating from the origin and x be a representative of l, i.e., $x \in l, x \neq o$. The set of all $f \in E^*$, satisfying the condition $\langle x, f \rangle \leq 0$, is said to be the *polar* of the ray l and is denoted by l^*. Evidently, this polar does not depend on the choice of a representative $x \in l$. Moreover, l^* is a closed half-space in E^* with the bounding hyperplane through the origin $o \in E^*$ and bd $l^* \| x^*$

for every nonzero $x \in l$. We remark that if $E = E^* = \mathbf{R}^n$ is the Euclidean space, then l^* is a half-space for which the ray l is its outward normal (Fig. 15).

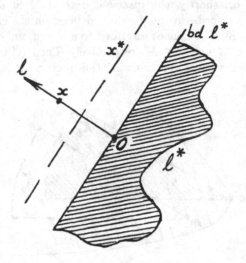

Fig. 15

Let now $K \subset E$ be a convex cone in E with the apex o. Its *polar* $K^* \subset E$ is the intersection of the polar half-spaces l^* for the rays $l \subset K$ emanating from the origin. In other words, K^* is the set of $f \in E^*$ which satisfy the condition $\langle x, f \rangle \leq 0$ for all $x \in K$ (Fig. 16):

$$K^* = \{f \in E^* : \langle x, f \rangle \leq 0 \text{ for all } x \in K\}.$$

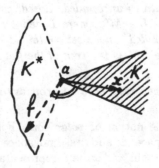

Fig. 16

This polar is a closed convex cone in E^* with the origin as its apex. Moreover, if the convex cone K is not closed, then $(\operatorname{cl} K)^* = K^*$. If in addition K is closed, then it is possible to reconstruct K with the help of K^*; indeed, $K = (K^*)^* = K^{**}$. Sometimes (if it is convenient) we consider the polar K^* of a cone $K \subset E$ with the apex a different from o:

$$K^* = \{f \in E^* : \langle x - a, f \rangle \leq 0 \text{ for all } x \in K\}.$$

It is clear that in this case $K^* = (-a + K)^*$, where $-a + K$ is the cone with apex o and hence its polar $(-a + K)^*$ is taken as above. But usually we consider the polar cone K^* only for a cone $K \subset E$ with the apex o.

Theorem 6.4: Let $K \subset E$ be a flat (of any dimension) through the origin. Then K is a convex cone with the apex o. A vector $f \in E^*$ belongs to K^* if and only if f is orthogonal to K (i.e., $\langle x, f \rangle = 0$ for all $x \in K$).   □

Theorem 6.5: Let $L \subset E$ be a closed half-space with the bounding hyperplane containing the origin, i.e., there exists a nonzero vector $f \in E^*$ such that $L = \{x : \langle x, f \rangle \leq 0\}$. Then L is a convex cone with the apex o. Its polar L^* consists of all vectors $\lambda f \in E$ with $\lambda \geq 0$. In other words, L^* is the ray emanating from the origin and passing through f, i.e., the outward normal of the half-space L (Fig. 15).   □

Theorem 6.6: A closed convex cone $K \subset E$ with the apex o is solid (i.e., K is a convex body) if and only if the cone K^* is pointed (i.e., K^* does not contain any line).   □

Theorem 6.7: Let $K_1, \cdots, K_s$ be closed, convex cones in E with the common apex at the origin. Then

$$(K_1 \cap \cdots \cap K_s)^* = \operatorname{cl} \operatorname{conv}(K_1^* \cup \cdots \cup K_s^*).   □$$

The following example shows that in Theorem 6.7 the closure operation on the right-hand side is essential.

EXAMPLE 6.8: In an orthogonal coordinate system x_1, x_2, x_3 in $E = E^* = \mathbf{R}^3$, let K_1 be the cone defined by the inequalities $2x_1x_3 - (x_2)^2 \geq 0, x_1 + x_3 \geq 0$ (Fig. 17). It touches the plane $x_3 = 0$ along the positive x_1-semiaxis. Furthermore, let K_2 be the half-space $x_1 \leq 0$. Then K_1, K_2 are convex, closed cones with the common apex at the origin, and $K_1 \cap K_2$ is the positive x_3-semiaxis. Consequently $(K_1 \cap K_2)^*$ is the polar cone of the positive x_3-semiaxis, i.e., $(K_1 \cap K_2)^*$ is the half-space $x_3 \leq 0$. On the other hand, K_1^* is the cone symmetric to K_1, i.e., $K_1^* = -K_1$, and the polar cone K_2^* is the positive x_1-semiaxis.

Finally, conv$(K_1^* \cup K_1^*)$ is the union of the *open* half-space $x_3 < 0$ and the x_1-axis. Thus conv$(K_1^* \cup K_2^*)$ is a nonclosed cone, whereas $(K_1 \cap K_2)^*$ is the closed half-space $x_3 \leq 0$. This means that

$$(K_1 \cap K_2)^* \neq \operatorname{conv}(K_1^* \cup K_2^*).   □$$

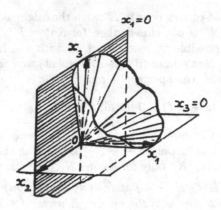

Fig. 17

The following Proposition was proved by M. K. Gavurin (private communication).

Theorem 6.9: Let $Q_1, \cdots, Q_s \subset E$ be convex, closed cones with common apex at the origin. If the cone $Q = \operatorname{conv}(Q_1 \cup \cdots \cup Q_s)$ is not closed, then there exist vectors $y_1 \in Q_1, \cdots, y_s \in Q_s$ not all equal to o such that $y_1 + \cdots + y_s = o$.

PROOF. Let $x_o \in (\operatorname{cl} Q) \setminus Q$. There exists a vector sequence $x^{(1)}, x^{(2)}, \cdots$ contained in Q such that $\lim_{k \to \infty} x^{(k)} = x_o$. Since $x^{(k)} \in Q$,

$$x^{(k)} = x_1^{(k)} + \cdots + x_s^{(k)}; \quad x_i^{(k)} \in Q_i \text{ for each } i \in \{1, \cdots, s\}.$$

We put $a_k = \max\left(\|x_1^{(k)}\|, \cdots, \|x_s^{(k)}\|\right)$. Since $x_o \in (\operatorname{cl} Q) \setminus Q$, i.e., $x_o \neq o$, we may suppose $a_k > 0$ for all $k = 1, 2, \cdots$

We are going to establish $a_k \to \infty$ as k increases. Indeed, admit that there exists a bounded subsequence of the sequence $a_1, a_2, \cdots$ Then, passing to the subsequence, we may suppose that $a_k < r$ for all $k = 1, 2, \cdots$, where r is a positive number. Consequently all points $x_i^{(k)}$ are situated in the ball of radius r centered at the origin ($i = 1, \cdots, s; \ k = 1, 2, \cdots$). Passing once more to a subsequence, we may suppose that for each $i = 1, \cdots, s$, there exists a limit $\lim_{k \to \infty} x_i^{(k)} = z_i$. Then $z_i \in Q_i$ for each $i = 1, \cdots, s$, since the cone Q_i is closed. Furthermore,

$$x_o = \lim_{k \to \infty} x^{(k)} = \lim_{k \to \infty} (x_1^{(k)} + \cdots + x_s^{(k)}) =$$

$$\lim_{k \to \infty} x_1^{(k)} + \cdots + \lim_{k \to \infty} x_s^{(k)} = z_1 + \cdots + z_s,$$

i.e., $x_o \in \operatorname{conv}(Q_1 \cup \cdots \cup Q_s) = Q$, contradicting the relation $x_o \in (\operatorname{cl} Q) \setminus Q$. This contradiction shows that $a_k \to \infty$ as k increases.

Denote the vector $\frac{1}{a_k}x_i^{(k)}$ by $y_i^{(k)}$. Then

$$\max\left(\|y_1^{(k)}\|,\cdots,\|y_s^{(k)}\|\right) = 1 \tag{1}$$

for every $k = 1, 2, \cdots$ We may suppose (passing to a subsequence if necessary) that for every $i = 1, \cdots, s$, there exists a limit $y_i = \lim_{k\to\infty} y_i^{(k)}$. It follows from (1) that at least one of the vectors $y_1, \cdots, y_s$ is distinct from o. Moreover, $y_i \in Q_i$ for each $i = 1, \cdots, s$, since the cone Q_i is closed. Finally,

$$
\begin{aligned}
y_1 + \cdots + y_s &= \lim_{k\to\infty}\left(y_1^{(k)} + \cdots + y_s^{(k)}\right) = \lim_{k\to\infty}\frac{1}{a_k}\left(x_1^{(k)} + \cdots + x_s^{(k)}\right) \\
&= \lim_{k\to\infty}\frac{1}{a_k}x^{(k)} = o,
\end{aligned}
$$

since $\lim_{k\to\infty} x^{(k)} = x_o$, $\lim_{k\to\infty} a_n = \infty$. ∎

EXAMPLE 6.10: Again, we consider Example 6.8. The cones $Q_1 = K_1^*$ and $Q_2 = K_2^*$ are closed, but the cone conv $(Q_1 \cup Q_2)$ is not closed. By virtue of Theorem 6.9, there are nonzero vectors $y_1 \in Q_1, y_2 \in Q_2$ such that $y_1 + y_2 = o$. Indeed, if e_1 is the unit vector of the positive x_1-axis, then $-e_1 \in Q_1, e_1 \in Q_2$, i.e., $y_1 = -e_1, y_2 = e_1$ are the desired vectors. This example illustrates the assertion of Theorem 6.9.

Theorem 6.11: Let $K \subset E$ be a closed, convex cone with the apex o and $F \neq \{o\}$ be an exposed face of K. Then there exists a (uniquely defined) exposed face F' of the cone K^* such that $F^* = \sup \operatorname{cone}_f K^*$ for every $f \in$ ri F' and, conversely, $(F')^* = \sup \operatorname{cone}_x F$ for every $x \in$ ri F. Thus we obtain a $1 - 1$-correspondence between the exposed faces of K distinct from $\{o\}$ and those in the cone K^*. □

Proof: For every nonzero $y \in E$, denote by l_y the ray emanating from o and passing through y. Similarly, for every nonzero $g \in E^*$, denote by l_g the ray emanating from o and passing through g.

Let now F be an exposed face of K and $x \in$ ri F. Since F is exposed, there exists a supporting half-space Π of K such that $F = K \cap$ bd Π. Denote by $h \in E^*$ a nonzero vector with $\Pi = l_h^*$. Then $h \in K^*$ and $\langle y, h \rangle = 0$ for $y \in F$, whereas $\langle y, h \rangle < 0$ for $y \in K \setminus F$ (since $F = K \cap l_h^*$).

We now fix a point $x \in$ ri F. Then l_x^* is a supporting hyperplane of K^*. Consequently $F' = K^* \cap$ bd l_x^* is an exposed face of the cone K^* consisting of the points $g \in K^*$ with $\langle x, g \rangle = 0$. Since $\langle x, h \rangle = 0$, the vector $h \neq o$ belongs to F' and hence $F' \neq \{o\}$. Let now $f \in$ ri F' and $f_1 \in$ ri F' be a vector such that $f \in]h, f_1[$. Then $\langle y, f \rangle = 0$ for $y \in F$ (since $\langle x, f \rangle = 0, x \in$ ri F, and $\langle y, f \rangle \leq 0$ for $y \in K$). Moreover, $\langle y, f \rangle < 0$ for $y \in K \setminus F$ (since $\langle y, h \rangle < 0$ and $\langle y, f_1 \rangle \leq 0$). Besides, $\langle x, g \rangle = 0$ for $g \in F'$ and $\langle x, g \rangle < 0$ for $g \in K^* \setminus F'$ (by the definition of the face F').

Finally, every supporting half-space Π of K^* at the point f has the form l_y^*, where $y \in K$ and $\langle y, f \rangle = 0$, i.e., $y \in F$. Consequently the vector $g \in E^*$

belongs to sup cone$_f K^*$ if and only if $g \in l_y^*$ for $y \in F$, i.e., $\langle y, g \rangle \leq 0$ for all $y \in F$. In other words,
sup cone$_f K^* = F^*$. Similarly, sup cone$_x K = (F')^*$.   ∎

EXAMPLE 6.12: In $\mathbf{R}^3$ with a coordinate system (x_1, x_2, x_3), consider the cone M defined by the inequalities $(x_1)^2 \leq 2x_2 x_3, x_2 \geq 0, x_3 \geq 0$. Hence M is a circular cone with positive x_2- and x_3-axis as generators and with its axis coinciding with the bisector of the first coordinate angle of the plane $x_1 = 0$, see Fig. 18. Furthermore, the ray defined by $x_2 = x_3 = 0, x_1 \geq 0$, will be denoted by l. The cone $K = $ conv $(M \cup l)$ is closed and convex. Moreover, the ray F, defined by $x_1 = x_3 = 0, x_2 \geq 0$, is a face of K, but this face is not exposed. We remark that F^* is not a supporting cone of the cone K^*. Since $K = $ conv $(M \cup l)$ is closed, the equality

$$K^* = (\text{ cl conv } (M \cup l))^* = M^* \cap l^*$$

holds, and this implies that K^* is the intersection of the cone $M^* = -M$ with the half-space $x_1 \leq 0$. Thus the cone F^* is the half-space $x_2 \leq 0$, and it is not a supporting cone of the cone K^* at a point $o \in K^*$.   □

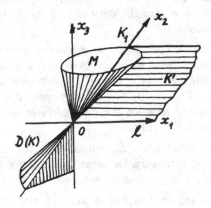

Fig. 18

Exercises

1. Prove that any closed, convex set K, having only one minimal face, is a convex cone.

2. Prove that if M is a half-bounded convex set (i.e., it does not contain any line), then each of its minimal faces is 0-dimensional.

3. Prove that an unbounded convex set $M \subset \mathbf{R}^n$ is half-bounded if and only if its inscribed cone is half-bounded.

4. Let $M_1, \cdots, M_s \subset \mathbf{R}^n$ be unbounded, convex sets and $N_1, \cdots, N_s$ be their translates. Prove that the inscribed cone of conv $N_1 \cup \cdots \cup N_s$ is a translate of the inscribed cone of conv $M_1 \cup \cdots \cup M_s$.

5. Prove that a closed, convex body $M \subseteq E$ is a convex cone if and only if each two of its faces have a nonempty intersection.

6. Prove that if M is a closed, convex set with $o \in M$, then the convex set cl $\left(\bigcup_{k>0} kM \right)$ is the supporting cone of M at the point o. Formulate a similar assertion for sup $\text{cone}_x M$ in the case when $x \in M$ does not coincide with the origin.

7. Give an example of a closed, convex cone $K \subset E$ with the apex o and a linear mapping $\pi : E \to E$ such that the cone $\pi(K)$ with the apex o is not closed.

8. Prove that the supporting cone sup $\text{cone}_x M$ of a convex body M at a point $x \in \text{bd } M$ is a half-space if and only if x is a regular boundary point of M.

9. Prove that if K is a closed, convex cone with an apex x, then each face of K is a cone with the apex x.

10. Prove that if K is a closed, convex cone with an apex x, then its supporting cone at any point $a \in \text{bd } K$ is a cone with the apex x, too.

11. Prove that if F is an exposed face of a closed, convex cone K, then for every point $x \in \text{ri } F$ the cone sup $\text{cone}_x K$ has aff F as its minimal face. Does this remain true for a nonexposed face?

12. Show that for every $k \in \{1, \cdots, n\}$ there is a closed, convex body $M \subset \mathbf{R}^n$ such that its inscribed cone is k-dimensional.

13. A compact, convex body $M \subset \mathbf{R}^n$ is said to be *nearly conic* if its inscribed cone K with an apex $a \in M$ is n-dimensional and, moreover, $M \subset U_h(K)$ for a positive real number h. Prove that if Γ is a flat through the point a and Γ has a non-empty intersection with int K, then $M \cap \Gamma$ is a nearly conic, convex body with the inscribed cone $K \cap \Gamma$.

14. Let $M \subset E$ be a compact, convex body with the origin in its interior and K be its supporting cone at a boundary point x. Prove that if a is a regular boundary point of K and Γ is a supporting hyperplane of K at a, then $\Gamma = f^*$ with $f \in \text{cl exp } M^*$.

15. Let $K \neq E$ be a closed, convex cone with the apex o and $l \subset K$ be a ray emanating from o. Prove that l^* is a supporting half-space of K^*. Prove the converse assertion.

16. Let $K \subset E$ be a closed, convex cone with $K \neq \{o\}$ and aff $K \neq E$. Prove that $(\text{aff } K)^*$ is the minimal face of the cone K^*.

17. Let K be a closed, convex cone with the apex o and F be a face of K. Prove that the face F is exposed if and only if F^* is a supporting cone for K^*.

18. Let $M \subset \mathbf{R}^n$ be an n-dimensional polytope. Prove that the cones (sup $\text{cone}_a M)^*$ taken over the vertices of M have pairwise no common interior point and the union of the cones coincides with $\mathbf{R}^n$.

§7 The Farkas Lemma and its generalization

A closed, convex cone $K \subset E = \mathbf{R}^n$ with the apex o is said to be *polyhedral* if it is the intersection of a finite number of closed half-spaces $\Pi_1, \cdots, \Pi_s$ with the bounding hyperplanes through o:

$$K = \Pi_1 \cap \cdots \cap \Pi_s;\ o \in \mathrm{bd}\ \Pi_i \ \text{for}\ i = 1, \cdots, s. \tag{1}$$

Theorem 7.1: For every polyhedral cone $K \subset E$ its polar cone K^* is polyhedral as well. $\square$

The following theorem contains an effective description of the polar cone K^* for every polyhedral cone $K \subset \mathbf{R}^n$. In Mathematical Programming and in Optimization Theory, this theorem is known as the *Farkas Lemma*. For extensions and generalizations we refer to [S-Wi] and [Zi].

Theorem 7.2: Let $K \subset E$ be a polyhedral cone with the apex o and (1) be its representation as the intersection of closed half-spaces with bounding hyperplanes through o. Denote by $e_1, \cdots, e_s$ the outward normals of these half-spaces, i.e., $e_i \in \Pi_i^*$ for $i = 1, \cdots, s$. A vector $f \in E^*$ belongs to the polar cone K^* if and only if f is a linear combination of the vectors $e_1, \cdots, e_s$ with nonnegative coefficients. $\square$

Consider a geometrical description of the Farkas Lemma. Suppose that the cone K is solid (i.e., n-dimensional) and there is no superfluous half-space in the representation (1), i.e., if we remove any of the half-spaces, then the intersection of the other ones does not coincide with K. Under these conditions, the intersection $K \cap \mathrm{bd}\Pi_i$ is an $(n-1)$-dimensional face of the cone K for each $i \in \{1, \cdots, s\}$, and the boundary of K is the union of these facets:

$$\mathrm{bd}\ K = (K \cap \mathrm{bd}\ \Pi_1) \cup \cdots \cup (K \cap \mathrm{bd}\ \Pi_s).$$

Let now l_i be the ray emanating from the origin and containing the vector e_i, i.e., $l_i = \Pi_i^* \subset E^*$ is the outward normal ray of the half-space $\Pi_i, i = 1, \cdots, s$. The Farkas Lemma affirms

$$K^* = \mathrm{conv}\ (l_1 \cup \cdots \cup l_s).$$

By this description, the Farkas Lemma follows immediately from Theorem 6.7.

For the sequel, it is convenient to suppose that $E = E^* = \mathbf{R}^n$ is the Euclidean space. Let K be a solid, closed, convex cone in $\mathbf{R}^n$ with the apex o. By $H(K)$ denote the set of all unit outward normals of the cone K at its regular boundary points. Thus $H(K)$ is a subset of the unit sphere $S^{n-1} \subset \mathbf{R}^n$. A vector $p \in S^{n-1}$ belongs to $H(K)$ if and only if there is a regular boundary point $a \in K$ such that the supporting half-space of K at a is given by $\{x : \langle x, p \rangle \le 0\}$. For example, if the vectors $e_1, \cdots, e_s$ in the Farkas Lemma are unit, then $H(K) = \{e_1, \cdots, e_s\}$, i.e., for K the corresponding set $H(K)$ is the set of all unit outward normals for its facets. We now formulate a generalization of the Farkas Lemma.

Theorem 7.3: Let $K \subset \mathbf{R}^n$ be a solid, closed, convex cone (distinct from $\mathbf{R}^n$ and not necessarily polyhedral) with the apex o. A vector f belongs to the polar cone K^* if and only if f is a linear combination of some vectors $e_1, \cdots, e_s \in \mathrm{cl}\, H(K), s \leq n$, with nonnegative coefficients. In other words,

$$K^* = \mathrm{conv} \bigcup_{e \in \, \mathrm{cl}\, H(K)} l(e),$$

where $l(e)$ is the ray emanating from the origin and containing e.   □

PROOF: Each vector from $H(K)$ belongs to the polar cone K^*. Hence each vector $e \in \mathrm{cl}\, H(K)$ also belongs to K^* (since K^* is closed). It follows that if $f = \lambda_1 e_1 + \cdots + \lambda_s e_s$ with $e_1, \cdots, e_s \in \mathrm{cl}\, H(K)$ and nonnegative coefficients $\lambda_1, \cdots, \lambda_s$, then f belongs to K^*.

We now establish the converse assertion. Fix an arbitrary nonzero vector $f \in K^*$. For every vector $g \in H(K)$, denote by $\Pi(g)$ the closed half-space $\{x : \langle x, g \rangle \leq 0\}$. Then $K = \bigcap_{g \in H(K)} \Pi(g)$. Furthermore, we denote by M the compact set $S^{n-1} \cap K$, and for every $k = 1, 2, \cdots$ we denote by N_k the compact set $S^{n-1} \setminus U_{1/k}(M)$. If $x \in N_k$, then there exists a vector $g \in H(K)$ such that $x \notin \Pi(g)$. In other words, the open half-spaces $\mathbf{R}^n \setminus \Pi(g)$, taken for all $g \in H(K)$, form an open covering of the compact set N_k. Hence there exists a finite collection of vectors $g_i^{(k)} \subset H(K), i = 1, \cdots, s_k$, such that

$$N_k \subset \bigcup_{i=1}^{s_k} \left(\mathbf{R}^n \setminus \Pi(g_i^{(k)}) \right),$$

i.e., $U_{1/k}(M) \supset V$, where

$$V = S \cap \bigcap_{i=1}^{s_k} \Pi \left(g_i^{(k)} \right).$$

This means that for every unit vector $x' \in V$ there exists a unit vector $x \in M$ with $\|x - x'\| \leq \frac{1}{k}$, i.e., the angle between x, x' (which is equal to $2\sin^{-1}(\frac{1}{2}\|x - x'\|)$, Fig. 19) is less than $\varepsilon_k = 2\sin^{-1}(1/2k)$.

Since the angle between the fixed vector f and every $x \in M \subset K$ is not less than $\frac{\pi}{2}$, we conclude that for every unit vector $x' \in V$ the angle between the vectors f, x' is *greater* than $\frac{\pi}{2} - \varepsilon_k$. We now denote by Q_k the convex cone consisting of all the vectors which form with f angles *not greater* than $\frac{\pi}{2} - \varepsilon_k$. Then the convex cones Q_k and V have no common points except for the origin. Consequently the cones can be separated, i.e., there exists a vector f_k with $\|f_k\| = \|f\|$ such that

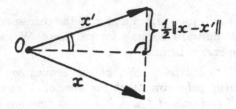

Fig. 19

$$\langle f_k, x \rangle \geq 0 \text{ for all } x \in Q_k, \tag{2}$$

$$\langle f_k, x \rangle \leq 0 \text{ for all } x \in V. \tag{3}$$

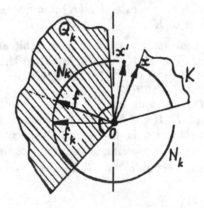

Fig. 20

By the inequality (2), the angle between f and f_k is not greater than ε_k (Fig. 20) and hence $\|f - f_k\| \leq \frac{1}{k}\|f\|$. By virtue of the Farkas Lemma, the inequality (3) means that

$$f_k = \sum \lambda_i^{(k)} g_i^{(k)} \text{ with } \lambda_i^{(k)} \geq 0$$

(summation over $i = 1, \cdots, s_k$). According to Theorem 1.2, we may suppose that this representation contains no more than n summands $\lambda_i^{(k)} g_i^{(k)}$. Moreover, adding summands $\lambda_i^{(k)} g_i^{(k)}$ with zero coefficients $\lambda_i^{(k)}$, we may suppose that the number of summands is exactly equal to n:

$$f_k = \lambda_1^{(k)} g_1^{(k)} + \cdots + \lambda_n^{(k)} g_n^{(k)}, \tag{4}$$

where $\lambda_i^{(k)} \geq 0, g_i^{(k)} \in H(K)$ for $i = 1, \cdots, n$. Conducting this construction for each $k = 1, 2, \cdots$, we obtain a sequence $f_1, f_2, \cdots$ such that the expansions (4) hold and, moreover,

$$\|f - f_k\| \leq \frac{1}{k}\|f\|, \text{ i.e., } \lim_{k\to\infty} f_k = f.$$

Since the cone K is n-dimensional, there exists a unit vector $x_o \in \text{int } K$. Let $r > 0$ be a number such that the ball of radius r centered at x_o is contained in K. Then for every nonzero vector $e \in H(K)$ the angle between e and x_o is not smaller than $\frac{\pi}{2} + \sin^{-1} r$ (Fig. 21) and hence

$$\langle e, x_o \rangle \leq \cos\left(\frac{\pi}{2} + \sin^{-1} r\right) = -r.$$

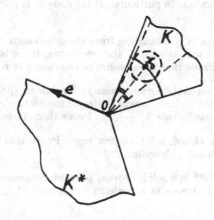

Fig. 21

In particular, $\langle g_i^{(k)}, x_o \rangle \leq -r$ for all $i = 1, \cdots, n$ and $k = 1, 2, \cdots$ Hence using (4), we obtain:

$$\langle f_k, x_o \rangle = \lambda_1^{(k)}\langle g_1^{(k)}, x_o \rangle + \cdots + \lambda_n^{(k)}\langle g_n^{(k)}, x_o \rangle \leq -r\left(\lambda_1^{(k)} + \cdots + \lambda_n^{(k)}\right).$$

Since $\langle f_k, x_o \rangle \geq -\|f_k\| = -\|f\|$, we conclude from the last inequality

$$\lambda_1^{(k)} + \cdots + \lambda_n^{(k)} \leq -\frac{1}{r}\langle f_k, x_o \rangle \leq \frac{\|f\|}{r},$$

and so far $|\lambda_i^{(k)}| \leq \frac{\|f\|}{r}$ for $i = 1, \cdots, n$. Consequently the sequence $\lambda_i^{(k)}$, $k = 1, 2, \cdots$, is bounded. Moreover, all the vectors $g_i^{(k)} \in H(K)$ are unit. Thus (passing to a subsequence, if necessary) we may suppose that for any $i = 1, \cdots, n$ there exist limits

$$\lim_{k \to \infty} g_i^{(k)} = e_i^*, \quad \lim_{k \to \infty} = \lambda_i^{(k)} = \lambda_i^* \geq 0.$$

We notice that $e_i^* \in \mathrm{cl}\, H(K)$, since $g_i^{(k)} \in H(K)$ for all k. It follows that

$$f = \lim_{k \to \infty} f_k = \lim_{k \to \infty} \left(\lambda_1^{(k)} g_1^{(k)} + \cdots + \lambda_n^{(k)} g_n^{(k)} \right) = \lambda_1^* e_1^* + \cdots + \lambda_n^* e_n^*. \quad \blacksquare$$

Exercises

1. Let $l_1, \cdots, l_s$ be rays in E emanating from the origin. Prove that $K = \mathrm{conv}\ (l_1 \cup \cdots \cup l_s)$ is either E or a polyhedral cone (in particular, K is closed).

2. Let $K \subset E$ be a pointed, solid, closed, convex cone. Prove that there exists a hyperplane Γ through an interior point of K such that the intersection $K \cap \Gamma$ is a compact, convex set. In particular, if the cone K is polyhedral, then $K \cap \Gamma$ is a polytope.

3. Let $l_1, \cdots, l_s$ be rays in E emanating from the origin such that $K = \mathrm{conv}\ (l_1 \cup \cdots \cup l_s)$ does not coincide with E and, moreover, there is no superfluous ray among $l_1, \cdots, l_s$. Prove that the number of the facets of K^* is equal to s.

4. Let $e_1, \cdots, e_n$ be an orthonormal basis in Euclidean space $E = E^* = \mathbf{R}^n$. Denote by K the polyhedral cone consisting of the vectors $x = \lambda_1 e_1 + \cdots + \lambda_n e_n$ with nonnegative coefficients $\lambda_1, \cdots, \lambda_n$. Prove that $K^* = -K$.

5. Let $K \subset \mathbf{R}^n$ be a closed, solid, convex cone. Prove that if $H(K)$ is infinite, then the cone K is not polyhedral.

6. Prove that if $K \subset \mathbf{R}^3$ is a solid, convex, pointed, rotational cone, then the set $H(K)$ is the circumference of a circle in the sphere $S^2 \subset \mathbf{R}^3$ (which is not a great circle of S^2).

7. Let $K \subset \mathbf{R}^n$ be a solid, closed, convex, pointed cone with the apex o and

$$Q = \mathrm{conv} \bigcup_{e \in H(K)} l(e).$$

Prove that $\mathrm{cl}\, Q$ coincides with K^*. Give an example for the case when Q is not closed.

8. Let $e \in \mathbf{R}^n$ be a nonzero vector and $\alpha > 0$ be a number smaller than $\frac{\pi}{2}$. Denote by Q_α the cone that consists of the vector o and all nonzero vectors which form with e angles not greater than α. Prove that $Q_\alpha^* = -Q_{\frac{\pi}{2} - \alpha}$. In particular, if $K = Q_{\frac{\pi}{4}}$, then the polar cone K^* coincides with $-K$.

9. Prove that $\dim(\mathrm{lin}\, H(K)) = m$ if and only if the cone K has an $(n - m)$-dimensional minimal face.

10. Let $K \subset \mathbf{R}^n$ be a cone with the apex o that contains the ball of radius r centered at a point x_o, where $\|x_o\| = 1$. Prove that for arbitrary unit vectors $f, g \in K^*$ the relation $\langle f, g \rangle \geq 2r^2 - 1$ holds.

11. Let $K \subset \mathbf{R}^n$ be a closed, convex cone with the apex o. Prove that a ray l, emanating from o, is an exposed face of K if and only if the unit vector $e \in l$ belongs to $H(K^*)$.

12. Let $K \subset \mathbf{R}^n$ be a closed, convex cone with the apex o and $M \subset K$ be the *minimal* (by inclusion) closed cone (nonconvex!) satisfying the condition conv $M = K$. Prove that $(M \cap S^{n-1}) \subset \mathrm{cl}\, H(K^*)$.

§8 Separable systems of convex cones

In this section we will give a short account of results which were obtained in the papers [Bt 3,4] (for Banach spaces in [Bt 9,10]). These results are applied in the Mathematical Theory of Optimal Control and in Mathematical Programming [Bt 11,12]. For us the theory of separation of convex cones is the main tool in the proof of a Helly-type theorem for H-convex sets [Bt 13] in §22.

Let $K_1, \cdots, K_s \subset \mathbf{R}^n$ be convex cones (not necessarily closed) with a common apex a. We say that the cones $K_1, \cdots, K_s$ are *separable* in $\mathbf{R}^n$ if there exists a hyperplane L passing through a that separates one of the cones $K_1, \cdots, K_s$ from the intersection of the other ones. This means that for an index i the cone K_i is contained in one closed half-space determined by L, whereas the intersection of the other cones is contained in the other closed half-space (Fig. 22).

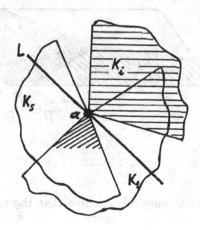

Fig. 22

<u>*Theorem 8.1:*</u> *Let* $K_1, \cdots, K_s \subset \mathbf{R}^n$ *be convex cones with a common apex* a. *The cones* $K_1, \cdots, K_s$ *are separable if and only if either* ri $K_1 \cap \cdots \cap$ ri $K_s = \emptyset$

or there exists a hyperplane that contains one of the cones and the intersection of the other ones.   □

PROOF: First we prove the part "if". Assume ri $K_1 \cap \cdots \cap$ ri $K_s = \emptyset$. Then there exists a positive integer $m < s$ such that

$$\text{ri } K_1 \cap \cdots \cap \text{ri } K_m \neq \emptyset, \quad \text{ri } K_1 \cap \cdots \cap \text{ri } K_m \cap \text{ri } K_{m+1} = \emptyset. \qquad (1)$$

According to Theorem 1.6,

$$\text{ri } (K_1 \cap \cdots \cap K_m) = \text{ri } K_1 \cap \cdots \text{ri } K_m.$$

Hence (1) can be rewritten in the form

$$\text{ri } (K_1 \cap \cdots \cap K_m) \cap \text{ri } K_{m+1} = \emptyset.$$

This means that the cones $K_1 \cap \cdots \cap K_m$ and K_{m+1} can be separated (Theorem 1.7). Consequently the intersection of all the cones $K_1, \ldots, K_s$ except for K_{m+1} can be separated from K_{m+1}, i.e., the system of the cones $K_1, \cdots, K_s$ is separable.

Assume now that there exists a hyperplane L that contains one of the cones and the intersection of the other ones, say $K_1 \subset L, K_2 \cap \cdots \cap K_s \subset L$ (Fig. 23). Then $K_1 \subset \Pi, K_2 \cap \cdots \cap K_s \subset \Pi'$, where Π, Π' are closed half-spaces which are determined by L, i.e., the system of cones $K_1, \cdots, K_s$ is separable.

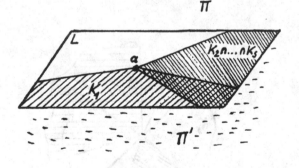

Fig. 23

We now prove the part "only if". Assume that the cones $K_1, \cdots, K_s$ are separable. Then we can choose an integer l, $2 \le l \le s$, such that there are l cones among $K_1, \cdots, K_s$ which are separable, while (in the case $l > 2$) every $l - 1$ of the cones are not separable. Without loss of generality we may suppose that the cones $K_1, \cdots, K_l$ are separable.

Consider the case $l = 2$, i.e., the cones K_1, K_2 are separable. In this case, $K_1 \subset \Pi, K_2 \subset \Pi'$, where Π, Π' are two closed half-spaces determined by a hyperplane $L = \text{bd } \Pi = \text{bd } \Pi'$. If $K_1 \subset L, K_2 \subset L$, then $K_1 \subset L, K_2 \cap \cdots \cap K_s \subset L$, i.e., the assertion "only if" is true. Otherwise, at least one of the cones K_1, K_2 (say K_1) is not contained in L. This means that ri $K_1 \subset \text{int } \Pi$, i.e., $(\text{ri } K_1) \cap (\text{ri } K_2) = \emptyset$ and consequently ri $K_1 \cap$ ri $K_2 \cap \cdots \cap$ ri $K_s = \emptyset$. Thus in the case $l = 2$, the assertion "only if" is true.

Consider the case $l > 2$. Then $K_1, \cdots, K_l$ are separable. Without loss of generality we may suppose that K_1 and $K_2 \cap \cdots \cap K_l$ are separable, i.e., $K_1 \subset \Pi, K_2 \cap \cdots \cap K_l \subset \Pi'$, where Π, Π' are the closed half-spaces with the common boundary $L = \text{bd } \Pi = \text{bd } \Pi'$. Moreover, the cones $K_2, \cdots, K_l$ are not separable. This means, by virtue of the part "if" which is already proved, that ri $K_2 \cap \cdots \cap$ ri $K_l \neq \emptyset$, i.e., according to Theorem 1.6, ri $(K_2 \cap \cdots \cap K_l) = $ ri $K_2 \cap \cdots \cap$ ri $K_l \neq \emptyset$.

If now $K_1 \subset L, K_2 \cap \cdots \cap K_l \subset L$, then $K_1 \subset L, K_2 \cap \cdots \cap K_s \subset L$, i.e., the assertion "only if" is true. Otherwise, i.e., if at least one of the cones $K_1, K_2 \cap \cdots \cap K_l$ is not contained in L, we conclude (as above) that ri $K_1 \cap (K_2 \cap \cdots \cap K_l) = \emptyset$, i.e., ri $K_1 \cap$ ri $K_2 \cap \cdots \cap$ ri $K_l = \emptyset$ and hence ri $K_1 \cap \cdots \cap$ ri $K_s = \emptyset$. Thus, at any rate, the assertion "only if" is true. ∎

Corollary 8.2: If the cones $K_1, \cdots, K_s$, maybe except for K_s, are n-dimensional, then int $K_1 \cap \cdots \cap$ int $K_{s-1} \cap K_s = \emptyset$ is a necessary and sufficient condition for separability. □

Theorem 8.1 gives a *geometrical* separability condition for a system of convex cones with a common apex. We are going now to establish an *algebraic* condition for separability. To understand how it may be formulated, consider at first the case $s = 2$.

EXAMPLE 8.3: Let $K_1, K_2 \subset \mathbf{R}^n$ be convex cones with a common apex a which are separable, i.e., there exists a hyperplane L through a such that K_1 is contained in one closed half-space P_1 determined by L, while K_2 is contained in the other closed half-space P_2 (Fig. 24). Let p_1 and $p_2 = -p_1$ be unit outward normals of these half-spaces. Then for every $x \in P_i$ the inequality $\langle p_i, x - a \rangle \leq 0$ holds. In particular, $\langle p_i, x - a \rangle \leq 0$ for every point $x \in K_i$ (since $K_i \subset P_i$), i.e., $p_i \in K_i^*, i = 1, 2$. Thus, if the cones K_1, K_2 are separable, then there exist nonzero vectors $p_1 \in K_1^*, p_2 \in K_2^*$ satisfying the condition $p_1 + p_2 = o$. It is easily shown that this necessary condition of separability is sufficient, too. □

The following Theorem generalizes the situation considered for two cones in Example 8.3.

Theorem 8.4: Convex cones $K_1, \cdots, K_s \subset \mathbf{R}^n$ with a common apex a are separable in $\mathbf{R}^n$ if and only if there exist vectors $p_1 \in K_1^*, \cdots, p_s \in K_s^*$ not all equal to o such that

$$p_1 + \cdots + p_s = o. \quad \square \tag{2}$$

PROOF: Without loss of generality we may suppose that $a = o$, i.e., the cones $K_1, \cdots, K_s$ have the common apex at the origin.

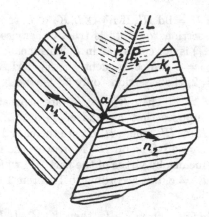

Fig. 24

First of all, we prove the part "if". Let $p_1 \in K_1^*, \cdots, p_s \in K_s^*$ be vectors not all equal to o such that (2) holds. Without loss of generality we may assume $p_s \neq o$. Denote by Π, Π' the half-spaces $\{x : \langle p_s, x \rangle \leq 0\}$, $\{x : \langle p_s, x \rangle \geq 0\}$, respectively. If $x \in K_s$, then $\langle p_s, x \rangle \leq 0$ (since $p_s \in K_s^*$). Therefore $K_s \subset \Pi$. Furthermore, if $x \in K_1 \cap \cdots \cap K_{s-1}$, then $\langle p_1, x \rangle \leq 0, \cdots, \langle p_{s-1}, x \rangle \leq 0$ and hence (by virtue of (2))

$$\langle p_s, x \rangle = \langle -p_1 - \cdots - p_{s-1}, x \rangle = -\langle p_1, x \rangle - \cdots - \langle p_{s-1}, x \rangle \geq 0.$$

Thus $K_1 \cap \cdots \cap K_{s-1} \subset \Pi'$. This means that the hyperplane $L = \Pi \cap \Pi'$ separates the cones $K_1 \cap \cdots \cap K_{s-1}$ and K_s, i.e., the cones $K_1, \cdots, K_s$ are separable.

Now we prove the part "only if". Let the cones $K_1, \cdots, K_s$ be separable. Then we can choose an integer l, $2 \leq l \leq s$, such that there are l cones among $K_1, \cdots, K_s$ which are separable, while (in the case $l > 2$) every $l-1$ of the cones are not separable. Without loss of generality we may suppose that $K_1, \cdots, K_l$ are separable.

In the case $l = 2$, i.e., when the cones K_1, K_2 are separable, there exist nonzero vectors $p_1 \in K_1^*, p_2 \in K_2^*$ such that $p_1 + p_2 = o$ (cf. Example 8.3). We now put $p_3 = \cdots = p_s = o$ (and therefore $p_i \in K_i^*$ for all $i = 3, \cdots, s$). Then (2) holds.

Consider the case $l > 2$. Then the cones $K_1, \cdots, K_l$ are separable. Without loss of generality we may suppose that K_1 and $K_2 \cap \cdots \cap K_l$ are separable, i.e., $K_1 \subset \Pi, K_2 \cap \cdots \cap K_l \subset \Pi'$, where Π, Π' are the half-spaces with a common boundary hyperplane. Moreover, the cones $K_2, \cdots, K_l$ are not separable. Hence, according to Theorem 8.1, ri $K_2 \cap \cdots \cap$ ri $K_l \neq \emptyset$. Now, by virtue of Theorem 1.6,

$$\text{cl } K_2 \cap \cdots \cap \text{cl } K_l = \text{cl } (K_2 \cap \cdots \cap K_l) \subset \Pi'.$$

Denote by p the unit outward normal of the half-space Π. Then $-p$ is the unit outward normal of the half-space Π'. The inclusions $K_1 \subset \Pi$, cl $K_2 \cap \cdots \cap$ cl $K_l \subset \Pi'$ mean that $p \in K_1^*$ and $-p \in (\text{cl } K_2 \cap \cdots \cap \text{cl } K_l)^*$. We consider two different possibilities:

a) The cone conv $(K_2^* \cup \cdots \cup K_l^*)$ is closed. Since $(\text{cl } K_i)^* = K_i^*$, we have (by Theorem 6.7)

$$\begin{aligned}
\text{conv } (K_2^* \cup \cdots \cup K_l^*) &= \text{cl conv } (K_2^* \cup \cdots \cup K_l^*) \\
&= \text{cl conv } (\text{cl } K_2^* \cup \cdots \cup \text{cl } K_l^*) \\
&= (\text{cl } K_2 \cap \cdots \cap \text{cl } K_1)^*.
\end{aligned}$$

Consequently $-p \in \text{conv } (K_2^* \cup \cdots \cup K_l^*)$. This means that there exist vectors $p_2 \in K_2^*, \cdots, p_l \in K_l^*$ such that $-p = p_2 + \cdots + p_l$. We put $p_1 = p \in K_1^*, p_{l+1} = \cdots = p_s = o$. Then the relation (2) holds and not all $p_1, \cdots, p_s$ are equal to o.

b) The cone conv $(K_2^* \cup \cdots \cup K_l^*)$ is not closed. Then there exist vectors $p_2 \in K_2^*, \cdots, p_l \in K_l^*$ not all equal to o such that $p_2 + \cdots + p_l = o$ (see Theorem 6.9). We now put $p_1 = o, p_{l+1} = \cdots = p_s = o$ (and therefore $p_i \in K_i^*$ for all $i = 1, \cdots, s$). Then (2) holds. ∎

Corollary 8.5: The cones $K_1, \cdots, K_s \subset \mathbf{R}^n$ with a common apex are separable if and only if the cones cl $K_1, \cdots,$ cl K_s are separable.

Indeed, since $(\text{cl } K_i)^* = K_i^*$, the separability condition contained in Theorem 8.4 is the same for $K_1, \cdots, K_s$ and for cl $K_1, \cdots,$ cl K_s. ∎

We now establish a new geometrical condition for nonseparability.

Theorem 8.6: Let $K_1, \cdots, K_s$ be convex cones in $\mathbf{R}^n$ with the common apex at the origin and ri $K_1 \cap \cdots \cap$ ri $K_s \neq \emptyset$. The cones $K_1, \cdots, K_s$ are not separable if and only if there exists a direct decomposition $\mathbf{R}^n = L_1 \oplus \cdots \oplus L_s$ such that $L_q \subset$ aff K_p for every $p \neq q$ (where it is possible that some of the subspaces L_i are trivial). □

PROOF: First we prove the part "if". Suppose that a decomposition $\mathbf{R}^n = L_1 \oplus \cdots \oplus L_s$ exists with $L_q \subset$ aff K_p for $p \neq q$. From this we have to deduce that the cones $K_1, \cdots, K_s$ are not separable. Assume the contrary, i.e., the cones $K_1, \cdots, K_s$ are separable. Then, by virtue of Theorem 8.1, there exist a hyperplane L and an index $q \in \{1, \cdots, s\}$ such that $K_q \subset L$ and the intersection of the cones $K_1, \cdots, K_s$, except for K_q, is also contained in L. Consequently the flat

$$\bigcap_{p \neq q} \text{aff } K_p = \text{aff } \left(\bigcap_{p \neq q} K_p \right)$$

is contained in L as well (cf. Theorem 1.6). This means that $L_q \subset \bigcap_{p \neq q} \text{aff } K_p \subset L$. Moreover, since $K_q \subset L$ (i.e., aff $K_q \subset L$), we conclude that $L_p \subset$ aff $K_q \subset$

L for every $p \neq q$. Thus *all* the subspaces $L_1, \cdots, L_s$ are contained in L, i.e., $\mathbf{R}^n \subset L$, contradicting that L is a hyperplane. Hence the cones $K_1, \cdots, K_s$ are not separable.

We now prove the part "only if", i.e., we suppose that $K_1, \cdots, K_s$ are not separable. All the more, the cones aff $K_1, \cdots$, aff K_s (with the common apex o) are not separable. Denote by H_q the intersection of the subspaces aff $K_1, \cdots$, aff K_s, except for aff K_q. Then $H_q \subset$ aff K_q for $p \neq q$. We have to prove that $H_1 + \cdots + H_s = \mathbf{R}^n$ (then there exist subspaces $L_1 \subset H_1, \cdots, L_s \subset H_s$ such that $\mathbf{R}^n = L_1 \oplus \cdots \oplus L_s$). To establish the equality $H_1 + \cdots + H_s = \mathbf{R}^n$, it suffices to show that $H_1^* \cap \cdots \cap H_s^* = \{o\}$ (cf. Theorem 6.7). Admit, on the contrary, that there exists a nonzero vector $y \in H_1^* \cap \cdots \cap H_s^*$. We remark that H_q^* is the vector sum of all the subspaces (aff $K_1)^*, \cdots,$ (aff $K_s)^*$, except for (aff $K_q)^*$ (by virtue of Theorem 6.7). Since $y \in H_1^*$, we conclude that $y = x_2 + \cdots + x_s$, where $x_p \in$ (aff $K_p)^*$ for $p = 2, \cdots, s$. Moreover, at least one of the vectors $x_2, \cdots, x_s$ is not equal to zero, because $y \neq o$. Without loss of generality we may suppose that $x_s \neq o$. Since now $y \in H_s^*$, we obtain $y = x_1' + \cdots + x_{s-1}'$, where $x_p' \in$ (aff $K_p)^*$ for $p = 1, \cdots, s-1$. Thus,

$$x_1' + (x_2' - x_2) + \cdots + (x_{s-1}' - x_{s-1}) - x_s = y - y = o,$$

i.e., we have vectors

$$x_1' \in (\text{aff } K_1)^*, (x_2' - x_2) \in (\text{aff } K_2)^*, \cdots,$$
$$(x_{s-1}' - x_{s-1}) \in (\text{aff } K_{s-1})^*, -x_s \in (\text{aff } K_s)^*,$$

not all equal to zero (since $x_s \neq o$) with zero sum. This means, by virtue of Theorem 8.4, that the flats aff $K_1, \cdots$, aff K_s are separable, contradicting what was said above. ∎

Exercises

1. Prove that *solid*, convex cones $K_1, \cdots, K_s$ with a common apex o are not separable if and only if (int $K_1) \cap \cdots \cap ($ int $K_s) \neq \emptyset$.

2. Let $K_1, \cdots, K_s$ be convex, solid cones in $\mathbf{R}^n$ having a common apex. Prove that these cones have a common interior point (i.e., int $K_1 \cap \cdots \cap$ int $K_s \neq \emptyset$) if and only if there exists a vector $b \in \mathbf{R}^n$ such that $\langle b, x \rangle < 0$ for any nonzero $x \in K_i^*$ and $i \in \{1, \cdots, s\}$.

3. Let $K_1, \cdots, K_s$ be convex, solid cones in $\mathbf{R}^n$ having a common apex. Prove that if the cones have a common interior point, then

$$(K_1 \cap \cdots \cap K_s)^* = \text{conv } (K_1^* \cup \cdots \cup K_s^*) \tag{3}$$

(without the closure operation on the right hand side).

4. Let $K_1, \cdots, K_s$ be convex cones in $\mathbf{R}^n$ with the common apex at the origin such that the cones $K_1, \cdots, K_{s-1}$ are solid. Prove that int $K_1 \cap \cdots \cap$ int $K_{s-1} \cap K_s = \{o\}$ if and only if there are vectors $a_1 \in K_1^*, \cdots, a_s \in K_s^*$ not all equal to o such that $a_1 + \cdots + a_s = o$ (Dubovitski-Miljutin Theorem [D-M]).

5. Prove that solid, convex cones $K_1, \cdots, K_s \subset E$ with the common apex o are nonseparable if and only if there exists a half-space $\Pi^* \subset E^*$ with the bounding hyperplane through o such that for every $q \in \{1, \cdots, s\}$ the cone K_q^* is contained in Π^* and $K_q^* \cap \mathrm{bd}\, \Pi^* = \{o\}$, see Fig. 25. Does this remain true for nonsolid cones?

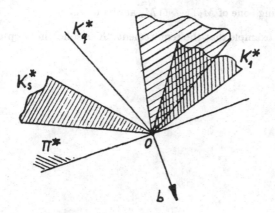

Fig. 25

6. Let the convex cones $K_1, \cdots, K_s$ with the common apex o be nonseparable. Prove that the cones $K_1, \cdots, K_{s-1}$ are nonseparable, too.

7. Let $K_1, \cdots, K_s$ be subspaces of E. Prove that the sets $K_1, \cdots, K_s$ (considered as cones with the common apex o) are separable if and only if there exists a line $L \subset E^*$ such that $L \subset K_1^* \cap \cdots \cap K_s^*$.

8. Prove that the separability of convex cones $K_1, \cdots, K_s \subset E$ with the common apex o means (in the "dual language") that there exist an index $q \in \{1, \cdots, s\}$ and a nonzero vector $f \in E^*$ such that $f \in K_q^*$ and $-f \in \mathrm{cl\; conv}\left(\bigcup_{p \neq q} K_p^*\right)$.

9. Prove Theorem 8.4 with the help of Exercise 8.

10. Let convex cones $K_1, \cdots, K_s \subset E$ with the common apex o possess the following property: there exists a hyperplane $\Gamma \subset E$ and an index l, $0 < l < s$, such that $K_1 \cap \cdots \cap K_l \subset \Pi$ and $K_{l+1} \cap \cdots \cap K_s \subset \Pi'$, where Π, Π' are half-spaces with the bounding hyperplane Γ. Prove that the cones $K_1, \cdots, K_s$ are separable.

11. Let $K_1, \cdots, K_s$ be convex cones with the common apex o satisfying ri $K_1 \cap \cdots \cap$ ri $K_s \neq \emptyset$. Prove that the cones $K_1, \cdots, K_s$ are separable if and only if the affine hulls aff $K_1, \cdots,$ aff K_s are separable.

12. Let $K_1, \cdots, K_s$ be solid, convex cones. Prove that the cones $K_1, \cdots, K_s$ are not separable if and only if $K_1^* + \cdots + K_s^*$ is a pointed cone. Does this remain true for nonsolid cones?

13. Prove that if convex cones $K_1, \cdots, K_s$ with common apex o are not separable, then the cone $K_1^* + \cdots + K_s^*$ is closed. Is the converse assertion true?

14. Prove that subspaces $K_1, \cdots, K_s \subset E$ are separable if and only if there exists a direct decomposition $E = L_1 \oplus \cdots \oplus L_s$ such that $L_p \subset K_q$ for $p \neq q$.

15. Let $M_1, \cdots, M_s$ be convex sets having the origin as a common boundary point. Prove that if the intersection $K = \bigcap\limits_{i=1}^{s} \text{sup cone}_o M_i$ is a solid cone, then K is the supporting cone of $M_1 \cap \cdots \cap M_s$ at the origin.

16. Show by an example that the requirement "K is solid" in the previous exercise is essential.

II. d-Convexity in normed spaces

In this chapter we consider basic properties of d-convex sets in finite-dimensional normed spaces. Central subjects are the support properties of d-convex sets (§10) and the properties of d-convex flats (§11). This chapter is of decisive importance for developing a machinery for solving combinatorial problems. Nevertheless, it is interesting for itself, because the family of d-convex sets has far-reaching analogies to the family of convex sets in $\mathbf{R}^n$. For example, the carrying flats, faces, inscribed cones, and supporting cones of d-convex sets are d-convex themselves; moreover, each boundary point of a d-convex body is contained in a d-convex supporting hyperplane. Questions referring to separability of d-convex sets are considered in section 13 of this chapter.

The notion of d-convexity was introduced by Menger [Me 1,2], de Groot [Gr], Aleksandrov-Zalgaller [A-Z], Soltan-Prisakaru [S-Pr] and other autors. But except for definitions, the first three papers contain almost no theorems on d-convex sets. The main results of the present chapter were obtained in papers [SP 6-9], [G-S-S], and [SP-SV 2]. For further basic developments around d-convexity and related concepts in metric and normed linear spaces, we refer to §9 in [D-G-K], [La 8,9], [SV 4], and §4 in [Man 3].

§9 The definition of d-convex sets

A real function $||x||$, given for $x \in \mathbf{R}^n$, is said to be a *norm* in the space $\mathbf{R}^n$ if it has the following properties:

(1) For each $x \in \mathbf{R}^n$, $||x|| \geq 0$ holds, and $||x|| = 0$ if and only if $x = o$.

(2) For every $x \in \mathbf{R}^n$ and an arbitrary real number λ, the relation $||\lambda x|| = |\lambda| \cdot ||x||$ holds.

(3) For arbitrary $x, y \in \mathbf{R}^n$, the inequality $||x + y|| \leq ||x|| + ||y||$ is satisfied.

The space $\mathbf{R}^n$, in which a norm is introduced, is called a (finite-dimensional) *normed space* or *Minkowski space*. For the sake of convenience and in contrast to the vector space $\mathbf{R}^n$ without norm, we denote the normed space by $\mathfrak{R}^n$.

The set $\sum$, consisting of the vectors $x \in \Re^n$ satisfying $\| x \| \leq 1$, is the
unit ball of this space. The unit ball is a bounded convex body (in the sense
of the linear convexity, which holds in the corresponding vector space $\mathbf{R}^n$)
symmetric about the point o.

Vice versa, let $\sum$ be a bounded convex body in the vector space $\mathbf{R}^n$ which
is centered at o. Then there exists exactly one norm in $\mathbf{R}^n$, for which $\sum$
represents the unit ball. This norm is defined in the following way. Let $x \in \mathbf{R}^n$
be distinct fromt o and x' denote the intersection point of the ray $[o, x)$ with
the boundary bd $\sum$ of $\sum$ (Fig. 26). Then there is a positive number $\lambda(x)$
satisfying $x = \lambda(x)\, x'$. We set $\| x \| = \lambda(x)$. It is easy to prove that $\lambda(x)$
satisfies the above axioms of the norm, and the unit ball for this (and only
this) norm coincides with the given convex body $\sum$.

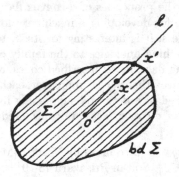

Fig. 26

Therefore the assignment of a norm (which transforms $\mathbf{R}^n$ into a normed
space $\Re^n$) is equivalent to the assignment of a compact, convex body cen-
tered at o. If the unit ball $\sum = \{x : \| x \| \leq 1\}$ is strictly convex, then the
corresponding normed space is also *strictly convex*. For example, the Eucli-
dean space $\mathbf{R}^n$ (whose norm is given by $\| x \| = \sqrt{\langle x, x \rangle}$) is a strictly convex
normed space. We set $d(x, y) = \| x - y \|$ and transform the normed space $\Re^n$
into a *metric space*, i.e., the function $d(x, y)$, called the *metric*, satisfies the
following axioms for arbitrary $x, y, z \in \Re^n$:

1° $d(x, y) \geq 0$ with the equality if and only if $x = y$,

2° $d(x, y) = d(y, x)$,

3° $d(x, y) \leq d(x, z) + d(z, y)$ (the *triangle inequality*).

In the sequel, we always consider the metric in $\Re^n$ which is induced by a
norm in the above manner. Now we assign the notion *d-convexity*. Since the
definition uses only the metric, its existence is ensured in each metric space.

But we will consider d-convexity only in $\Re^n$ (with the metric induced by the norm).

For $a, b \in \mathbf{R}^n$, the set $\{x : d(a, b) = d(a, x) + d(x, b)\}$ is a d-*segment* with endpoints a, b denoted by $[a, b]_d$. (Denoting by $[a, b]$ a linear segment with endpoints a, b we maintain.)

Furthermore, a set $M \subset \Re^n$ is d-*convex* if for any points $a, b \in M$ the d-segment $[a, b]_d$ is completely contained in M. The empty set is d-convex and also $\Re^n$ is d-convex.

There is another description of d-convexity, namely: the set $M \subset \Re^n$ is d-convex if, for any three points $a, b \in M, x \in \Re^n$, the equality $d(a, b) = d(a, x) + d(x, b)$ implies that $x \in M$. Obviously, this definition of d-convexity is equivalent to the preceding one.

These definitions imply that each d-convex set is also linearly convex (since $[a, b] \subset [a, b]_d$ for any $a, b \in \mathbf{R}^n$). The converse is not true.

EXAMPLE 9.1: Consider the space $\mathbf{R}^2$ with a coordinate system (x_1, x_2) and introduce the norm $\| x \| = \max\{|x_1|, |x_2|\}$. The unit ball $\sum$ of this space $\Re^2$ is the square described by the inequalities $|x_1| \leq 1, |x_2| \leq 1$. This unit ball is linearly convex, but not d-convex. Indeed, the points $a = (1, 1), b = (-1, 1)$ belong to $\sum$, but not the point $z = (0, 2)$. At the same time, $d(a, b) = 2, d(a, z) = d(z, b) = 1$, i.e., $z \in [a, b]_d$ (cf. Fig. 27). □

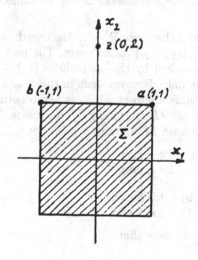

Fig. 27

In addition, a definition of d-convexity is obtainable by using the term of the length of simple arcs which connect points of the set M. (As usual, the length of simple arcs in $\Re^n$ is defined by means of the metric induced by the norm.)

<u>*Theorem 9.2:*</u> *The set $M \subset \Re^n$ is d-convex if and only if for arbitrary points $a, b \in M$ the following holds: each simple arc of length $\| a - b \|$ with endpoints a, b is completely contained in M.* $\square$

PROOF: Assume that for any $a, b \in M$ each simple arc of length $\| a - b \|$ with the endpoints a, b is contained in M. From this we conclude that M is d-convex. Let therefore $a, b \in M$ and $z \in \Re^n$ be points satisfying $d(a, b) = d(a, z) + d(z, b)$. The segments $[a, z]$ and $[z, b]$ are simple arcs of length $\| a - z \| = d(a, z)$ and $\| z - b \| = d(z, b)$, respectively. Thus the simple arc $l = [a, z] \cup [z, b]$ has the length $d(a, z) + d(z, b)$, i.e., $d(a, b)$. Therefore $l \subset M$ and hence $z \in M$. Consequently the set M is d-convex.

On the other hand, let M be a d-convex set and p denote a simple arc having $a, b \in M$ as its endpoints and $\| a - b \| = d(a, b)$ as its length. We choose an arbitrary point $z \in p$ and denote by p_1, p_2 the two parts in which p is partitioned by z. Since p_1 is a simple arc with endpoints a and z, its length ϱ_1 is not smaller than $\| a - z \| = d(a, z)$. Analogously, $\varrho_2 \geq \| z - b \| = d(z, b)$. Hence the length $\varrho_1 + \varrho_2$ of the whole arc p is not smaller than $d(a, z) + d(z, b)$. At the same time, the assumption implies that the length of p equals just $d(a, b)$. Hence $d(a, b) \geq d(a, z) + d(z, b)$, and the triangle inequality implies $d(a, b) = d(a, z) + d(z, b)$. Because M is assumed to be d-convex, this means $z \in M$ and $p \subset M$. ∎

The following example shows that a d-segment is not, in general, a d-convex set, i.e., the family of the d-segments is not contained in the family of all d-convex sets.

EXAMPLE 9.3: Consider the space $\Re^3$ with the coordinate system (x_1, x_2, x_3) and $\| x \| = \max\{|x_1|, |x_2|, |x_3|\}$ as its norm. The unit ball $\sum$ of this space is a cube which is described by the inequalities $|x_1| \leq 1, |x_2| \leq 1, |x_3| \leq 1$. We will show that the only d-convex *body* in this space is given by $\Re^3$ itself. Since each body contains a subset which is homothetic to $\sum$, it suffices to show that a d-convex set M containing $\sum$ is necessarily equal to $\Re^3$. Thus assume that a d-convex set M contains $\sum$. We set

$$a_k = \begin{cases} (k, 1, 0) & \text{for } k \text{ even,} \\ (k, 0, 1) & \text{for } k \text{ odd,} \end{cases}$$

$$b_k = \begin{cases} (k, -1, 0) & \text{for } k \text{ even,} \\ (k, 0, -1) & \text{for } k \text{ odd,} \end{cases}$$

see Fig. 28. It is easy to show that

$$d(a_k, b_k) = 2, \, d(a_k, a_{k+1}) = d(a_k, b_{k+1}) = d(b_k, a_{k+1}) = d(b_k, b_{k+1}) = 1,$$

and therefore $a_{k+1}, b_{k+1} \in [a_k, b_k]_d$.

Since $a_1, b_1 \in \sum \subset M$ and M is d-convex, the points a_k, b_k belong to M. Furthermore, since any d-convex set is linearly convex, this implies that the positive x_1-semiaxis is contained in M.

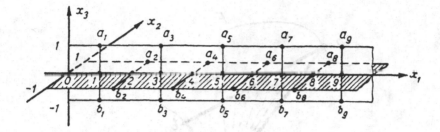

Fig. 28

The analogous reasoning can be applied to each other semiaxis. Therefore every coordinate axis is contained in M. Thus, by convexity arguments, $M = \Re^3$.

Furthermore, if for $a, b \in \Re^3$ the vector $a - b$ is not parallel to a spatial diagonal of the cube $\sum$, then the d-segment $[a, b]_d$ has interior points, i.e., it is a body. Since such a body is bounded, it cannot coincide with $\Re^3$. Hence in $\Re^3$ there are d-segments which are not d-convex.   □

Theorem 9.4: The intersection of an arbitrary family of d-convex sets is a d-convex set.   □

PROOF: Let $\{M_\alpha\}$ denote a family of d-convex sets, where α runs over an index set A. We set $M = \cap M_\alpha$. If $a, b \in M$, then for all $\alpha \in A$ the relation $a, b \in M_\alpha$ holds, and by the d-convexity of M_α we have $[a, b]_d \subset M_\alpha$. Since this inclusion holds for each $\alpha \in A$, we have $[a, b]_d \subset M$. Thus the set M is d-convex.   ■

The smallest d-convex set containing a set $F \subset \mathbf{R}^n$ is said to be *d-convex hull* of F. The theorem above shows that the d-convex hull exists for an arbitrary set $F \subset \Re^n$. It is the intersection of the d-convex sets containing F. We denote the d-convex hull of F by $\mathrm{conv}_d F$.

Now we are concerned with the question whether Theorem 1.2 can be transferred to the case of the d-convex hull.

At first we look at some other variants of the theorem. Let $F \subset \mathbf{R}^n$ be an arbitrary set. Denote by $I(F)$ the union of the segments $[a, b]$ with endpoints $a, b \in F$. The transition from F to $I(F)$ we call the *process of segment joining*. Furthermore, we can consider the set $I(I(F))$ obtained by the two-fold application of this process etc. We remark that the sets $I(F), I(I(F)), \cdots$, constructed by successive application of segment joining, are contained in $\mathrm{conv}\, F$. Theorem 1.2 says that, starting with F, a finite number of iterations of this process yields $\mathrm{conv}\, F$. For example, the two-fold application gives the set of the triangles $[a, b, c]$ with $a, b, c \in F$ (Fig. 29). Analogously, the n-fold application yields the union of all n-dimensional simplices $[a_o, a_1, \cdots, a_n]$ with

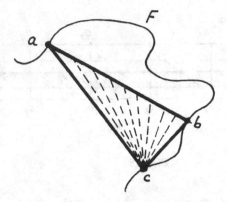

Fig. 29

the vertices in F. In accordance with Theorem 1.2, this is already sufficient for obtaining conv F from F (see Exercise 3 in §1).

The process of segment joining can be introduced for d-segments in $\Re^n$, too. Analogously, we write $I_d(F)$ for the set obtained from $F \subset \Re^n$ by this process. Hence one is motivated to ask whether from an arbitrary set $F \subset \Re^n$ the d-convex hull is obtainable by a finite number of corresponding iterations. It is easy to see that the question has a negative answer. For showing this, we denote by diam M the *diameter* of the set M, i.e., the upper bound of the distances $d(x,y)$ with $x,y \in M$. If $x \in [a,b]_d$, where $a,b \in F$, then $d(a,x)+d(x,b) = d(a,b)$ and hence at least one of the distances $d(a,x), d(x,b)$ is not larger than $\frac{1}{2}d(a,b)$. For instance, let $d(a,x) \leq \frac{1}{2}d(a,b) \leq \frac{1}{2}$ diam F. Analogously, if $y \in [c,d]_d$ for $c,d \in F$, then one can write $d(c,y) \leq \frac{1}{2}d(c,d) \leq \frac{1}{2}$ diam F. Hence we obtain

$$d(x,y) \leq d(x,a) + d(a,c) + d(c,y) \leq 2 \text{ diam } F$$

(Fig. 30). Up to now the following is obtained: if the set F has diameter h, then the set obtained from F by a single process of d-segment joining has a diameter not larger than $2h$; the next step of d-segment joining yields a set

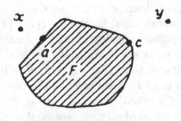

Fig. 30

of diameter not larger than $4h$, and so on. Hence a finite number of iterations of the process of d-segment joining yields again a bounded set. So, starting with $\sum$ from Example 9.3, a finite number of such iterations cannot give the d-convex hull $\mathrm{conv}_d \sum = \Re^3$ of $\sum$ (see also Example 10.4 below).

On the other hand, countably many iterations of the process of d-segment joining yield the d-convex hull, i.e., we obtain the following

__Theorem 9.5:__ *Let $M_o \subset \Re^n$ be an arbitrary set and M_i denote the set obtained from M_{i-1} by the process of d-segment joining $(i = 1, 2, \cdots)$. Then the set $M' = \bigcup_{i=1}^{\infty} M_i$ is the d-convex hull of M_o.* $\square$

<u>PROOF:</u> Obviously, by $M_i \subset \mathrm{conv}_d M_o$ for each $i = 1, 2, \cdots$, we have $M' \subset \mathrm{conv}_d M_o$. To prove the converse inclusion, it suffices to verify the d-convexity of M' (since $\mathrm{conv}_d M_o$ is the smallest d-convex set containing M_o). For any $a, b \in M'$ there are some i and j such that $a \in M_i, b \in M_j$. With $M_o \subset M_1 \subset M_2 \subset \cdots$ and denoting by k the largest integer from i, j, this implies $a, b \in M_k$. By the introduced construction, $[a, b]_d \subset M_{k+1} \subset M'$. Hence M' is d-convex. ∎

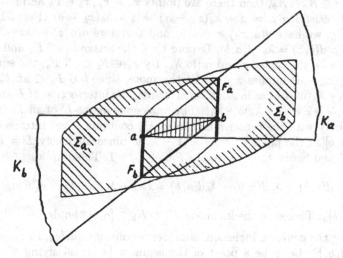

Fig. 31

The following theorem gives a geometric description of d-segments. To formulate it, we need some further notation. For $a, b \in \Re^n$, denote by $\sum_a, \sum_b$ the balls of the radius $r = d(a, b)$ with centers a and b, respectively. Furthermore, we write F_a for the face of b in $\sum_a$ and, similarly, F_b for the face of a in $\sum_b$ (Fig. 31). Now by K_a we denote the cone with the apex a consisting of the points $a + \lambda(x - a)$, where $x \in F_a, \lambda \geq 0$, and by K_b the analogous cone with the apex b and representation $b + \lambda(y - b)$, where $y \in F_b, \lambda \geq 0$. It is easy to

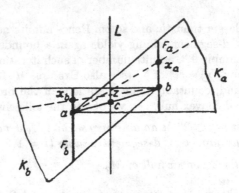

Fig. 32

see that K_a, K_b are symmetric with respect to the midpoint of the segment $[a, b]$.

Theorem 9.6: In the introduced notation, $[a, b]_d = K_a \cap K_b$. □

PROOF: If $z \in K_a \cap K_b$, then there are points $x_a \in F_a, x_b \in F_b$ and numbers $\lambda_a, \lambda_b \geq 0$ satisfying $z = a + \lambda_a(x_a - a) = b + \lambda_b(x_b - b)$ (Fig. 32). By $x_a \in$ bd $\sum_a$, we have $d(a, x_a) = d(a, b)$ and therefore $d(a, z) = \lambda_a \cdot d(a, b)$. Analogously, $d(b, z) = \lambda_b \cdot d(a, b)$. Denote by L the image of aff F_a under the homothety with the center a and ratio λ_a. By $x_a \in F_a \subset$ aff F_a, the equality $z = a + \lambda_a(x_a - a)$ implies $z \in L$. Furthermore, since $b \in F_a \subset$ aff F_a, the point $c = a + \lambda_a(b - a)$ lies in L. In other words, the intersection of L and the segment $[a, b]$ is $c = (1 - \lambda_a)a + \lambda_a b$. The analogous image L' of aff F_b (under the homothety with the center b and ratio λ_b) contains z and intersects the segment $[a, b]$ in the point $c' = \lambda_b + (1 - \lambda_b)b$. Since obviously $L = L'$, we have $c = c'$ and hence $\lambda_a = 1 - \lambda_b$, i.e., $\lambda_a + \lambda_b = 1$. This implies

$$d(a, z) + d(z, b) = \lambda_a d(a, b) + \lambda_b d(a, b) = (\lambda_a + \lambda_b)d(a, b) = d(a, b),$$

i.e., $z \in [a, b]_d$. Therefore the inclusion $K_a \cap K_b \subset [a, b]_d$ holds.

For showing the converse inclusion, consider a point $y \in [a, b]_d$, i.e., $d(a, b) = d(a, y) + d(y, b)$. Let p be a point of the segment $[a, b]$, satisfying $d(a, p) = d(a, y)$. Then

$$d(p, b) = d(a, b) - d(a, p) = d(a, b) - d(a, y) = d(y, b).$$

We put $\mu = \frac{d(a,y)}{d(a,b)} = \frac{d(a,p)}{d(a,b)}$ and consider the points $f = a + \frac{1}{\mu}(y - a), g = a + \frac{1}{1-\mu}(b - y)$. (For $\mu = 0$ or $\mu = 1$ the relation $y \in K_a \cap K_b$ is obvious. Therefore we assume $0 < \mu < 1$.) We have

$$d(a, f) \quad = \quad \| f - a \| = \frac{1}{\mu} \| y - a \| = \frac{1}{\mu}d(a, y) = d(a, b),$$

$$d(a,g) \;=\; \parallel g - a \parallel = \frac{1}{1-\mu} \parallel b - y \parallel = \frac{1}{1-\mu} d(y,b) = d(a,b).$$

This means that the points f and g belong to the boundary of the ball $\sum_a$. Furthermore, the equality $\mu f + (1-\mu)g = b$ holds, i.e., f and g belong to the face F_a of $b \in \sum_a$. Therefore $f,g \in K_a$, leading also to $y \in K_a$ (since $y = a + \mu(f-a)$). Analogously, we can derive that $y \in K_b$, i.e., $y \in K_a \cap K_b$. Hence $[a,b]_d \subset K_a \cap K_b$. ∎

Exercises

1. In a Minkowski space $\Re = \Re^n$, denote by $\sum_{a,r}$ the set of all points $x \in \Re$ satisfying the condition $\parallel x - a \parallel \le r$ for some $a \in \Re$ and $r \ge 0$. Thus the set $\sum_{a,r}$ is centered at a and homothetic to the unit ball $\sum$ with the ratio r (for $r \ne 1$). Prove that if $a \notin M \subset \Re$, then the distance $d(a,M)$ between a and the set M (i.e., inf $\parallel a - y \parallel$ with y running through M) is equal to sup $\{ r \ge 0 : \sum_{a,r} \cap M = \emptyset \}$.

2. Prove that if $M \subset \Re$ is a flat and $a \in \Re$, then there exists a point $b \in M$ such that $\parallel a - b \parallel = d(a,M)$.

3. Let $\Re$ denote an n-dimensional Minkowski space with the unit ball $\sum$ and $\Re^*$ be the n-dimensional Minkowski space with the unit ball $\sum^*$. Furthermore, let $x \in \Re$ be a point and $x^* \in \Re^*$ be its polar hyperplane. Prove that $\parallel x \parallel_\Re \cdot d_{\Re^*}(o, x^*) = 1$.

4. Let $\Re$ be the n-dimensional Minkowski space whose unit ball $\sum$ is described (in a coordinate system $x_1, \cdots, x_n$) by the inequalities $|x_i| \le 1, i = 1, \cdots, n$. Prove that for every point $x = (x_1, \cdots, x_n) \in \Re$ its norm $\parallel x \parallel$ is equal to max $(|x_1|, \cdots, |x_n|)$.

5. Prove that for the Minkowski plane $\Re^2$, considered in Example 9.1, the d-convex hull of its unit ball $\sum$ is the square with the vertices $(\pm 2, 0)$, $(0, \pm 2)$.

6. Consider the Minkowski space $\Re^3$ introduced in Example 9.3. Prove that the only d-convex sets in this space $\Re^3$ are: (i) the whole space $\Re^3$; (ii) the lines parallel to spatial diagonals of the cube $\sum$, as well as segments, intervals, semiintervals and rays lying on these lines; (iii) points; (iv) the empty set.

7. Again we refer to the Minkowski space $\Re^3$ introduced in Example 9.3. Prove that for every two distinct points $a, b \in \Re^3$, with $a - b$ not parallel to a spatial diagonal of the cube $\sum$, the relation $\mathrm{conv}_d(\{a,b\}) = \Re^3$ holds.

8. Prove that if for a Minkowski space $\Re$ its unit ball $\sum$ is strictly convex, then d-convexity in $\Re$ coincides with linear convexity.

9. Let $\Re$ be the n-dimensional Minkowski space whose unit ball $\sum$ is a regular cross-polytope, i.e., the convex hull (in the sense of linear convexity) of $2n$ points lying on the coordinate semiaxes at the distance 1 from the origin. Prove that for every compact set $M \subset \Re$ its d-convex hull coincides with its circumscribed coordinate parallelotope (described by inequalities $a_i \le x_i \le b_i, i = 1, \cdots, n$).

10. Describe all d-convex sets in the Minkowski plane whose unit ball $\sum$ is described in a coordinate system (x_1, x_2) by

$$x_1^2 + x_2^2 \leq 1, \ |x_1| \leq \frac{1}{2}.$$

11. Describe all d-convex sets in the Minkowski plane having the regular hexagon as its unit ball.

12. Prove that for every pair of points a, b of an arbitrary Minkowski plane $\Re^2$ the d-segment $[a, b]_d$ is a d-convex set, i.e., $\text{conv}_d\{a, b\} = [a, b]_d$.

13. Let $M_1, \cdots, M_s \subset \Re^n$. Prove that if $M_i \cup M_j$ is d-convex for every $i, j = 1, \cdots, s$, then the set $M_1 \cup \cdots \cup M_s$ is d-convex.

§10 Support properties of d-convex sets

Theorem 10.1: _Let_ $\{M_i\}$ _be an increasing sequence of d-convex sets in $\Re^n$._ _Then the set_ $N = \bigcup\limits_{i=1}^{\infty} M_i$ _is d-convex._ $\square$

PROOF: If $a, b \in N$, then $a \in M_i$, $b \in M_j$ for certain i, j. Thus $a, b \in M_k$, where k is the largest number from i, j. Since M_k is d-convex, we have $[a, b]_d \subset M_k$ and hence $[a, b]_d \subset N$. ∎

Theorem 10.2: _If M is a d-convex set in $\Re^n$, then_ ri M _and_ cl M _are also d-convex sets._ $\square$

PROOF: For an arbitrary point $a \in$ ri M, denote by M_k the set obtained from M by the homothety with the center a and ratio $1 - \frac{1}{k}$, $k = 1, 2, 3, \cdots$ Since this transformation transfers d-convexity, the sets $M_1 \subset M_2 \subset \cdots$ are d-convex. By Theorem 10.1, also $M' = \bigcup\limits_{i=1}^{\infty} M_k$ is d-convex, and this set is equal to ri M.

Let now N_k be the set obtained from M by the homothety with the center a and ratio $1 + \frac{1}{k}$, $k = 1, 2, 3, \cdots$. Again, the sets N_k and $N' = \bigcap\limits_{i=1}^{\infty} N_k$ are d-convex (cf. Theorem 9.4), and with $N' = $ cl M the theorem is proved. ∎

Corollary 10.3: _Let F be an arbitrary set in $\Re^n$. Then the inclusion_ $\text{conv}_d(\text{cl } F) \subset \text{cl } (\text{conv }_d F)$ _holds._ $\square$

In fact, since $\text{conv}_d F$ is d-convex, Theorem 10.2 implies d-convexity of cl $(\text{conv}_d F)$ and since this closed set contains F, we have cl $F \subset$ cl $(\text{conv}_d F)$. By d-convexity of the set cl $(\text{conv}_d F)$, the inclusion conv_d (cl F) $\subset$ cl $(\text{conv}_d F)$ follows. ∎

We remark that in the case of linear convexity, for any bounded set $F \subset \mathbf{R}^n$ the equality conv (cl F) $=$ cl (conv F) holds. For the case of d-convexity, the analogous equality is false, even if F is a compact set. This is shown by the following

EXAMPLE 10.4: In the space $\mathbf{R}^3$ with a coordinate system (x_1, x_2, x_3) consider the set $\sum = \mathrm{conv}\,(P_1 \cup P_2 \cup Q)$, where Q is described by the conditions $x_3 = 0$, $x_1^2 + x_2^2 = 3$ and P_1, P_2 are balls of radius 1 with the centers $(0, 0, 1)$ and $(0, 0, -1)$, respectively. (In Fig. 33, a meridian of the rotational body $\sum$ is shown, i.e., the intersection of $\sum$ and the plane $x_2 = 0$.) Now consider the normed space $\Re^3$ obtained from $\mathbf{R}^3$ by taking $\sum$ as unit ball. Denoting by L the plane $x_3 = 0$, it is easy to see that if the vector $a - b$ is parallel to L or the angle between $a - b$ and L is not smaller than $\frac{\pi}{3}$ (in the Euclidean metric with the norm $\| x \| = \sqrt{x_1^2 + x_2^2 + x_3^2}$), then the segments $[a, b]$ and $[a, b]_d$ coincide. But if this angle is positive and smaller than $\frac{\pi}{3}$, then the corresponding d-segment $[a, b]_d$ is the parallelogram (cf. Theorem 9.6) which has one interior angle $\frac{\pi}{3}$, the segment $[a, b]$ as its longer diagonal, and two sides parallel to L (cf. Fig. 34 and Fig. 31). Now let C be a set described by the inequalities $x_1^2 + x_2^2 < 3, |x_3| \le 2$ (i.e., C is a cylinder without its generators) and $M = C \cup Q$. By the above description of d-segments in $\Re^3$, it is easy to see that M is d-convex and, moreover, $M = \mathrm{conv}_d \sum$ (see also Example 11.4). Hence for the compact set $F = \sum$, the d-convex set $\mathrm{conv}_d(\mathrm{cl}\,\sum) = \mathrm{conv}_d \sum$ is not closed, i.e., $\mathrm{conv}_d(\mathrm{cl}\,\sum) \ne \mathrm{cl}\,(\mathrm{conv}_d \sum)$. Finally, we remark that here it is impossible to obtain $\mathrm{conv}_d \sum$ from $\sum$ by a finite number of iterations of the process of d-segment joining, since $\mathrm{conv}_d \sum$ is not compact. (Indeed, after finitely many iterations of this process compact sets are transferred to compact sets.)

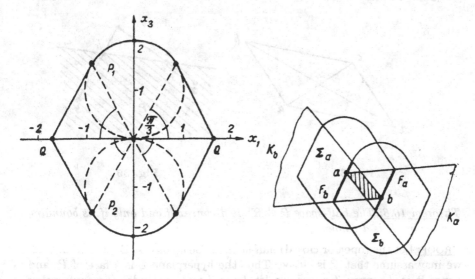

Fig. 33   Fig. 34

Theorem 10.5: Let $M \subset \mathbf{R}^n$ be a d-convex set and $a \in M$. Then the supporting cone of M at a is also a d-convex set.   □

PROOF: Let M_k be the set obtained from M by the homothety with the center a and ratio k, where $k = 1, 2, \cdots$ Then $M_1 \subset M_2 \subset \cdots$ and the sets M_k are d-convex. By Theorem 10.1, $M' = \bigcup_{k=1}^{\infty} M_k$ is d-convex. Theorem 10.2 implies that also cl M' is d-convex, and this set coincides with the supporting cone of M at a.   ■

Corollary 10.6: The carrying flat aff M of any d-convex set $M \subset \Re^n$ is d-convex.   □

It is sufficient to apply the previous assertion for the case $a \in$ ri M.

Theorem 10.7: Each face of any closed d-convex set $M \subset \Re^n$ is d-convex.   □

PROOF: Let F be an arbitrary face of M and $a, b \in F$, $c \in [a, b]_d$. Thus $d(a, b) = d(a, c) + d(c, b)$. We set $x = \frac{1}{2}a + \frac{1}{2}b$ and $d = 2x - c$, see Fig. 35, i.e., x is the midpoint of the segment $[a, b]$ and $[c, d]$. Then $a - d = c - b, d - b = a - c$, and $d(a, d) = d(c, b), d(d, b) = d(a, c)$. Hence $d(a, d) + d(d, b) = d(c, b) + d(a, c) = d(a, b)$, i.e., $d \in [a, b]_d$. Since $a, b \in F \subset M$, the d-convexity of M implies $[a, b]_d \subset M$. In particular, we have $c, d, x \in M$. Let now F_x denote the face of x. Since x is the midpoint of $[a, b]_d$ and $[c, d]_d$, each of the points a, b, c, d belongs to F_x. Finally, the relation $x \in [a, b]_d \subset F$ implies $F_x \subset F$ (Theorem 2.1), i.e., $c \in F$. Since c is an arbitrary point of $[a, b]_d$, we have $[a, b]_d \subset F$, i.e., F is d-convex.   ■

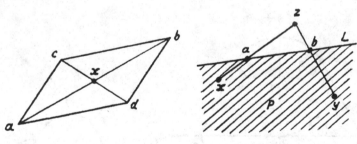

Fig. 35   Fig. 36

Theorem 10.8: The half-space $P \subset \Re^n$ is d-convex if and only if its bounding hyperplane L is d-convex.   □

PROOF: Let the (open or closed) half-space P be d-convex. By Theorem 10.2, we may assume that P is closed. Thus the hyperplane L is a face of P, and hence it is d-convex. Conversely, let L be d-convex. Under the assumption that P is not d-convex, there exist points $x, y \in P$ and $z \notin P$ with $d(x, z) + d(z, y) = d(x, y)$. Denote by a, b the intersection points of the hyperplane L

with the segments $[x, z]$ and $[z, y]$, respectively (Fig. 36, the points a and x as well as the points b and y can coincide). With $a \in [x, z], b \in [y, z]$ we have $d(x, z) = d(x, a) + d(a, z), d(z, y) = d(z, b) + d(b, y)$. Thus $d(x, y) = d(x, a) + d(a, z) + d(z, b) + d(b, y)$. The triangle inequality implies

$$d(x, y) \leq d(x, a) + d(a, b) + d(b, y),$$

yielding $d(a, b) \geq d(a, z) + d(z, b)$. Therefore $d(a, b) = d(a, z) + d(z, b)$, contradicting the d-convexity of L (since $a, b \in L$ and $z \notin L$). Hence the closed half-space P, defined by L, is d-convex, and Theorem 10.2 implies that int P is d-convex. ■

Theorem 10.9: Let $M \subset \Re^n$ be a d-convex body. For each point $x_o \in$ bd M, there is a d-convex supporting hyperplane of M through x_o.   □

PROOF: Let K_o be the supporting cone of M at x_o. The set K_o is d-convex with $x_o \in$ bd K_o (cf. Theorem 10.5). Let now x_1 be an arbitrary boundary point of K_o with $x_o \neq x_1$ and K_1 be the supporting cone of K_o at x_1. Then K_1 is also d-convex and the line through x_1, x_o belongs to the boundary of K_1. Now we choose a point $x_2 \in$ bd K_1, which is not contained in the line, and denote by K_2 the supporting cone of K_1 at x_2. The set K_2 is d-convex, and the two-dimensional plane through x_o, x_1, x_2 is contained in bd K_2. Continuing these arguments, we obtain a d-convex body K_{n-1} whose boundary contains the hyperplane L through points $x_o, x_1, \cdots, x_{n-1}$. Thus K_{n-1} is a closed half-space bounded by L. By Theorem 10.8, L is d-convex. It remains to remark that, because of $M \subset K_o \subset K_1 \subset \cdots \subset K_{n-1}$, the hyperplane L is a supporting hyperplane of M through x_o. ■

In the assertion above we demanded that M be a _body_. It should be noticed that this cannot be omitted. In Example 9.3, a spatial diagonal of the cube $\sum$ is closed and d-convex, but does not have any d-convex supporting hyperplane (since there is no d-convex hyperplane in $\Re^3$).

Theorem 10.10: Each closed, d-convex body $M \subset \Re^n$, not identical with $\Re^n$ can be represented as the intersection of a family of closed, d-convex half-spaces.

PROOF: Let L be a supporting hyperplane of the body M. By Π_L denote the closed half-space with respect to L which satisfies $M \subset \Pi_L$. We will show that $M = \bigcap \Pi_L$, where the intersection is taken over the d-convex supporting hyperplanes of M (correspondingly, each half-space Π_L is d-convex, cf. Theorem 10.8). Obviously, we have $M \subset \bigcap \Pi_L$. For showing the converse inclusion, let a, b be points with $a \notin M$ and $b \in$ int M. Then the intersection $[a, b] \cap (\text{bd } M)$ is a unique point c. Let now $\varepsilon > 0$ be a real number such that the ball Q with the center b and radius ε is contained in int M. We set $\delta = \lambda\varepsilon$, where λ is the ratio of the lengths of $[a, c]$ and $[a, b]$, see Fig. 37. Let now x be an arbitrary point satisfying $d(c, x) < \delta$; then the ray $[a, x)$ intersects the ball, i.e., $(\text{int } M) \cap [a, x)$ is nonempty. By Theorem 2.5, there exists a regular boundary point $x_o \in$ bd M satisfying $d(c, x_o) < \delta$, cf. Fig.

38. Since x_o is a regular boundary point of M, the supporting cone K_o of M at x_o is a half-space, i.e., $K_o = \Pi_{L_o}$, where L_o is the supporting hyperplane of M through x_o. By Theorems 10.5 and 10.8, L_o is d-convex. Let now $c_o \in [a, x_o) \cap \operatorname{int} M$. By $c_o \notin L_o$, the relations $c_o \in \operatorname{int} M \subset \operatorname{int} K_o$ and $x_o \in]a, c_o[$ (with $x_o \in L_o$) imply that a and c_o lie on opposite sides of L_o, i.e., $a \notin K_o$. Thus if $a \notin M$, then $a \notin \Pi_{L_o}$ for some d-convex hyperplane L_o and therefore $a \notin \bigcap \Pi_L$ (the intersection taken over the d-convex supporting hyperplanes). Hence $\bigcap \Pi_L \subset M$. ∎

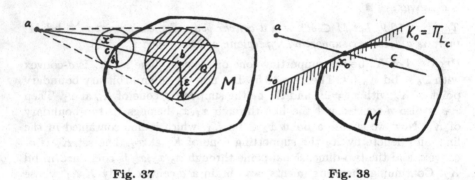

Fig. 37   **Fig. 38**

Theorem 10.11: Let K be a cone inscribed in an unbounded, d-convex set $M \subset \Re^n$. Then K is d-convex. □

PROOF: For $a \in \operatorname{ri} M$ and each $x \in M$, we denote by M_x the translate of M with $a - x$ as translation vector. Since M is d-convex, the same holds for each $M_x, x \in M$, and for $Q = \bigcap_{x \in M} M_x$. We will show the coincidence of the cone K, inscribed to M at a, and Q, which will complete the proof.

Let K_x denote the cone inscribed in M at $x \in M$, see Fig. 39. The cone K is the translate of K_x with respect to the translation vector $a - x$. Since $K_x \subset M$, we have $M_x \supset K$ (for arbitrary $x \in M$) and therefore $Q \supset K$ holds.

Let now $b \notin K, b \in \operatorname{aff} M$. Then the ray $[a, b)$ is not completely contained in M. For $\{y\} = \operatorname{rbd} M \cap [a, b)$, we choose a point $x \in [a, y]$ satisfying $d(x, y) < d(a, b)$, see Fig. 40. It is easy to see that $b \notin M_x$, leading to $b \notin Q$. Hence the conditions $b \notin K, b \in \operatorname{aff} M$ imply $b \notin Q$. On the other hand, if $b \notin \operatorname{aff} M$, then for each $x \in M$ the inclusion $b \notin M_x$ holds, i.e., in this case $b \notin Q$. Thus $Q \subset K$. ∎

Theorem 10.12: For each closed, unbounded, d-convex set $M \subset \Re^n$ the asymptotic scope $\tilde{M}$ is d-convex. □

PROOF: For $a \in M$, let K be the cone with the apex a which is inscribed to M. We choose an arbitrary point $x \in K$ with $x \neq a$ and consider the lines

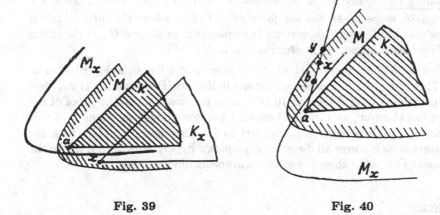

Fig. 39   Fig. 40

through the points from M which are parallel to the vector $x - a$. The union of the lines is denoted by M_x. We set $M' = \bigcap \operatorname{cl} M_x$, where the intersection is taken over all points $x \in K, x \neq a$. We will show $M' = \tilde{M}$ and, in addition, that this set is d-convex. For a point $b \notin \operatorname{cl} M_x$, linear convexity of $\operatorname{cl} M_x$ implies the following: There is a hyperplane L such that L contains M_x in one of its closed half-spaces, but this half-space (denoted in the sequel by Π_L) does not contain b. Since Π_L contains lines parallel to $x - a$ (this vector is parallel to L), the intersection of M and some hyperplane parallel to L contains a ray with starting point $y \in M$, parallel to $x - a$. (This ray lies in the cone inscribed in M with respect to y and is therefore contained in M.) Hence the intersection of M and each hyperplane parallel to L is either empty or unbounded. Therefore we can assume that L is an asymptotic hyperplane of M (if necessary, L can be translated in the direction of Π_L). Thus $b \notin \Pi_L$ implies $b \notin \tilde{M}$, and $\tilde{M} \subset \operatorname{cl} M_x$ is proved. Since this inclusion holds for an arbitrary point $x \in K$ with $a \neq x$, we have also $\tilde{M} \subset M'$.

For showing the converse inclusion, assume $c \notin M$, i.e., there is an asymptotic hyperplane L such that $M \subset \Pi_L, c \notin \Pi_L$. Let L' be the hyperplane through a, parallel to L. Then $L' \cap M$ is unbounded and contains a ray l' with a as its starting point. Since there is some $x' \in K, x' \neq a$, such that $l' = [a, x')$, we have $l' \subset K$. Furthermore, l' is parallel to L. Hence there exists a line which is parallel to l', contains a point sufficiently near to c, and satisfies the following: its intersection with Π_L (and consequently with M) is empty. Thus $c \notin \operatorname{cl} M_{x'}$ and $c \notin M'$; this implies $M' \subset \tilde{M}$.

With $M' = \tilde{M}, M' = \bigcap \operatorname{cl} M_x$ and Theorems 9.4 and 10.2, one can conclude d-convexity of M from d-convexity of each set M_x. It is easy to see that $M_x = \bigcup_k M_k$, where M_k is the image of M under the translation by $k(a - x), k = 1, 2, \cdots$ Since $M_1 \subset M_2 \subset \cdots$, the d-convexity of M_x is obvious by Theorem 10.1. ∎

Corollary 10.13: Let M be an unbounded d-convex body in $\Re^n$. Then the asymptotic scope of M has the form $\tilde{M} = \bigcap \Pi_L$, where the intersection is taken over all d-convex asymptotic hyperplanes L of M and Π_L is the closed half-space with respect to L which contains M.   □

In fact, let $b \notin \tilde{M}$, i.e., $b \notin \text{cl } M_x$ for a point $x \neq a$ from the inscribed cone K. Then there exists a *d*-convex, closed half-space P containing cl M_x, but not containing b (cf. Theorem 10.10). As in the first part of the proof of the preceding theorem, we can conclude that the bounding hyperplane L of P is an asymptotic hyperplane with respect to M. Therefore $b \notin \bigcap \Pi_L$ (with the intersection taken over all *d*-convex asymptotic hyperplanes of M). Hence the inclusion $\bigcap_L \subset \tilde{M}$ is shown, and the converse is obvious.   ■

Exercises

1. Let a, b be arbitrary points in a normed space $\Re^n$. Prove that the diameter of the *d*-segment with endpoints a, b is equal to $\| a - b \|$, i.e., diam $[a, b]_d = \| a - b \|$.

2. Prove that for every bounded set $A \subset \Re^n$ the relation

   $$\text{diam } I_d(A) \leq 2 \text{ diam } A$$

   holds.

3. Let Φ be a convex set in the (x_1, x_3)-plane which is bounded by four circular arcs and two segments, as in the Fig. 41. Let the angle φ be smaller than ψ. Denote by $\sum$ the body in $\mathbf{R}^3$ obtained from Φ by rotation around the x_3-axis. Prove that in the space $\Re^3$ with the unit ball $\sum$ the relation $\text{conv}_d \sum = \sum$ holds.

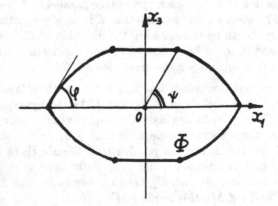

Fig. 41

4. Prove that if the unit ball $\sum \subset \Re^n$ is a *d*-convex set, then diam $\text{conv}_d M = $ diam M for every bounded set $M \subset \Re^n$.

5. Let $\Re^n$ be a Minkowski space whose unit ball $\sum$ is d-convex. Furthermore, let $M \subset \Re^n$ be a *diametrically maximal* bounded set, i.e., diam $(M \cup \{x\}) >$ diam M for every $x \notin M$. Prove that M is d-convex.

6. Let $\Re^n, n \geq 3$, be a Minkowski space with the following property: there exists a subspace L with dim $L > 1$ such that the intersection $L \cap \sum$ is a polytope in L. Prove that the unit ball $\sum \subset \Re^n$ is not d-convex.

7. Show by a counterexample that the following assertion (which is related to Theorem 10.2) is false: if $M \subset \Re^n$ is a linearly convex set such that ri M and cl M are d-convex, then M is d-convex.

8. Prove by a counterexample that the following assertion (converse to Theorem 10.7) is false: if $M \subset \Re^n$ is a linearly convex set and all its proper faces are d-convex, then M is d-convex.

9. Prove the following assertion (converse to Theorem 10.9): if $M \subset \Re^n$ is a linearly convex body such that for every $a \in$ bd M there exists a d-convex supporting hyperplane through a, then M is d-convex. Is it essential that M is a body?

10. Prove that a closed, linearly convex set $M \subset \Re^n$ is d-convex if and only if for every $a \in M$ the cone sup $\text{cone}_a M$ is d-convex.

11. Let $M \subset \Re^n$ be a d-convex set. We say that $a \in M$ is a *d-extreme point* of M if it is impossible to find points $x, y \in M$ such that $a \in$ ri $([x,y]_d)$. Prove that $a \in M$ is a d-extreme point of M if and only if a is an extreme point of M (in the sense of linear convexity). It follows that for every compact, d-convex set M the equality $M = \text{conv}_d(\text{ext}_d M)$ holds.

12. Let $M \subset \mathbf{R}^2$ be a regular octagon. Construct $\sum \subset \mathbf{R}^2$ such that in the corresponding Minkowski plane (with unit circle $\sum$) the equality $\text{conv}_d \sum = M$ holds.

13. Show by a counterexample that the following assertion is false: let $Q = \text{conv}_d \sum$, where $\sum$ is the unit ball of a Minkowski space $\Re^n$. A closed, convex cone K is d-convex if and only if there exists a point $a \in$ bd Q such that, up to a translation, $K = \text{sup cone}_a Q$. Show that the above assertion is false even for $Q = \sum$.

§11 Properties of d-convex flats

In the following, we will deduce some necessary and sufficient criteria for d-convexity of a subspace $L \subset \Re^n$. Let $\sum$ be the unit ball of $\Re^n$. The face of an arbitrary point $x \in$ bd $\sum$ with respect to $\sum$ is denoted by Φ_x. We say that a subspace $L \subset \Re^n$ is *equatorial* (with respect to $\sum$) if for every point $x \in L \cap$ bd $\sum$ the face Φ_x of x is *contained in L*.

Theorem 11.1: A subspace $L \subset \Re^n$ is d-convex if and only if it is equatorial.
$\square$

PROOF: Assume that L is not equatorial, i.e., Φ_x is not contained in L for a point $x \in (\text{bd} \sum) \cap L$. Then there are points $y, z \in \Phi_x \subset \text{bd} \sum$ such that x is an interior point of the segment $[y, z]$ and $y, z \notin L$. We can assume that $d(x, y) = d(x, z)$, i.e., $x = \frac{1}{2}y + \frac{1}{2}z$ (Fig. 42). If we set $a = 2x$, then $d(o, a) = 2d(o, x) = 2$, $d(o, y) = 1$, $d(y, a) = d(o, z) = 1$, i.e., $d(o, a) = d(o, y) + d(y, a)$ and hence $y \in [o, a]_d$. By $o, a \in L$ and $y \notin L$, the subspace is not d-convex.

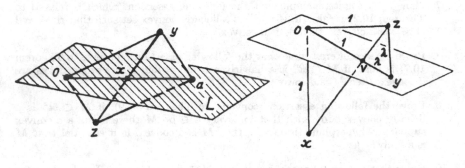

Fig. 42   Fig. 43

Conversely, let L be a subspace which is not d-convex. Then there exist points $y \in L$ and $z \notin L$ such that $d(o, y) = d(o, z) + d(z, y)$. In addition, we may suppose that $d(o, z) = \| z \| = 1$. For $\{v\} = (\text{bd} \sum) \cap [o, y]$ (Fig. 43), the equality $\| v \| = 1$ holds. From $d(o, y) = d(o, z) + d(z, y) = 1 + d(z, y)$, $d(o, y) = d(o, v) + d(v, y) = 1 + d(v, y)$ it follows that $d(z, y) = d(v, y) = \lambda > 0$. Setting $x = \frac{1}{\lambda}(y - z)$, we have $\| x \| = \frac{1}{\lambda} \| y - z \| = \frac{1}{\lambda}d(z, y) = 1$, i.e., $x \in \text{bd} \sum$. Furthermore,

$$\frac{\lambda}{1 + \lambda}x + \frac{1}{1 + \lambda}z = \frac{1}{1 + \lambda}y = v.$$

Hence v is an interior point of $[z, x] \subset \sum$ and this means that z and x lie in the face Φ_v of the point $v \in (\text{bd} \sum) \cap L$. Since $z \notin L$, the face Φ_v is not contained in L, i.e., L is not equatorial. ∎

Corollary 11.2: A one-dimensional subspace $l \subset \Re^n$ is d-convex if and only if $l \cap (\text{bd} \sum)$ is a pair of extreme points of $\sum$. □

Corollary 11.3 For a strictly convex unit ball $\sum \subset \Re^n$, an arbitrary subspace $L \subset \Re^n$ (and hence an arbitrary flat) is d-convex. In other words, for this case d-convexity and linear convexity are identical. Conversely, if both the types of convexity coincide, then the unit ball $\sum \subset \Re^n$ is strictly convex. □

EXAMPLE 11.4: Consider the normed space $\Re^3$ described in Example 10.4. By L denote the plane $x_3 = 0$ and by l the line $x_1 = x_2 = 0$. Each plane through l intersects the unit ball in the figure shown in Fig. 33; this figure is bounded by four segments and two circular arcs. The four segments are the only faces of $\sum$ which are not points. Hence the only d-convex planes of $\Re^3$ are the 2-flats parallel or orthogonal to L (in the sense of Euclidean norm, as in Example 10.4). A line is d-convex if and only if it is parallel to L or forms an angle with L not smaller than $\frac{\pi}{3}$. Hence, by Theorem 10.10, any closed, d-convex *body* in $\Re^3$ has the form $G \oplus N$, where G denotes a closed, convex figure in the plane $x_3 = 0$ and $N \subset R$ is a segment, a ray or the complete line of real numbers. Again we observe that the smallest closed d-convex set containing $\sum$ (i.e., the set cl $(\text{conv}_d \sum)$) is the cylinder cl C (see Example 10.4) and therefore

$$\text{conv}_d{\textstyle\sum} \supset \text{int } C. \quad \square$$

EXAMPLE 11.5: Consider the convex hull of the points $(\pm 1, 0, 0), (0, \pm 1, 0),$ $(0, 0, \pm 1)$, i.e. the regular octahedron, as the unit ball of the normed space $\Re^3$. By Theorem 11.1, each d-convex plane (line) in $\Re^3$ has to be parallel to a coordinate plane (line). In accordance with Theorem 10.10, the only compact d-convex bodies in $\Re^3$ are the parallelotopes with edges parallel to the coordinate axes. $\square$

EXAMPLE 11.6: Let the norm in $\Re^n$ (with a coordinate system $(x_1, x_2, \cdots,$ $x_n)$) be introduced by $\| x \| = \max(|x1i|, \cdots, |x_n|)$. The unit ball $\sum$ of $\Re^n$ is the cube described by the inequalities $|x_1| \leq 1, \cdots, |x_n| \leq 1$. It is easy to show that for $n \geq 3$ there is no d-convex k-flat with $2 \leq k \leq n-1$ in this normed space. A line in $\Re^n$ is d-convex if and only if it is parallel to a main diagonal of the cube $\sum$, i.e., parallel to a vector of the form $(\pm 1, \cdots, \pm 1)$ for any combination of signs. $\square$

Theorem 11.7: For each normed space $\Re^n$ there exist one-dimensional d-convex subspaces $l_1, \cdots, l_n$ such that $\Re^n = l_1 \oplus \cdots \oplus l_n$. $\square$

PROOF: By Theorem 2.2, the convex hull of the extreme points of the unit ball $\sum$ coincides with $\sum$. Hence we can find n extreme points $a_1, \cdots, a_n$ not lying in an $(n-1)$-dimensional subspace. These points are correspondingly contained in one-dimensional subspaces $l_1, \cdots, l_n$ which are d-convex (cf. Corollary 11.2). Then $l_1 \oplus \cdots \oplus l_n = \mathbf{R}^n$.   ■

We remark that it is possible that there are no more than n one-dimensional d-convex subspaces (see Example 11.5).

A normed space $\Re^n$ is said to be an α-*space* if for each bounded set $F \subset \Re^n$ the d-convex hull $\text{conv}_d F$ also is bounded.

Theorem 11.8: The space $\Re^n$ is an α-space if and only if there exist d-convex hyperplanes $L_1, L_2, \cdots, L_n \subset \Re^n$ whose intersection is a point. $\square$

PROOF: Let $\Re^n$ be an α-space. Then this space contains bounded, d-convex bodies (as e.g. $\text{conv}_d \sum$). In accordance with Theorem 10.2, $\Re^n$ contains compact, d-convex bodies. Let $M \subset \text{int } \sum$ be a compact, d-convex body. By Theorem 10.10, the representation $M = \bigcap_\gamma P_\gamma$ holds, where each P_γ is a closed, d-convex half-space and γ runs through a suitable index set. Let $a \in$ int M. For any $x \in$ bd $\sum$, there is some γ and some translate P'_γ of P_γ such that $a \in$ int P'_γ and $x \notin P_\gamma$. By the compactness of bd $\sum$ one can find a finite index set $\gamma_1, \cdots, \gamma_k$ such that $N = P'_{\gamma_1} \cap \cdots \cap P'_{\gamma_k}$ does not contain any boundary point of $\sum$, i.e., $N \subset$ int $\sum$. Moreover, a is an interior point of N. In other words, N is a d-convex polytope contained in int $\sum$. By Theorem 10.8, the bounding hyperplanes $L'_{\gamma_1}, \cdots, L'_{\gamma_k}$ of the supporting half-spaces $P'_{\gamma_1}, \cdots, P'_{\gamma_k}$ of N are d-convex. We assume that each $(n+1)$-tuple from $\{L'_{\gamma_1}, \cdots, L'_{\gamma_k}\}$ has an empty intersection (if necessary, this can be ensured by suitably small translations). Then an arbitrary vertex of N is the intersection of an n-tuple from $\{L'_{\gamma_1}, \cdots, L'_{\gamma_k}\}$.

Conversely, let $L_1, \cdots, L_n \subset \Re^n$ be d-convex hyperplanes whose intersection is a point. We may assume that this point is o. The intersection of each $(n-1)$-tuple from $\{L_1, \cdots, L_n\}$ is a line. Thus we can choose the corresponding n lines as the axes of a coordinate system $(x_1, \cdots, x_n)$ in $\mathbf{R}^n$, where the condition $x_i = 0$ describes the hyperplane L_i. The half-space given by $x_i \leq \lambda$ (or $x_i \geq \lambda$) is d-convex (cf. Theorem 10.8). Hence the n-dimensional parallelotope defined by $|x_1| \leq \lambda, \cdots, |x_n| \leq \lambda$ is a d-convex set. For a sufficiently large λ, an arbitrarily given bounded set F is contained in this parallelotope and therefore $\text{conv}_d F$ is also bounded. Hence $\Re^n$ is an α-space. ∎

Corollary 11.9: Each two-dimensional normed space $\Re^2$ is an α-space. □

In fact, by Theorem 11.7 the existence of d-convex lines $l_1, l_2 \subset \Re^2$ with $l_1 \oplus l_2 = \Re^2$ is ensured. Since these lines are hyperplanes with respect to $\Re^2$, Theorem 11.8 gives the assertion. ∎

Let $\Re^n$ be a normed space with unit ball $\sum$. Then if L is a nontrivial, linear subspace of $\mathbf{R}^n$, the intersection $L \cap \sum$ is a nonempty, compact subset of $\Re^n$, distinct from $\{o\}$. Thus for two arbitrary nontrivial subspaces $L_1, L_2 \subset \Re^n$ the number

$$\Theta(L_1, L_2) = \varrho(L_1 \cap \textstyle\sum, L_2 \cap \textstyle\sum) \tag{1}$$

is defined, where ϱ is the Hausdorff distance. Since for each subspace $L \subset \Re^n$ the set $L \cap \sum$ is uniquely determined and vice versa, $\Theta(L_1, L_2)$ is a metric in the set G of all nontrivial subspaces of $\Re^n$. The *topology* determined by this metric in G does not depend on the choice of the corresponding norm, i.e., it is defined by the vector space $\mathbf{R}^n$. The set G_k of all k-dimensional linear subspaces of $\Re^n$ is contained in G, and therefore G_k, together with G, is a metric space (with the metric (1)). To avoid misunderstandings, we will write $G_k(\Re^n)$ instead of G_k if further normed spaces (not identical with $\Re^n$) are taken into consideration. Each of these metric (or topological) spaces $G, G_1, G_2, \cdots, G_{n-1}$ is compact, i.e., these spaces have the property that from an arbitrary open cover of them a finite cover can be chosen.

Furthermore, by E_k (or $E_k(\Re^n)$), if it is necessary to emphasize the space $\Re^n$) we denote the set of all d-convex, k-dimensional subspaces of $\Re^n$. Thus $E_k \subset G_k$, i.e., E_k is a subspace of the metric space G_k (with metric (1)).

Theorem 11.10: The set E_{n-1} (with the metric (1)) is compact.   □

PROOF: Since any closed subspace of a compact space is compact, it remains to show that E_{n-1} is a closed subspace of G_{n-1}, i.e., $G_{n-1} \setminus E_{n-1}$ is open in G_{n-1}. Let L be from $G_{n-1} \setminus E_{n-1}$, i.e., L be an $(n-1)$-dimensional subspace which is not d-convex. We have to prove that all $(n-1)$-subspaces, which are (in the sense of (1)) sufficiently near to L, are also not d-convex. Since L is not d-convex, there are points $x \in L, y \notin L$ satisfying $y \in [o, x]_d$, i.e., $d(o, x) = d(o, y) + d(y, x)$. We set $a = x - \frac{1}{2}y, b = \frac{1}{2}y$ (Fig. 44) and obtain $\frac{1}{2}a + \frac{1}{2}b = \frac{1}{2}x$, i.e., $\frac{1}{2}x \in L$ is the midpoint of the segment $[a, b]$. By $b \notin L$, the points a und b belong to different open half-spaces with respect to L. Thus,

$$d(o, b) + [d(b, y) + d(y, x)] = d(o, y) + d(y, x) = d(o, x);$$

$$d(o, b) + d(b, x) \geq d(o, x),$$

and therefore $d(b, x) \geq d(b, y) + d(y, x)$. The triangle inequality implies

$$d(b, x) = d(b, y) + d(y, x). \tag{2}$$

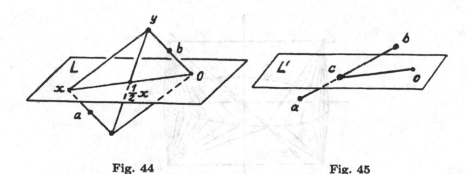

Fig. 44   Fig. 45

Furthermore, $d(b, x) = \parallel x - b \parallel = \parallel x - \frac{1}{2}y \parallel = \parallel a \parallel = d(o, a)$ and, analogously, $d(b, y) = d(o, b), d(y, x) = d(b, a)$. Therefore (2) implies

$$d(o, a) = d(o, b) + d(b, a). \tag{3}$$

Again, since a and b belong to different open half-spaces regarding L, the same holds for a hyperplane $L' \subset G_{n-1}$ sufficiently near to L. Hence $L' \cap [a, b]$ is an interior point of the segment $[a, b]$, denoted by c, see Fig. 45. With (3) we have

· $d(o,a) = d(o,b) + d(b,c) + d(c,a)$;

$d(o,a) \leq d(o,c) + d(c,a)$,

and therefore $d(o,c) \geq d(o,b) + d(b,c)$. By the triangle inequality, this yields

$d(o,c) = d(o,b) + d(b,c)$,

i.e., $b \in [o,c]_d$. But $o, c \in L'$ and $b \notin L'$; therefore L' is not d-convex. Thus all subspaces sufficiently near to L and belonging to G_{n-1} are not d-convex. ∎

EXAMPLE 11.11: We will show that there exist a normed space $\Re^n$ and a positive integer $k < n-1$ such that E_k is not compact (i.e., not closed in G_k). In $\mathbf{R}^3$ with a coordinate system (x_1, x_2, x_3), consider the set Q defined by the equalities $x_1^2 + x_2^2 = 1$, $x_3 = 0$ and the segments $S_1 = [a,b], S_2 = [c,d]$ with $a = (1,0,1), b = (1,0,-1), c = (-1,0,1), d = (-1,0,-1)$, see Fig. 46. Furthermore, we take $\sum = \text{conv}\{Q \cup S_1 \cup S_2\}$ as the unit ball of a normed 3-space $\Re^3$. It is easy to see that the points of Q, except for $(\pm 1, 0, 0)$, are extreme points of $\sum$. Thus by Corollary 11.2 all one-dimensional subspaces from aff Q, except for the x_1-axis, are d-convex. This shows that the set E_1 of all one-dimensioal d-convex subspaces of $\Re^3$ is not closed in G_1 (and hence it is not compact). □

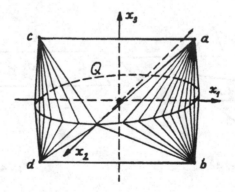

Fig. 46

Theorem 11.12: Let $\Re^n$ *be a normed space,* $\sum$ *be its unit ball, and L be a* d-convex subspace. We set $\sum_* = \sum \cap L$ and consider L as a normed space $\mathbf{L}$ with the unit ball $\sum_*$. Then the set $M \subset L$ is d-convex in $\Re^n$ if and only if it is d-convex in $\mathbf{L}$. □

PROOF: For $x \in L$, the point $[o,x] \cap \text{bd} \sum$ coincides with $[0,x) \cap \text{rbd} \sum_*$. Therefore the norms of x coincide in both the spaces $\Re^n$ and $\mathbf{L}$. Hence for

$x, y \in L$ the d-segment $[x, y]_d \subset \Re^n$ is also contained in L (since L is d-convex). Thus the d-segments with the endpoints x and y are coinciding in both the spaces. ∎

Theorem 11.13: Let $L \subset \Re^n$ be an arbitrary d-convex subspace of dimension k, $2 \le k \le n-1$. Let $F(L)$ denote the set of the subspaces $L \cap \Gamma$, where Γ runs over all d-convex subspaces of dimension $n-1$ not containing L. Then cl $F(L) \subset E_{k-1}(\mathbf{R}^n)$. $\square$

PROOF: In view of Theorem 11.12, let $\mathbf{L}$ be a normed space with unit ball $\sum_*$. By d-convexity of $L \cap \Gamma$ in $\Re^n$ (and by $L \cap \Gamma \subset L$), Theorem 11.12 implies that $L \cap \Gamma$ is d-convex in $\mathbf{L}$, too. Hence $F(L)$ is contained in the set $E_{k-1}(\mathbf{L})$ of all d-convex $(k-1)$-subspaces of $\mathbf{L}$. By Theorem 11.10, the set $E_{k-1}(\mathbf{L})$ is compact, i.e., it is closed in $G_{k-1}(\mathbf{L})$, and therefore we have cl $F(L) \subset E_{k-1}(\mathbf{L})$. Now Theorem 11.12 implies $E_{k-1}(\mathbf{L}) \subset E_{k-1}(\mathbf{R}^n)$. ∎

Finally, we formulate two theorems about the d-convex hull of the unit ball in a Minkowski space $\Re^n$.

Theorem 11.14: The unit ball of a normed space $\Re^n$ satisfies the condition conv$_d \sum = \sum$ *if and only if the body $\sum$ is equatorial, i.e, each its regular supporting hyperplane is d-convex, i.e., the corresponding parallel subspace is equatorial with respect to $\sum$.* $\square$

The idea of the proof is sketched in the Exercises 1–7.

Corollary 11.15: For $n \ge 3$, let the unit ball $\sum$ of $\Re^n$ be an n-polytope. Then the condition conv$_d \sum = \sum$ *cannot be satisfied.* $\square$

PROOF: To prove this, it is sufficient to embed $\sum$ into the projective augmentation $\mathbb{P}^n$ of the Euclidean n-space $\mathbf{R}^n$. By Theorem 11.14, each pair of parallel facets of the centrally symmetric, convex n-polytope $\sum$ has to be parallel to an $(n-1)$-subspace which is equatorial with respect ot $\sum$. Thus the number p of parallel facet pairs of $\sum$ cannot be larger than the number q of equatorial $(n-1)$-subspaces of $\sum$, i.e., $p \le q$. Projecting the $(n-2)$-skeleton of bd $\sum$ from the origin into the hyperplane at infinity denoted by $\mathbb{P}^{n-1}$, we obtain a projective arrangement A of hyperplanes (actually, of $(n-2)$-flats) with respect to $\mathbb{P}^{n-1}$, where the set of equatorial $(n-1)$-subspaces of $\sum$ and the set of $(n-2)$-flats in A are in a $1-1$-correspondence. Analogously, there is a $1-1$-correspondence between parallel facet pairs of $\sum$ and $(n-1)$-cells of A. Since the system of outward facet normals of $\sum$ has to span $\mathbf{R}^n$ and the normals of the equatorial $(n-1)$-subspaces of $\sum$ have to do the same, the $(n-2)$-flats of A cannot have a common point. It is known that for such projective arrangements of q flats of dimension $n-2$ in $\mathbb{P}^{n-1}$, the lower bound on the number p of $(n-1)$-cells is exactly attained if A forms a "near pencil", i.e., if $q-n+2$ of the $(n-2)$-flats meet at a projective $(n-3)$-flat, see [Sha] and [Mar 1]. This yields $p \ge 2^{n-2}(q-n+2)$, contradicting the above relation $p \le q$ for $n \ge 3$. ∎

It should be remarked that an alternate proof is obtainable by means of Theorem 1 from [Mar 3].

These combinatorial arguments do not work for $n = 2$. In fact, for the two-dimensional case the condition $\text{conv}_d \sum = \sum$ is realizable if $\sum$ is a polygon. An example is given by a regular hexagon $\sum$.

Furthermore, let $M \subset \mathbf{R}^n$ be a compact, convex body centered at the origin. Consider the following problem: is there a unit ball $\sum \subset \mathbf{R}^n$ transforming $\mathbf{R}^n$ into a Minkowski space such that $\text{conv}_d \sum = M$? The answer is not always positive, cf. Exercises 6, 8, 9, and 10.

Exercises

1. Suppose that the unit ball $\sum \subset \Re^2$ is a polygon (centered at the origin). Denote by $l_1, \cdots, l_s$ the one-dimensional subspaces passing through the vertices of $\sum$. For a compact set $M \subset \Re^2$ consider two supporting lines of M parallel to l_i and denote by Π_i (Fig. 47) the closed strip between them, $i = 1, \cdots, s$. Prove that $\text{conv}_d M = \Pi_1 \cap \cdots \cap \Pi_s$.

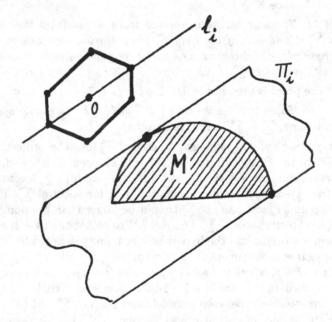

Fig. 47

2. Generalize the result of Exercise 1 for the case when the unit ball $\sum \subset \Re^2$ is not a polygon.

3. In the notation of Exercise 1, prove that the unit ball $\sum \subset \Re^2$ is *d*-convex if and only if it has two 1-faces parallel to l_i for every $i \in \{1, \cdots, s\}$. In other words, $\text{conv}_d \sum = \sum$ if and only if $\sum$ is equatorial, i.e., for each 1-face of $\sum$ the correspondingly parallel line through the origin contains two vertices of $\sum$.

4. Prove that if the unit ball $\sum \subset \Re^2$ is a regular polygon with $4p + 2$ vertices, $p = 1, 2, \cdots$, then $\operatorname{conv}_d \sum = \sum$.

5. Prove that the unit ball $\sum$ of a Minkowski plane $\Re^2$ satisfies the condition $\operatorname{conv}_d \sum = \sum$ if and only if $\sum$ is equatorial, i.e., for every regular supporting line of $\sum$ the correspondingly parallel line through the origin intersects $\operatorname{bd} \sum$ in two extreme points of $\sum$.

6. Prove that for the polygon $M \subset \mathbf{R}^2$ in Fig. 48 there exists no $\sum \subset \mathbf{R}^2$ such that in the Minkowski plane $\Re^2$ with unit ball $\sum$ the equality $\operatorname{conv}_d \sum = M$ holds. Prove the analogous assertion for the set M given in Fig. 49.

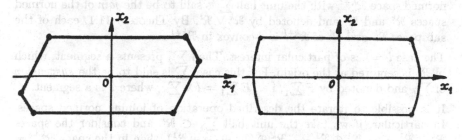

Fig. 48   **Fig. 49**

7. Let M be a compact, convex body in an n-dimensional Minkowski space $\Re^n$. Consider all the equatorial hyperplanes Γ of the unit ball $\sum \subset \Re^n$ such that two supporting hyperplanes of $\operatorname{conv}_d M$ parallel to Γ are regular, and denote by Π_Γ the closed strip between them. Prove that

$$\operatorname{conv}_d M = \bigcap_\Gamma \Pi_\Gamma.$$

8. Let $M \subset \Re^n$ be a compact, convex body centered at the origin and Γ be a regular supporting hyperplane of M. Admit that there exists a body $\sum \subset \Re^n$ centered at the origin such that in the Minkowski space with $\sum$ as unit ball the equality $\operatorname{conv}_d \sum = M$ holds. Prove that under this assumption $\sum = M$ and $\Gamma \cap \sum \neq \emptyset$. In particular, if x is a boundary point of M which simultaneously is regular and exposed, then $x \in \sum$.

9. Let M be the regular cross-polytope in $\mathbf{R}^n$. Describe the unit ball $\sum$ such that in the corresponding Minkowski space $\Re^n$ the equality $\operatorname{conv}_d \sum = M$ holds.

10. Give an example of a compact, convex body $M \subset \mathbf{R}^3$ centered at the origin such that there exists no $\sum \subset \mathbf{R}^3$ for which $\operatorname{conv}_d \sum = M$ in the Minkowski space $\Re^3$ with unit ball $\sum$.

11. Prove that for a Minkowski space $\Re^n$ the condition $\operatorname{diam} \operatorname{conv}_d \sum < \infty$ holds if and only if there exist d-convex hyperplanes $L_1, \cdots, L_n$ through the origin satisfying $L_1 \cap \cdots \cap L_n = \{o\}$.

12. Having a number q, construct a Minkowski space $\Re^n$ with $q < \operatorname{diam} \operatorname{conv}_d \sum < \infty$.

§12 The join of normed spaces

Let $\mathbf{R}^{k+l}$ be a direct sum of its subspaces $\mathbf{R}^k$ and $\mathbf{R}^l$, i.e., $\mathbf{R}^{k+l} = \mathbf{R}^k \oplus \mathbf{R}^l$. Furthermore, let $\sum_1$ and $\sum_2$ denote compact, convex bodies (centered at o) in $\mathbf{R}^k$ and $\mathbf{R}^l$, respectively. Thus with $\sum_1$ and $\sum_2$ as unit balls the normed spaces $\Re^k$ and $\Re^l$ are obtained from $\mathbf{R}^k$ and $\mathbf{R}^l$, respectively. We set

$$\sum = \operatorname{conv}\{\textstyle\sum_1 \cup \sum_2\} = \sum_1 \vee \sum_2$$

and obtain a compact, convex body in $\mathbf{R}^{k+l}$ centered at the origin. The normed space $\Re^{k+l}$ with the unit ball $\sum$ is said to be the *join* of the normed spaces $\Re^k$ and $\Re^l$ and denoted by $\Re^k \vee \Re^l$. By Theorem 11.1, each of the subspaces $\Re^k$ and $\Re^l$ of $\Re^{k+l}$ is *d*-convex in $\Re^{k+l}$.

The case $l = 1$ is of particular interest. Then $\sum_2$ presents a segment, which in $\Re^1$ is centered at the origin. For this case, $\sum$ is said to be the *suspension* of $\sum_1$ and denoted by $E\sum_1$, i.e., $E\sum_1 = I \vee \sum_1$, where I is a segment.

It is possible to iterate the described operation of joining normed spaces. In particular, if we take the unit ball $\sum_1 \subset \Re^k$ and consider the spaces $\Re^k \oplus \Re^1, \Re^k \oplus \Re^1 \oplus \Re^1, \cdots, \Re^k \oplus (\Re^1 \oplus \cdots \oplus \Re^1)$, then in the space $\Re^{k+t} = \Re^k \oplus (\Re^1 \oplus \cdots \oplus \Re^1)$ we obtain the unit ball $E^t \sum_1 = E(E^{t-1}\sum_1)$. For example, if $\Re^n = \Re^1 \oplus \cdots \oplus \Re^1$ is a join of one-dimensional subspaces, then its unit ball is an n-dimensional cross-polytope or, in other words, an $(n-1)$-fold suspension over a segment, see Fig. 50.

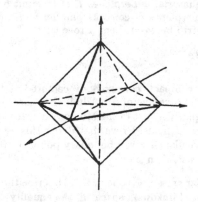

Fig. 50

Theorem 12.1: Let the normed space $\Re^{k+l}$ be the join of the spaces $\Re^k$ and $\Re^l$. If the points $a, b, x \in \Re^{k+l}$ satisfy the inclusions $a - x \in \Re^k$, $b - x \in \Re^l$, then $x \in [a, b]_d$, i.e., $d(a, b) = d(a, x) + d(x, b)$. $\square$

<u>PROOF:</u> Let $\sum_1, \sum_2$ denote the unit balls of the normed spaces $\Re^k$ and $\Re^l$, respectively. Since $a - x \in \Re^k$, we have $a - x = \lambda_1 e_1$ with $e_1 \in \operatorname{bd}\sum_1$

and $\lambda_1 \geq 0$. Analogously, $x - b = \lambda_2 e_2$ with $e_2 \in$ bd $\sum_2, \lambda_2 \geq 0$. Since $\| e_1 \| = \| e_2 \| = 1$, the equalities

$$d(a, x) = \| a - x \| = \| \lambda_1 e_1 \| = \lambda,$$

$$d(x, b) = \| x - b \| = \| \lambda_2 e_2 \| = \lambda_2$$

hold. Furthermore,

$$a - b = (a - x) + (x - b) = \lambda_1 e_1 + \lambda_2 e_2 = (\lambda_1 + \lambda_2)e,$$

where

$$e = \frac{\lambda_1}{\lambda_1 + \lambda_2} e_1 + \frac{\lambda_2}{\lambda_1 + \lambda_2} e_2.$$

By $e \in [e_1, e_2]_d \subset$ bd $\sum$ we have $\| e \| = 1$, and therefore

$$d(a, b) = \| a - b \| = \| (\lambda_1 + \lambda_2)e \| = \lambda_1 + \lambda_2 = d(a, x) + d(x, b). \quad \blacksquare$$

The following notions are introduced in view of the next theorem. Again, $\Re^{k+l}$ is written for the join of $\Re^k$ and $\Re^l$. Let now π_1 denote the parallel projection of $\Re^{k+l}$ along $\Re^l$ onto $\Re^k$ and π_2 the corresponding parallel projection of $\Re^{k+l}$ onto $\Re^l$. Furthermore, we denote by $A + B$ the vector sum of $A \subset \Re^k, B \subset \Re^l$, i.e., the set of all points $x + y$ with $x \in A, y \in B$. It is easily shown that $A + B = \pi^{-1}(A) \cap \pi_2^{-1}(B)$ for any nonempty $A \subset \Re^k, B \subset \Re^l$. Let conv$_d M$ denote the d-convex hull of a set $M \subset \Re^{k+l}$ with respect to $\Re^{k+l}$. We remark that for an arbitrary set $A \subset \Re^k$ the inclusion conv$_d A \subset \Re^k$ holds (since the subspace $\Re^k$ is d-convex). Analogously, $B \subset \Re^l$ implies conv$_d B \subset \Re^l$.

Theorem 12.2: For an arbitrary set $M \subset \Re^{k+l}$ the relation conv$_d M =$ conv$_d \pi_1(M) +$ conv$_d \pi_2(M)$ *holds.* $\square$

PROOF: Without loss of generality we may assume that $o \in M$ (if necessary, a translation is used). For $x \in M$, let $y = \pi_1(x)$. Then $y \in \Re^k, x - y \in \Re^l$ and by Theorem 12.1,

$$d(o, x) = d(o, y) + d(y, x).$$

Hence $y \in$ conv$_d M$, and this shows $\pi_1(M) \subset$ conv$_d M$. Then

$$\text{conv}_d \pi_1(M) \subset \text{conv}_d(M) \tag{1}$$

and, analogously,

$$\text{conv}_d \pi_2(M) \subset \text{conv}_d(M). \tag{2}$$

Let now $z \in$ conv$_d \pi_1(M) +$ conv$_d \pi_2(M)$, i.e., $z = a_1 + a_2$ with $a_1 \in$ conv$_d \pi_1(M), a_2 \in$ conv$_d \pi_2(M)$. Thus, by Theorem 12.1, we have

$$d(a_1, a_2) = d(a_1, z) + d(z, a_2),$$

and since $a_1, a_2 \in \mathrm{conv}_d M$ (cf. (1) and (2)), $z \in \mathrm{conv}_d M$. Therefore the inclusion

$$\mathrm{conv}_d M \supset (\mathrm{conv}_d \pi_1(M) + \mathrm{conv}_d \pi_2(M)).$$

holds. For showing the converse inclusion, consider the set $C = \pi_1^{-1}(\mathrm{conv}_d \pi_1(M))$. This is a cylinder spanned by the flats which are parallel to $\Re^l$ and contain points from $\mathrm{conv}_d \pi_1(M)$. Show the *d*-convexity of this cylinder. Let $z \in [a, b]_d$ for $a, b \in C$. We set $z' = \pi_1(z)$. Then there exist points $x, y \in C$ such that $d(x, y) = d(x, z') + d(z', y)$. Now we set $x' = \pi_1(x)$, $y' = \pi_1(y)$ and obtain, by Theorem 12.1,

$$d(x, z') = d(x, x') + d(x', z'),$$

$$d(z', y) = d(z', y') + d(y', y).$$

Thus,

$$d(x, y) = d(x, x') + d(x', z') + d(z', y') + d(y', y).$$

On the other hand, the triangle inequality implies

$$d(x, y) \leq d(x, x') + d(x', y') + d(y', y).$$

Comparing, we obtain

$$d(x', y') = d(x', z') + d(z', y').$$

From $x', y' \in \pi_1(C) = \mathrm{conv}_d \pi_1(M)$ it follows that $z' \in \mathrm{conv}_d \pi_1(M) \subset C$ and therefore $z \in C$. Thus $[a, b]_d \subset C$ and the set C is *d*-convex. The obvious inclusion $M \subset C$ implies $\mathrm{conv}_d M \subset C$. Analogously, the inclusion $\mathrm{conv}_d M \subset D$ with $D = \pi_2^{-1}(\mathrm{conv}_d \pi_2(M))$ is provable. Hence $\mathrm{conv}_d M \subset (C \cap D)$. Since $C \cap D = \mathrm{conv}_d \pi_1(M) + \mathrm{conv}_d \pi_2(M)$, the converse inclusion holds, too. ∎

EXAMPLE 12.3: Example 11.5 refers to a normed space $\Re^3$ which is the join of three one-dimensional spaces (the coordinate axes). By Theorem 12.2, in this space $\Re^3$ only the sets of the form $A + B + C$ are *d*-convex, where A, B, C are convex subsets of the coordinate axes. In particular, the only compact, *d*-convex sets are parallelotopes with edges parallel to the coordinate axes. The analogous inequalities hold in the space $\Re^n$ having the *n*-dimensional cross-polytope as its unit sphere (i.e., in the space representable as join of one-dimensional subspaces). □

The next theorem on the joining operation might be interesting for itself.

Theorem 12.4: Let $\Re^n$ be a normed space with unit ball $\sum$. We set $\alpha = \frac{1}{2}\mathrm{diam}(\mathrm{conv}_d \sum)$. Then for each bounded set $M \subset \Re^n$ the relation

$$\operatorname{diam}\,(\operatorname{conv}_d M) \le \alpha \operatorname{diam}\,M$$

holds.   □

PROOF: It suffices to consider only the case $\alpha < \infty$. Then the ball $\sum$ has a bounded d-convex hull and hence each bounded set $M \subset \Re^n$ has a bounded d-convex hull. Furthermore, by means of homotheties, it is enough to show that diam $M = 2$ implies the inequality diam $(\operatorname{conv}_d M) \le 2\alpha$. Therefore we assume diam $M = 2$.

Let M' denote the convex hull of M in the linear sense, i.e., $M' = \operatorname{conv}\,M$. Then diam $M' = $ diam $M = 2$. With respect to an arbitrary point $a \in \operatorname{ri}\,M'$, let $\varepsilon > 0$ be a number such that $U_\varepsilon(a) \cap \operatorname{aff}\,M' \subset M'$ (where $U_\varepsilon(a)$ is a ball in $\Re^n$, having the center a and radius ε). Since $U_\varepsilon(a) \cap \operatorname{aff}\,M'$ has the diameter 2ε and is enclosed in M', we have $\varepsilon \le 1$. Furthermore, for each point $x \in M'$ the inequality $d(a,x) \le 2 - \varepsilon$ holds (since a point $b \in \operatorname{bd}\,U_\varepsilon(a)$ exists for which $a \in [x,b]$, see Fig. 51). From this we obtain that diam $(M' \cup U_\varepsilon(a)) = 2$. In fact, for $x \in M', y \in U_\varepsilon(a)$ we have

$$d(x,y) \le d(x,a) + d(a,y) \le 2 - \varepsilon + \varepsilon = 2;$$

but if $x,y \in U_\varepsilon(a)$, then $d(x,y) \le 2\varepsilon \le 2$. If we set $M'' = \operatorname{cl}\,\operatorname{conv}\,(M' \cup U_\varepsilon(a))$, then M'' is a closed, convex *body* satisfying $M'' \supset M$ and diam $M'' = 2$.

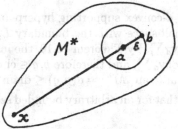

Fig. 51

Let $p,q \in \operatorname{cl}\,\operatorname{conv}_d M''$ be points with $d(p,q) = \operatorname{diam}\,(\operatorname{conv}_d M'')$. In view of suitable translations, it is possible to assume $p = -q$, i.e., o is the center of the segment $[p,q]$.

Now let L', L'' denote an arbitrary pair of parallel, d-convex supporting hyperplanes of the body $\operatorname{cl}\,\operatorname{conv}\sum$. By Γ', Γ'' denote the correspondingly parallel supporting hyperplanes of the body $\operatorname{cl}\,\operatorname{conv}_d M''$. Of course, the intersections $\Gamma' \cap M'', \Gamma'' \cap M''$ are nonempty (otherwise, one could translate at least one of the hyperplanes Γ', Γ'' in such a manner that the strip enclosed, and hence $\operatorname{conv}_d M''$, could become smaller). For $x' \in \Gamma' \cap M'', x'' \in \Gamma'' \cap M''$

we have $d(x', x'') \leq$ diam $M'' = 2$, and therefore at least one of the hyperplanes Γ', Γ'' has a point in common with $\sum$ (otherwise $d(x', x'') > 2$ would
hold). For example, let $\Gamma' \cap \sum$ be nonempty. Then Γ' is contained in the
strip bounded by the hyperplanes L', L''. Clearly, the points p, q lie in the
half-space P' bounded by Γ' with cl conv$_d M'' \subset P'$. Therefore they lie also
in the half-space Π bounded by one of the hyperplanes L', L'', since $\Pi \supset P'$
(see Fig. 52). But then, by symmetry with respect to o, both the points p, q
are contained in the strip determined by the hyperplanes L' and L''.

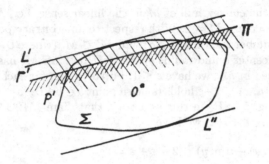

Fig. 52

Thus for an arbitrary d-convex supporting hyperplane L of cl conv$_d \sum$, the
points p, q lie in the half-space with the boundary L, which is a supporting
one of the body cl conv$_d \sum$. By Theorem 10.10, the intersection of these half-
spaces is given by cl conv$_d \sum$, and therefore $p, q \in$ cl conv$_d \sum$. Thus we have
diam (conv$_d M$) $\leq$ diam (conv$_d M''$) $= d(p, q) \leq$ diam (conv$_d \sum$) $= 2\alpha$. ∎

Theorem 12.4 implies that for an arbitrary bounded set $M \subset \Re^n$ the following
holds:

$$\frac{\text{diam (conv}_d M)}{\text{diam } M} \leq \frac{\text{diam (conv}_d \sum)}{\text{diam } \sum}.$$

Furthermore, for an arbitrary number $\alpha' \in [1, \alpha]$ one can find a bounded set
$M \subset \Re^n$ satisfying

$$\frac{\text{diam (conv}_d M)}{\text{diam } M} = \alpha'.$$

In fact, if we consider the set $M = \sum' \cap$ conv$_d \sum$ (where $\sum'$ is a ball of
radius r, $1 \leq r \leq \alpha$, centered at o), then by $\sum \subset M \subset$ conv$_d \sum$ the equalities
conv$_d M =$ conv$_d \sum$ and diam (conv$_d M$) $=$ diam (conv$_d \sum$) $= 2\alpha$ hold. On
the other hand, it is easy to see that diam $M = 2r$ and therefore

$$\frac{\text{diam (conv}_d M)}{\text{diam } M} = \frac{\alpha}{r}.$$

If r is running through $[1, \alpha]$, then the same holds for $\frac{\alpha}{r}$.

Theorem 12.5: _Let_ $\Re^{k+l}$ _be the join of normed spaces_ $\Re^k$ _and_ $\Re^l$. _Denote by_ $\sum_1, \sum_2, \sum$ _the unit balls of_ $\Re^k, \Re^l$, _and_ $\Re^{k+l}$, _respectively. Then_

$$\mathrm{diam}\,(\mathrm{conv}_d\textstyle\sum) = \mathrm{diam}\,(\mathrm{conv}_d\textstyle\sum_1) + \mathrm{diam}\,(\mathrm{conv}_d\textstyle\sum_2). \quad \square$$

PROOF: According to Theorem 12.2,

$$\mathrm{conv}_d\textstyle\sum = \mathrm{conv}_d\pi_1(\textstyle\sum) + \mathrm{conv}_d\pi_2(\textstyle\sum) = \mathrm{conv}_d\textstyle\sum_1 + \mathrm{conv}_d\textstyle\sum_2.$$

By Theorem 12.1, this implies the above equality. ∎

The proved assertions allow some further interpretations. We set

$$\alpha(\Re^n) = \sup_M \frac{\mathrm{diam}\,(\mathrm{conv}_d M)}{\mathrm{diam}\,M},$$

where M runs through the bounded subsets of $\Re^n$. Theorem 12.4 implies that this upper bound is attained by the unit ball $\sum$, i.e.,

$$\alpha(\Re^n) = \frac{1}{2}\mathrm{diam}\,(\mathrm{conv}_d\textstyle\sum).$$

The normed space $\Re^n$ is an α-space (cf. Theorem 11.8) if and only if $\alpha(\Re^n) < \infty$. Theorem 12.5 shows that

$$\alpha(\Re^{k+l}) = \alpha(\Re^k) + \alpha(\Re^l).$$

EXAMPLE 12.6: Let $\Re^n$ be the join of one-dimensional subspaces, i.e., its unit ball $\sum$ is an n-dimensional cross-polytope. Taking these one-dimensional subspaces as axes of a coordinate system $(x_1, \cdots, x_n)$, we obtain the following equality for an arbitrary vector $x = (x_1, \cdots, x_n) \in \Re^n$ (cf. Theorem 12.1):

$$\| x \| = \sum_{i=1}^{n} |x_i|.$$

By Theorem 12.5, we have diam $(\mathrm{conv}_d \sum) = 2n$, i.e., $\alpha(\Re^n) = n$. This is easily verified, since the set $\mathrm{conv}_d \sum$ is a cube which is described in $\Re^n$ by the inequalities

$$|x_1| \leq 1, \cdots, |x_n| \leq 1,$$

cf. Theorem 12.2. $\square$

Theorem 12.7: _Let_ $\Re^n = L_1 \oplus L_2$, _where_ L_1, L_2 _denote d-convex subspaces of_ $\Re^n$. _The normed space_ $\Re^n$ _is the join of the subspaces_ L_1, L_2 _if and only if the union_ $L_1 \cup L_2$ _contains every one-dimensional d-convex subspace of_ $\Re^n$.
$\square$

PROOF: Assume that every one-dimensional, d-convex subspace of $\Re^n$ is contained in $L_1 \cup L_2$. According to Corollary 11.2, the unit ball $\sum$ of $\Re^n$ has no extreme points outside of $L_1 \cup L_2$. Denoting by N the set of extreme points of $\sum$, we have with $N_1 = N \cap L_1, N_2 = N \cap L_2$ the equality $N = N_1 \cup N_2$. Hence Theorem 2.2 implies

$$\sum = \text{conv } N = \text{conv } (N_1 \cup N_2) = \text{conv } (\textstyle\sum_1 \cup \sum_2),$$

where $\sum_1 = \text{conv } N_1$ and $\sum_2 = \text{conv } N_2$. It is easy to see that $\sum_1 = L_1 \cap \sum, \sum_2 = L_2 \cap \sum$, and therefore $\sum = \text{conv } ((\sum \cap L_1) \cup (\sum \cap L_2))$. This shows that $\Re^n$ is the join of L_1 and L_2.

Conversely, if $\Re^n$ is the join of subspaces L_1 and L_2, then all extreme points of $\sum$ lie in $L_1 \cup L_2$, from which (with Corollary 11.2) the necessity is obtained. ∎

Exercises

1. Consider the space $\Re^{k+l} = \Re^k \vee \Re^l$ with the unit ball $\sum = \sum_1 \vee \sum_2$. Prove that the segments $[a, b]$ with $a \in \sum_1$ and $b \in \text{rbd } \sum_2$ have pairwise no common interior points.

2. In the notation of Exercise 1, prove that the union of the segments $[a, b]$ (with $a \in \sum_1$ and $b \in \text{rbd } \sum_2$) coincides with $\sum$.

3. In the notation of Exercise 1, prove that bd $\sum$ is the union of the segments $[a, b]$, where $a \in \text{rbd } \sum_1$, $b \in \text{rbd } \sum_2$.

4. Prove that the k-fold suspension over an l-dimensional cross-polytope is a $(k+l)$-dimensional cross-polytope.

5. Prove that the unit ball $\sum = \sum_1 \vee \cdots \vee \sum_s$ of the Minkowski space $\Re_1 \vee \cdots \vee \Re_s$ is the union of all simplices with the vertices $a_1 \in \sum_1, \cdots, a_s \in \sum_s$. In what case is such a simplex situated in bd $\sum$?

6. Let $\Re^{k+l} = \Re^k \vee \Re^l$. Prove that for any nonempty sets $A \subset \Re^k, B \subset \Re^l$ the relation $A + B = \pi_1^{-1}(A) \cap \pi_2^{-1}(B)$ holds, where π_1, π_2 are projections of $\Re^{k+l}$ onto $\Re^k$ and $\Re^l$, in each case parallel to the other subspace.

7. Prove the following assertion which is converse to Theorem 12.1: Let $\Re^{k+l} = \Re^k \oplus \Re^l$ and $\sum, \sum_1, \sum_2$ be the unit balls of $\Re^{k+l}, \Re^k$, and $\Re^l$, respectively. If for any $a \in \Re^k, b \in \Re^l$ the equality $\| a + b \| = \| a \| + \| b \|$ holds, then $\sum = \sum_1 \vee \sum_2$.

8. Prove that a subspace $L \subset \Re^k \vee \Re^l$ is equatorial if and only if $L = L_1 + L_2$, where $L_1 \subset \Re^k, L_2 \subset \Re^l$ are equatorial.

9. Prove that if $\Re^{k+l} = \Re^k \vee \Re^l$, then for any $a \in \Re^k, b \in \Re^l$ the equality $[o, a + b]_d = [o, a]_d + [o, b]_d$ holds.

10. Prove that if $\Re^{k+l} = \Re^k \vee \Re^l$, then for any d-convex sets $A \subset \Re^k$, $B \subset \Re^l$ with $o \in A$, $o \in B$ the equality $\text{conv}_d(A \cup B) = \bigcup_{a \in A, b \in B} [a, b]_d$ holds.

11. In the notation of Exercise 6, prove the following implication: if $\text{conv}_d M = \text{conv}_d \pi_1(M) + \text{conv}_d \pi_2(M)$ for every $M \subset \Re^{k+l}$, then $\Re^{k+l} = \Re^k \vee \Re^l$ (i.e., $\sum = \sum_1 \vee \sum_2$).

§13 Separability of d-convex sets

Theorem 13.1: _Let M_1, M_2 be d-convex sets in a two-dimensional Minkowski space $\Re^2$, satisfying_ ri $M_1 \cap$ ri $M_2 = \emptyset$. _Then M_1 and M_2 are strictly separable in $\Re^2$ by some d-convex line._   □

PROOF: If M_1 is a point (outside of ri M_2), then the assertion follows from Theorem 10.10 (or 10.9) for dim $M_2 = 2$ and from Theorem 11.7 (with Corollary 10.6) for dim $M_2 \leq 1$. Thus we can assume that M_1, M_2 are not points. By ri $M_1 \cap$ ri $M_2 = \emptyset$ and Theorem 1.7, the sets M_1, M_2 are strictly separable (in the linear sense) by a line $l \subset \mathbf{R}^2$. In other words, the inclusions $M_1 \subset P_1$ and $M_2 \subset P_2$ hold, where P_1, P_2 are the closed half-planes regarding $l \subset \Re^2$ and at least one of the sets M_1, M_2 is not contained in l.

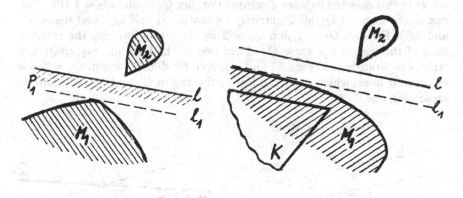

Fig. 53   Fig. 54

Let $M_1 \cap (P_1 \setminus l) \neq \emptyset$. By $M_1 \subset P_1$, there exists a line $l_1 \parallel l$ which is either a supporting line or an asymptotic line of M_1 (see Fig. 53 and Fig. 54, respectively), which has M_1, M_2 on different sides (it is possible that l_1 and l coincide). If l_1 is an asymptotic line of M_1, then the cone inscribed into M_1 is a (possibly straight) angle with one leg parallel to l, see Fig. 54. For this case, Theorems 10.11 and 10.9 imply d-convexity of l. The same holds if M_2 has an asymptotic line parallel to l. Therefore we can restrict our considerations to the following case: M_1 and M_2 have parallel supporting lines l_1, l_2 (possibly

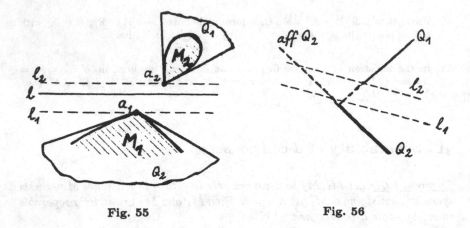

Fig. 55   Fig. 56

with $l_1 = l_2$) and these sets lie on different sides of the strip Π enclosed by l_1 and l_2 (Fig. 55), with the possible subcase $M_2 \subset l_2$.

For $a_1 \in l_1 \cap M_1, a_2 \in l_2 \cap M_2$, let Q_1, Q_2 denote the supporting cones of the figures M_1, M_2 at the points a_1 and a_2, respectively. Then, by Theorem 10.5, Q_1 and Q_2 are d-convex sets on different sides of Π with $Q_1 \not\subset l_1$. Each of the figures Q_1, Q_2 is either one-dimensional or a (possibly straight) angle, where in this case the legs are d-convex. For $\dim Q_1 = \dim Q_2 = 1$ (Fig. 56), one of the lines aff Q_1, aff Q_2 strictly separates Q_1 and Q_2 (and hence M_1 and M_2). For $\dim Q_1 = 1, \dim Q_2 = 2$ we denote by m_1, m_2 the carrying lines of the legs of the angle Q_2. Then one of the lines m_1, m_2, aff Q_1 is a strictly separating one (Figs. 57 and 58). And for $\dim Q_1 = \dim Q_2 = 2$, one of the four lines, which carry the legs of the angles Q_1 and Q_2, is a strictly separating one.   ■

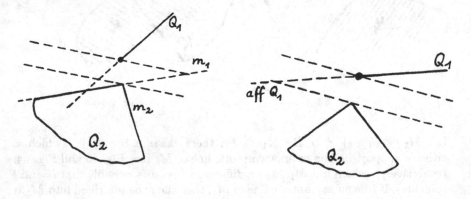

Fig. 57   Fig. 58

Theorem 13.2: _The following properties of a normed space_ $\Re^n$ _are equivalent:_

(1) _If arbitrary d-convex sets_ $M_1, M_2 \subset \Re^n$ _satisfy the condition_ ri $M_1 \cap$ ri $M_2 = \emptyset$, _then they are strictly separable by a d-convex hyperplane in_ $\Re^n$.

(2) _The sum of two arbitrary d-convex subspaces is a d-convex subspace._

(3) _The space_ $\Re^n$ _is the join of some of its d-convex subspaces, each of which is either two-dimensional or strictly convex._ $\square$

PROOF: We prove the theorem by the sequence $(1) \Rightarrow (2) \Rightarrow (3) \Rightarrow (1)$.
$(1) \Rightarrow (2)$: Let $\Re^n$ have the property (1). We remark that (1) can be transferred to an arbitrary d-convex subspace $\mathbb{L} \subset \Re^n$. Indeed, if $M_1, M_2 \subset \mathbb{L}$ are d-convex sets with ri $M_1 \cap$ ri $M_2 = \emptyset$, then there exists a d-convex hyperplane Γ strictly separating M_1 and M_2 in $\Re^n$. It is clear that Γ cannot contain $\mathbb{L}$ (since otherwise M_1, M_2 would lie in Γ), and therefore $\Gamma \cap \mathbb{L}$ is a d-convex hyperplane of the space $\mathbb{L}$ which strictly separates M_1 and M_2 (Fig. 59).

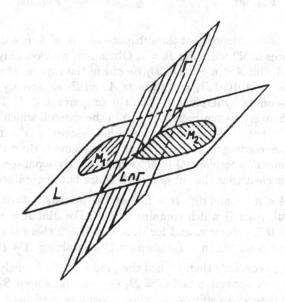

Fig. 59

Now we prove that for any d-convex subspace $A \subset \Re^n$, dim $A \le n - 2$, there exists a d-convex subspace $A' \supset A$ with dim $A' = $ dim $A + 1$. In fact, let l be a one-dimensional subspace not contained in A (the existence of l is ensured by Theorem 11.7) and l_1 be one of the rays on l with starting point o, see Fig. 60.

Clearly, also l_1 is d-convex. Since ri $A \cap$ ri $l_1 = \emptyset$, there is a d-convex hyperplane Γ' strictly separating A and l_1. Obviously, $A \subset \Gamma'$. Thus there exists an $(n-1)$-dimensional subspace $\Gamma' \supset A$. Applying the above reasoning to the space Γ', we analogously obtain a d-convex $(n-2)$-subspace $\Gamma'' \supset A$ (if dim $A <$ dim $\Gamma' - 1$), and so on.

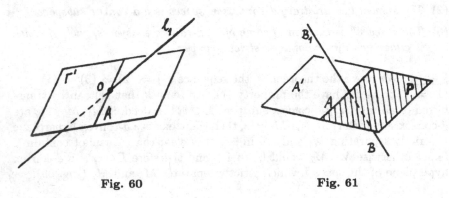

Fig. 60   Fig. 61

Now we prove the d-convexity of the subspace $A + B$, if A, B are arbitrary d-convex subspaces of $\Re^n$ with dim $B = 1$. Obviously, we have only to consider the case $B \not\subset A$, dim $A < n - 1$. Let B_1 be one of the rays on B with starting point o. Because of ri $A \cap$ ri $B_1 = \emptyset$, the sets A and B_1 are strictly separable in $\Re^n$ by some d-convex hyperplane L with the property $A \subset L$. The aforesaid implies that there is a d-convex flat $A' \supset A$ in the space L which satisfies dim $A' =$ dim $A + 1$. Let P be one of the closed half-spaces in A' with respect to A, see Fig. 61. According to Theorem 10.8, P is d-convex. By ri $P \cap$ ri $B = \emptyset$, there is a d-convex hyperplane Q in $\Re^n$ which strictly separates P and B in this space. It is clear that the subspaces A, B are both contained in Q.

Thus, if dim $A < n - 1$ and dim $B = 1$, then there exists a d-convex, $(n-1)$-dimensional subspace Q which contains $A + B$. For dim $A = n - 2$ we see that $Q = A + B$ is d-convex, and for dim $A < n - 2$ there is (by the same reasoning) some d-convex $(n-2)$-subspace Q' containing $A + B$, and so on.

Now it is easy to conclude that $\Re^n$ has the property (2). Namely, let A, C be arbitrary d-convex subspaces and $C = B_1 \oplus \cdots \oplus B_k$, where $B_1, \cdots, B_k$ are one-dimensional d-convex subspaces (whose existence is ensured, by Theorem 11.7). The assertions proved above imply that the subspaces $A + B_1, (A + B_1) + B_2, (A + B_1 + B_2) + B_3, \cdots$ are d-convex. Thus $A + C$ is a d-convex subspace.

(2) $\Rightarrow$ (3): Assume that $\Re^n$ has the property (2). Let $\Re^n$ be represented as the join of d-convex subspaces $\mathbb{L}_1, \cdots, \mathbb{L}_k$, each of which itself is not a join. (The case $k = 1$ is possible, for which $\Re^n$ is not a join.) For dim $\mathbb{L}_i = 1$, the space $\mathbb{L}_i$ is strictly convex, and the case dim $\mathbb{L}_i = 2$ is permitted by (3). Hence it remains to verify the d-convexity of $\mathbb{L}_i$ for dim $\mathbb{L}_i > 2$.

Thus, let $\mathbb{L}$ be a normed space with dim $\mathbb{L} = s \geq 3$ which is not a join and has the property (2). Denote by $l_1, \cdots, l_s$ arbitrary one-dimensional, d-convex subspaces not contained in a hyperplane of $\mathbb{L}$ (Theorem 11.7) and by $e_1, \cdots, e_s$ nonzero vectors correspondingly parallel to $l_1, \cdots, l_s$. A nonempty set $M \subset \{1, \cdots, s\}$ is said to be *realizable* if the following holds: there is a one-dimensional d-convex subspace of $\mathbb{L}$ whose direction can be represented by $\lambda_1 e_1 + \cdots + \lambda_s e_s$, where $\lambda_i \neq 0$ for and only for $i \in M$.

Now we show the following implication: *If a set M is realizable, then its nonempty subsets are also realizable.* Without loss of generality we may show this only for $\{1, \cdots, k\}$, assuming that $\{1, \cdots, q\}$ with $q > k$ is realizable. Let m be a one-dimensional, d-convex subspace having a direction vector of the form $\lambda_1 e_1 + \cdots + \lambda_q e_q$, where $\lambda_i \neq 0$ for each $i \in \{1, \cdots, q\}$. By (2), the subspaces $m + l_{k+1} + \cdots + l_s$ and $l_1 + l_2 + \cdots + l_k$ are d-convex, and hence their intersection is also d-convex. But this intersection is a one-dimensional subspace with a direction vector of the form $\lambda_1 e_1 + \cdots + \lambda_k e_k$, i.e., the set $\{1, \cdots, k\}$ is realizable.

Our next assertion to be verified is given by the following implication: *if sets M_1 and M_2 with nonempty intersection are realizable, then also the set $M_1 \cup M_2$*. For example, let M_1, M_2 be given by $M_1 = \{1, \cdots, k\}, M_2 = \{p, p+1, \cdots, q\}$ with $p \leq k$ and $q \leq s$. The above implication guarantees the realizability of the set $\{k, k+1, \cdots, q\}$. Let m', m'' be one-dimensional, d-convex subspaces whose direction vectors can be represented by $\lambda_1 e_1 + \cdots + \lambda_k e_k$ and $\mu_k e_k + \cdots + \mu_1 e_q$, respectively with nonzero coefficients. By a suitable multiplication, the equality $\lambda_k = \mu_k$ is established. By (2), the subspaces $m' + l_{k+1} + \cdots + l_q$ and $m'' + l_1 + \cdots + l_{k-1}$ and also their intersection are d-convex. But this intersection is a one-dimensional subspace, whose direction vector is parallel to $\lambda_1 e_1 + \cdots + \lambda_k e_k + \mu_{k+1} e_{k+1} + \cdots + \mu_q e_q$. Therefore the set $\{1, 2, \cdots, q\} = M_1 \cup M_2$ is realizable.

Let now N be a set consisting of the integer 1 and, in addition, of the integers $i \in \{2, 3, \cdots, q\}$ for which the set $\{1, i\}$ is realizable. By the assertions above, N is realizable. We assume that N is not identical with the whole set $\{1, \cdots, s\}$. Thus let $N = \{1, \cdots, k\}$ for some k satisfying $1 \leq k < s$. Since $\mathbb{L}$ is not the union of the subspaces $l_1 + \cdots + l_k$ and $l_{k+1} + \cdots + l_s$, there is a one-dimensional, d-convex subspace m^* lying in neither of the subspaces (cf. Theorem 12.7). Hence the direction of m^* can be represented by a vector of the form $\nu_1 e_1 + \cdots + \nu_s e_s$, where at least one of the coefficients $\nu_1, \cdots, \nu_k$ and at least one of the coefficients $\nu_{k+1}, \cdots, \nu_s$ have to be $\neq 0$. Hence M^* is realizable, when M^* denotes the set of all indices i with $\nu_i \neq 0$ (where $M^* \cap N \neq \emptyset$ and $M^* \not\subset N$). But then the set $M^* \cup N$ is realizable. This contradicts the assumption that N contains all indices i for which $\{1, i\}$ is realizable. Hence $N = \{1, \cdots, s\}$.

Thus an *arbitrary* subset of $\{1, \cdots, s\}$ is realizable.

Let now $Q \subset \mathbb{L}$ be a *d*-convex, $(s - 2)$-dimensional subspace. We will prove that an *arbitrary* hyperplane containing Q is *d*-convex. Suppose the contrary. The set of all hyperplanes which, in each case, contain Q is homeomorphic to a circumference, and the set of *d*-convex ones among them is a closed subset of that circumference (cf. Theorem 11.10). By the assumption, this closed subset and the whole circumference do not coincide. In other words, there are two *d*-convex hyperplanes L_1, L_2 containing Q such that there is no *d*-convex hyperplane in one of the dihedral angles between them. We choose one-dimensional, *d*-convex subspaces $l'_1 \subset L_1, l'_2 \subset L_2$ not contained in Q (by Theorem 11.7, such subspaces exist). In addition (and also by Theorem 11.7), we choose one-dimensional, *d*-convex subspaces $l'_3, \cdots, l'_s$ such that $Q = l'_3 + \cdots + l'_s$. The corresponding direction vectors $e'_1, \cdots, e'_s$ form a basis of the space $\mathbb{L}$. The assertions on realizability imply the existence of a one-dimensional, *d*-convex subspace p which is defined by a vector $\lambda_1 e'_1 + \lambda_2 e'_2 + \lambda_3 e'_3$, where $\lambda_1, \lambda_2, \lambda_3 \neq 0$ (cf. Fig. 62). Here the condition $s \geq 3$ is used.

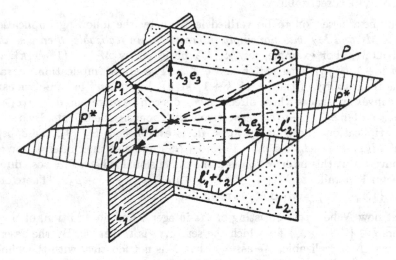

Fig. 62

We set $p_1 = (p + l'_2) \cap L_1, p_2 = (p + l'_1) \cap L_2, p^o = (p_1 + p_2) \cap (l'_1 + l'_2)$. The one-dimensional subspaces p_1, p_2, p^o are *d*-convex, and they are defined by the vectors $\lambda_1 e_1 + \lambda_3 e'_3$, $\lambda_2 e'_2 + \lambda_3 e'_3$, and $\lambda_1 e'_1 - \lambda_2 e'_2$, respectively. But the vectors $\lambda_1 e'_1 + \lambda_2 e'_2 + \lambda_3 e'_3$ and $\lambda_1 e'_1 - \lambda_2 e'_2$ lie in *different* pairs of vertical angles defined by the hyperplanes L_1 and L_2. Hence the *d*-convex hyperplanes $Q + p$ and $Q + p^o$ lie also in different pairs of vertical angles. This contradicts the choice of L_1 and L_2, and therefore an arbitrary hyperplane through Q has to be *d*-convex.

Now let $l_1, \cdots, l_s$ be one-dimensional, *d*-convex subspaces satisfying $l_1 + \cdots + l_s = \mathbb{L}$ (Theorem 11.7). Let Q_{ij} denote the sum of all subspaces except for l_i, l_j. The $(s-2)$-dimensional subspaces Q_{ij} are d-convex. The aforesaid relations imply that an arbitrary hyperplane containing some Q_{ij} is d-convex. In other words, if l is an arbitrary one-dimensional subspace (not necessarily d-convex), then all the subspaces $l + Q_{ij}$ $(i \neq j)$ are d-convex. But we have $l = \bigcap_{i \neq j} (l + Q_{ij})$, which implies that an arbitrary one-dimensional subspace $l \subset \mathbb{L}$ is d-convex. In view of Corollary 11.2, this yields the strict convexity of $\mathbb{L}$.

(3) $\Rightarrow$ (1): Let $\Re^n$ be the join of the subspaces $\mathbb{L}_1, \cdots, \mathbb{L}_k$, where each of the subspaces is either two-dimensional or strictly convex. Furthermore, let $M, N \subset \Re^n$ be d-convex sets with ri $M \cap$ ri $N = \emptyset$. By Theorem 12.2, $M = M_1 + \cdots + M_k$ and $N = N_1 + \cdots + N_k$, where M_i, N_i denote d-convex sets of the subspace $\mathbb{L}_i$. By ri $M =$ ri $M_1 + \cdots +$ ri M_k and ri $N =$ ri $N_1 + \cdots +$ ri N_k we conclude that ri $M_i \cap$ ri $N_i = \emptyset$ for some i. As in the case dim $\mathbb{L}_i = 2$ (cf. Theorem 13.1), also in the case of strict convexity of $\mathbb{L}_i$ (see Theorem 1.7 and Corollary 11.3) we have the following implication: if ri $M_i \cap$ ri $N_i = \emptyset$, then there exists a d-convex hyperplane $\Gamma_i \subset \mathbb{L}_i$ strictly separating M_i and N_i. But then $\mathbb{L}_1 + \cdots + \mathbb{L}_{i-1} + \Gamma_i + \mathbb{L}_{i+1} + \cdots + \mathbb{L}_k$ is a d-convex hyperplane in $\Re^n$ strictly separating M and N. Hence $\Re^n$ has the property (1). ∎

Theorem 13.2 shows that there exist normed spaces such that ri $M \cap$ ri $N = \emptyset$ does not imply the existence of a d-convex hyperplane strictly separating M and N. The following example gives an illustration.

EXAMPLE 13.3: In Euclidean 3-space $\mathbf{R}^3$ with an orthornomal coordinate system (x_1, x_2, x_3) we consider the cube C defined by the inequalities $|x_i| \leq 1, i = 1, 2, 3$. Furthermore, we consider two supporting planes through every edge of the cube, enclosing an angle of $\frac{\pi}{6}$ with one of the neighbouring facets of C. The intersection of the 24 half-spaces, defined by these supporting planes and containing C in each case, is a convex polyhedron (Fig. 63) centered at o and having the 14 vertices

$$(\pm 1, \pm 1, \pm 1), (\pm (1 + \frac{\sqrt{3}}{3}), 0, 0), (0, \pm(1 + \frac{\sqrt{3}}{3}), 0), (0, 0, \pm(1 + \frac{\sqrt{3}}{3})).$$

Considering this polyhedron as unit ball $\sum$ of a normed space $\Re^3$, this space has (according to Theorem 11.1) only six d-convex, two-dimensional subspaces, which are determined by the equalities

$$x_1 \pm x_2 = 0, \; x_1 \pm x_3 = 0, \; x_2 \pm x_3 = 0.$$

Now let K_1 be the cone determined by

$$x_1 + x_2 \geq 0, x_1 - x_2 \geq 0$$

and, analogously, K_2 be the cone defined by

$$-x_1 + x_3 \geq 0, -x_1 - x_3 \geq 0,$$

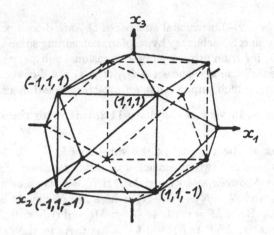

Fig. 63

cf. Fig. 64. By Theorem 10.8, both the cones are d-convex. The intersection of their interiors is empty. The point $(0,0,0)$ is the common apex of K_1 and K_2; since the x_3-axis belongs to K_1 and the x_2-axis belongs to K_2, the plane $x_1 = 0$ is the only linearly convex plane by which K_1 and K_2 are (strictly) separated. But this plane is not d-convex, i.e., K_1 and K_2 are *not separable* in $\Re^3$ by any d-convex plane.   □

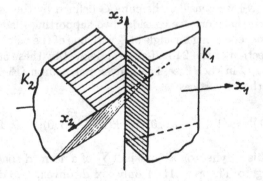

Fig. 64

In view of a generalization of the separability of d-convex sets we will give a further definition. Let M_1, M_2 be nonempty, d-convex sets in $\Re^n$. The sets M_1, M_2 are said to be *d-separable* if there exist d-convex cones $K_1, K_2 \subset \Re^n$ such that the following conditions are satisfied:

(1) $M_1 \subset K_1, M_2 \subset K_2, \operatorname{ri} K_1 \cap \operatorname{ri} K_2 = \emptyset$;

(2) in view of (1), the cones K_1, K_2 are maximal.

It easy to see that, in the case of linear convexity, the corresponding cones K_1, K_2 are given by two closed half-spaces, defined by a hyperplane. This shows that the d-separability is a natural generalization of the usual separability of convex sets.

Theorem 13.4: If for d-convex sets $M, N \subset \Re^n$ the condition ri $M \cap$ ri $N = \emptyset$ holds, then the sets are d-separable.   □

PROOF: By Theorem 10.2, we can assume that the sets M and N are closed. Let Γ be a hyperplane (possibly not d-convex) strictly separating M and N. Thus, with P_1, P_2 denoting the closed half-spaces defined by Γ, we have $M \subset P_1, N \subset P_2$, and at least one of the sets M, N is not contained in Γ. Assume $M \not\subset \Gamma$ and let Π_1 denote the smallest closed half-space containing M and contained in P_1. The bounding hyperplane Γ_1 of Π_1 is parallel to Γ. Analogously, we construct a half-space Π_2 for N with bounding hyperplane $\Gamma_2 \parallel \Gamma$.

If $M \cap \Gamma_1 = \emptyset$, then M is not bounded (since M is closed) and Γ_1 is an asymptotic hyperplane of M. Let $l = [a, x)$ be a ray with $l \subset M$ and $l \| \Gamma_1$. Then cl $M_x \subset \Pi_1$ and cl $M_x \cap \Gamma_1 \neq \emptyset$, where M_x is as in the proof of Theorem 10.12. Since the set cl $M_x \supset M$ is not contained in Γ, we have ri cl $M_x \cap$ ri $N = \emptyset$. Thus the possibility of replacing M by the d-convex set cl M_x (if necessary) shows that we can assume $M \cap \Gamma_1 \neq \emptyset$. Analogously, $N \cap \Gamma_2 \neq \emptyset$ is obtained. Hence we have $M \subset \Pi_1, N \subset \Pi_2$, and $M \not\subset \Gamma$.

Let now $x_1 \in M \cap \Gamma_1, x_2 \in N \cap \Gamma_1$. By Q_1 we denote the supporting cone of M at x_1 and, analogously, by Q_2 that of N at x_2. Then $Q_1 \subset \Pi_1, Q_2 \subset \Pi_2$, and $Q_1 \not\subset \Gamma$. Hence ri $Q_1 \cap$ ri $Q_2 = \emptyset$.

Denote by Ξ the set of d-convex cones K satisfying $M \subset K$ and ri $K \cap$ ri $Q_2 = \emptyset$, in each case. The set Ξ is nonempty; e.g., $Q_1 \in \Xi$. Furthermore, if $\{K_\alpha\}$ is an increasing sequence of cones contained in Ξ, then also the cone
$$K' = \text{cl}\left(\bigcup_\alpha K_\alpha\right)$$
is from Ξ. By Theorems 10.1 and 10.2, K' is d-convex. Moreover, starting with an arbitrary cone in this sequence, all the successors K_α have the same dimension. Therefore ri $K' = \bigcup_\alpha$ ri K_α, and this implies ri $K' \cap$ ri $Q_2 = \emptyset$ and $K' \in \Xi$. The lemma of Zorn [Zo] shows that there is at least one maximal element K_1 in the set Ξ.

Analogously, we can replace the d-convex cone Q_2 (satisfying $M_2 \subset Q_2$, ri $K_1 \cap$ ri $Q_2 = \emptyset$) by a maximal cone K_2 with the same properties.   ■

EXAMPLE 13.5: In $\mathbf{R}^3$ with a coordinate system (x_1, x_2, x_3), let $\sum$ denote the convex hull of the points $(\pm 2, 0, 0), (0, \pm 2, 0), (\pm 1, 0, \pm 1), (0, \pm 1, \pm 1)$, see Fig. 65. We take $\sum$ as the unit ball of a normed space $\Re^3$. The only d-convex planes in $\Re^3$ are those parallel to the plane $x_3 = 0$. Let M_1 be the positive semiaxis x_1 and M_2 be the positive semiaxis x_2. The sets M_1, M_2

are *d*-convex and ri $M_1 \cap$ ri $M_2 = \emptyset$. Consider now the *d*-convex cones K_1, K_2 which separate M_1, M_2. It is easy to see that (up to the order of indices) these cones are uniquely determined: they coincide with the half-space $x_3 \leq 0$ and the half-space $x_3 \geq 0$, respectively. But these cones do not realize a *strict* separation of M_1 and M_2 (since neither of the sets has a common point with ri K_1 and ri K_2). This illustrates that Theorem 13.4 cannot be sharpened by the condition of strict separability of the sets M_1 and M_2.   □

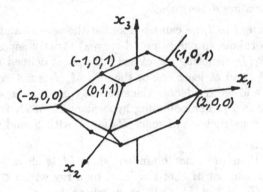

Fig. 65

Exercises

1. Let $\mathfrak{R}^3$ denote the normed 3-space introduced in Example 10.4. Construct *d*-convex sets M_1 and M_2 satisfying ri $M_1 \cap$ ri $M_2 = \emptyset$ which are not strongly separable by a *d*-convex plane.

2. Prove that if for any compact, *d*-convex sets $A, B \subset \mathfrak{R}^n$ the sum $A + B$ is *d*-convex, then $\mathfrak{R}^n = \mathfrak{R}_1 \vee \cdots \vee \mathfrak{R}_s$, where each $\mathfrak{R}_i$ is either one-dimensional or two-dimensional.

3. Let $\mathfrak{R}^3$ denote the normed 3-space from Example 10.4. Show that there exist two compact, *d*-convex sets $A, B \subset \mathfrak{R}^3$ such that the sum $A + B$ is not *d*-convex.

4. Let a, b be different points in a Minkowski plane $\mathfrak{R}^2$. Prove that the *d*-segment $[a, b]_d$ is either the linear segment $[a, b]$ or a parallelogram with two opposite vertices a, b.

5. Prove that for any points $a, b \subset \mathfrak{R}^2$ the inclusion $[o, a]_d + [o, b]_d \supset [o, a + b]_d$ holds.

6. Prove that for any *d*-convex sets $A, B \subset \mathfrak{R}^2$ the sum $A + B$ is *d*-convex.

7. Let $A, B \subset \mathfrak{R}^2$ be *d*-convex sets with ri $A \cap$ ri $B = \emptyset$. As usual, we denote by $A - B$ the set of all points $x = a - b$, where $a \in A$, $b \in B$. Prove that $A - B$ is *d*-convex and $o \notin$ ri $(A - B)$. Using this, give a new proof of Theorem 13.1.

8. Show by an example that, in general, for $a, b \in \Re^n$ the inclusion $[o, a]_d + [o, b]_d \supset [o, a + b]_d$ is false.

9. Is the inclusion, indicated in the previous exercise, true if $\Re^n = \Re_1 \vee \cdots \vee \Re_s$ with each $\Re_i$ one-dimensional or two-dimensional?

10. Prove that if $\Re^n = \Re_1 \vee \cdots \vee \Re_s$ with each $\Re_i$ one-dimensional, then for any d-convex sets $A, B \subset \Re^n$ the sum $A + B$ is d-convex.

11. Prove that if for any compact set $M \subset \Re^2$ the 1-fold joining of the segments $[a, b]_d$, $a, b \in M$, gives the d-convex hull $\mathrm{conv}_d M$, then $\Re^2$ is the join of two one-dimensional spaces.

§14 The Helly dimension of a set family

First of all, we recall the statement of the classical Helly Theorem.

Theorem 14.1: Let $M_1, \cdots, M_s$ be convex sets in $\mathbf{R}^n, s \geq n + 2$. If each $n + 1$ of these sets have a nonempty inersection, then

$$M_1 \cap \cdots \cap M_s \neq \emptyset. \quad \square$$

There are several different proofs of this Theorem and a number of applications in the surveys [D-G-K] and [E]; see also [H-D-K], [Hl] and [Ra]. One more proof of this Theorem is given below, in §22, where the Helly Theorem is obtained as a particular case of a more general result [Bt 13].

EXAMPLE 14.2: Let $T \subset \mathbf{R}^n$ be an r-dimensional simplex and $a_1, \cdots, a_{r+1}$ be its vertices. By M_i denote the $(r-1)$-dimensional face of T opposite to the vertex $a_i, i = 1, \cdots, r + 1$. Then each r of the sets $M_1, \cdots, M_{r+1}$ have a nonempty intersection, but $M_1 \cap \cdots \cap M_{r+1} = \emptyset$. $\quad \square$

This example shows that in the Helly Theorem the number $n + 1$ cannot be replaced by a smaller one.

EXAMPLE 14.3: Let $x_1, \cdots, x_n$ be cartesian coordinates in $\mathbf{R}^n$. Denote by P the family of the *coordinate parallelotopes*, i.e., parallelotopes with facets parallel to the coordinate hyperplanes. In other words, each parallelotope $M \in P$ is defined by a system of inequalities $a_1 \leq x_1 \leq b_1, \cdots, a_n \leq x_n \leq b_n$, where $a_i \leq b_i$ are real numbers, $i = 1, \cdots, n$. It is possible to say that P is the family of all compact, d-convex sets in the Minkowski space $\Re^n$ with the regular cross-polytope as its unit ball. It is easily shown that the following assertion is true: *Let $M_1, \cdots, M_s \in P$, where $s > 2$. If each two of the parallelotopes $M_1, \cdots, M_s$ have a nonempty intersection, then*

$$M_1 \cap \cdots \cap M_s \neq \emptyset.$$

Indeed, let M_i be defined by inequalities

$$a_1^{(i)} \le x_1 \le b_1^{(i)}, \cdots, a_n^{(i)} \le x_n \le b_n^{(i)}.$$

We fix an index $j = 1, \cdots, n$. By π_j we denote the projection of $\mathbf{R}^n$ onto the x_j-axis (parallel to the other axes). Then $\pi_j(M_1), \cdots, \pi_j(M_s)$ are segments in the x_j-axis, and each two of them have a nonempty intersecion. Consequently, by virtue of the Helly Theorem for $\mathbf{R}^1$, the segments $\pi_j(M_1), \cdots, \pi_j(M_s)$ have a common point x_j'. In other words,

$$a_j^{(1)} \le x_j' \le b_j^{(1)}, \cdots, a_j^{(s)} \le x_j' \le b_j^{(s)}.$$

This is true for every index $j = 1, \cdots, n$, and therefore the point $x' = (x_1', \cdots, x_n')$ belongs to each parallelotope $M_i (i = 1, \cdots, s)$, i.e., $M_1 \cap \cdots \cap M_s \ne \emptyset$. $\square$

This example shows that if we limit ourselves by considering only the family P (more narrow than the family of all convex sets in $\mathbf{R}^n$), then the integer $n + 1$ in the Helly Theorem may be replaced by 2.

For a better understanding of the aforesaid it is convenient to use the notion of *Helly dimension* of a family of sets.

DEFINITION 14.4: Let F be an infinite family of sets and m be an integer. We say that F has the property $\cap_m$ if for every collection $M_1, \cdots, M_s$ of sets from F ($s > m + 1$) the following assertion is true: if each $m + 1$ sets of this collection have a common point, then $M_1 \cap \cdots \cap M_s \ne \emptyset$.

Theorem 14.1 shows that the family C_n of all convex subsets of $\mathbf{R}^n$ possesses the property $\cap_n$. Example 14.2 shows that C_n does not have the property $\cap_m$ for $m < n$. In other words, n is the *least* of the integers m such that C_n possesses the property $\cap_m$. Example 14.3 shows that the family of the coordinate parallelotopes has the property $\cap_1$.

For the sake of convenience, we say that a family F of sets possesses the property $\cap_o$ if for any sets $M_1, \cdots, M_s \in F$, $s \ge 2$, the intersection $M_1 \cap \cdots \cap M_s$ is nonempty.

EXAMPLE 14.5: Denote by K the family of all translates of a half-space $P \subset \mathbf{R}^n$. It is obvious that K possesses the property $\cap_o$. We remark that the intersection of *all* the sets of the family K is empty. $\square$

Now we can introduce the Helly dimension of a set family F.

DEFINITION 14.6: Let F be an infinite family of sets. The *Helly dimension* him F of the family F is the smallest of integers $m > 0$ such that F possesses the property $\cap_m$. If F does not possess the property $\cap_m$ for any $m \ge 0$, then him $F = \infty$. $\square$

Thus, according to the classical Helly Theorem, for the family C_n of all convex subsets in $\mathbf{R}^n$ we have him $C_n = n$. This justifies the term "Helly dimension" for him F, using this term not only for C_n, but for any infinite family of sets, too. Furthermore, the family P of all coordinate parallelotopes satisfies the condition him $P = 1$. Obviously, for any subfamily $F \subset C_n$ we have $0 \leq$ him $F \leq n$.

The term *Helly dimension* was introduced in [SP 7]. In the literature, this notion is also known as Helly number (the difference between both numbers is one); see [D-G-K], [E], and [SV 4] for general references and futher developments. The following assertion is proved in the paper [SP 7].

Theorem 14.7: Let F be an infinite family of sets such that $0 <$ him $F < \infty$. Then him F is the largest of the integers m for which there exist sets $M_1, \cdots, M_{m+1} \in F$ such that each m of them have a nonempty intersection and $M_1 \cap \cdots \cap M_{m+1} = \emptyset$. $\square$

PROOF: We say that an integer m is *admissible* if there exist sets $M_1, \cdots, M_{m+1} \in F$ such that each m of them have a nonempty intersection and $M_1 \cap \cdots \cap M_{m+1} = \emptyset$. The Theorem affirms that him F is the largest of the admissible integers.

First of all, we prove that there exists at least one admissible integer. Since the inequality him $F > 0$ holds, the family F does not have the property $\cap_o$, i.e., there is a finite subfamily of F with empty intersection. Let r be the *least* of the integers q, for which there exist $q+1$ sets belonging to F with empty intersection. Then there are sets $M_1, \cdots, M_{r+1} \in F$ with $M_1 \cap \cdots \cap M_{r+1} = \emptyset$. At the same time, each r of the sets $M_1, \cdots, M_{r+1}$ have a common point (since r is the *least* of the integers q). This means that r is admissible.

Furthermore, it is clear that if m is admissible, then him $F \geq m$. Consequently, by virtue of him $F < \infty$, there exists the largest admissible integer p. Thus him $F \geq p$.

Finally, let $N_1, \cdots, N_s \in F$ be sets every $p+1$ of which have a nonempty intersection. Then each $p+2$ of them have a nonempty intersection, too (otherwise p would not be the largest of the admissible integers). Analogous reasoning shows that every $p+3$ of the sets $N_1, \cdots, N_s$ have a nonempty intersection, etc. After all, we conclude that $N_1 \cap \cdots \cap N_s \neq \emptyset$. This means that him $F \leq p$. ∎

Let now $\Re^n$ be an n-dimensional normed space. We will consider the following families of sets which are closely connected to the metric d given in $\Re^n$:

the family V_d of all d-convex sets in $\Re^n$,

the family $V_d^{(cl)}$ of all closed, d-convex sets in $\Re^n$,

the family $V_d^{(b)}$ of all d-convex *bodies* in $\Re^n$.

With respect to these families, we introduce certain combinatorial dimensions of the space $\Re^n$, namely:

$$\text{him } \Re^n = \text{him } V_d; \quad \text{him }^{(cl)}\Re^n = \text{him } V_d^{(cl)}; \quad \text{him }^{(b)}\Re^n = \text{him } V_d^{(b)}.$$

In [SP 7], the first of these invariants is said to be the *Helly dimension* of the space $\Re^n$. (We remark that all these invariants can also be considered in arbitrary metric spaces. But this would go beyond the limits of the present book.)

Since the inclusions $\mathbf{V}_d \supset \mathbf{V}_d^{(\mathrm{cl})} \supset \mathbf{V}_d^{(b)}$ hold, the introduced invariants satisfy the inequalities

$$\mathrm{him}\,\Re^n \geq \mathrm{him}^{(\mathrm{cl})}\Re^n \geq \mathrm{him}^{(b)}\Re^n.$$

EXAMPLE 14.8: Consider the space $\Re^3$ from Example 10.4 (see also Example 11.4). We choose the d-convex set M described in Example 10.4 and denote by M_1, M_2, M_3 the translates of M with the translation vectors $p_i = (\sqrt{3}\cos\alpha_i, \sqrt{3}\sin\alpha_i, 0)$, $i = 1, 2, 3$, where $\alpha_1 = 0, \alpha_{2,3} = \pm\frac{2}{3}\pi$. The images of M_1, M_2, M_3 under orthogonal projection onto the plane $x_3 = 0$ are shown in Fig. 66. Finally, we denote by M_o the plane $x_3 = 2$. All the sets M_o, M_1, M_2, M_3 are d-convex, each three of them have nonempty intersection, and $M_o \cap M_1 \cap M_2 \cap M_3 = \emptyset$. Hence $\mathrm{him}\,\Re^3 = 3$. At the same time, $\mathrm{him}^{(\mathrm{cl})}\Re^3 < 3$. Indeed, let N_1, N_2, N_3 be convex bodies obtained from $\mathrm{cl}\,M$ by the translations with the vectors $q_i = (2\cos\alpha_i, 2\sin\alpha_i, 0), i = 1, 2, 3$, for the same α_i as above. The orthogonal projections of N_1, N_2, N_3 into the plane $x_3 = 0$ are shown in Fig. 67. Each two of the bodies N_1, N_2, N_3 have a common point, while $N_1 \cap N_2 \cap N_3 = \emptyset$. Consequently $\mathrm{him}^{(\mathrm{cl})}\Re^3 \geq 2$. In fact, $\mathrm{him}^{(\mathrm{cl})}\Re^3 = 2$. $\square$

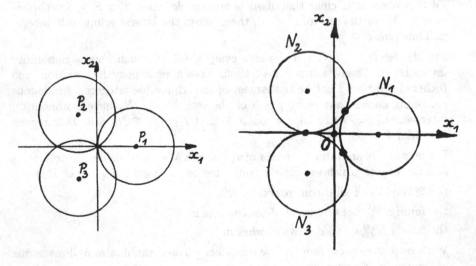

Fig. 66   Fig. 67

EXAMPLE 14.9: In $\mathbf{R}^3$ with a coordinate system (x_1, x_3, x_3), let $\sum$ denote the convex polyhedron with the 14 vertices

$$(\pm 1, 0, 0), (0, \pm 1, 0), (0, 0, \pm 1), (\pm\tfrac{2}{5}, \pm\tfrac{2}{5}, \pm\tfrac{2}{5})$$

(all combinations of the signs). This polyhedron is obtained from an octa-hedron (with the first six vertices) by the erection of a triangular pyramid onto each of its facets, cf. Fig. 68. Taking $\sum$ as unit ball, we transform the space $\mathbf{R}^3$ into a normed space $\Re^3$. According to Theorem 11.1, the only two-dimensional, d-convex subspaces of $\Re^3$ are the coordinate planes $x_1 = 0, x_2 = 0, x_3 = 0$. Consequently the only d-convex bodies in $\Re^3$ are solid coordinate parallelotopes (i.e., parallelotopes whose facets are parallel to the coordinate planes). It follows (cf. Example 14.3), that $\mathrm{him}^{(b)}\Re^3 = 1$. At the same time, $\mathrm{him}^{(\mathrm{cl})}\Re^3 > 1$. Indeed, denote by M_1 the plane $x_1 = 1$, by M_2 the plane $x_2 = 0$, and by M_3 the line $x_1 = x_2 = x_3$. All the sets M_1, M_2, M_3 are closed and d-convex, and they have pairwise common points, but $M_1 \cap M_2 \cap M_3 = \emptyset$. Thus $\mathrm{him}^{(\mathrm{cl})}\Re^3 \geq 2$. (Actually, $\mathrm{him}^{(\mathrm{cl})}\Re^3 = 2$.) $\quad\square$

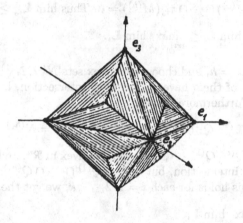

Fig. 68

The Examples 14.8 and 14.9 show that $\mathrm{him}\,\Re^n$, $\mathrm{him}^{(b)}\Re^n$ and $\mathrm{him}^{(\mathrm{cl})}\Re^n$ are *different* dimensional invariants of the normed space $\Re^n$.

Theorem 14.10: *Let* $\Re^n = \mathbb{L}_1 \oplus \cdots \oplus \mathbb{L}_k$ *be the join of spaces* $\mathbb{L}_1, \cdots, \mathbb{L}_k$. *Then*

$$\mathrm{him}\,\Re^n = \max_{i=1,\cdots,k} \mathrm{him}\,\mathbb{L}_i,$$

$$\mathrm{him}^{(\mathrm{cl})}\Re^n = \max_{i=1,\cdots,k} \mathrm{him}^{(\mathrm{cl})}\mathbb{L}_i,$$

$$\mathrm{him}^{(b)}\Re^n = \max_{i=1,\cdots,k} \mathrm{him}^{(b)}\mathbb{L}_i. \quad\square$$

<u>PROOF:</u> Denote by $\sum$ the unit ball of $\Re^n$ and set $\sum_i = \sum \cap \mathbf{L}_i$. Then $\sum_i$ is the unit ball of a normed space $\mathbf{L}_i$ and the equality

$$\sum = \text{conv} \left(\sum_1 \cup \cdots \cup \sum_k \right)$$

holds. Furthermore, denote by $\mathbf{L}_i^\triangle$ the direct sum of all the subspaces $\mathbf{L}_1, \cdots, \mathbf{L}_k$ except for $\mathbf{L}_i$ and by π_i the projection of $\Re^n$ onto $\mathbf{L}_i$, parallel to the subspace $\mathbf{L}_i^\triangle$. Then, by Theorem 12.2, for each d-convex set $M \subset \Re^n$ the equality $M = \pi_1(M) + \cdots + \pi_k(M)$ holds. Besides, $M_i = \pi_i(M)$ is a d-convex set in the space $\mathbf{L}_i$. Conversely, if M_i is a d-convex set of the space $\mathbf{L}_i$ ($i = 1, \cdots, k$), then $M = M_1 + \cdots + M_k$ is d-convex in the space $\Re^n$ and $\pi_i(M) = M_i$ holds.

We set him $\Re^n = h$. Since $h > 0$, we can choose d-convex sets $M^{(o)}, M^{(1)}, \cdots, M^{(h)}$ in $\Re^n$ such that each h of them have a nonempty intersection, but $M^{(o)} \cap M^{(1)} \cap \cdots \cap M^{(h)} = \emptyset$. Then, for any index i, each h of the sets $\pi_i(M^{(o)}), \cdots, \pi_i(M^{(h)})$ have a nonempty intersection, but there exists an $i = i_o$ such that $\pi_{i_o}(M^{(o)}) \cap \cdots \cap \pi_{i_o}(M^{(h)}) = \emptyset$. Thus him $\mathbf{L}_{i_o} \geq h$ and therefore

$$\text{him } \Re^n = h \leq \text{him } \mathbf{L}_i \leq \max_{i=1,\cdots,k} \text{him } \mathbf{L}_i. \tag{1}$$

We now set him $\mathbf{L}_i = h_i$ and choose d-convex sets $N^{(o)}, N^{(1)}, \cdots, N^{(h_i)}$ in $\mathbf{L}_i$ such that each h_i of them have a nonempty intersection, but $N^{(o)} \cap N^{(1)} \cap \cdots \cap N^{(h_i)} = \emptyset$. Furthermore, we set

$$Q^{(o)} = N^{(o)} + \mathbf{L}_o^\triangle, \; Q^{(1)} = N^{(1)} + \mathbf{L}_1^\triangle, \cdots, Q^{(h_i)} = N^{(h_i)} + \mathbf{L}_i^\triangle.$$

Then all the sets $Q^{(o)}, Q^{(1)}, \cdots, Q^{(h_i)}$ are d-convex in $\Re^n$, and each h_i of them have a nonempty intersection, but $Q^{(o)} \cap Q^{(1)} \cap \cdots \cap Q^{(h_i)} = \emptyset$. Hence him $\Re^n \geq h_i$. Since this holds for each $i = 1, 2, \cdots, k$, we get the inequality

$$\text{him } \Re^n \geq \max_{i=1,\cdots,k} \text{him } \mathbf{L}_i. \tag{2}$$

The inequalities (1) and (2) imply the first equality in Theorem 14.10. The other two equalities can be verified in an analogous manner. It should be noticed that if $M^{(j)}$ is closed, then also all the sets $\pi_i(M^{(j)})$ are closed; and if $N^{(j)}$ is closed, then also $N^{(j)} + \mathbf{L}_i^\triangle$ is closed. Furthermore, if $M^{(j)}$ is a body in $\Re^n$, then $\pi_i(M^{(j)})$ is a body in $\mathbf{L}_i$; and in just the same way, if $N^{(j)}$ is a body in $\mathbf{L}_i$, then also $N^{(j)} + \mathbf{L}_i^\triangle$ is a body in $\Re^n$.

Corollary 14.11: Let $\Re^n$ be the join of subspaces $\mathbf{L}_1, \cdots, \mathbf{L}_k$. Then

$$\text{him}^{(b)} \Re^n \leq \text{him}^{(\text{cl})} \Re^n \leq \text{him } \Re^n \leq \max_{i=1,\cdots,k} \dim \mathbf{L}_i. \quad \Box$$

Corollary 14.12: Let $\Re^n$ be the join of subspaces $\mathbf{L}_1, \cdots, \mathbf{L}_k$, where $\dim \mathbf{L}_1 \geq 1$ and the subspaces $\mathbf{L}_2, \cdots, \mathbf{L}_k$ are one-dimensional (i.e., the unit ball of $\Re^n$ is a $(k-1)$-fold suspension over the unit ball of $\mathbf{L}_1$). Then him $\Re^n$ = him $\mathbb{L}_1$; him$^{(\text{cl})}\Re^n$ = him$^{(\text{cl})}\mathbb{L}_1$. $\square$

This follows from Theorem 14.10: for one-dimensional subspaces $\mathbb{L}_i$ we have him $\mathbb{L}_i$ = him$^{(\text{cl})}\mathbb{L}_i$ = 1 (since in a one-dimensional space each convex set is d-convex) and for the space $\mathbb{L}_1$ with dim $\mathbb{L}_1 > 1$ the inequality him$^{(\text{cl})}\mathbb{L}_1 \geq 1$ holds (since, by Theorem 11.7, the space $\mathbb{L}_1$ contains one-dimensional, d-convex subspaces). We remark that for the dimension him$^{(b)}$ an analogous equality is, in general, false. Namely, if $\mathbb{L}_1$ is the space as in Example 9.3, then dim $\mathbb{L}_1 > 1$, but him$^{(b)}\mathbb{L}_1 = 0$. Therefore him$^{(b)}(\mathbb{L}_1 \oplus \cdots \oplus \mathbb{L}_k) = 1 >$ him$^{(b)}\mathbb{L}_1$ (where the spaces $\mathbb{L}_2, \cdots, \mathbb{L}_k$ are one-dimensional).

Furthermore, Corollary 14.12 implies that for every r, $1 \leq r \leq n$, there exists a normed space $\Re^n$ with him $\Re^n = r$.

EXAMPLE 14.13: For a normed space $\Re^n$ with norm $\| x \| = \sum_{i=1}^{n} |x_i|$ (cf. Example 14.3), we have him $\Re^n = 1$, i.e., an arbitrary finite family of d-convex sets has a nonempty intersection if each two of the sets have a common point (since the d-convex sets are coordinate parallelotopes). $\square$

REMARK 14.14: Theorem 14.10 shows that if a Minkowski space $\Re$ is the join of two (or more) Minkowski spaces of smaller dimensions, then him $\Re <$ dim $\Re$ (the analogous inequality holds for him$^{(\text{cl})}$ and him$^{(b)}$). It is natural to ask whether the converse assertion is true, i.e., whether the relation him $\Re <$ dim $\Re$ implies that $\Re$ is a join of several lower-dimensional Minkowski spaces. Furthermore, in view of the Examples 11.6 and 12.6 another question can be posed. In these examples we have the *polar* Minkowski spaces $\Re, \Re^*$ (of dimension n), i.e., their unit balls $\sum, \sum^*$ are polars of each other. For these spaces we have him $\Re = 1$, him $\Re^* = 1$, i.e., him $\Re =$ him $\Re^*$. The question is whether this is true for any pairs of polar Minkowski spaces. In Chapter IV we will see that both these questions have *negative* answers (cf. Exercises 12 and 13 in §25).

Exercises

1. a) Let sets $M_1, \cdots, M_{n+2} \subset \mathbf{R}^n$ be convex. Prove that if each $n + 1$ of them have a nonempty intersection, then there are points $b_1, \cdots, b_{n+2}$ such that $b_i \in M_j$ for $i \neq j$.

   b) Let $b_1, \cdots, b_{n+2} \in \mathbf{R}^n$. Prove that there are nonzero real numbers $\lambda_1, \cdots, \lambda_{n+2}$ such that
   $$\lambda_1 b_1 + \cdots + \lambda_{n+2} b_{n+2} = o, \quad \lambda_1 + \cdots + \lambda_{n+2} = 0.$$

   c) Without loss of generality we may suppose that for an integer q, $1 \leq q \leq n + 1$, the above numbers $\lambda_1, \cdots, \lambda_{n+2}$ satisfy the conditions
   $$\lambda_1 > 0, \cdots, \lambda_q > 0; \lambda_{q+1} < 0, \cdots, \lambda_{n+2} < 0 \quad \lambda_1 + \cdots + \lambda_q = 1.$$
   Prove that the point
   $$p = \lambda_1 b_1 + \cdots + \lambda_q b_q = -\lambda_{q+1} b_{q+1} - \cdots - \lambda_{n+2} b_{n+2}$$
   belongs to each set $M_i, i = 1, \cdots, n+2$. This means that $M_1 \cap \cdots \cap M_{n+2} \neq \emptyset$.

d) Using the previous partial exercise, prove the Helly Theorem for arbitrary dimension n. This proof was suggested by Radon [Ra] (for details, see [D-G-K]).

2. We sketch another proof of the Helly Theorem (this time by induction on the dimension n). Suppose that for dimension $n-1$ the Helly Theorem is established and consider the situation as in Exercise 1 (a). Admit $M_1 \cap \cdots \cap M_{n+2} = \emptyset$. We choose the points $b_1, \cdots, b_{n+2}$ as in Exercise 1 (a) and denote by F_i the convex hull of all the points $b_1, \cdots, b_{n+2}$, except for b_i ($i = 1, \cdots, n+2$). Then $F_1, \cdots, F_{n+2}$ are compact, convex sets such that $F_i \subset M_i$ and, moreover, all the sets $F_1, \cdots, F_{n+2}$ except F_i have a common point b_i. Since $F_i \subset M_i$, the intersection $F_1 \cap \cdots \cap F_{n+2}$ is empty, i.e., two compact, convex sets $\Phi = F_1 \cap \cdots \cap F_{n+1}$ and F_{n+2} have empty intersection. Consequently there is a hyperplane L strongly separating these convex sets, i.e., $\Phi \subset P_1$, $M_{n+2} \subset P_2$, where P_1, P_2 are the open half-spaces defined by L. In particular, $F_1 \cap \cdots \cap F_{n+1} \cap L = \emptyset$. Prove that every n of the convex sets

$$L \cap F_1, L \cap F_2, \cdots, L \cap F_{n+1}$$

have a nonempty intersection. Deduce from this that all these sets, lying in the $(n-1)$-dimensional space L, have a nonempty intersection, i.e., $F_1 \cap \cdots \cap F_{n+1} \cap L \neq \emptyset$, contradicting the choice of L. This contradiction shows that $F_1 \cap \cdots \cap F_{n+2} \neq \emptyset$, and hence $M_1 \cap \cdots \cap M_{n+2} \neq \emptyset$. Further argue as in Exercise 1 (d).

3. Prove that if $M_1, \cdots, M_s \subset \mathbf{R}^n$, $s > n + 1$, are convex bodies every $n + 1$ of which have a common interior point, then int $M_1 \cap \cdots \cap$ int $M_s \neq \emptyset$.

4. Prove the Helly Theorem for an infinite family of sets in the following form: Let A be an infinite set of indices and $M_\alpha, \alpha \in A$, be closed, convex sets in $\mathbf{R}^n$, at least one of which is compact. If each $n + 1$ of the sets M_α have a common point, then $\bigcap_{\alpha \in A} M_\alpha \neq \emptyset$.

5. Show by examples that the requirements "all M_α are closed" and "at least one them is compact" in the previous exercise are essential.

6. Prove the Helly Theorem for an infinite family of flats: Let A be an infinite set of indices and $M_\alpha, \alpha \in A$, be flats in $\mathbf{R}^n$. If each $n + 1$ of the flats M_α have a nonempty intersection, then $\bigcap_{\alpha \in A} M_\alpha$ is nonempty.

7. A convex body $M \subset \mathbf{R}^n$ is said to be a *cylinder* if there exists a direct decomposition $\mathbf{R}^n = L_1 \oplus L_2$ such that $M = L_1 \oplus M_2$, where $M_2 \subset L_2$ is a compact, convex set. Prove that if A is an infinite set of indices and for each $\alpha \in A$ the convex body $M_\alpha \subset \mathbf{R}^n$ is a cylinder, then the Helly Theorem for this family is true, i.e., $\bigcap_{\alpha \in A} M_\alpha \neq \emptyset$ if every $n + 1$ of the cylinders M_α have a common point.

8. Prove that if for infinite set families F_1, F_2 the inclusion $F_1 \subset F_2$ holds, then him $F_1 \leq$ him F_2.

9. In a coordinate system (x_1, x_2) in $\mathbf{R}^2$, let F be the family of the convex sets $M \subset \mathbf{R}^2$ each of which contains a ray directed as the positive x_1-axis and is contained between two lines parallel to the x_1-axis. Prove that him $F = 1$.

10. Let $M_o \subset \mathbf{R}^2$ be defined by the inequality $x_2 \geq (x_1)^2$ and F be the family of sets, each of which is obtained from M_o by either a translation or a homothety with a positive ratio. Prove that the family F possesses the property $\cap_o$, i.e., him $F = 0$.

11. Prove that if $M \subset \mathbf{R}^n$ is an unbounded, convex set with an n-dimensional inscribed cone, then the family F of the translates of M satisfies the condition him $F = 0$.

12. Let F be the family of the circumferences in the plane $\mathbf{R}^2$. Prove that him $F = 3$. Generalize this to $\mathbf{R}^n$. Give another generalization, considering the family of the curves of second order in the plane.

13. Prove that for the Minkowski space of Example 10.4 the equalities $\mathrm{him}^{(\mathrm{cl})} \Re^3 = 2$ and $\mathrm{him}^{(b)} \Re^3 = 2$ hold.

14. Prove that for the Minkowski space of Example 14.9 the equalities $\mathrm{him}\, \Re^3 = 2$ and $\mathrm{him}^{(\mathrm{cl})} \Re^3 = 2$ hold.

15. Prove that if in a Minkowski space $\Re^n$ the unit ball $\sum$ is a polytope, then him $\Re^n = \mathrm{him}^{(\mathrm{cl})} \Re^n$.

16. Prove the Helly Theorem for an infinite family of d-convex sets in the following form. Let $\Re^n$ be a Minkowski space and $h = \mathrm{him}^{(\mathrm{cl})} \Re^n$. Let, furthermore, A be an infinite set of indices, and $M_\alpha, \alpha \in A$, be closed, d-convex sets in $\Re^n$, at least one of which is compact. If every $h + 1$ of the sets M_α have a common point, then $\bigcap_{\alpha \in A} M_\alpha \neq \emptyset$.

§15 d-Star-shaped sets

M.A. Krasnosel'ski [Ks] proved a theorem on star-shaped sets. We recall its statement in a generalized form, as it is contained in the book of K. Leichtweiss [Lei]. (For more recent developments around Krasnosel'ski's theorem the reader is referred to the surveys §E 2 in [C-F-G] and [E]).

Let $M \subset \mathbf{R}^n$ be a compact set. This set is said to be *star-shaped* if there exists a point $x_o \in M$ such that for every $x \in M$ the segment $[x_o, x]$ is contained in M. In other words, every point $x \in M$ is *visible* in M from the point x_o. Krasnosel'ski's theorem affirms that if every $n + 1$ points of M are visible in M from a point $x_o \in M$, then *all* the points of M are visible in M from a point of the set M, i.e., M is star-shaped.

The proof of this theorem, offered by Krasnosel'ski for a closed region in $\mathbf{R}^2$ bounded by a finite number of segments (Fig. 69), was given in a simple and general form in [Lei]. The proof is based on the classical Helly theorem.

In this section, we prove a theorem (and several of its consequences) which generalizes the Krasnosel'ski theorem and makes it more precise, see also

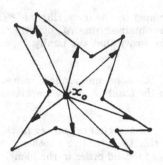

Fig. 69

[B-M-SP]. Namely, we consider *d-star-shaped sets* in a Minkowski space $\Re^n$, which are obtained when we replace usual segments by *d*-segments. More precisely, a set $M \subset \Re^n$ is said to be *d-star-shaped* if there exists a point $x_o \in M$ such that for every $x \in M$ the *d*-segment $[x_o, x]_d$ is contained in M. In other words, every point $x \in M$ is *d-visible* in M from x_o. To establish a necessary and sufficient condition for *d*-star-shapedness (which generalizes Krasnosel'ski's theorem), we introduce some auxiliary notions.

Let a be a point of a set $M \subset \mathbf{R}^n$ (in particular, a boundary point). A ray l emanating from a is said to have a *free direction* (with respect to M) if there exists a point $b \in l$ distinct from a such that the interval $]a, b[$ does not contain any points of M, i.e., $[a, b[\cap M = \{a\}$. Otherwise, the ray l has a *nonfree direction*.

The union of all rays emanating from $a \in M$ which have nonfree directions (with respect to M) is said to be the cone of nonfree directions and is denoted by $\mathrm{nof}_a M$. If, in particular, M is convex, then the cone cl $(\mathrm{nof}_a M)$ coincides with the supporting cone sup $\mathrm{cone}_a M$.

Let now M be a set in a Minkowski space $\Re^n$. We say that M possesses the property of *internal-local d-conicity* if for every point $a \in M$ the set cl (conv $(\mathrm{nof}_a M)$) is *d*-convex.

EXAMPLE 15.1: Let $\Re^2$ be the Minkowski plane whose unit ball is the square with the vertices $(\pm 1, 0), (0, \pm 1)$. In this plane, *d*-segments coincide with coordinate rectangles $a_1 \leq x_1 \leq b_1, a_2 \leq x_2 \leq b_2$. Let now M be a closed region whose boundary is the union of a finite number of segments parallel to the coordinate axes (Fig. 70). The cones of nonfree directions are: the whole plane (for the point $a_1 \in M$); half-planes (for the points a_2, a_3); right angles (for the points a_4, a_5); the angles of the value $\frac{3\pi}{2}$ (for the points a_6, a_7). It is easily shown that cl (conv $(\mathrm{nof}_a M)$) is either the plane or a half-plane, or a right angle; consequently the set M is internal-local *d*-conic at each of its points.   □

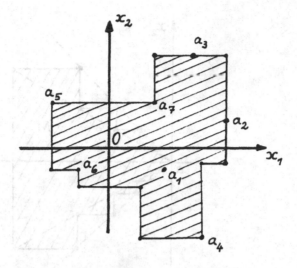

Fig. 70

EXAMPLE 15.2: The closed region M in Fig. 71 is bounded by a line which is the union of infinitely many segments and the point $a = (0,0)$. The set M is not internal-local d-conic (with respect to the same Minkowski metric), since the cone of nonfree directions $\mathrm{nof}_a M$ at a is the right angle between the bisectors of the first and the second quadrants, and the condition conv $(\mathrm{nof}_a M) = \mathrm{conv}_d\,(\mathrm{nof}_a M)$ is not satisfied.

EXAMPLE 15.3: Let $\Re^3$ be the Minkowski space whose unit ball is the octahedron with vertices $(\pm 1, 0, 0)$, $(0, \pm 1, 0)$, $(0, 0, \pm 1)$. Let, furthermore, K_1 be a rotational cone which is contained in the half-space $x_1 \le 0$ and touches the (x_2, x_3)-plane along the positive x_3- axis. We denote by K_2 the negative x_3-axis and put $M = K_1 \cup K_2$. Then for the point $a = (0,0,0)$ the convex hull conv $(\mathrm{nof}_a M)$ is *not closed*, while $\mathrm{conv}_d M$ is the *closed* half-space $x_1 \le 0$. This shows that the closure operation in the definition of local-internal d-convexity is essential.   □

We now describe one more condition for a set $M \subset \Re^n$. Let $M \subset \Re^n$. We say that M possesses the property of *external-local d-conicity* if for every $a \in M$ and each ray l, that emanates from a and has a free direction, there is a positive number ε such that for each positive $\mu < \lambda < \varepsilon$ the relation $[a + \lambda v, a + \mu v]_d \cap M = \emptyset$ holds, where v is a vector directed along l.

We now establish Kranosel'ski's theorem for Minkowski spaces (a first version of this theorem is contained in [B-SP3]).

Theorem 15.4: Let $M \subset \Re^n$ be a compact set that is internal- and external-local d-conic. The number him $\Re^n$ *is denoted by* h. *The set M is d-star-shaped if and only if for every $h+1$ points $x_1, \cdots, x_{h+1} \in M$ there is a point $y \in M$*

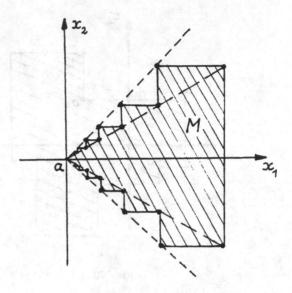

Fig. 71

such that the points $x_1, \cdots, x_{h+1}$ are d-visible in M from y, i.e., $[y, x_i]_d \subset M$ for $i = 1, \cdots, h + 1$. □

PROOF: For every point $a \in M$, we put

$$C(a) = \text{cl conv}_d(M \cap \text{nof}_a M). \tag{1}$$

We will establish that the *d*-convex set

$$N = \bigcap_{a \in M} C(a)$$

is nonempty and every point $x_o \in N$ has the desired property, i.e., each point $x \in M$ is *d*-visible in M from x_o:

$$[x, x_o]_d \subset M \text{ for each } x \in M.$$

Let $x_1, \cdots, x_{h+1}$ be points of M. Then (according to the condition of the theorem) there is a point $y \in M$ such that

$$[y, x_i]_d \subset M, i = 1, \cdots, h + 1.$$

Hence $y \in [y, x_i]_d \subset \text{nof}_{x_i} M$, and $y \in M$, i.e., $y \in C(x_i)$, $i = 1, \cdots, h + 1$. Thus every $h + 1$ of the sets $C(a), a \in M$, have a common point y. By virtue of Helly's theorem for *d*-convex sets (see the definition of the number him $\Re^n$ in §14) each *finite* family of *d*-convex sets $C(a)$ has a nonempty intersection.

Let now $Q \subset \Re^n$ be a Euclidean ball containing M in its interior. For every $z \in$ bd Q we consider the maximal, closed Euclidean ball $U(z)$ centered at z, which contains no point of M in its interior. We take a point $a \in M \cap$ bd $U(z)$ and denote by $P(a)$ the closed half-space satisfying

$$P(a) \cap U(z) = \{a\}, \tag{2}$$

see Fig. 72. It follows from (2) that $\mathrm{nof}_a M \subset P(a)$ and, consequently,

$$
\begin{aligned}
C(a) &= \mathrm{cl}\ \mathrm{conv}_d(M \cap \mathrm{nof}_a M) \subset \mathrm{cl}\ \mathrm{conv}_d(\mathrm{nof}_a M) \\
&= \mathrm{cl}\ \mathrm{conv}\ (\mathrm{nof}_a M) \subset P(a)
\end{aligned}
\tag{3}
$$

(by virtue of the condition of internal-local d-conicity). We conclude that the set $C(a)$ has no common point with the ray r_z which emanates from z and has the direction of the outward normal to the ball Q. This means that $N \cap r_z = \emptyset$. Since this is true for all points $z \in$ bd Q, the inclusion $N \subset B$ holds. Furthermore, since B is compact, $C(a)$ is closed, and every finite family of the sets $C(a)$ has a nonempty intersection; hence the *whole* family $C(a), a \in M$, has a nonempty intersection, i.e., $N \neq \emptyset$.

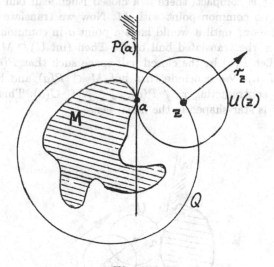

Fig. 72

We now prove that $N \subset M$. Suppose the contrary, i.e., there exists a point $z \in N$ not belonging to M. Then, denoting by $U(z)$ the maximal, closed Euclidean ball centered at z which contains no point of M in its interior (Fig. 73), we take a point $a \in M \cap$ bd $U(z)$. As above, we consider the half-space $P(a)$ satisfying (3). We conclude that $\mathrm{nof}_a M \subset P(a)$, and consequently the inclusion (3) holds. This means that $z \in N \subset P(a)$, contradicting the inclusion $z \in$ int $U(z)$. This contradiction shows that $N \subset M$.

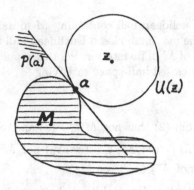

Fig. 73

Furthermore, we establish that for any points $x_o \in N$ and $x \in M$ the *linear* segment $[x_o, x]$ is contained in M, i.e., M is star-shaped in the linear sense. Suppose the contrary, i.e., there is a point $w \in [x_o, x]$ which does not belong to M. Since M is compact, there is a closed Euclidean ball B centered at w, which has no common point with M. Now we translate the ball B in the direction $x - x_o$ until it would have a point a in common with M (Fig. 74). We denote the translated ball by U. Then $(\text{int } U) \cap M = \emptyset$ and $a \in (\text{bd}U) \cap M$. Let $P(a)$ be the closed half-space such that $P(a) \cap \text{int } U = \emptyset$ and $a \in \text{bd } P(a)$. We conclude that $\text{nof}_a M \subset P(a)$ and hence (cf. (3)) $C(a) \subset P(a)$, contradicting $x_o \notin P(a), x_o \in N \subset C(a)$. This contradiction shows that M is star-shaped in the linear sense.

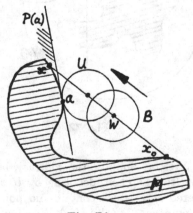

Fig. 74

Thus from the condition of internal-local *d*-conicity of M we have deduced that the whole set M is visible (in the linear sense) from every point $x_o \in N$. Now, with the help of the property of external-local *d*-conicity, we prove that M is *d*-visible (from each point $x_o \in M$), i.e., for any points $x_o \in N$, $x \in M$ the *d*-segment $[x_o, x]_d$ is contained in M.

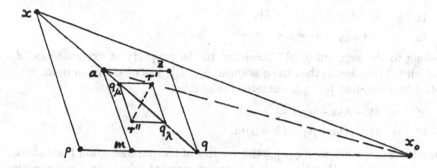

Fig. 75

Assume that a point $p \in [x_o, x]_d$ exists which does not belong to M (Fig. 75). Then all the points close enough to p do not belong to M. Let ρ be a positive number such that the point $q = p + \rho(x_o - p)$ is not contained in M. We denote by a the point from $[q, x]$ which is *nearest* to q and belongs to M, and by v the vector $q - a$. By virtue of the external-local *d*-conicity, an $\varepsilon > 0$ exists such that $[a + \lambda v, a + \mu v]_d \cap M = \emptyset$ for every positive $\mu < \lambda < \varepsilon$. We notice that $[x_o, a] \subset M$ (according to the linear star-shapedness of M which was established above).

Furthermore, since $p \in [x_o, x]_d$ and therefore $q \in [x_o, x]_d$, the following equalities hold:

$$\| x - x_o \| = \| x - p \| + \| p - x_o \| = \| x - p \| + \| p - q \| + \| q - x_o \|,$$

$$\| x - x_o \| = \| x - q \| + \| q - x_o \|.$$

Consequently,

$$\| x - p \| + \| p - q \| = \| x - q \|, \tag{4}$$

i.e., $p \in [q, x]_d$. We now denote by k the number $\frac{\|q-a\|}{\|q-x\|}$ and by g the homothety with the center q and ratio k. Then $g(x) = a$, and the point $m = g(p)$ belongs to the segment $[q, p]$. By virtue of $g(x) = a, g(p) = m, g(q) = q$, the equation (4) implies

$$\| a - m \| + \| m - q \| = \| a - q \| .$$

This means that $m \in [a, q]_d$. Furthermore, for the point $z = q + a - m$ (symmetric to m with respect to the midpoint of $[q, a]$) we have

$$\| q - z \| + \| z - a \| = \| m - a \| + \| q - m \| = \| q - a \|,$$

i.e., $z \in [q, a]_d$. For $0 < \mu < \lambda < 1$ the points

$$q_\lambda = a + \lambda(q - a) = a + \lambda v,$$
$$q_\mu = a + \mu(q - a) = a + \mu v$$

belong to the segment $[q, a]$. According to the property of external-local d-conicity, if $\mu < \lambda < \varepsilon$, then the d-segment $[q_\lambda, q_\mu]_d$ has no common point with M. The d-segment $[q_\lambda, q_\mu]_d$ contains the points

$$r' = (1 - \lambda)a + \mu q + (\lambda - \mu)z,$$
$$r'' = (1 - \lambda)a + \mu q + (\lambda - \mu)m,$$

which are (together with q_λ, q_μ) the vertices of the parallelogram homothetic to conv $(amqz)$ with ratio $\lambda - \mu$. Hence the segment $[r', r'']$ does not contain any point from M. However the line passing through the points r', r'' meets the segment $[x_o, a]$ at the point which, for a fixed $\lambda < \varepsilon$ and $\mu \to 0$, tends to the common point of the segments $[x_o, a]$ and $[a + \lambda(m - a), a + \lambda(z - a)]$. Consequently, for a fixed $\lambda < \varepsilon$ and μ small enough, the segment $[r', r'']$ *contains* a point of the segment $[x_o, a] \subset M$ (we recall that M is linearly visible from x_o, and $a \in M$), i.e., $[r', r''] \cap M \neq \emptyset$, contradicting the aforesaid. This contradiction shows that $[x_o, x]_d \subset M$. ∎

As an immediate consequence of the theorem just established we obtain the classical Krasnosel'ski theorem in the formulation taken from the book of K. Leichtweiss [Lei].

Corollary 15.5: A compact set $K \subset \mathbf{R}^n$ *is star-shaped if and only if for every* $n + 1$ *points* $x_1, \cdots, x_{n+1} \in M$ *there is a point* $y \in M$ *such that* $x_1, \cdots, x_{n+1}$ *are visible in* M *from* y, *i.e.,* $[y, x_i] \subset M$ *for* $i = 1, \cdots, n + 1$. □

Indeed, in Euclidean space $\mathbf{R}^n$ (when the considered d-convexity coincides with the linear convexity) the conditions of internal-local and external-local conicity are trivially satisfied and him $\mathbf{R}^n = n$. ∎

As a second consequence, we consider the case of a compact, polyhedral region $M \subset \Re^n$, i.e., $M = $ cl (int M) and the boundary of M is the union of a finite number of $(n - 1)$-dimensional convex polytopes which are called the *facets* of M.

Corollary 15.6: Let $M \subset \Re^n$ *be a compact, polyhedral region each facet of which has a d-convex affine hull. The number* him $\Re^n$ *we denote by h. The set M is d-star-shaped if and only if every $h + 1$ points of M are d-visible in M from a point of the set M.* ∎

Indeed, let $a \in M$. Then for a sufficiently small ball B centered at a the equality $M \cap B = (\text{nof}_a M) \cap M$ holds. Moreover, $\text{nof}_a M$ is a polyhedral

cone (in general, not convex) such that each of its $(n-1)$-dimensional faces
is situated in a *d*-convex hyperplane. It follows that for every $x \in \Re^n$ the
d-segment $[a, x]_d$ either is contained in $\mathrm{nof}_a M$, or its relative interior has no
common point with $\mathrm{nof}_a M$. This means that the properties of internal-local
and external-local *d*-conicity are satisfied for M. It remains to apply Theorem
15.4. ∎

Corollary 15.7: Let $M_1, \cdots, M_s \subset \Re^n$, $s > h+1$, be sets such that

(i) the union of every $h+1$ of them is *d*-star-shaped, and

(ii) the set $M = M_1 \cup \cdots \cup M_s$ possesses the properties of external-local and
     internal-local *d*-conicity.

Then M is d-star-shaped. □

Indeed, let $x_1, \cdots, x_{h+1} \in M$. For every $i \in \{1, \cdots, h+1\}$ we choose an index
q_i such that $x_i \in M_{q_i}$. Then

$$x_1, \cdots, x_{h+1} \in M_{q_1} \cup \cdots \cup M_{q_{h+1}} \tag{5}$$

(if some of the indices $q_1, \cdots, q_{h+1}$ coincide, then we add some other sets
M_i, in order to have a union of $h+1$ sets on the right-hand side of (5)). By
assumption, the set $N = M_{q_1} \cup \cdots \cup M_{q_{h+1}}$ is *d*-star-shaped, i.e., the points
$x_1, \cdots, x_{h+1}$ are *d*-visible in N from a point $y \in N$. So far, $x_1, \cdots, x_{h+1}$ are
d-visible in M from the point y (since $N \subset M$). Thus every $h+1$ points of
M are *d*-visible in M from a point $y \in M$. It remains to apply Theorem 15.4.
∎

If, in particular, we consider the Euclidean space $\mathbf{R}^n$ (replacing *d*-convexity
by linear convexity), then we obtain the following consequence.

Corollary 15.8: Let $M_1, \cdots, M_s \subset \mathbf{R}^n$, $s > n+1$. If the union of every $n+1$
of the sets $M_1, \cdots, M_s$ is star-shaped, then $M_1 \cup \cdots \cup M_s$ is also star-shaped.
□

This result was obtained by K. Kolodziejczyk [Ko 2].

Exercises

1. Prove that the set M in Example 15.2 is *d*-star-shaped.

2. Let $M_1, \cdots, M_s$ be *d*-convex sets. Prove that if $M_1 \cap \cdots \cap M_s \neq \emptyset$, then the set
   $M_1 \cup \cdots \cup M_s$ is *d*-star-shaped.

3. Let $\Re^2$ be the Minkowski plane as in Example 15.1 and $M \subset \Re^2$ be a compact,
   linearly convex figure. Consider the circumscribed rectangle Q of M with sides
   parallel to the coordinate axes. Let, finally, $x_o \in M$ and l_1, l_2 be lines through
   x_o parallel to the coordinate axes. Denote by a, b, c, d the intersection points
   of the lines l_1, l_2 with the boundary of Q. Prove that the set M is *d*-visible
   from the point x_o if and only if the points a, b, c, d belong to M. With the help
   of this, find a necessary and sufficient condition for the *d*-star-shapedness of a
   linearly convex set $M \subset \Re^2$. Generalize this for the *n*-dimensional case.

4. Denote by $\sum$ the set which in an orthonormal coordinate system (x_1, x_2, x_3) is defined by the inequalities

$$|x_1|^{\frac{2}{|x_3|}} + |x_2|^{\frac{2}{|x_3|}} \leq 1 - x_3^2 \text{ for } x_3 \neq 0,$$

$$|x_1| \leq 1, |x_2| \leq 1 \text{ for } x_3 = 0.$$

Prove that $\sum$ is a convex body in $\mathbf{R}^3$ which contains the contour of the square $|x_1| \leq 1$, $|x_2| \leq 1$, $x_3 = 0$ in its boundary, while every boundary point a of $\sum$ with $x_3 \neq 0$ is exposed (i.e., there exists a supporting plane Γ of $\sum$ such that $\Gamma \cap \sum = \{a\}$).

5. Construct another convex body $\sum$ in $\mathbf{R}^3$ which is centrally symmetric with respect to the origin and possesses the properties as in Exercise 4, i.e., $\sum$ contains the contour of the square $|x_1| \leq 1$, $|x_2| \leq 1$, $x_3 = 0$ in its boundary, while every boundary point of $\sum$ with $x_3 \neq 0$ is exposed.

6. Let $\Re^3$ be a normed space with the unit ball $\sum$ as in the Exercises 4 or 5. Denote by M the rectangle with vertices $(\pm 1, 0, -1)$, $(\pm 1, 0, 1)$. Prove that every finite set contained in M is d-visible in M from a point $y \in M$. Prove that M is not d-star-shaped (i.e., there is no point $x_o \in M$ from which the whole set M is d-visible).

7. Let $\Re^3$ be a Minkowski space as in Example 10.4 and $\sum$ be its unit ball. Prove that $\sum$ is external-local d-conic, but not internal-local d-conic.

8. Let $\Re^2$ be the Minkowski plane whose unit ball is the rhombus with the vertices $(\pm 1, 0)$, $(0, \pm 1)$. Denote by Q_o the first quadrant of the plane and by Q_k its translate $a_k + Q_o$, where $a_k = \left(-\frac{1}{k^2}; \frac{1}{k}\right)$. Finally, we put

$$P = Q_o \cup \left(\bigcup_{k=1}^{\infty} Q_k \right)$$

and denote by M the intersection of the set P with the square $|x_1| \leq 2$, $|x_2| \leq 2$. Prove that M is internal-local d-conic, but not external-local d-conic. This example and the example in Exercise 7 show that the requirements of internal-local d-conicity and external-local d-conicity are independent from each other.

9. Prove that the set M considered in the previous exercise is d-star-shaped.

10. Let $M_1, \cdots, M_s$ be d-convex polytopes in a Minkowski space $\Re^n$ with him $\Re^n = h$. Prove that if the union of every $h + 1$ of these polytopes is d-star-shaped, then $M_1 \cup \cdots \cup M_s$ is also d-star-shaped.

11. Show by an example that there exists a d-star-shaped body $M \subset \Re^n$ such that for a point x_o, from which the body is d-visible, the inclusion $\text{conv}_d\{x_o, x\} \subset M$ is not satisfied for all points $x \in M$.

III. H-convexity

The preceding chapter contains a theorem about a representation of any d-convex body as the intersection of d-convex half-spaces. A natural question emerges for the possibility of neglecting a norm (by which d-convex half-spaces are introduced) in order to find other ways to describe half-spaces whose intersections determine certain classes of convex sets.

A simple possibility is based on the fixation of a subset H of the unit sphere $S^{n-1} \subset \mathbf{R}^n$ and on the consideration of only those half-spaces whose outward normals belong to the set H. This is the basic idea of H-convexity, cf. [Bt 5].

By analogy with the preceding chapter, we will study the properties of H-convex sets, to develop a machinery for solving problems of combinatorial geometry. Moreover, having in view properties of linearly convex sets in $\mathbf{R}^n$ or d-convex sets in $\Re^n$, the H-convex sets are interesting for themselves. By suitable comparisons, we will also clarify some questions concerning d-convex sets.

The first three paragraphs of this chapter are devoted to algebraic tools which are needed for the investigation of H-convexity. These algebraic tools, mainly consisting of the functional md, were introduced in [Bt 6]. In the second part of the chapter we will give a Helly-type theorem for H-convex sets [Bt 6] and some of its applications [Bt 8], cf. also [Bt-SP 3,4].

§16 The functional md for vector systems

In this section we will consider the *minimal dependence*, which is the main algebraic apparatus in this chapter (and the following ones).

We say that vectors $a_1, \cdots, a_{m+1} \in \mathbf{R}^n$ are *minimally dependent* if they are the vertices of an m-dimensional simplex T that has the origin in its relative interior (Fig. 76). In particular, two vectors a_1, a_2 are minimally dependent if they have opposite directions, i.e., the origin belongs to the interval $]a_1, a_2[$.

We now turn to an algebraic description of this geometrical definition. The vectors

$$a_1, \cdots, a_{m+1} \in R^n$$

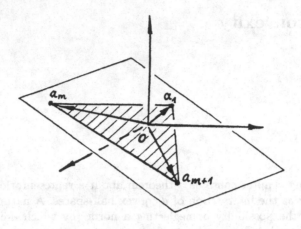

Fig. 76

are said to be *positively dependent* if there are positive numbers $\lambda_1, \cdots, \lambda_{m+1}$ such that $\lambda_1 a_1 + \cdots + \lambda_{m+1} a_{m+1} = o$.

Theorem 16.1: If vectors $a_1, \cdots, a_{m+1} \in \mathbf{R}^n$ are minimally dependent, then they are positively dependent and every m of them are linearly independent. $\square$

PROOF: Since $a_1, \cdots, a_{m+1}$ are minimally dependent, the origin o belongs to the relative interior of the m-dimensional simplex T with the vertices $a_1, \cdots, a_{m+1}$. Consequently,

$$o = \lambda_1 a_1 + \cdots + \lambda_{m+1} a_{m+1}, \tag{1}$$

where the numbers $\lambda_1, \cdots, \lambda_{m+1}$ are positive and $\lambda_1 + \cdots + \lambda_{m+1} = 1$. It follows that the vectors $a_1, \cdots, a_{m+1}$ are positively dependent.

Admit that some m of the vectors $a_1, \cdots, a_{m+1}$ (say, $a_1, \cdots, a_m$) are linearly dependent. Then the subspace $L = \text{lin}\,(a_1, \cdots, a_m)$ (where lin denotes the linear hull) has dimension $\leq m - 1$. By virtue of (1), the vector

$$a_{m+1} = -\frac{1}{\lambda_{m+1}}(\lambda_1 a_1 + \cdots + \lambda_m a_m)$$

belongs to the subspace L, too. Thus all the points $a_1, \cdots, a_{m+1}$ are lying in L, contradicting dim $T = m$.   ∎

Theorem 16.2: If vectors $a_1, \cdots, a_{m+1} \in \mathbf{R}^n$ are positively dependent and some m of them are linearly independent, then the vectors $a_1, \cdots, a_{m+1}$ are minimally dependent. $\square$

PROOF: Without loss of generality we may suppose that the vectors $a_1, \cdots, a_m$ are linearly independent and the positive dependence between the vectors $a_1, \cdots, a_{m+1}$ has the form (1), where

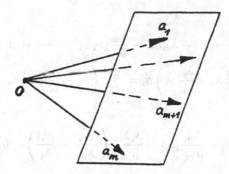

Fig. 77

$$\lambda_1 > 0, \cdots, \lambda_{m+1} > 0; \quad \lambda_1 + \cdots + \lambda_{m+1} = 1. \tag{2}$$

Admit $a_{m+1} \in \text{aff}\,(a_1, \cdots, a_m)$ (Fig. 77), i.e., $a_{m+1} = \mu_1 a_1 + \cdots + \mu_m a_m$, where $\mu_1 + \cdots + \mu_m = 1$. Then, according to (1),

$$\lambda_1 a_1 + \cdots + \lambda_m a_m + \lambda_{m+1}(\mu_1 a_1 + \cdots + \mu_m a_m) = o.$$

This is a *nontrivial* depence between the vectors $a_1, \cdots, a_m$ (since the sum of the coefficients is equal to 1), contradicting the linear independence of $a_1, \cdots, a_m$. Thus $a_{m+1} \notin \text{aff}\,(a_1, \cdots, a_m)$, i.e., $a_1, \cdots, a_m, a_{m+1}$ are the vertices on an m-dimensional simplex T. The relations (1), (2) show now that the origin o belongs to the relative interior of T, i.e., the vectors $a_1, \cdots, a_{m+1}$ are minimally dependent. ∎

Consequence 16.3: Let $a_1, \cdots, a_{m+1}$ be positively dependent vectors in $\mathbf{R}^n$. If some m of them are linearly independent, then each m of the vectors $a_1, \cdots, a_{m+1}$ are linearly independent.

Theorem 16.4: Let $a_o, a_1, \cdots, a_m \in \mathbf{R}^n$ be positively dependent vectors and $a_o \neq o$. If the vectors $a_1, \cdots, a_m$ are linearly dependent, then it is possible to choose among $a_o, a_1, \cdots, a_m$ less than $m + 1$ positively dependent vectors such that a_o is contained in the set of the chosen vectors. □

<u>PROOF:</u> Let $\lambda_o a_o + \lambda_1 a_1 + \cdots + \lambda_m a_m = o$ be a positive dependence between the vectors. We may suppose $\lambda_o = 1$. Let, furthermore, $\nu_1 a_1 + \cdots + \nu_m a_m = o$ with at least one of the numbers $\nu_1, \cdots, \nu_m$ distinct from 0. Without loss of generality we may suppose (changing the enumeration of the vectors $a_1, \cdots, a_m$ if necessary) that

$$\frac{|\nu_1|}{\lambda_1} \le \frac{|\nu_2|}{\lambda_2} \le \cdots \le \frac{|\nu_m|}{\lambda_m}.$$

Then $\nu_m \neq 0$. Consequently,

$$-a_o = \lambda_1 a_1 + \cdots + \lambda_m a_m$$

$$= \lambda_1 a_1 + \cdots + \lambda_m a_m - \frac{\lambda_m}{\nu_m}(\nu_1 a_1 + \cdots + \nu_m a_m)$$

$$= \sum_{i=1}^{m}\left(\lambda_i - \frac{\lambda_m}{\nu_m}\nu_i\right)a_i = \sum_{i=1}^{m-1}\left(\lambda_i - \frac{\lambda_m}{\nu_m}\nu_i\right)a_i.$$

Furthermore,

$$\lambda_i - \frac{\lambda_m}{\nu_m}\nu_i \geq \lambda_i - \frac{\lambda_m}{|\nu_m|}|\nu_i| = \frac{\lambda_i \lambda_m}{|\nu_m|}\left(\frac{|\nu_m|}{\lambda_m} - \frac{|\nu_i|}{\lambda_i}\right) \geq 0.$$

Thus,

$$a_o + \sum_{i=1}^{m-1}\left(\lambda_i - \frac{\lambda_m}{\nu_m}\nu_i\right)a_i = o,$$

where the coefficients are non-negative. Removing the summands with null coefficients (if there are), we obtain a positive dependence between a_o and some of the vectors $a_1, \cdots, a_{m-1}$. ∎

Theorem 16.5: Let $a_o, a_1, \cdots, a_m \in \mathbf{R}^n$ be positively dependent vectors, each of which is distinct from o. If they are not minimally dependent, then it is possible to choose from them less than $m+1$ minimally dependent vectors such that a_o is contained in the set of the chosen vectors.   □

PROOF: Since the vectors $a_o, a_1, \cdots, a_m$ are not minimally dependent, the vectors $a_1, \cdots, a_m$ are linearly dependent (by virtue of Theorem 16.2). According to Theorem 16.4 it is possible to choose from $a_o, a_1, \cdots, a_m$ less than $m+1$ positively dependent vectors in such a way that a_o belongs to the chosen vectors. Admit, for example, that $a_o, a_1, \cdots, a_k$ are positively dependent, where $k < m$. If the vectors $a_1, \cdots, a_k$ are linearly independent, then the vectors $a_o, a_1, \cdots, a_k$ are minimally dependent (Theorem 16.2), and the aim is achieved. If, however, the vectors $a_1, \cdots, a_k$ are linearly dependent, then it is possible to apply the same method once more, etc. ∎

Let $H \subset \mathbf{R}^n$ be a system (finite or infinite) of nonzero vectors. By md H we denote the largest of the integers m such that in H there is a system of $m+1$ minimally dependent vectors. If H does not contain any system of minimally dependent vectors, then md $H = 0$. Usually we will suppose that $\mathbf{R}^n$ is a Euclidean space and consider a vector system H consisting of unit vectors, i.e., $H \subset S^{n-1}$, where S^{n-1} is the unit sphere of $\mathbf{R}^n$.

We say that a vector system $H \subset \mathbf{R}^n$ is *one-sided* if there exists a closed half-space Π with its bounding hyperplane through the origin such that $H \subset \Pi$. The system is named *strictly one-sided* if there exists a half-space Π with $o \in \mathrm{bd}\,\Pi$ such that $H \subset \mathrm{int}\,\Pi$. Thus in the case when $\mathbf{R}^n$ is an Euclidean space, a vector system $H \subset \mathbf{R}^n$ is one-sided if and only if there exists a vector $q \neq o$ such that $\langle q, e \rangle \leq 0$ for each $e \in H$. Furthermore, H is strictly one-sided if and only if there exists a vector $q \neq o$ such that $\langle q, e \rangle < 0$ for each $e \in H$.

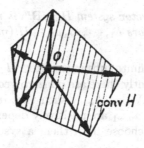

Fig. 78

Theorem 16.6: A vector system $H \subset \mathbf{R}^n$ is one-sided if and only if $o \notin$ int conv H. In other words, H is not one-sided if and only if the origin o is an interior point of conv H (Fig. 78).   □

PROOF: If H is one-sided, then H is contained in a half-space with bounding hyperplane through the origin, and consequently $o \notin$ int conv H. Conversely, if $o \notin$ int conv H, then there exists a hyperplane L through o such that H is contained in one of the closed half-spaces with respect to L, i.e., the vector system is one-sided. ■

Theorem 16.7: Let $H \subset \mathbf{R}^n$ be a vector system which is not one-sided and $a \in \mathbf{R}^n$. Then there exist vectors $b_1, \cdots, b_s \in H$ and positive real numbers $\lambda_1, \cdots, \lambda_s$ such that $a = \lambda_1 b_1 + \cdots + \lambda_s b_s$.   □

PROOF: Since $o \in$ int conv H, there exists a positive number ν such that $\nu a \in$ conv H. Consequently $\nu a = \nu_1 b_1 + \cdots + \nu_s b_s$ for some $b_1, \cdots, b_s \in H$ and positive numbers $\nu_1, \cdots, \nu_s$. Denoting the positive number $\frac{\nu_i}{\nu}$ by λ_i for $i = 1, \cdots, s$, we obtain $a = \lambda_1 b_1 + \cdots + \lambda_s b_s$. ■

Theorem 16.8: If a vector system $H \subset \mathbf{R}^n$ is not one-sided, then there exists a finite subsystem $H_o \subset H$ which is also not one-sided.   □

PROOF: Since H is not one-sided, $o \in$ int conv H. Let $T \subset \mathbf{R}^n$ be an n-dimensional simplex such that $o \in$ int T and $T \subset$ conv H. Denote by $c_1, \cdots, c_{n+1}$ the vertices of T. Since $c_i \in$ conv H, there exists a *finite* subset $H_i \subset H$ such that $c_i \in$ conv H_i. Then the finite set $H_o = H_1 \cup \cdots \cup H_{n+1}$ satisfies the requirement, since $T \subset$ conv H_o and hence $o \in$ int conv H_o. ■

Theorem 16.9.: If a vector system $H \subset \mathbf{R}^n$ is strictly one-sided, then md $H = 0$.   □

PROOF: Let $q \in \mathbf{R}^n$ be a vector such that $\langle q, a \rangle < 0$ for each $a \in H$. Then for any positive linear combination $\lambda_1 a_1 + \cdots + \lambda_k a_k$ of vectors $a_1, \cdots, a_k \in H$ we have $\langle q, \lambda_1 a_1 + \cdots + \lambda_k a_k \rangle < 0$, i.e., $\lambda_1 a_1 + \cdots + \lambda_k a_k \neq o$. Hence the vectors $a_1, \cdots, a_k$ are not minimally dependent. Thus there is no system of minimally dependent vectors in H and consequently md $H = 0$. ■

Theorem 16.10: If a vector system $H \subset \mathbf{R}^n$ is not one-sided, then md $H >$ 0, i.e., *there exist vectors* $a_1, \cdots, a_{m+1} \in H$ $(m \geq 1)$ *which are minimally dependent.*   □

PROOF: There exists a finite subsystem $H_o \subset H$ that is not one-sided, i.e., $o \in$ int conv H_o. Consequently there are nonzero vectors $a_1, \cdots, a_{m+1} \in H_o \subset H$ and positive numbers $\lambda_1, \cdots, \lambda_{m+1}$ such that $o = \lambda_1 a_1 + \cdots + \lambda_{m+1} a_{m+1}$, i.e., the vectors $a_1, \cdots, a_{m+1}$ are positively dependent. By virtue of Theorem 16.5, it is possible to choose from them a system of minimally dependent vectors.   ■

Exercises

1. Let vectors $a_1, \cdots, a_{m+1} \in \mathbf{R}^n$ be positively dependent and several k of them (where $k < m$) be linearly dependent. Prove that the vectors $a_1, \cdots, a_{m+1}$ are not minimally dependent.

2. Let $M \subset \mathbf{R}^3$ be the regular octahedron and H be the set of the unit outward normals of its two-dimensional faces. Prove that md $H = 3$. Generalize this to the n-dimensional cross-polytope (i.e., the polytope with the vertices $e_1, \cdots, e_n, -e_1, \cdots, -e_n$, where $e_1, \cdots, e_n$ is an orthonormal basis in $\mathbf{R}^n$).

3. Let $M \subset \mathbf{R}^n$ be an n-dimensional simplex and H be the set of the outward normals of its facets. Prove that md $H = n$.

4. Let $M \subset \mathbf{R}^3$ be a regular prism with r-sided basis, where $r \neq 4$. Denote by H the set of the unit outward normals of its facets. Prove that md $H = 2$.

5. Show that if $H \subset H_1 \subset S^{n-1}$, then md $H \leq$ md H_1.

6. Prove that if $H \subset S^{n-1}$ is contained in a closed hemisphere, then md $H \leq n-1$. Show by examples that all the values md $H = 0, 1, \cdots, n-1$ are possible.

7. Let $L_o, L_1, \cdots, L_n$ be subspaces of $\mathbf{R}^n$ such that $L_o \subset L_1 \subset \cdots \subset L_n = \mathbf{R}^n$ and dim $L_i = i$, $i = 0, 1, \cdots, n$. For each $i \in \{1, \cdots, n\}$, denote by P_i an *open* half-space in L_i which is bounded by L_{i-1}. Prove that the set $H' = P_1 \cup P_2 \cup \cdots \cup P_n$ satisfies the condition md $H' = 0$ and, at the same time, H' is not strictly one-sided. This example shows that the converse to Theorem 16.9 is false.

8. Prove that the vector system H', described in Exercise 7, is "universal", i.e., if a vector system $H \subset \mathbf{R}^n$ with $o \notin H$ is not strictly one-sided and satisfies md $H = 0$, then H is contained (up to a motion with o as fixed point) in H'.

9. Let $H \subset \mathbf{R}^n$ be a vector system and C be its spanned cone, i.e., C is the set of the vectors $\lambda_1 a_1 + \cdots + \lambda_n a_n$ with $a_i \in H$ and $\lambda_i \geq 0$, $i = 1, \cdots, n$. Prove that if m is the maximal dimension of a subspace contained in C, then md $H \leq m$. Is it possible that md $H < m$?

10. Let $H \subset \mathbf{R}^n$ be a vector system, C its spanned cone, and L the maximal subspace contained in C. Prove that md $H =$ md $(H \cap L)$.

11. Let $\mathbf{R}^n = L_1 \oplus \cdots \oplus L_s$ and $H = \{a_1, \cdots, a_p\} \subset \mathbf{R}^n$ be a system of minimally dependent vectors. Prove that if $H \subset L_1 \cup \cdots \cup L_s$, then there exists an index i such that $H \subset L_i$.

12. Let $\mathbf{R}^n = L_1 \oplus \cdots \oplus L_s$, and $H \subset \mathbf{R}^n$ be a vector system such that $H \subset L_1 \cup \cdots \cup L_s$. Prove that md $H = \max(\mathrm{md}\,(H \cap L_1), \cdots, \mathrm{md}\,(H \cap L_s))$.

13. Let $H \subset \mathbf{R}^n$ be a non-one-sided vector system and $a \in \mathbf{R}^n$ be a nonzero vector. Prove that there exist vectors $e_1, \cdots, e_k \subset H$ such that the system $a, e_1, \cdots, e_k$ is minimally dependent.

14. Prove that md is an affine invariant, i.e., if $H \subset \mathbf{R}^n$ and $f : \mathbf{R}^n \to \mathbf{R}^n$ is a nondegenerate affine transformation, then md $f(H) = $ md H.

15. Let $H \subset \mathbf{R}^3$ be a centrally symmetric vector system with md $H = 2$. Prove that there exists a direct decomposition $\mathbf{R}^3 = L \oplus N$, dim $L = 2$, such that $H \subset L \cup N$.

16. Prove that if a vector system $\{a_o, \cdots, a_n\} \subset \mathbf{R}^n$ is not one-sided, then the vectors $a_o, \cdots, a_n$ are minimally dependent.

17. Let $\mathbf{R}^n = L_1 \oplus \cdots \oplus L_s$. Prove that if a vector system $H \subset L_1 \cup \cdots \cup L_s$ is not one-sided, then for each $i = 1, \cdots, s$ the vector system $H \cap L_i$ is not one-sided in L_i. Is the opposite assertion true?

§17 The ε-displacement Theorem

A vector system $\{e'_1, \cdots, e'_p\}$ is said to be an *ε-displacement* of a system $\{e_1, \cdots, e_p\}$ if $\| e'_1 - e_1 \| < \varepsilon, \cdots, \| e'_p - e_p \| < \varepsilon$.

Theorem 17.1: Let $H_o = \{e_1, \cdots, e_p\} \subset \mathbf{R}^n$ be a non-one-sided vector system. Then there exists a number $\varepsilon > 0$ such that for every vector system $H'_o = \{e'_1, \cdots, e'_p\}$, obtained from H_o by an ε-displacement, the inequality md $H'_o \geq$ md H_o holds. $\square$

PROOF: The origin o is an interior point of the polytope $M_o = \mathrm{conv}\, H_o$. Let $r > 0$ be a number such that the ball $\sum_r$ of radius r, centered at the origin, is contained in int M_o. There exists a number $\varepsilon_1 > 0$ such that for $\varepsilon < \varepsilon_1$ the polytope $M'_o = \mathrm{conv}\, H'_o$ contains $\sum_r$ in its interior, too.

Furthermore, let $R > 0$ be a number such that M_o is contained in the interior of the ball $\sum_R$ of radius R, centered at the origin. There exists a number $\varepsilon_2 > 0$ such that for $\varepsilon < \varepsilon_2$ the polytope M'_o is contained in int $\sum_R$, too.

Since md $H_o = m > 0$, there are $m + 1$ minimally dependent vectors in H_o. Without loss of generality we may suppose that the vectors $e_1, \cdots, e_{m+1}$ are minimally dependent. Then the m-dimensional simplex T with the vertices $e_1, \cdots, e_{m+1}$ contains the origin in its relative interior. There exists an $\varepsilon_3 > 0$ such that for $\varepsilon < \varepsilon_3$ the points $e'_1, \cdots, e'_{m+1}$ are not situated in any $(m - 1)$-dimensional flat, i.e., the vectors $e'_1, \cdots, e'_{m+1}$ are the vertices of an m-dimensional simplex T'. We remark that, in general, aff T' does not contain the origin.

Let $f : \mathrm{aff}\, T \to \mathrm{aff}\, T'$ be the affine mapping which maps $e_1, \cdots, e_{m+1}$ correspondingly to $e'_1, \cdots, e'_{m+1}$. Each point $x \in T$ has the form $x =$

$\lambda_1 e_1 + \cdots + \lambda_{m+1} e_{m+1}$, where the numbers $\lambda_1, \cdots, \lambda_{m+1}$ are non-negative and $\lambda_1 + \cdots + \lambda_{m+1} = 1$. Moreover, if $x \in \text{ri } T$, then the numbers $\lambda_1, \cdots, \lambda_{m+1}$ are positive. The image of x is the point

$$f(x) = \lambda_1 f(e_1) + \cdots + \lambda_{m+1} f(e_{m+1}) = \lambda_1 e_1' + \cdots + \lambda_{m+1} e_{m+1}',$$

and therefore

$$\begin{aligned}
\|f(x) - x\| &= \|\lambda_1(e_1' - e_1) + \cdots + \lambda_{m+1}(e_{m+1}' - e_{m+1})\| \\
&< \lambda_1 \varepsilon + \cdots + \lambda_{m+1}\varepsilon = \varepsilon.
\end{aligned}$$

In particular, $\| f(o) \| = \| f(o) - o \| < \varepsilon$. Furthermore, $f(o) \in \text{ri } T'$, since $o \in \text{ri } T$.

Let $\rho > 0$ be a number such that the m-dimensional ball of radius ρ, centered at the origin and situated in aff T, is contained in ri T. There exists a number $\varepsilon_4 > 0$ such that for $\varepsilon < \varepsilon_4$ the m-dimensional ball of radius ρ, centered at $f(o)$ and situated in aff T', is contained in ri T'.

Let now ε be a positive number which is smaller than each of the numbers $\varepsilon_1, \varepsilon_2, \varepsilon_3, \varepsilon_4, \frac{\rho r}{R+r}$. Denote by L the $(n-m)$-dimensional subspace that is the orthogonal complement of aff T', and by $p : R^n \to L$ the orthogonal projection. Then $p(M_o')$ is an $(n-m)$-dimensional polytope in L which contains the origin in its interior. There are two possible cases which will be considered separately: $pf(o) = o$ and $pf(o) \neq o$.

If $pf(o) = o$, then the flat aff T' passes through o. Since

$$\| f(o) \| < \varepsilon < \frac{R+r}{r}\varepsilon < \frac{R+r}{r}\frac{\rho r}{R+r} = \rho,$$

the point o is situated at a distance less than ρ from $f(o)$ and consequently $o \in \text{ri } T'$ (since $\varepsilon < \varepsilon_4$). This means that the vectors $e_1', \cdots, e_{m+1}'$ are minimally dependent, and hence md $H_o' \geq m$.

Suppose now that $pf(o) \neq o$. Denote by l the ray emanating from the origin with the direction opposite to $pf(o)$. Let y be the intersection point of the ray l and the boundary of the polytope $p(M_o') \subset L$. Then $\| y \| > r$ (since $\varepsilon < \varepsilon_1$, i.e., $\sum_r \subset \text{int } M_o'$, and hence the $(n-m)$-dimensional ball $p\sum_r \subset L$ of radius r centered at the origin is contained in ri $p(M_o')$). Consequently $pf(o) + \tau y = o$, where $\tau > 0$ is less than $\frac{\varepsilon}{r}$ (since $\| pf(o) \| \leq \|f(o)\| < \varepsilon$ and $\| y \| > r$).

Denote by Γ the face of minimal dimension of $p(M_o')$ that contains the point y. Let $a_1, \cdots, a_s$ be the vertices of this face. It is possible to choose from $a_1, \cdots, a_s$ some points (say $a_1, \cdots, a_q$, where $q \leq s$) such that $a_1, \cdots, a_q$ are the vertices of a simplex $Q \subset \Gamma$ and $y \in \text{ri } Q$. Then $y = \mu_1 a_1 + \cdots + \mu_q a_q$, where $\mu_1, \cdots, \mu_q$ are positive numbers with $\mu_1 + \cdots + \mu_q = 1$.

Since $a_1, \cdots, a_q$ are vertices of the polytope $p(M_o')$, there exist vertices $b_1, \cdots, b_q$ of M_o' such that $pb_i = a_i$, $i = 1 \cdots, q$. The vertices $b_1, \cdots, b_q$ belong to the system H_o', and

$$p(f(o) + \tau(\mu_1 b_1 + \cdots + \mu_q b_q)) = pf(o) + \tau(\mu_1 a_1 + \cdots + \mu_q a_q)$$
$$= pf(o) + \tau y = o.$$

This means that the vector

$$c = f(o) + \tau(\mu_1 b_1 + \cdots + \mu_q b_q)$$

belongs to the kernel of the projection p, i.e., c is parallel to the flat aff T'. Since $\varepsilon < \varepsilon_2$ and $\varepsilon < \frac{\rho r}{R+r}$, we obtain

$$\| c \| \leq \| f(o) \| + \tau(\mu_1 \| b_1 \| + \cdots + \mu_q \| b_q \|) < \varepsilon + \tau(\mu_1 R + \cdots + \mu_q R)$$

$$= \varepsilon + \tau R < \varepsilon + \frac{\varepsilon}{r} R < \rho.$$

Consequently the distance of the point

$$f(o) - c = -\tau(\mu_1 b_1 + \cdots + \mu_q b_q) \in \text{aff } T'$$

from $f(o)$ is less than ρ, i.e., $(f(o) - c) \in \text{ri } T'$ (since $\varepsilon < \varepsilon_4$). This means that there are positive numbers $\nu_1, \cdots, \nu_{m+1}$ satisfying

$$f(o) - c = \nu_1 e'_1 + \cdots + \nu_{m+1} e'_{m+1}, \quad \nu_1 + \cdots + \nu_{m+1} = 1.$$

Comparing with the previous equality, we find

$$\nu_1 e'_1 + \cdots + \nu_{m+1} e'_{m+1} + \tau(\mu_1 b_1 + \cdots + \mu_q b_q) = o,$$

i.e., we obtain a positive dependence between the vectors

$$e'_1, \cdots, e'_{m+1}, b_1, \cdots, b_q, \tag{1}$$

which are contained in H'_o.

We show that the vectors (1), except for the last one, are linearly independent, i.e., the vectors (1) are minimally dependent. Indeed, let

$$\alpha_1 e'_1 + \cdots + \alpha_{m+1} e'_{m+1} + \beta_1 b_1 + \cdots + \beta_{q-1} b_{q-1} = o \tag{2}$$

be a linear dependence. Since $pe'_1 = \cdots = pe'_{m+1} = pf(o)$, we obtain from (2), using the projection p,

$$(\alpha_1 + \cdots + \alpha_{m+1})pf(o) + \beta_1 a_1 + \cdots \beta_{q-1} a_{q-1} = o. \tag{3}$$

But the vectors $pf(o), a_1, \cdots, a_q$ are minimally dependent, and consequently the vectors $pf(o), a_1, \cdots, a_{q-1}$ are linearly independent. Hence $\beta_1 = \cdots \beta_{q-1} = 0$, i.e., the dependence (2) takes the form

$$\alpha_1 e'_1 + \cdots + \alpha_{m+1} e'_{m+1} = o. \tag{4}$$

Furthermore, $e'_1, \cdots, e'_{m+1}$ are the vertices of the simplex T' such that $o \notin$ aff T' (since $pf(o) \neq o$), and consequently they are linearly independent, i.e., it follows from (4) that $\alpha_1 = \cdots = \alpha_{m+1} = 0$.

Since the number of minimally dependent vectors (1) is equal to $m + 1 + q > m + 1$, we have md $H'_o > m$.

Thus in any case md $H'_o \geq m$, i.e., md $H'_o \geq$ md H_o.   ∎

EXAMPLE 17.2: We show that in the case of a *one-sided* system H_o the assertion of Theorem 17.1 could be false. Let e_1, e_2 be a basis in a two-dimensional plane $L \subset \mathbf{R}^3$. We put

$$H_o = \{e_1, -e_1, e_2, -e_2, e_1 + e_2, -e_1 - e_2\}.$$

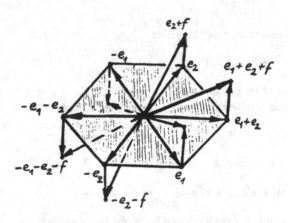

Fig. 79

Clearly, md $H_o = 2$ and, besides, the system H_o is one-sided in $\mathbf{R}^3$. Let, furthermore, the system H'_o have the form (Fig. 79)

$$H'_o = \{e_1 + f, -(e_1 + f), e_2 + f, -(e_2 + f), e_1 + e_2 + f, -(e_1 + e_2 + f)\},$$

where f is a nonzero vector orthogonal to L. Then the vectors $e_1 + f, e_2 + f, e_1 + e_2 + f$ contained in H'_o are linearly independent. It follows that md $H'_o = 1$. At the same time, the system H'_o is otained from H_o by an ε-displacement, where $\varepsilon = \|f\|$. Thus an arbitrarily small displacement of the system H_o yields a vector system H'_o such that md $H'_o <$ md H_o.   □

The following theorem is obtained by P. Živalevich [Ži].

Theorem 17.3: If a vector system $H \subset R^n$ is not one-sided, then md $H =$ md cl H.   □

PROOF: Let md cl $H = m$. We choose in cl H minimally dependent vectors $e_1, \cdots, e_{m+1}$. Furthermore, we choose in cl H vectors $e_{m+2}, \cdots, e_p$ such that the vector system

$$H_o = \{e_1, \cdots, e_{m+1}, \cdots, e_p\}$$

is not one-sided. Then md $H_o = m$. Let $\varepsilon > 0$ be a positive number, as in Theorem 17.1. Since $e_1, \cdots, e_p \in \mathrm{cl}\, H$, there is a vector system $H_o' = \{e_1', \cdots, e_p'\} \subset H$ such that $\|e_1 - e_1'\| \leq \varepsilon, \cdots, \|e_p - e_p'\| \leq \varepsilon$. According to Theorem 17.1, we have md $H_o' \geq$ md $H_o = m =$ md cl H. Consequently md $H \geq$ md cl H (since $H \supset H_o'$). The opposite inequality is obvious, since $H \subset \mathrm{cl}\, H$. ∎

Exercises

1. A finite vector system $H = \{e_1, \cdots, e_p\}$, which is not one-sided, possesses the following property: For every $\varepsilon > 0$ there exists a system $H' = \{e_1', \cdots, e_p'\}$ obtained from H by an ε-displacement such that md $H' \geq m$. Is it true that md $H \geq m$?

2. A finite vector system $H = \{e_1, \cdots, e_p\}$, which is not one-sided, possesses the following property: For every $\varepsilon > 0$ there exists a system $H' = \{e_1', \cdots, e_p'\}$ obtained from H by an ε-displacement such that md $H' \leq m$. Is it true that md $H \leq m$?

3. Let $H \subset \mathbf{R}^n$ be a vector system which is not one-sided, and $a_1, \cdots, a_{m+1} \in H$ be minimally dependent vectors. Furthermore, let $L = \mathrm{lin}\, (a_1, \cdots, a_{m+1})$ and $\pi : \mathbf{R}^n \to \mathbf{R}^n/L$ be the natural projection. We denote by k the number md $\pi(H \setminus L)$. Prove that for every $\varepsilon > 0$ there exists an ε-displacement $H \to H'$ such that md $H' \geq m + k$ and $H' \cap L = H \cap L$.

4. Let $H \subset \mathbf{R}^n$ be a vector system which is not one-sided. Prove that for every $\varepsilon > 0$ there exists an ε-displacement $H \to H'$ such that md $H' = n$.

5. Let $L \subset \mathbf{R}^n$ be a p-dimensional subspace and $\pi : \mathbf{R}^n \to \mathbf{R}^n/L$ be the natural projection. Furthermore, let $H_o = \{a_1, \cdots, a_{n+2}\}$ be a vector system such that the vectors $a_1, \cdots, a_{p+1} \subset L$ are minimally dependent and $\pi(a_{p+2}), \cdots, \pi(a_{n+2})$ are minimally dependent in $\mathbf{R}^n/L$. Prove (without using Theorem 17.1) that there exists a number $\varepsilon > 0$ such that for every system H_o', obtained from H_o by an ε-displacement, the relation md $H_o' \geq p$ holds.

6. A vector system $H_o = \{e_1, \cdots, e_n, f_1, \cdots, f_k\} \subset \mathbf{R}^n$ $(1 \leq k \leq n)$ is said to be *simplicial* if $e_1, \cdots, e_n$ is a basis in R^n and there are numbers $p_o = 0 < p_1 < \cdots < p_{k-1} < p_k = n$ such that the following conditions are satisfied: *(i)* the vectors $e_1, \cdots, e_{p_1}, f_1$ are minimally dependent, *(ii)* for every $i \in \{2, \cdots, k\}$ the vectors $\pi_i(e_{p_{i-1}+1}), \cdots, \pi_i(e_{p_i}), \pi_i(f_i)$ are minimally dependent in L_i/L_{i-1}, where $L_i = \mathrm{lin}\, (e_1, \cdots, e_{p_i})$ and $\pi_i : L_i \to L_{i-1}$ is the natural projection. Let $H \subset \mathbf{R}^n$ be a vector system which is not one-sided. Prove that if H is minimal (i.e., each of its proper subsystems is one-sided), then the system H is *simplicial*.

7. Prove (without using Theorem 17.1) that if $H_o \subset \mathbf{R}^n$ is a *simplicial* system (cf. the preceding exercise), then there exists a number $\varepsilon > 0$ such that for every system H_o', obtained from H by an ε-displacement, the relation md $H_o' \geq$ md H_o holds.

8. Give a new proof of Theorem 17.1, using that every vector system, which is not one-sided, contains a minimal one.

9. Prove that any minimal, non-one-sided vector system in $\mathbf{R}^n$ consists of no more that $2n$ vectors. In what case does it consist of $2n$ vectors?

10. Describe all minimal vector systems in $\mathbf{R}^2$ which are not one-sided.

11. Prove that every minimal, non-one-sided vector system in $\mathbf{R}^3$ coincides, up to an affine transformation, with $H(M)$, where M is either a cube, or a regular prism with triangular basis, or a regular pyramid over a square, or the regular tetrahedron.

12. Prove that if H is a minimal, non-one-sided vector system in $\mathbf{R}^3$ and q is the number of its vectors, then md $H = 7 - q$.

13. In connection with the previous exercise, the conjecture arises that if H is a minimal, non-one-sided vector system in $\mathbf{R}^n$ and q is the number of its vectors, then md $H = (2n + 1) - q$. Show by an example that this conjecture is false. What is the least integer n for which such an n-dimensional counterexample exists?

14. Let $H \subset S^3$ be a minimal, non-one-sided, non-splittable vector system (i.e., there is no direct decomposition $\mathbf{R}^n = L_1 \oplus L_2$, dim $L_1 > 0$, dim $L_2 > 0$, such that $H \subset L_1 \cup L_2$). Prove that md $H = 9 - q$, where q is the number of vectors in H.

15. Let $H = \{a_1, \cdots, a_p\}$ be a vector system which is not one-sided. Prove that there exist non-negative continuous real functions $\lambda_1(x), \cdots, \lambda_p(x)$ on $x \in \mathbf{R}^n$ such that for any $x \in \mathbf{R}^n$ the following equality holds:

$$x = \lambda_1(x)a_1 + \cdots + \lambda_p(x)a_p.$$

16. Is it true that every simplicial vector system in $\mathbf{R}^n$ is not one-sided?

17. Show by an example that the following assertion is false: If $H_o = \{e_1, \cdots, e_n, f_1, \cdots, f_k\}$ is a simplicial vector system in $\mathbf{R}^n$ (cf. Exercise 6), then it is minimal, i.e., each of its proper subsystems is one-sided.

18. Let $\{e_1, \cdots, e_p\}$ be a minimal vector system in $\mathbf{R}^n$ which is not one-sided. Prove that for any $i \in \{1, \cdots, p\}$ the convex hull of $\{e_1, \cdots, e_p\} \setminus \{e_i\}$ is a facet of the n-polytope $M = \text{conv} \{e_1, \cdots, e_p\}$. Prove that each facet of M has such a form.

19. Prove that if $H = \{a_1, \cdots, a_q\} \subset \mathbf{R}^n$ is a minimal vector system which is not one-sided, then H is a *positive basis* in $\mathbf{R}^n$, i.e., each $x \in \mathbf{R}^n$ is uniquely representable as a linear combination of the vectors $a_1, \cdots, a_q$ with nonnegative coefficients.

20. Is the requirement "H is not one-sided" in Theorem 17.3 essential?

21. Let H be a finite, one-sided vector system in $\mathbf{R}^n$ such that md $H > 0$. Prove that for every $\varepsilon > 0$ there exists a vector system H' obtained from H by an ε-displacament such that md $H = 0$.

§18 Lower semicontinuity of the functional md

Let $M \subset \mathbf{R}^n$ be a convex body. We recall that a boundary point a of M is said to be *regular* if there exists only one supporting hyperplane of M through a. The closed half-space P containing M and bounded by this hyperplane is said to be the *supporting half-space* of M at the regular boundary point a. The set of all unit vectors $x \in S^{n-1}$, which are outward normals of the body M at regular boundary points, will be denoted by $H(M)$. If, for example, M is a convex polytope, then $H(M)$ is a finite subset of the unit sphere S^{n-1}. For brevity, the number md $H(M)$ will be denoted by md M.

To indicate a connection of the vector system $H(M)$ with properties of the polar body M^*, we will suppose that $\mathbf{R}^n$ is the Euclidean space and introduce the *normalization* of vector systems. Let $Q \subset \mathbf{R}^n$ be a system of nonzero vectors. By norm Q we denote the set of all *normed* vectors from Q, i.e., $y \in$ norm Q if and only if there exists a vector $x \in Q$ such that $y = \frac{1}{\|x\|}x$.

Theorem 18.1: For every compact, convex body $M \subset \mathbf{R}^n$ with the origin in its interior, the equality $H(M) =$ norm exp M^* holds.   □

PROOF: Let $p \in H(M)$, i.e., p is the unit outward normal of a supporting half-space Π at a regular boundary point $a \in$ bd M. For the supporting hyperplane $\Gamma =$ bd Π, the corresponding point Γ^* belongs to exp M^* (cf. §2). Consequently p and Γ^* are lying on the same ray emanating from the origin (Fig. 80), i.e., $p =$ norm Γ^*. This proves the inclusion $H(M) \subset$ norm exp M^*. Conducting this reasoning in the opposite direction, we obtain the converse inclusion.   ■

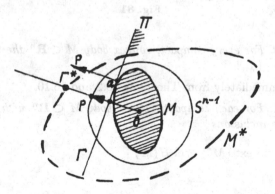

Fig. 80

Theorem 18.2 For a convex body $M \subset \mathbf{R}^n$ the set $H(M) \subset S^{n-1}$ is not one-sided if and only if M is compact.   □

PROOF: If M is compact and $o \in$ int M, then M^* is a compact, convex body with $o \in$ int M^*, i.e., $o \in$ int (conv exp M^*). By virtue of Theorem 16.6, this means that the vector system exp M^* is not one-sided. Consequently, according to Theorem 18.1, the vector system $H(M)$ is also not one-sided.

On the other hand, admit that the body M is noncompact, i.e. unbounded. Then there exists a ray $l \subset M$ emanating from $x_o \in M$, i.e., $y = x_o + \lambda q \in M$ for each $\lambda \geq 0$, where $q \neq o$ is a vector directed along the ray l, see Fig. 81. Let now $p \in H(M)$, i.e., p is the outward normal of the body M at a regular boundary point $a \in$ bd M. Then $M \subset \{x : \langle p, x - a \rangle \leq 0\}$ and consequently $\langle p, y - a \rangle = \langle p, x_o + \lambda q - a \rangle \leq 0$. Since this is true for any $\lambda \geq 0$, we conclude that $\langle p, q \rangle \leq 0$. Thus $\langle p, q \rangle \leq 0$ for every $p \in H(M)$, i.e., the vector system $H(M)$ is one-sided. ∎

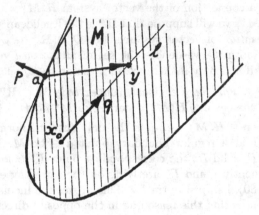

Fig. 81

Corallary 18.3: For every compact, convex body $M \subset \mathbf{R}^n$ the inequality md $M > 0$ *holds.*   □

This follows immediately from Theorems 18.2 and 16.10.

Corollary 18.4: For each compact, convex body $M \subset \mathbf{R}^n$ with the origin in its interior the inclusions

$$H(M) \subset \text{norm ext } (M^*) \subset \text{cl } H(M)$$

hold.   □

This follows immediately from Theorems 18.1 and 2.6.

Theorem 18.5: The functional md*, defined on the family of all compact, convex bodies, is an affine invariant.*   □

PROOF: Let $M \subset \mathbf{R}^n$ be a compact, convex body and $f : \mathbf{R}^n \to \mathbf{R}^n$ be a nondegenerate linear mapping. We may suppose that the origin is an interior point of M and hence (according to Theorem 18.1) md $M = $ md $H(M) = $ md exp (M^*).

If now $N = f(M)$ is the image of M, then $M^* = f^*(N^*)$, and consequently

md exp $(M^*) = $ md exp $(f^*(N^*))$.

But exp $(f^*(N^*)) = f^*($exp $(N^*))$, since f^* is a nondegenerate linear mapping and the set of exposed points is, evidently, an affine invariant. Hence,

md exp $(f^*(N^*)) = $ md $f^*($exp $(N^*))$.

Finally, for vector systems the functional md is affine invariant (since the minimal dependence is). So, we conclude

md $f^*($exp $(N^*)) = $ md (exp $(N^*)) = $ md $H(N) = $ md N. ∎

We remark that the assertion of Theorem 18.5 is not evident, since outward normals of the body M, in general, do not pass to outward normals of the body $N = f(M)$.

We now reformulate the lower semicontinuity property of the operator exp (cf. §5) in terms of the functional md.

Theorem 18.6: *The functional* md, *defined on the family of all compact, convex bodies, possesses the lower semicontinuity property. This means that for any compact, convex body $M \subset \mathbf{R}^n$ there exists a positive real number δ such that for each compact, convex body $N \subset \mathbf{R}^n$ with $\varrho(M,N) < \delta$ the inequality* md $N \geq $ md M *holds. In other words, let $M = \lim\limits_{k \to \infty} M_k$, where M, M_k are compact, convex bodies. If* md $M_k \leq r$ *for all $k = 1, 2, \cdots$, then* md $M \leq r$.
□

PROOF: We may suppose that $o \in $ int M. Let md $M = m$. We choose regular boundary points $a_1, \cdots, a_{m+1}$ of the body M such that the unit outward normals $q_1, \cdots, q_{m+1}$ of M at these points are minimally dependent. Furthermore, let $q_{m+2}, \cdots, q_p$ be unit outward normals of M at regular boundary points $a_{m+2}, \cdots, a_q$ of the body M such that the vector system $\{q_1, \cdots, q_{m+1}, \cdots, q_p\}$ is not one-sided (Theorems 18.2 and 16.8). According to Theorem 18.1, there exists a vector system $H_o = \{e_1, \cdots, e_{m+1}, \cdots, e_p\} \subset$ exp (M^*) such that norm $H_o = \{q_1, \cdots, q_m, \cdots, q_p\}$. Consequently the system H_o is not one-sided and we have md $H_o = m$. Let $\varepsilon > 0$ be a number as in Theorem 17.1 (for the system H_o). By virtue of Theorem 5.2, there exists a positive number $\delta > 0$ such that if $\varrho(M, N) < \delta$, then exp $M^* \subset U_\varepsilon($exp $(N^*))$ and hence $H_o \subset U_\varepsilon($exp $(N^*))$. Thus, if $\varrho(M, N) < \delta$, then there exist vectors $e_1', \cdots, e_p' \in $ exp (N^*) such that $\| e_1' - e_1 \| < \varepsilon, \cdots, \| e_p' - e_p \| < \varepsilon$, i.e., the system $H_o' = \{e_1', \cdots, e_p'\}$ is obtained from H_o by an ε-displacement. According to the definition of the number ε, this means that md $H_o' \geq m$. Consequently md exp $(N^*) \geq m$, and we conclude (by virtue of Theorem 18.1) that md $H(N) \geq m$, i.e., md $N \geq m =$ md M. ∎

Exercises

1. Prove that for any n-dimensional parallelotope $M \subset \mathbf{R}^n$ the equality md $M = 1$ holds.

2. Prove that for any n-dimensional simplex $T \subset \mathbf{R}^n$ the equality md $T = n$ holds.

3. Prove that for the n-dimensional regular cross-polytope $M \subset \mathbf{R}^n$ the equality md $M = n$ holds.

4. Prove that for every n-dimensional convex polytope $M \subset \mathbf{R}^n$ the equality

   $$H(M) = \text{norm ext } (M^*)$$

   holds.

5. Let $M \subset \mathbf{R}^n$ be a compact, convex body and $a_1, \cdots, a_q$ be its regular boundary points such that the intersection of the supporting half-spaces of M at these points is a polytope N ("circumscribed polytope" of M). Prove that md $N \leq$ md M.

6. Prove that for every compact, convex body with $o \in \text{int } M$ the relation md $M = \text{md exp}(M^*)$ holds.

7. Assume that a convex body M satisfies the condition md $M > 0$. Is the body M necessarily compact?

8. Give an example of a compact, convex body $M \subset \mathbf{R}^n$ for which the inclusion $H(M) \subset \text{norm ext } (M^*)$ is strict.

9. Give an example of a compact, convex body $M \subset \mathbf{R}^n$ for which the inclusion norm ext $(M^*) \subset \text{cl } H(M)$ is strict.

10. We say that a convex body $C \subset \mathbf{R}^n$ is a k-simplicial cylinder if there exists a decomposition $\mathbf{R}^n = L_1 \oplus L_2$ such that $C = T + L_2$, where $T \subset L_1$ is a simplex and dim $T = \dim L_1 = k$. Prove that for a compact, convex body $M \subset \mathbf{R}^n$ the number md M is the largest of the integers k such that there exist regular boundary points $a_1, \cdots, a_{k+1} \in \text{bd } M$ for which the intersection of corresponding supporting half-spaces is a k-simplicial cylinder.

11. Give another proof of Theorem 18.5 by using the previous exercise.

12. Let $M \subset \mathbf{R}^n$ be the cube described in an orthonormal coordinate system $(x_1, \cdots, x_n)$ by the inequalities $0 \leq x_i \leq 1, i = 1, \cdots, n$. Furthermore, let $M_k = M \cap P_k$, where P_k is the closed half-space described by the inequality $x_1 + \cdots + x_n \geq \frac{1}{k}$ $(k = 1, 2, \cdots)$. Prove that $M = \lim_{k \to \infty} M_k$ and, moreover, md $M = 1$, md $M_k = n$ for $k = 1, 2, \cdots$. This example shows that the functional md does not possess the upper semicontinuity property, i.e., the equalities $\lim_{k \to \infty} M_k = M$ and md $M_k = r$ for all k do not imply md $M \geq r$.

§19 The definition of H-convex sets

In the Euclidean vector space $\mathbf{R}^n$, let H denote an arbitrary subset of the unit sphere $S^{n-1} \subset \mathbf{R}^n$. Each half-space of the form $\{x : \langle f, x \rangle \leq \lambda\}, f \in H, \lambda \in \mathbf{R}$, is said to be an H-convex half-space, and each set representable as intersection of a family of H-convex half-spaces is said to be H-convex, cf. [Bt 5]. We say that the whole space $\mathbf{R}^n$ is also H-convex.

The definition shows that each H-convex set is closed and linearly convex. The converse is only true when $H = S^{n-1}$. In other words, for $H \neq S^{n-1}$ there exists a linearly convex (closed) set which is not H-convex.

EXAMPLE 19.1: If a vector $f_o \in S^{n-1}$ does not belong to H, then the hyperplane $L = \{x : \langle f_o, x \rangle = \lambda_o\}$ is not H-convex. In fact, L is not contained in the half-space $\{x : \langle f, x \rangle \leq \lambda\}$ if the vector $f \in H$ is different from $\pm f_o$. Furthermore, any half-space $\{x : \langle f_o, x \rangle \leq \lambda\}$ is not H-convex, since $f_o \notin S^{n-1}$. If even the remaining half-spaces $\{x : \langle f_o, x \rangle \geq \lambda\}$ are H-convex (i.e., $-f_o \in H$), the hyperplane L cannot be represented as intersection of such half-spaces.

Theorem 19.2: Let $M \subset \mathbf{R}^n$ be a convex body distinct from $\mathbf{R}^n$. If $H(M) \subset H \subset S^{n-1}$, then M is H-convex. $\quad\square$

PROOF: Let a be a regular boundary point of M and Π_a be the corresponding supporting half-space. Then the unit outward normal p_a of the half-space Π_a belongs to $H(M)$. Since $H(M) \subset H$, the vector p_a belongs to H, i.e., the half-space Π_a is H-convex. Thus the supporting half-space of M at any of its regular boundary points is H-convex. The body M coincides with the intersection of the supporting half-spaces at its regular boundary points, i.e., M is the intersection of a family of H-convex half-spaces. Thus M is H-convex. ∎

Theorem 19.3: For any H-convex body $M \subset \mathbf{R}^n$ distinct from $\mathbf{R}^n$ the inclusion $H(M) \subset \operatorname{cl} H$ holds. $\quad\square$

PROOF: Let $p \in H(M)$. Then there exists a regular boundary point a of M such that p is the unit outward normal of M at the point a. We fix an $\varepsilon > 0$. Since a is _regular_, there exists a point $b \notin M$ (close enough to a) with the following property: If Π is a closed half-space with the unit outward normal q such that $\Pi \supset M, b \notin \Pi$, then $\| p - q \| < \varepsilon$ (Fig. 82). Let now Π_1 be an H-convex half-space with $\Pi_1 \supset M, b \notin \Pi_1$ (such a half-space exists, since M is H-convex and $b \notin M$). For the unit outward normal q_1 of Π_1 the relation $\| p - q_1 \| < \varepsilon$ holds (according to the choice of b). Moreover, since the half-space Π_1 is H-convex, we have $q_1 \in H$. Thus, for every ε there exists a point $q_1 \in H$ such that $\| p - q_1 \| < \varepsilon$. This means that $p \in \operatorname{cl} H$. ∎

Theorem 19.4: A compact, convex body $M \subset \mathbf{R}^n$ is H-convex if and only if $\overline{H(M) \subset \operatorname{cl} H}$ holds. $\quad\square$

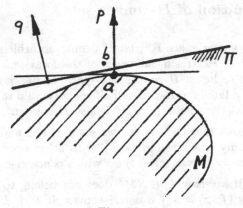

Fig. 82

<u>PROOF</u>: By virtue of Theorem 19.3, it suffices to prove only the part "if". Let M be a compact, convex body with $H(M) \subset \mathrm{cl}\, H$. We have to prove that M is H-convex. Let x be a point not contained in M. Then there exists a regular boundary point a of M such that the corresponding supporting half-space Π does not contain x, see Fig. 83. The unit outward normal e of Π belongs to $H(M)$ (since Π is the supporting half-space of Π at its *regular* boundary point a), and consequently $e \in \mathrm{cl}\, H$. By virtue of the compactness of M, there exists a vector $e' \in H$ (close enough to e) and a closed half-space Π' with the outward normal e' such that $M \subset \Pi'$, $x \notin \Pi'$. Since $e' \in H$, the half-space Π' is H-convex. Thus x does not belong to the intersection of *all* H-convex half-spaces which contain M. This is true for every point $x \notin M$, and therefore the intersection and M coincide. Hence, M is H-convex. ∎

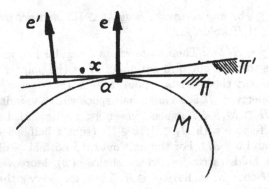

Fig. 83

Theorem 19.5: *Admit that the set $H \subset S^{n-1}$ is closed. Then a convex body $M \subset \mathbf{R}^n$ is H-convex if and only if $H(M) \subset H$.*   □

This is an immediate consequence of Theorems 19.2 and 19.3.   ∎

We remark that the convex body M in Theorem 19.5 is not supposed to be compact.

Exercises

1. Let $H \subset S^{n-1}$ be a nonempty set. Prove that if $M \subset \mathbf{R}^n$ is H-convex, then each translate of M is H-convex, too.

2. Prove that if $M \subset \mathbf{R}^n$ is H-convex, then its image under a homothety with any positive ratio is also H-convex.

3. Consider the assertion which is contrary to Theorem 19.2: "If M is an H-convex body distinct from $\mathbf{R}^n$, then $H(M) \subset H$." Show by an example that this assertion is false.

4. According to Theorem 19.4, a compact, convex body M with $H(M) \subset \operatorname{cl} II$ is H-convex. Does this remain true for non-compact, convex bodies?

5. Prove that a compact, convex body $M \subset \mathbf{R}^n$ is H-convex if and only if it is $(\operatorname{cl} H)$-convex.

6. Prove that if the set $H \subset S^{n-1}$ is one-sided, then each nonempty H-convex set is unbounded.

7. Prove that if $H \subset S^{n-1}$ is not one-sided, then there exist compact, H-convex bodies.

8. Let $M \subset \mathbf{R}^n$ be a compact, convex body with $o \in \operatorname{int} M$. Denote the set norm $\exp (M^*)$ by H. Prove that M is H-convex.

9. Let $M \subset \mathbf{R}^n$ be a compact, convex body with $o \in \operatorname{int} M$. Prove that if M is H-convex, then norm $\exp (M^*) \subset \operatorname{cl} H$.

10. Prove that $\{o\}$ is an H-convex set if and only if the set H is not one-sided.

§20 Upper semicontinuity of the H-convex hull

From the definition of H-convexity it follows immediately that if M_α is an H-convex set for each $\alpha \in A$, then so is $\bigcap_{\alpha \in A} M_\alpha$. This obvious proposition gives the permission to define the H-*convex hull* of any set $M \subset \mathbf{R}^n$. Namely, the intersection of all H-convex sets containing $M \subset \mathbf{R}^n$ is said to be the H-*convex hull* of M. We denote it by $\operatorname{conv}_H M$.

Obviously, $\operatorname{conv}_H M$ is the *minimal* H-convex set which contains M. On the other hand, the set $\operatorname{conv}_H M$ is (if it does not coincide with $\mathbf{R}^n$) the intersection of all H-convex half-spaces which contain M.

We remark that $\mathrm{conv}_H M$ contains M together with its closure $\mathrm{cl}\, M$ (since all H-convex sets are closed). In other words, $\mathrm{conv}_H M = \mathrm{conv}_H(\mathrm{cl}\, M)$.

<u>EXAMPLE 20.1:</u> Let $H = \{e_1, \cdots, e_n, -e_1, \cdots, -e_n\}$, where $e_1, \cdots, e_n$ is a basis of $\mathbf{R}^n$. Then for every compact, convex body $X \subset \mathbf{R}^n$ its H-convex hull is the circumscribed parallelotope of X, whose facets are orthogonal to the vectors $e_1, \cdots, e_n$, respectively (Fig. 84). $\square$

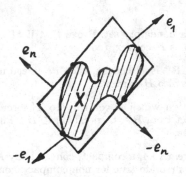

Fig. 84

The following proposition establishes a connection between the notions H-*convex hull* and *vector sum*.

<u>Theorem 20.2:</u> *Let $M_1, M_2 \subset \mathbf{R}^n$ and $H \subset S^{n-1}$ be a nonempty set. Then the inclusion*

$$\mathrm{conv}_H M_1 + \mathrm{conv}_H M_2 \subset \mathrm{conv}_H(M_1 + M_2)$$

holds. $\square$

<u>PROOF:</u> Let $x \notin \mathrm{conv}_H(M_1 + M_2)$. Then there exists an H-convex half-space P such that $M_1 + M_2 \subset P$ and $x \notin P$. Denote by e the unit outward normal of P. We may suppose that for every $\varepsilon > 0$ the half-space $P - \varepsilon e$ does not contain the set $M_1 + M_2$ (Fig. 85). Let, furthermore, P_1, P_2 be analogous half-spaces for M_1, M_2 with the same outward normal e. In other words, $M_1 \subset P_1$ and for every $\varepsilon > 0$ the half-space $P_1 - \varepsilon e$ does not contain M_1 (analogously for M_2). Then $P = P_1 + P_2$. Since the half-spaces P_1, P_2 are H-convex, the inclusions

$$\mathrm{con}_H M_1 \subset P_1, \quad \mathrm{conv}_H M_2 \subset P_2$$

hold. It follows from $x \notin P$ and $P = P_1 + P_2$ that $x \notin P_1 + P_2$, so much $x \notin \mathrm{conv}_H M_1 + \mathrm{conv}_H M_2$.   ∎

<u>EXAMPLE 20.3:</u> We indicate an example which shows that the converse inclusion, i.e.

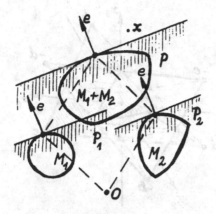

Fig. 85

$$\text{conv}_H M_1 + \text{conv}_H M_2 \supset \text{conv}_H(M_1 + M_2), \tag{1}$$

is, in general, false. Consider the vector system

$$H = \left\{ \frac{1}{\sqrt{2}}(\pm e_1 \pm e_3), \frac{1}{\sqrt{2}}(\pm e_2 \pm e_3) \right\}$$

in R^3 (all combinations of signs), where e_1, e_2, e_3 is an orthonormal basis. It is easily shown that the tetrahedron T_1 with the vertices $a(0; 2; 1)$, $b(0; -2; 1)$, $c(2; 0; -1)$, $d(-2; 0; -1)$ is H-convex, and the tetrahedron T_2 with the vertices $-a, -b, -c, -d$ is H-convex, too (Fig. 86). For example, the face of T_1 with the vertices a, b, c is situated in the plane whose normal is (up to a sign) the vector product

$$[a - b, a - c] = [4e_2, -2e_1 + 2e_2 + 2e_3] = 8(e_3 + e_1),$$

i.e., the corresponding unit normals are $\pm \frac{1}{\sqrt{2}}(e_1 + e_3) \in H$.

At the same time, the vector sum $T_1 + T_2$ has a two-dimensional face which is the sum of the segments $[a, b] \subset T_1$ and $[-c, -d] \subset T_2$. This face is situated in the plane $x_3 = 2$ and has the unit outward normals $\pm e_3 \notin H$. Thus the polytope $T_1 + T_2$ is not H-convex and, consequently,

$$\text{conv}_H T_1 + \text{conv}_H T_2 = T_1 + T_2 \neq \text{conv}_H(T_1 + T_2).$$

This means that for $M_1 = T_1$ and $M_2 = T_2$ the inclusion (1) does not hold. Moreover, this example shows that, in general, the vector sum of H-convex sets is not H-convex. $\square$

Theorem 20.4: Let $M \subset \mathbf{R}^n$ be a compact, H-convex set and ε be a positive real number. Then there exists a real number $\delta > 0$ such that

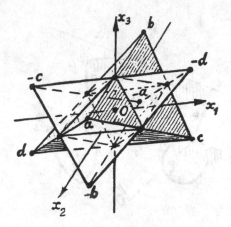

Fig. 86

$$\mathrm{conv}_H U_\delta(M) \subset U_\varepsilon(M). \quad \square$$

<u>PROOF</u>: Denote by B the boundary of the compact, convex body cl $U_\varepsilon(M)$.
Let $a \in B$. Since M is H-convex and $a \notin M$, there exists a closed half-space
$P(a)$ which is H-convex with $P(a) \supset M$, $a \notin P(a)$. We fix such a half-space
$P(a)$ for every $a \in B$. The complementary half-space $Q(a) = \mathbf{R}^n \setminus P(a)$ is
open and $a \in Q(a)$. This means that the sets $Q(a)$ form an open covering of
the set B. Since B is compact, there is a finite set $\{a_1, \cdots, a_p\} \subset B$ such that
$B \subset Q(a_1) \cup \cdots \cup Q(a_p)$. Consequently the convex body $P(a_1) \cap \cdots \cap P(a_p)$
is contained in $U_\varepsilon(M)$. Denote by $e(a_i)$ the unit outward normal of the half-
space $P(a_i)$, $i = 1, \cdots, p$. Then $e(a_i) \in H$, since the half-space $P(a_i)$ is
H-convex.

Since the intersection $P(a_1) \cap \cdots \cap P(a_p)$ is contained in the open, convex
set $U_\varepsilon(M)$, there exists a positive number δ such that the intersection

$$(P(a_1) + \delta e(a_1)) \cap \cdots \cap (P(a_p) + \delta e(a_p)) \tag{2}$$

is contained in $U_\varepsilon(M)$, too (Fig. 87). It is clear that

$$U_\delta(M) \subset (P(a_1) + \delta e(a_1)) \cap \cdots \cap (P(a_p) + \delta e(a_p)).$$

Furthermore, since the intersection (2) is H-convex, the inclusion

$$\mathrm{conv}_H U_\delta(M) \subset (P(a_1) + \delta e(a_1)) \cap \cdots \cap (P(a_p) + \delta e(a_p)) \subset U_\varepsilon(M)$$

is true.   ∎

Theorem 20.4 expresses the *upper semicontinuity property* of the H-convex
hull. It is possible to formulate another form of this property which uses the
notion of *upper limit* of a set sequence.

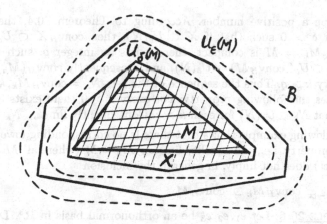

Fig. 87

Let $N_1, N_2 \cdots$ be a sequence of sets in $\mathbf{R}^n$. The *upper limit* $\overline{\lim}_{k \to \infty} N_k$ of the sequence is the set of the points $x \in \mathbf{R}^n$ such that for every $\varepsilon > 0$ the intersection $U_\varepsilon(x) \cap N_k$ is nonempty for an infinite set of indices k. In other words, $x \in \overline{\lim}_{k \to \infty} N_k$ if for every $\varepsilon > 0$ and every integer m the intersection

$$U_\varepsilon(x) \cap \left(\bigcup_{k > m} N_k \right)$$

is nonempty.

It is well known that the upper limit $\overline{\lim}_{k \to \infty} N_k$ of every set sequence $N_1, N_2, \cdots$ is a *closed* set. Moreover, let $N \subset R^n$ be a closed set. Then, for the validity of the inclusion $\overline{\lim}_{k \to \infty} N_k \subset N$, it is sufficient that for every $\varepsilon > 0$ there exists an integer p such that $N_k \subset U_\varepsilon(N)$ for each $k > p$. Now we can formulate another form of the upper semicontinuity for the H-convex hull.

Theorem 20.5: Let $H \subset S^{n-1}$ be a *non-one-sided* set. Let, furthermore, $\lim_{k \to \infty} M_k = M$, where M, M_k are compact, convex sets. Then

$$\overline{\lim}_{k \to \infty} \mathrm{conv}_H M_k \subset \mathrm{conv}_H M. \quad \square$$

PROOF: Let $B \subset R^n$ be a ball containing the compact set M. Since H is not one-sided, there exists a finite subset $\{e_1, \cdots, e_p\} \subset H$ which is not one-sided (Theorem 16.8). Let $\Pi_1, \cdots, \Pi_p$ be supporting half-spaces of the ball B with the outward normals $e_1, \cdots, e_p$. The intersection $Q = \Pi_1 \cap \cdots \cap \Pi_p$ is compact, since $H(Q) = \{e_1, \cdots, e_p\}$ is not one-sided (Theorem 18.2), i.e., Q is an H-convex, compact body. Furthermore, since $M \subset B \subset Q$, the inclusion $\mathrm{conv}_H M \subset \mathrm{conv}_H Q = Q$ holds. It follows that the set $\mathrm{conv}_H M$ is compact. We denote this compact set by N.

Let ε be a positive number. According to Theorem 20.4, there exists a number $\delta > 0$ such that if $X \subset U_\delta(N)$, then $\mathrm{conv}_H X \subset U_\varepsilon(N)$. Since $\overline{\lim}_{k\to\infty} M_k = M$ is compact, there exists an integer q, such that $M_k \subset U_\delta(M) \subset U_\delta(\mathrm{conv}_H M) = U_\delta(N)$, and consequently $\mathrm{conv}_H(M_k) \subset U_\varepsilon(M)$, for every $k > q$. Thus the sequence of the sets $N_k = \mathrm{conv}_H M_k, k = 1, 2, \cdots$, possesses the following property: For every $\varepsilon > 0$ there exists an integer q such that $N_k \subset U_\varepsilon(N)$ for each $k > q$. Consequently $\overline{\lim}_{k\to\infty} N_k \subset N$. ∎

The following example shows that there is no corresponding lower semicontinuity property, i.e., the equality $\lim_{k\to\infty} M_k = M$ (where all M_k and M are compact) does not imply, in general, the inclusion

$$\underline{\lim}_{k\to\infty} \mathrm{conv}_H M_k \supset \mathrm{conv}_H M. \tag{3}$$

EXAMPLE 20.6: Let e_1, e_2, e_3 be an orthonormal basis in $\mathbf{R}^3$. Denote by H the set of the vectors $f = (f_1, f_2, f_3) \in S^2$ satisfying the condition $|\langle f, e_1 \rangle| \leq |\langle f, e_2 \rangle|$, i.e., $|f_1| \leq |f_2|$ (Fig. 88). We use the symbol M for the segment $[-e_2, e_2]$. Let $P = \{x : \langle f, x \rangle \leq \lambda\}$ be an H-convex half-space containing M. Then $\lambda \geq 0$, since $o \in M$. Furthermore, $\langle f, -e_2 \rangle \leq \lambda, \langle f, e_2 \rangle \leq \lambda$, i.e., $|\langle f, e_2 \rangle| \leq \lambda$. Since $f \in H$, we have $|\langle f, e_1 \rangle| \leq |\langle f, e_2 \rangle| \leq \lambda$, i.e., $e_1 \in P$. Thus e_1 belongs to any H-convex half-space containing M, and hence $e_1 \in \mathrm{conv}_H M$. It follows that $M \neq \mathrm{conv}_H M$.

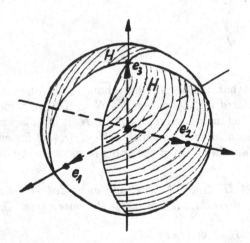

Fig. 88

For $k = 1, 2, \cdots$, we now put

$$f_k^{(+)} = \frac{ke_1 + (k+1)e_2 + (k^2+k)e_3}{k^2+k+1},$$

$$f_k^{(-)} = \frac{-ke_1 + (k+1)e_2 + (k^2+k)e_3}{k^2+k+1}.$$

It is easily shown that $\pm f_k^{(+)} \in H, \pm f_k^{(-)} \in H$. Moreover, $\pm e_2 \in H$. Let M_k be the intersection of the half-spaces

$$\left\{x : \langle f_k^{(+)}, x \rangle \leq 0\right\}, \left\{\langle f_k^{(+)}, x\rangle \geq 0\right\}, \left\{x : \langle f_k^{(-)}, x\rangle \leq 0\right\},$$

$$\left\{x : \langle f_k^{(-)}, x\rangle \geq 0\right\}. \{x : \langle e_2, x\rangle \leq 1\}, \{x : \langle e_2, x\rangle \geq -1\}.$$

The set M_k is H-convex. The intersection of the first and the second of the indicated half-spaces is the plane $L_k^{(+)}$ passing trough o and being orthogonal to the vector $f_k^{(+)}$. The intersection of the third and the fourth half-spaces is the plane $L_k^{(-)}$ passing through o and being orthogonal to the vector $f_k^{(-)}$. Therefore the intersection of the four half-spaces is the line passing through o and being parallel to the vector product $[f_k^{(+)}, f_k^{(-)}]$, i.e., parallel to the vector $e_2 - \frac{1}{k}e_3$ (Fig. 89).

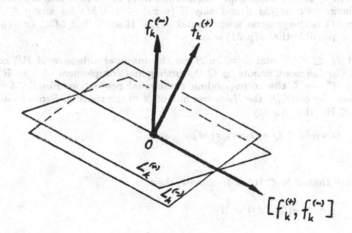

Fig. 89

It follows that M_k (i.e., the intersection of all six half-spaces) is the segment $[-e_2 + \frac{1}{k}e_3, e_2 - \frac{1}{k}e_3]$. Hence, $\lim_{k\to\infty} M_k = M$. Since M_k is H-convex for all $k = 1, 2, \cdots$, we have

$$\lim_{k\to\infty} \operatorname{conv}_H M_k = \lim_{k\to\infty} M_k = M \neq \operatorname{conv}_H M.$$

This shows that the operation conv_H has no lower semicontinuity property.
□

Exercises

1. Prove that if H is symmetric with respect to the origin and M is centrally symmetric, then the set $\text{conv}_H M$ is also centrally symmetric.

2. Let $H \subset S^{n-1}$ consist of the vectors $\frac{1}{\sqrt{n}}(\pm e_1 \pm e_2 \pm \cdots \pm e_n)$ (with all combinations of signs). What is the H-convex hull of the unit ball?

3. Let $\sum \subset \mathbf{R}^n$ be an equatorial body centered at the origin (see Theorem 11.14) and $H = H(\sum)$. Prove that for every H-convex sets M_1, M_2 their vector sum $M_1 + M_2$ is also H-convex.

4. Prove that, in the notation of the previous exercise, for arbitrary sets $M_1, M_2 \subset \mathbf{R}^n$ the inclusion (1) holds.

5. Prove that, again in the notation of the third exercise, for every sequence $M_1, \cdots, M_k$ of convex sets, convergent to a convex set M, the inclusion (3) holds. In other words, the H-convex hull is continuous.

6. Let $M \subset \mathbf{R}^n$ be a nonempty set distinct from $\mathbf{R}^n$. Prove that a point $x_o \in \mathbf{R}^n$ belongs to $\text{conv}_H M$ if and only if $\langle p, x_o \rangle \leq s(p, M)$ for any $p \in H$, where $s(p, M)$ is the greatest lower bound of $\lambda \in \mathbf{R}$ such that $M \subset \{x : \langle p, x \rangle \leq \lambda\}$ (it is possible that $s(p, M) = \infty$).

7. Let $H \subset S^{n-1}$ and $L = \text{lin } H$ be the minimal subspace of $\mathbf{R}^n$ containing H. Furthermore, denote by Q the orthogonal complement of L in $\mathbf{R}^n$ and by $\pi : \mathbf{R}^n \to L$ the corresponding orthogonal projection. Finally, for $N \subset L$ denote by $\text{conv}'_H N$ the H-convex hull of N in the space L. Prove that for every $M \subset \mathbf{R}^n$ the equality

   $$\text{conv}_H M = Q + \text{conv}'_H \pi(M)$$

   holds.

8. Prove that if $M \subset \mathbf{R}^n$ is H-convex, then

   $$\bigcap_{\varepsilon > 0} \text{conv}_H (U_\varepsilon(M)) = M.$$

9. Prove that for every nonempty, bounded set $M \subset \mathbf{R}^n$ its H-convex hull $\text{conv}_H M$ is bounded if and only if H is not one-sided.

10. Let $M_1, M_2, \cdots$ be a sequence of convex sets in $\mathbf{R}^n$. Prove that if $\varlimsup_{k \to \infty} M_k$ is nonempty and contained in a bounded, H-convex set M, then

    $$\varlimsup_{k \to \infty} \text{conv}_H M_k \subset M.$$

11. Let M be the body defined in an orthonormal coordinate system (x_1, x_2, x_3) by the inequalities

    $$(x_1)^2 \leq 2x_2 x_3,\ 0 \leq x_3 \leq 1,\ x_2 \geq 0$$

    (Fig. 90), and let $H = H(M)$. Prove that for $\varepsilon = \frac{1}{3}$ there is no $\delta > 0$ such that $\text{conv}_H U_\delta(M) \subset U_\varepsilon(M)$. This example shows that for unbounded H-convex bodies the conclusion of Theorem 20.4 is, in general, incorrect.

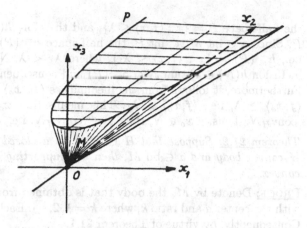

Fig. 90

12. Let $H = \{e_1, e_2, -e_1, -e_2\}$, where e_1, e_2 is an orthonormal basis in $\mathbf{R}^2$. Furthermore, let $M_k = [o, ke_1 + e_2]$ and M denote the ray emanating from the origin and containing e_1. Prove that $\overline{\lim}_{k \to \infty} M_k = M$ and, nevertheless, the inclusion $\overline{\lim}_{k \to \infty} \mathrm{conv}_H M_k \subset M$ is false. This example shows that, in general, the conclusion in Exercise 10 does not hold for an unbounded, H-convex set M.

§21 Supporting cones of H-convex bodies

Theorem 21.1: Let $H \subset S^{n-1}$ be a closed set. For every increasing sequence $M_1 \subset M_2 \subset \cdots$ of H-convex bodies, the body $N = \mathrm{cl}\left(\bigcup_k M_k\right)$ is H-convex, too.

PROOF: Let $x_o \notin N$. We fix a point $a \in \mathrm{int}\, M_1$ and choose a homothety h with the center a and ratio $q > 1$ such that $x_o \notin h(N)$. In particular, $x_o \notin h(M_k)$ for every $k = 1, 2, \cdots$. Consequently for every $k = 1, 2, \cdots$ there exists a vector $f_k \in H$ and a real number λ_k such that the half-space $P_k = \{x : \langle f_k, x \rangle \leq \lambda_k\}$ contains the body $h(M_k)$, but $x_o \notin P_k$, i.e., $\langle f_k, x_o \rangle > \lambda_k$. The sequence $\lambda_1, \lambda_2, \cdots$ is bounded, since $\lambda_k < \langle f_k, x_o \rangle \leq |x_o|$ and $\lambda_k \geq \langle f_k, a \rangle \geq -|a|$. Therefore we may suppose (passing to a subsequence, if necessary) that there exist limits

$$\lim_{k \to \infty} \lambda_k = \lambda, \qquad \lim_{k \to \infty} f_k = f,$$

where f is a unit vector. Moreover, $f \in H$, since H is closed.

If $x \in h\left(\bigcup_k M_k\right)$, then $\langle f_k, x \rangle \leq \lambda_k$ for all k except, maybe, a finite number of indices. Hence $\langle f, x \rangle \leq \lambda$. This means that the body $h\left(\bigcup_k M_k\right)$ is contained in the half-space $P = \{x : \langle f, x \rangle \leq \lambda\}$, and therefore $h(N) = h(\text{cl } (\bigcup_k M_k)) \subset P$. Since $a \in \text{int } M_1 \subset \text{int } P$, the half-space $h^{-1}(P)$ is contained in int P, i.e., $h^{-1}(P) = \{x : \langle f, x \rangle \leq \lambda_1\}$, where $\lambda_1 < \lambda$. Now it follows from the inclusion $h(N) \subset P$ that $N \subset h^{-1}(P)$ and consequently $\text{conv}_H N \subset h^{-1}(P)$. Furthermore, it follows from the inequalities $\langle f_k, x_o \rangle > \lambda_k, k = 1, 2, \cdots$, that $\langle f, x_o \rangle \geq \lambda$, i.e., $\langle f, x_o \rangle > \lambda_1$. This means that $x_o \notin h^{-1}(P)$, i.e., $x_o \notin \text{conv}_H N$. Thus, if $x_o \notin N$, then $x_o \notin \text{conv}_H N$, i.e., $\text{conv}_H N = N$. ∎

<u>Theorem 21.2</u>: *Suppose that $H \subset S^{n-1}$ is a closed set. If $M \subset \mathbf{R}^n$ is an H-convex body and $a \in \text{bd } M$, then the supporting cone $\text{supcone}_a M$ is H-convex.*

PROOF: Denote by M_k the body that is obtained from M by the homothety with the center a and ratio k, where $k = 1, 2, \cdots$. Each body M_k is H-convex. Consequently, by virtue of Theorem 21.1,

$$\text{sup cone }_a M = \text{cl } \left(\bigcup_k M_k \right)$$

is also H-convex. ∎

EXAMPLE 21.3: We now show that the requirement "H is closed" in Theorem 21.2 is essential. Let e_1, e_2 be an orthonormal basis in $\mathbf{R}^2$. We put $H = S^1 \setminus \{e_2\}$. Then the unit ball $B \subset \mathbf{R}^2$ is an H-convex body and e_2 is its boundary point. The half-space $P = \{x : \langle x, e_2 \rangle \leq 1\}$ is the supporting cone of M at the point e_2. But this supporting cone is not H-convex (Fig. 91).

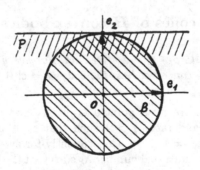

Fig. 91

EXAMPLE 21.4: We now show that the requirement "M is a body" in Theorem 21.2 is essential. Let e_1, e_2, e_3 be an orthonormal basis in $\mathbf{R}^3$. We put

$$f(\alpha) = (\cos^2\alpha \sin\alpha)e_1 - (\sin^2\alpha\cos\alpha)e_2 + \sqrt{1 - \sin^2\alpha \cos^2\alpha} \, e_3$$

and denote by H the set of all vectors $\pm f(\alpha)$, where $\alpha \in [0, \frac{\pi}{2}]$. The set H is closed and consists of two closed curves each of which has $\pm e_3$ as angle

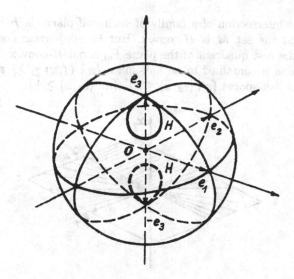

Fig. 92

points. The tangent rays of the curve H at these points are parallel to the x_1- and x_2-axes (Fig. 92). Denote by M the set of all vectors $x \in \mathbf{R}^3$ which satisfy the conditions $\| x - e_1 \| \le 1, \| x - e_2 \| \le 1, \langle x, e_3 \rangle = 0$ (Fig. 93). In

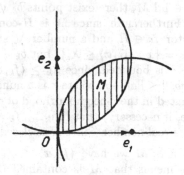

Fig. 93

other words, M is the intersection of two circles with radius 1 and the centres e_1, e_2 situated in the plane $P_o = \{x : \langle x, e_3 \rangle = 0\}$. The plane $P_o = \mathrm{aff}\ M$ is H-convex, since $\pm e_3 \in H$. Furthermore, if $0 < \alpha < \frac{\pi}{2}$, then the plane $\{x : \langle x, f(\alpha) \rangle = \mathrm{const}\}$ is H-convex, and its intersection with P_o is a line that is parallel to the vector product $[e_3, f(\alpha)] = \gamma(e_1 \sin \alpha + e_2 \cos \alpha)$, where $\gamma = \sin \alpha \cos \alpha \ne 0$. It follows that a half-plane $Q \subset P_o$ is H-convex if and only if its boundary line is parallel to a vector from the first quadrant.

Certainly, the intersection of a family of such half-planes is H-convex, too. It follows that the set M is H-convex. But its supporting cone with the apex o, i.e., the first quadrant of the plane P_o, is not H-convex. Indeed, this supporting cone is contained in no half-space $\{x : \langle f, x \rangle \leq \lambda\}$ with $f \in H$, except for the half-spaces $\{x : \langle e_3, x \rangle \leq \lambda\}$, $\{x : \langle e_3, x \rangle \geq \lambda\}$.

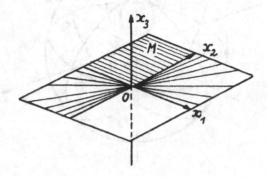

Fig. 94

Theorem 21.5: Suppose that $H \subset S^{n-1}$ is a closed set. Then for any H-convex set $M \subset R^n$ and any point $a \in \mathrm{bd}\, M$ there exists an H-convex half-space $P \supset M$ such that its boundary hyperplane passes through a.   □

PROOF: Since $a \in \mathrm{bd}\, M$, there exist points $b_k \notin M, k = 1, 2, \cdots$, such that $a = \lim_{k \to \infty} b_k$. Furthermore, since M is H-convex, for every $k = 1, 2, \cdots$ there exist a vector $f_k \in H$ and a number λ_k such that M is contained in the half-space $P_k = \{x : \langle f_k, x \rangle \leq \lambda_k\}$, but $b_k \notin P_k$, i.e., $\langle f_k, b_k \rangle > \lambda_k$. The sequence $\lambda_1, \lambda_2, \cdots$ is bounded, since $\lambda_k \geq \langle f_k, a \rangle \geq -||a||$, and, moreover, $\lambda_k < \langle f_k, b_k \rangle \leq ||b_k|| \leq ||a|| + r$, where r is a number such that all the points $b_1, b_2, \cdots$ are situated in the r-neighbourhood of a. We may suppose (passing to a subsequence, if necessary) that $\lim_{k \to \infty} f_k = f$, $\lim_{k \to \infty} \lambda_k = \lambda$, where $||f|| = 1$ and λ is a real number. Since H is closed, the inclusion $f \in H$ holds.

For every point $x \in M$ we have $\langle f_k, x \rangle \leq \lambda_k$, $k = 1, 2, \cdots$, and hence $\langle f, x \rangle \leq \lambda$. This means that M is contained in the half-space $P = \{x : \langle f, x \rangle \leq \lambda\}$. This half-space is H-convex, since $f \in H$. Furthermore, for every $k = 1, 2, \cdots$ the inequality $\langle f_k, b_k \rangle > \lambda_k$ holds and hence, passing to the limit, we obtain the relation $\langle f, a \rangle \geq \lambda$. On the other hand, $\langle f, a \rangle \leq \lambda$, since $a \in M$. Consequently $\langle f, a \rangle = \lambda$, i.e., $a \in \mathrm{bd}\, P$.   ∎

Theorem 21.6. Suppose that the set $H \subset S^{n-1}$ is closed and symmetric with respect to the origin. Then for any H-convex set $M \subset R^n$ and any point $a \in \mathrm{bd}\, M$ there exists (at least) one H-convex supporting hyperplane of M passing through a. In particular, if M is a body and a is its regular boundary point, then the supporting hyperplane of M passing through a is H-convex.

PROOF: According to Theorem 21.5, there is an H-convex half-space P containing M such that $a \in$ bd P. The unit outward normal e of the half-space P belongs to H. Consequently, $-e \in H$. This means that the second closed half-space P' with the boundary bd P' = bd P is H-convex, too. Hence the hyperplane $\Gamma = P \cap P'$ is H-convex as well, i.e., Γ is an H-convex supporting hyperplane of M passing through the point a.   ∎

Corollary 21.7: Let $H \subset S^{n-1}$ be a closed set and $K \subset \mathbf{R}^n$ be an H-convex, solid cone. A vector $y \in \mathbf{R}^n$ belongs to the polar cone K^* if and only if y is a linear combination of some vectors $e_1, \cdots, e_n \in K^* \cap H$ with non-negative coefficients.   □

Indeed, $H(K) \subset$ cl H (Theorem 19.3). Therefore cl $H(K) \subset$ cl $H = H$. Moreover, $H(K) \subset K^*$, i.e, cl $H(K) \subset K^*$. Hence cl $H(K) \subset K^* \cap H$. If now $y \in K^*$, then, according to Theorem 7.3, $y = \lambda_1 e_1 + \cdots + \lambda_n e_n$, where all the coefficients are nonnegative and $e_1, \cdots, e_n \in$ cl $H(K)$. Consequently $e_1, \cdots, e_n \in K^* \cap H$.   ∎

EXAMPLE 21.8: We show that if the cone K is not a body (i.e., dim $K < n$), then the assertion of Corollary 21.7 is, in general, false. Let e_1, e_2, e_3 be an orthonormal basis in $\mathbf{R}^3$. Denote by H the set that contains $e_1, -e_3$ and all the vectors

$$f(\gamma) = -\frac{\sqrt{3}}{4}(1 - \cos\gamma)e_1 + \left(\frac{1}{2}\sin\gamma\right)e_2 + \frac{1}{4}(3 + \cos\gamma)e_3, 0 \leq \gamma \leq 2\pi.$$

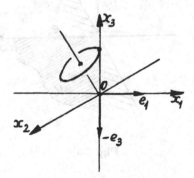

Fig. 95

Then H is the union of the vectors $e_1, -e_3$ and the circumference that consists of all the vectors $f(\gamma)$ (Fig. 95). The set H is closed and not one-sided. The two-dimensional plane $L = \{x : \langle x, e_3 \rangle = 0\}$ is H-convex, since $-e_3 \in H$ and $e_3 = f(0) \in H$. Furthermore, the H-convex half-space

$$\Pi(\gamma) = \{x : \langle x, f(\gamma) \rangle \leq 0\}$$

is described by the inequality

$$-\frac{\sqrt{3}}{4}(1-\cos\gamma)x_1 + \left(\frac{1}{2}\sin\gamma\right)x_2 + \frac{1}{4}(3+\cos\gamma)x_3 \le 0.$$

Consequently for $0 < \gamma < \pi$ the intersection $L \cap \Pi(\gamma)$ is an *H*-convex set that consists of the points $(x_1, x_2, 0)$ satisfying the condition

$$-\frac{\sqrt{3}}{4}(1-\cos\gamma)x_1 + \left(\frac{1}{2}\sin\gamma\right)x_2 \le 0, \quad \text{i.e.,} \quad x_2 \le \left(\frac{\sqrt{3}}{2}\tan\frac{\gamma}{2}\right)x_1$$

(Fig. 96). The *H*-convex cone

$$K = L \cap \left(\bigcap_{0 < \gamma < \pi} \Pi(\gamma) \right)$$

coincides with the angle described by $x_1 \ge 0, x_2 \le 0, x_3 = 0$. The vector $y = e_2$ belongs to K^*. But it is impossible to represent y as a nonnegative linear combination of vectors from $K^* \cap H$.

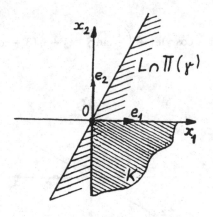

Fig. 96

Indeed, K^* is defined by the inequalities $x_1 \le 0$, $x_2 \ge 0$. Hence K^* is contained in the half-space $P = \{x : \langle x, e_1 \rangle \le 0\}$ and all the vectors from $K^* \cap H$ except for $\pm e_3$ belong to the *open* half-space int P. It follows that the vector $y = e_3 \in$ bd P is not representable in the form $y = \lambda_1 e_1 + \cdots + \lambda_s e_s$, where $e_1, \cdots, e_s \in K^* \cap H$ and the coefficients are non-negative.

<u>*Theorem 21.9:*</u> *Let $H \subset S^{n-1}$ be a closed set and $K_1, \cdots, K_s$ be H-convex, solid cones with a common apex. If the cones $K_1, \cdots, K_s$ are separable and each $s - 1$ of them are not separable, then $s \le$ md $H + 1$.*   □

PROOF: According to Theorem 8.4, there exist vectors $p_1 \in K_1^*, \cdots, p_s \in K_s^*$ not all equal to o such that

$$p_1 + \cdots + p_s = o. \tag{1}$$

Since each $s - 1$ of the cones $K_1, \cdots, K_s$ are not separable, every vector p_i is not equal to o (again by Theorem 8.4). Furthermore, according to Corollary 21.7

$$p_i = \sum_j \lambda_j^{(i)} e_j^{(i)}$$

(finite sum), where all the vectors $e_j^{(i)} \neq 0$ are contained in $K_i^* \cap H$ and the coefficients are positive. Now (1) may be rewritten in the form

$$\sum_i \sum_j \lambda_j^{(i)} e_j^{(i)} = o.$$

This equality shows that the vectors $e_j^{(i)}$ are positively dependent. We choose from $e_j^{(i)}$ a minimally dependent system of vectors (this is possible by virtue of Theorem 16.5). Hence we obtain a minimal dependence

$$\sum_i^{\triangle} \sum_j^{\triangle} \mu_j^{(i)} e_j^{(i)} = o \tag{2}$$

where the numbers $\mu_j^{(i)}$ are positive and the symbol $\triangle$ above shows the summation over the chosen vectors. Every chosen vector $e_j^{(i)}$ belongs to $K_i^* \cap H$. Therefore the vector $f_i = \sum_j^{\triangle} \mu_j^{(i)} e_j^{(i)}$ belongs to K_i^*. Since the cone K_i is solid, the polar cone K_i^* is pointed (Theorem 6.6). Consequently there exists a supporting half-space P_i of K_i^* such that the supporting hyperplane bd P_i has only the point o in common with K_i^*, i.e., the whole cone K_i^*, except the point o, is contained in int P_i. This means that all the vectors $e_j^{(i)}$ belong to int P_i. Consequently, if i is an index such that there is at least one vector $e_j^{(i)}$ on the left-hand side of (2), then $f_i \neq o$. The equality (2) becomes $\sum_i^{\triangle} f_i = o$. If the number of summands were less than s, then the corresponding cones would be separated (Theorem 8.4), contradicting the situation. Therefore in the equality $\sum_i^{\triangle} f_i = o$ the number of summands is equal to s, and hence the number of summands in (2) (denote this number by q) is *not less* than s. Since $e_j^{(i)} \in K_i^* \cap H \subset H$ and (2) is a minimal dependence, we conclude that there are $q \geq s$ minimally dependent vectors in H. Thus, md $H \geq s-1$. ∎

Theorem 21.10: Suppose that the set $H \subset S^{n-1}$ is closed and symmetric with respect to the origin. Then every maximal face of an H-convex body is H-convex, too. □

PROOF: Let M be an H-convex body and N be a maximal face of M. For an arbitrary point $a \in$ ri N we consider an H-convex supporting hyperplane Γ

of M through the point a, cf. Theorem 21.6. The intersection $\Gamma \cap M$ coincides with N (since the face N is maximal). Since both the sets Γ, M are H-convex, we conclude that N is H-convex. ∎

Exercises

1. Show by a counterexample that the following assertion is false: if $K \subset \mathbf{R}^n$ is a closed, convex cone such that all its proper faces are H-convex, then K is H-convex. Is there a counterexample with a solid cone K?

2. Show by a counterexample that the following assertion is false: if H is closed and M is an H-convex set, then the flat aff M is H-convex.

3. Show by an example that the condition "H is closed" in Theorem 21.5 is essential (even if M is an H-convex body).

4. Show by an example that the condition "H is centrally symmetric with respect to the origin" in Theorem 21.6 is essential.

5. Decide whether the following assertion is true: Let $H \subset S^{n-1}$ be closed and not one-sided. If for any boundary point a of a closed, convex set M there is an H-convex supporting half-space of M at the point a, then M is H-convex. What happens if M is a closed, convex body?

6. Let the set $H \subset S^{n-1}$ be closed, non-one-sided and symmetric with respect to the origin. Prove that if M is an H-convex body, then each of its maximal faces is H-convex, too.

7. Let H be as in Example 21.4 and

$$M = \{x : \langle e_1, x \rangle \le 0, \langle e_2, x \rangle \ge 0, \langle e_3, x \rangle = 0\},$$

see Fig. 94. Prove that M is H-convex, but its maximal face

$$N = \{x : \langle e_2, x \rangle = 0, \langle e_3, x \rangle = 0, \langle e_1, x \rangle \le 0\}$$

is not H-convex. This example shows that the conclusion of Exercise 6 is false if M is not a body.

8. Let the set $H \subset S^{n-1}$ be non-one-sided and symmetric with respect to the origin. Prove that if M is a *compact*, H-convex set (not necessarily a body), then each of its maximal faces is H-convex. Does it hold for non-maximal faces?

9. Is the condition "H is closed" in Corollary 21.7 essential?

10. Let $H \subset S^{n-1}$ be a set with md $H = m$. Construct $m+1$ separable, H-convex cones $K_1, \cdots, K_{m+1}$ with a common apex such that every m of them are not separable.

11. Prove that md H is the maximal of the integers s such that there exist $m+1$ separable, H-convex cones with a common apex every m of which are not separable.

12. We say that convex cones $K_1, \cdots, K_s \subset \mathbf{R}^n$ are *intersectionally free* if there are translates of the cones with empty intersection. Prove that convex cones with common apex are intersectionally free if and only if they are separable.

§22 The Helly Theorem for H-convex sets

Theorem 22.1: Let $H \subset S^{n-1}$ be a set that is not one-sided. The Helly dimension of the family C_H of the H-convex sets is equal to md H.   □

PROOF: We put md $H = m$ and choose minimally dependent vectors $n_1, \cdots, n_{m+1} \in H$. Furthermore, we put (Fig. 97)

$$P_i = \{x : \langle n_i, x \rangle \leq -1\}, i = 1, \cdots, m+1. \tag{1}$$

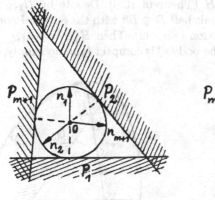

| Fig. 97 | Fig. 98 |

First of all, we prove that $P_1 \cap \cdots \cap P_{m+1} = \emptyset$. Indeed, let $\lambda_1, \cdots, \lambda_{m+1}$ be positive numbers such that

$$\lambda_1 n_1 + \cdots + \lambda_{m+1} n_{m+1} = o. \tag{2}$$

Up to a common positive multiplier, these numbers are uniquely defined. Assume that there is a point $x \in P_1 \cap \cdots \cap P_{m+1}$, i.e., $\langle n_i, x \rangle \leq -1$ for all $i = 1, \cdots, m+1$. Then

$$\langle \lambda_1 n_1 + \cdots + \lambda_{m+1} n_{m+1}, x \rangle = \lambda_1 \langle n_1, x \rangle + \cdots + \lambda_{m+1} \langle n_{m+1}, x \rangle$$
$$\leq -\lambda_1 - \cdots - \lambda_{m+1} < 0,$$

contradicting (2). Thus, $P_1 \cap \cdots \cap P_{m+1} = \emptyset$.

Furthermore, each m of the half-spaces $P_1, \cdots, P_{m+1}$ have a nonempty intersection. Indeed, the equation $\langle n_i, x \rangle = -1$ determines a hyperplane $\Gamma_i = \mathrm{bd} P_i \subset R^n$ with normal n_i (Fig. 98). Every m of the vectors $n_1, \cdots, n_{m+1}$ are linearly independent (since $n_1, \cdots, n_{m+1}$ are minimally dependent; cf. Theorem 16.1). Therefore the intersection of all hyperplanes $\Gamma_1, \cdots, \Gamma_{m+1}$, except for Γ_i, is nonempty.

Thus there are points $b_1, \cdots, b_{m+1}$ such that $b_i \in \Gamma_j \subset P_j$ for $i \neq j$, i.e., b_i belongs to all the half-spaces $P_1, \cdots, P_{m+1}$ except for P_i. This means that for every m of the H-convex half-spaces $P_1, \cdots, P_{m+1}$ their intersection is nonempty, while $P_1 \cap \cdots \cap P_{m+1} = \emptyset$, i.e., him $C_H \geq m$ (Theorem 14.7). In other words, him $C_H \geq \mathrm{md}\, H$.

We now prove the opposite inequality him $C_H \leq \mathrm{md}\, M$. Set him $C_H = k$. Then there are H-convex sets $G_1, \cdots, G_{k+1}$ such that every k of them have a nonempty intersection, while $G_1 \cap \cdots \cap G_{k+1} = \emptyset$. Let b_i be a point belonging to all $G_1, \cdots, G_{k+1}$, except for G_i, i.e., $b_i \in G_j$ for $j \neq i$. We remark that there exists a compact, H-convex body $Q \subset R^n$. Indeed, let $\{a_1, \cdots, a_p\}$ be a non-one-sided, finite subset of H (Theorem 16.8). Denote by $\Pi_1, \cdots, \Pi_p$ the supporting half-spaces of the unit ball $B \subset R^n$ with the outward normals $a_1, \cdots, a_p$ and by Q their intersection (Fig. 99). Then $H(Q) = \{a_1, \cdots, a_p\}$ and, according to Theorem 18.2, the body Q is compact (and nonempty, since $Q \supset B$).

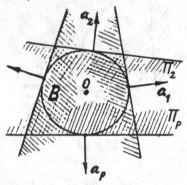

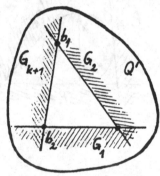

Fig. 99   Fig. 100

Applying a homothety whose ratio is large enough (with a suitable center), we can obtain an H-convex, compact body Q' that contains the constructed points $b_1, \cdots, b_{k+1}$ (Fig. 100). (We remark that if a set is H-convex, then any of its images under a homothety with positive ratio is H-convex, too.) We now put

$$F_i = Q' \cap G_i, \quad i = 1, \cdots, k+1,$$

(Fig. 101). Then the sets $F_1, \cdots, F_{k+1}$ are compact, H-convex bodies and $b_i \in F_j$ for $i \neq j$. It follows that each k of the bodies $F_1, \cdots, F_{k+1}$ have a nonempty intersection. At the same time, $F_1 \cap \cdots \cap F_{k+1} = \emptyset$, since $F_i \subset G_i$ for $i = 1, \cdots, k+1$.

Since the bodies $F_1, \cdots, F_{k+1}$ are compact and $F_1 \cap \cdots \cap F_{k+1} = \emptyset$, there exists a number $\varepsilon > 0$ such that $U_\varepsilon(F_1) \cap \cdots \cap U_\varepsilon(F_{k+1}) = \emptyset$. According to Theorem 20.4, there exists a number $\delta > 0$ such that

$\operatorname{conv}_H U_\delta(F_i) \subset U_\varepsilon(F_i)$, $i = 1, \cdots, k+1$.

Thus $Q_i = \operatorname{conv}_H U_\delta(F_i)$ is a compact, H-convex body for every $i = 1, \cdots k+1$, and $Q_1 \cap \cdots \cap Q_{k+1} = \emptyset$ (since $Q_i \subset U_\varepsilon(F_i)$). Moreover, $F_i \subset \operatorname{int} Q_i$ (since Q_i contains the body $U_\delta(F_i)$). In particular, $b_i \in \operatorname{int} Q_j$ for $i \neq j$.

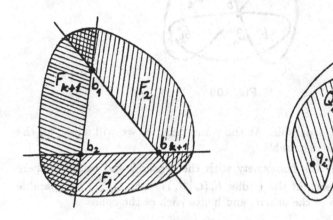

Fig. 101   Fig. 102

For every $i = 1, \cdots, k+1$, we fix an interior point q_i of the body Q_i. By $Q_i^{(\lambda)}$ denote the body obtained from Q_i by the homothety with the center q_i and ratio $1 + \lambda$ (Fig. 102). If λ is small enough, then the intersection $Q_1^{(\lambda)} \cap \cdots \cap Q_{k+1}^{(\lambda)}$ is empty. Hence there exists a number $\mu > 0$ such that this intersection is empty for $\lambda < \mu$, while $Q_1^{(\mu)} \cap \cdots \cap Q_{k+1}^{(\mu)} \neq \emptyset$, i.e., there is a point $b \in Q_1^{(\mu)} \cap \cdots \cap Q_{k+1}^{(\mu)}$ (Fig. 103). It is easily shown that b is a *boundary point* of the body $Q_i^{(\mu)}$ for each $i = 1, \cdots, k+1$. Indeed, assume that $b \in \operatorname{int} Q_j^{(\mu)}$ for an index j. Then there exists a point $c \in [b, b_j]$ distinct from b that belongs to $\operatorname{int} Q_j^{(\mu)}$. Moreover, the point c belongs to $\operatorname{int} Q_i^{(\mu)}$ for every $i \neq j$ (since $b \in Q_i^{(\mu)}$, $b_j \in \operatorname{int} Q_i \subset \operatorname{int} Q_i^{(\mu)}$). Thus $c \in \operatorname{int} Q_i^{(\mu)}$ for $i = 1, \cdots, k+1$, and therefore $Q_1^{(\lambda)} \cap \cdots \cap Q_{k+1}^{(\lambda)} \neq \emptyset$ with $\lambda = \mu - \varepsilon$ for $\varepsilon > 0$ small enough. But this contradicts the definition of μ. This shows that $b \in \operatorname{bd} Q_i^{(\mu)}$ for each $i = 1, \cdots, k+1$.

Moreover, every of the bodies $Q_1^{(\mu)}, \cdots, Q_{k+1}^{(\mu)}$ is separable from the intersection of the other ones (since $\operatorname{int} Q_1^{(\mu)} \cap \cdots \cap \operatorname{int} Q_{k+1}^{(\mu)} = \emptyset$).

Let $C_1, \cdots, C_{k+1}$ be the supporting cones of $Q_1^{(\mu)}, \cdots, Q_{k+1}^{(\mu)}$ at their common boundary point b. The cones $C_1, \cdots, C_{k+1}$ with common apex b are solid and each k of them have a nonempty intersection which is solid, too (since $b_i \in \operatorname{int} Q_j^{(\mu)} \subset \operatorname{int} C_j$ for $i \neq j$). This means that every k of the cones

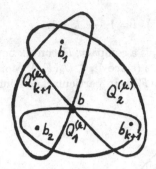

Fig. 103

$C_1, \cdots, C_{k+1}$ are not separable. At the same time, as we will show, all the cones $C_1, \cdots, C_{k+1}$ are separable.

Indeed, denote by h_r the homothety with the center b and ratio r (where $r = 1, 2, \cdots$). Then each of the bodies $h_r(Q_1^{(\mu)},) \cdots, h_r(Q_{k+1}^{(\mu)})$ is separable from the intersection of the others, and hence each of the cones

$$\bigcup_r h_r(Q_1^{(\mu)}), \cdots, \bigcup_r h_r(Q_{k+1}^{(\mu)}) \qquad (3)$$

with the common apex b is separable from the intersection of the others. In other words, the system (3) of convex cones is separable. Consequently their closures

$$\mathrm{cl}\left(\bigcup_r h_r(Q_1^{(\mu)})\right), \cdots, \mathrm{cl}\left(\bigcup_r h_r(Q_{k+1}^{(\mu)})\right)$$

are separable, as well (Corollary 8.5), i.e., the supporting cones $C_1, \cdots, C_{k+1}$ are separable.

Finally, for every index $i = 1, \cdots, k+1$ the body $Q_i^{(\mu)}$ is homothetic to Q_i with positive ratio, and hence $Q_i^{(\mu)}$ is H-convex. It follows that $Q_i^{(\mu)}$ is (cl H)-convex, and consequently the cone C_i is (cl H)-convex (Theorem 21.2). Applying Theorem 21.9, we conclude that $k+1 \le \mathrm{md\ cl}\ H + 1$, i.e., him $C_H \le \mathrm{md\ cl}\ H$. At last, $\mathrm{md\ cl}\ H = \mathrm{md}\ H$ (Theorem 17.3), i.e., him $C_H \le \mathrm{md}\ H$. ∎

Corollary 22.2: Let $H \subset S^{n-1}$ be a non-one-sided set. The Helly dimension of the family $C_H^{(c)}$ of all compact, H-convex sets is equal to $\mathrm{md}\ H$. □

PROOF: Since $C_H^{(c)} \subset C_H$, we conclude that him $C_H^{(c)} \le$ him $C_H = \mathrm{md}\ H$. Thus it is sufficient to prove that him $C_H^{(c)} \ge$ him $C_H = \mathrm{md}\ H = m$. Let P_i, b_i be as at the beginning of the previous proof (i.e., $b_i \in P_j$ for $i \ne j$), and Q' be a compact, H-convex body containing all the points $b_1, \cdots, b_{m+1}$. Then the sets $F_1 = P_1 \cap Q', \cdots, F_{m+1} = P_{m+1} \cap Q'$ are compact, H-convex sets with empty intersection such that every m of them have a common point. Consequently him $C_H^{(c)} \geq m = \mathrm{md}\, H$ (Theorem 22.1). ∎

EXAMPLE 22.3 In Theorem 22.1, the requirement that H is not one-sided (i.e., H is not contained in a closed hemisphere of S^{n-1}) is essential. Indeed, let e_1, e_2 be an orthonormal basis in $\mathbf{R}^2$. We put

$$H = \{x \in S^1 : \langle x, e_2 \rangle < 0\} \cup \{e_1\}.$$

The set H is one-sided and $\mathrm{md}\, H = 0$. At the same time him $C_H = 1$. In fact, a closed, convex set $M \subset \mathbf{R}^2$ distinct from $\mathbf{R}^2$ is H-convex if and only if either M is a left half-plane (with the boundary parallel to e_2), or M satisfies the following two conditions: $i) M$ does not contain a line parallel to e_2; $ii) M$ contains a ray whose direction is defined by e_2 (Fig. 104). Consequently, if every two of the H-convex sets $M_1, \cdots, M_s$ have a common point, then $M_1 \cap \cdots \cap M_s \neq \emptyset$ (since, denoting by l the line $x_2 = c$ where $c > 0$ is large enough, every two of the one-dimensional convex sets $M_1 \cap l, \cdots, M_s \cap l$ have a common point). This means that him $C_H \leq 1$. But the equality him $C_H = 0$ does not hold, since there are two disjoint H-convex sets. Thus him $C_H = 1 \neq \mathrm{md}\, H$. □

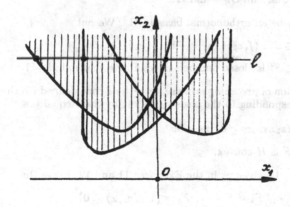

Fig. 104

Exercises

1. Prove a Helly-type theorem for an infinite family of H-convex sets in the following form: let A be a set of indices and $M_\alpha, \alpha \in A$, be H-convex sets, at least one of which is compact. If every $\mathrm{md}\, H + 1$ of the sets M_α have a common point, then $\bigcap_\alpha M_\alpha \neq \emptyset$.

2. Let $H \subset S^{n-1}$ be a *closed* set (maybe, one-sided). Prove that him $C_H = $ md H.

3. Let $M \subset \mathbf{R}^n$ be a convex body, $m = $ md M, and $M_1, \cdots, M_s$, $s \geq m + 2$, be translates of M. Prove that if every $m + 1$ of the bodies $M_1, \cdots, M_s$ have a common point, then $M_1 \cap \cdots \cap M_s \neq \emptyset$. (In the next chapter we will establish an assertion which, in a sense, is converse.)

4. Let $H \subset S^{n-1}$ be a set which is non-one-sided and symmetric with respect to the origin. Denote by L the family of all H-convex flats. Prove that him $L = $ md H.

5. Deduce the classical Helly Theorem from Theorem 22.1.

6. Prove that for the family of all H-convex cones the Helly dimension is equal to md H (where $H \subset S^{n-1}$ can be one-sided).

7. Show that the following position of three translates M_1, M_2, M_3 of a regular cross-polytope $M \subset \mathbf{R}^n$ is possible: the cross-polytopes M_1, M_2, M_3 have pairwise common points, but $M_1 \cap M_2 \cap M_3 = \emptyset$. Thus if F denotes the family of all translastes of M, then him $F \neq 1$. Can you evaluate him F?

8. Let $M \subset \mathbf{R}^3$ be a regular prism with n-sided basis and $M_1, \cdots, M_s$ be its translates, $s \geq 4$. Prove that if every three of the bodies $M_1, \cdots, M_s$ have a common point, then $M_1 \cap \cdots \cap M_s \neq \emptyset$.

9. Let $L \subset \mathbf{R}^n$ be a subspace and $H \subset S^{n-1} \cap L$ be a set which is not one-sided in L. Prove that him $C_H = $ md H.

10. Let e_1, e_2, e_3 be an orthonormal basis in $\mathbf{R}^3$. We put

$$H = \{f \in S^2 : \langle f, e_2 + e_3 \rangle = -1\} \cup \{e_2\}.$$

The set $H \subset S^2$ is closed and one-sided. Prove that md $H = 1$.

11. In the notation of preceding exercise, let F be a cone defined (in the coordinate system corresponding to the basis e_1, e_2, e_3) by the inequalities

$$(x_1)^2 \leq 2x_2 x_3, \quad x_2 \geq 0, \quad x_3 \geq 0.$$

Prove that F is H-convex.

12. Let H and F be given as in the Exercises 11 and 12. Prove that

$$F_1 = F + e_1, \quad F_2 = F - e_1, \quad F_3 = \{x : \langle e_2, x \rangle \leq 0\}$$

are H-convex sets such that every two of them have a common point and $F_1 \cap F_2 \cap F_3 = \emptyset$. Thus him $C_H \geq 2$ (in fact, him $C_H = 2$), i.e., him $C_H \neq $ md H. This example shows that the requirement "H is not one-sided" in Theorem 22.1 is essential.

13. Let $L \subset \mathbf{R}^3$ be a two-dimensional plane. Construct a finite set $H \subset S^2$ which is not one-sided and has the following properties:

   (i) For the family C_H of all H-convex sets in $\mathbf{R}^3$ the relation him $C_H = 3$ holds.

   (ii) For the family $C_H^{(L)}$ of H-convex sets contained in L the relation him $C_H^{(L)} = 1$ holds.

14. Let r, k be integers with $0 \le r < k < n$. Prove that there exists a one-sided set $H \subset S^{n-1}$ such that md $H = r$ and him $C_H = k$, where C_H denotes the family of the H-convex sets in $\mathbf{R}^n$.

15. Prove that him $C_H \ge$ md H for any set $H \subset S^{n-1}$ (where H may be even one-sided and nonclosed).

16. Prove that for every integer r satisfying $1 \le r \le n$ there exists a compact, convex body $M \subset \mathbf{R}^n$ such that the family $T(M)$ of all its translates satisfies him $T(M) = r$.

17. Let $K_1, \cdots, K_s$, $s > n+1$, be convex cones in $\mathbf{R}^n$ with a common apex. Prove that if there exist vectors $p_i \in K_i^*$, $i = 1, \cdots, s$, not all equal to o such that $p_1 + \cdots + p_s = o$, then some $n+1$ of the cones have a common interior point.

18. Let $K_1, \cdots, K_s$, $s >$ md $H+1$, be H-convex cones in $\mathbf{R}^n$ with a common apex. Prove that if there exist vectors $p_i \in K_i^*$, $i = 1, \cdots, s$ not all equal to o such that $p_1 + \cdots + p_s = o$, then some md $H+1$ of the cones have a common interior point.

§23 Some applications of H-convexity

In this section we indicate several applications of the Helly-type theorem for H-convex sets. The first theorem is carefully proved, the other ones are only formulated, since they are direct generalizations of some classical theorems of Combinatorial Geometry (obtained by considering H-convexity instead of the usual one). In the exercises, the proofs are shortly sketched.

Let $M \subset \mathbf{R}^n$ be a compact, convex body. By cov M we denote the smallest of the positive integers m with the following property: Let $X \subset \mathbf{R}^n$ be an arbitrary set; if every $m+1$ points of X are contained in a translate of M, then the whole set X is contained in a translate of M.

To clarify the sense of the number cov M, we consider simple examples.

EXAMPLE 23.1: Let $M \subset \mathbf{R}^2$ be a disk. It is well known that if every three points of a set $X \subset \mathbf{R}^2$ can be covered by a translate of M, then the whole set X can be covered by a translate of M. This means that cov $M = 2$. Analogously, for any ball $M \subset \mathbf{R}^n$ the equality cov $M = n$ holds. □

EXAMPLE 23.2: Let $M \subset \mathbf{R}^2$ be the square defined (in an orthogonal cartesian coordinate system x_1, x_2) by $0 \le x_1 \le 1, 0 \le x_2 \le 1$. Two points $a = (a_1, a_2)$, $b = (b_1, b_2)$ in $\mathbf{R}^2$ can be covered simultaneously by a translate of M if and only if $|a_1 - b_1| \le 1$, $|a_2 - b_2| \le 1$. It follows that if every *two* points of a set $X \subset \mathbf{R}^2$ can be covered by a translate of M, then the whole set X is contained in a translate of M. In other words, cov $M = 1$. Analogously, cov $M = 1$ for every n-dimensional parallelotope $M \subset \mathbf{R}^n$. □

We remark that for the bodies considered in Examples 23.1 and 23.2, the equality cov $M =$ md M holds. The following theorem (cf. [B-Ch 2]) affirms that this is true in general.

Theorem 23.3: *For any compact, convex body $M \subset \mathbf{R}^n$, the equality* cov $M =$ md M *holds. In other words, if every* md $M+1$ *points of a set $X \subset \mathbf{R}^n$ are contained in a translate of M, then the whole set X is contained in a translate of M (and the number* md $M+1$ *cannot be diminished).*

PROOF: First we establish the inequality cov $M \le$ md M. Denote the number md M by m. For an arbitrary point $x \in \mathbf{R}^n$, we denote by N_x the set of all vectors $v \in \mathbf{R}^n$ such that $x \in (v + M)$. In other words, $N_x = x - M$ (Fig. 105). This means that N_x is a translate of the body $N_o = -M$. We put $H = H(N_o)$. Then the body N_o is H-convex (Theorem 19.2) and each body N_x (i.e., a translate of N_o) is also H-convex. Now we notice that md $N_o =$ md $M = m$, since N_o is symmetric to M with respect to the origin. Consequently md $H =$ md $H(N_o) =$ md $N_o = m$. Let now $X \subset \mathbf{R}^n$ be a set. The set X is contained in a translate of M if and only if there exists a vector v which belongs to N_x for any $x \in X$ or, what is the same, $\bigcap_{x \in X} N_x \ne \emptyset$.

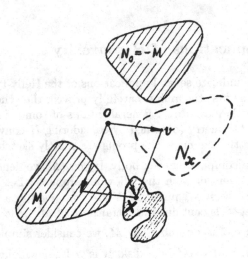

Fig. 105

Assume that every $m+1$ points of X are contained in a translate of M, i.e., every $m+1$ of the sets N_x, where $x \in X$, have a nonempty intersection. It follows from Theorem 22.1 that every *finite* collection of the bodies N_x, $x \in X$, has a nonempty intersection (since each N_x belongs to C_H, and him $C_H =$ md $H = m$). Moreover, by virtue of the compactness of each body N_x, the *whole* family of the bodies N_x, $x \in X$, has a nonempty intersection, i.e., $\bigcap_{x \in X} N_x \ne \emptyset$. This means that X is contained in a translate of M. Thus the number md $M = m$ possesses the following property: if every $m+1$ points of a set $X \subset \mathbf{R}^n$ are contained in a translate of M, then the whole set X is contained in a translate of M. Consequently cov $M \le m$, i.e., cov $M \le$ md M.

For establishing the converse inequality cov $M \geq$ md $M = m$, let $a_1, \cdots, a_{m+1}$ be regular boundary points of M such that the outward unit normals $p_1, \cdots, p_{m+1}$ of M at the points are minimally dependent. This means that every m of the vectors $p_1, \cdots, p_{m+1}$ are linearly independent and, moreover, there exist numbers $\lambda_1, \cdots, \lambda_{m+1}$ such that

$$\lambda_1 p_1 + \cdots + \lambda_{m+1} p_{m+1} = o; \ \lambda_1 > 0, \cdots, \lambda_{m+1} > 0. \tag{1}$$

First of all, we prove that every m of the points $a_1, \cdots, a_{m+1}$ are contained in a translate of the open set int M. For simplicity, we concentrate upon the points $a_1, \cdots, a_m$ (other cases are analogous). Since the vectors $p_1, \cdots, p_m$ are linearly independent, there exists a vector q such that

$$\langle q, p_1 \rangle < 0, \cdots, \langle q, p_m \rangle < 0$$

(Fig. 106). This means that $a_i + \varepsilon q \in$ int M $(i = 1, \cdots, m)$ if $\varepsilon > 0$ is small enough (since a_i is a *regular* boundary point). In other words, $a_i \in -\varepsilon q +$ int M for all $i = 1, \cdots, m$, i.e., the points $a_1, \cdots, a_m$ are contained in a translate of int M. Similarly, every m of the points $a_1, \cdots, a_{m+1}$ are contained in a translate of int M.

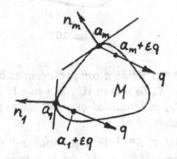

Fig. 106

Denote by h_k the homothety with center at a fixed point $c \in$ int M and ratio $k > 1$. Every m of the points $h_k(a_1), \cdots, h_k(a_{m+1})$ are contained in a translate of int M if the ratio $k > 1$ is close enough to 1. We fix such a ratio k and denote the set $\{h_k(a_1), \cdots, h_k(a_{m+1})\}$ by X. Thus every m points of the set X are contained in a translate of int M and hence in a translate of M.

To establish the inequality cov $M \geq$ md M, it is sufficient to prove that the set X is contained in no translate of M. Indeed, assume that $X \subset v + M$ for a vector $v \in R^n$, i.e.,

$$h_k(a_i) = v + x_i, \ x_i \in M$$

for each $i = 1, \cdots, m + 1$. Since p_i is the outward normal of the body $h_k(M)$ at its regular boundary point $h_k(a_i)$ and $x_i \in M \subset \operatorname{int} h_k(M)$ (Fig. 107), the inequality

$$\langle -v, p_i \rangle = \langle x_i - h_k(a_i), p_i \rangle < 0$$

is true, $i = 1, \cdots, m + 1$. Consequently,

$$\langle -v, \lambda_1 p_1 + \cdots + \lambda_{m+1} p_{m+1} \rangle < 0,$$

contradicting (1). ∎

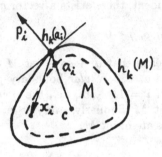

Fig. 107

Theorem 23.4: Let $M \subset R^n$ be a compact, convex body with md $M = m$. Let, furthermore, $X_1, \cdots, X_s$ be sets in $\mathbf{R}^n$, $s > m+1$. If the union of every $m+1$ of the sets $X_1, \cdots, X_s$ is contained in a translate of M, then $X_1 \cup \cdots \cup X_s$ is contained in a translate of M. □

Theorem 23.5: Let $H \subset S^{n-1}$ be a set with md $H = m$ which is not one-sided. Let, furthermore, $M_1, \cdots, M_s$ be H-convex sets, $s > m + 1$. If a set $X \subset R^n$ is contained in a translate of the intersection of each $m + 1$ of the sets $M_1, \cdots, M_s$, then X is contained in a translate of $M_1 \cap \cdots \cap M_s$. □

Theorem 23.6: Let $M \subset R^n$ be a compact, convex body with md $M = m$. Then there exists a point $q \in \operatorname{int} M$ such that each chord $[a, b]$ of M passing through q (Fig. 108) satisfies the conditions

$$|a - q| \geq \frac{1}{m + 1} |a - b|, \quad |b - q| \geq \frac{1}{m + 1} |a - b|. \quad \square$$

Theorem 23.7: Each compact, convex body $M \subset R^n$ contains a ball of radius $\frac{1}{m+1} \triangle$, where $m = $ md M and $\triangle$ is the minimal width of the body M. □

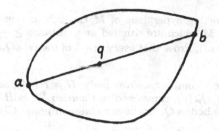

Fig. 108

Exercises

1. Let $M \subset \mathbf{R}^3$ be a pyramid whose basis is a parallelogram. Prove that cov $M = 2$. This means that if every three points of a set $X \subset \mathbf{R}^3$ are contained in a translate of M, then the whole set X is contained in a translate of M.

2. Let $M \subset \mathbf{R}^3$ be a pyramid whose basis is not a parallelogram. Prove that cov $M = 3$.

3. Let $M \subset \mathbf{R}^n$ be a compact, convex body and md $M = m$. Prove that there exists an m-dimensional simplex $T \subset \mathbf{R}^n$ such that each of its proper faces is contained in a translate of M, whereas T is not contained in any translate of M.

4. Let $M \subset \mathbf{R}^n$ be a compact, convex body. Prove that cov $M = n$ if and only if there exist regular boundary points $a_1, \cdots, a_{n+1}$ of M such that the intersection of the supporting half-spaces of M at these points is an n-dimensional simplex.

5. Let $M \subset \mathbf{R}^n$ be a compact, convex body, $H = H(M)$, and $X \subset \mathbf{R}^n$ be a set. Denote by N_X the set of all $v \in \mathbf{R}^n$ such that $X \subset v + M$. Prove that the set N_X is $(-H)$-convex (where $-H$ is the set symmetric to H with respect to the origin).

6. Using the result of the previous exercise, give a proof of Theorem 23.4.

7. Let $M \subset \mathbf{R}^n$ be a compact, convex body. For every point $a \in M$, denote by h_a the homothety with center a and ratio $\frac{m}{m+1}$, where m is a positive integer. Prove that for every points $a_1, \cdots, a_{m+1} \in M$ the point $\frac{1}{m+1}(a_1 + \cdots + a_{m+1})$ belongs to each of the bodies $h_{a_1}(M), \cdots, h_{a_{m+1}}(M)$.

8. Let $M \subset \mathbf{R}^n$ be a compact, convex body, $H = H(M)$, and $m = \mathrm{md}\ M$. Prove that each body $h_a(M)$ (obtained from M by the homothety with center a and ratio $\frac{m}{m+1}$) is H-convex. Prove that each $m + 1$ of the bodies $h_a(M)$, $a \in M$, have a common point. Deduce from this that *all* the bodies $h_a(M)$, $a \in M$, have a common point. Give a proof of Theorem 23.6.

9. Prove by an example that the number $\frac{1}{m+1}$ in Theorem 23.6 cannot be increased.

10. Let $M \subset \mathbf{R}^n$ be a compact, convex body and $\triangle$ denote the minimal width (thickness) of M. Furthermore, let a be a regular boundary point of M, and

Γ be the supporting hyperplane of M through a. Denote by $Q_m(a)$ the set of all points of M which are situated at a distance $\geq \frac{1}{m+1}\Delta$ from Γ (with a positive integer m). Prove that every $m+1$ of the sets $Q_m(a)$ have a nonempty intersection.

11. Let $M \subset \mathbf{R}^n$ be a compact, convex body, $H = H(M)$, and $m = \text{md } M$. Prove that each body $Q_m(a)$, considered in Exercise 10, is H-convex. Deduce from this that *all* the bodies $Q_m(a)$ have a common point. Give a proof of Theorem 23.7.

12. Prove by an example that the number $\frac{1}{m+1}$ in Theorem 23.7 cannot be replaced by a larger one.

13. Let $M_1, \cdots, M_s$ be convex, half-bounded sets in $\mathbf{R}^n$, $s > n+1$. Prove that if the convex hull of the union of every $n+1$ of these sets is half-bounded, then the set conv $(M_1 \cup \cdots \cup M_s)$ is half-bounded, too (cf. [B-SP 4]).

14. Let $M_1, \cdots, M_s$ be H-convex, half-bounded sets in $\mathbf{R}^n$, $s > \text{md } H + 1$. Prove that if the convex hull of the union of every md $H + 1$ of these sets is half-bounded, then the set conv $(M_1 \cup \cdots \cup M_s)$ is half-bounded, too (cf. [B-SP 4]).

15. Let $H \subset S^{n-1}$ be a set which is not one-sided and $a_1, \cdots, a_s \in \mathbf{R}^n$. Prove that there exists a point $q \in \mathbf{R}^n$ such that for every H-convex half-space P with $q \in P$ the number of the points $a_1, \cdots, a_s$ contained in P is not smaller than
$$\frac{s}{\text{md } H+1}.$$

16. Let $H \subset S^{n-1}$ be a set which is not one-sided, and $Q \subset \mathbf{R}^n$ be a compact set whose k-dimensional volume is equal to V. Prove that there exists a point $q \in \mathbf{R}^n$ such that for every H-convex half-space P containing q the k-dimensional volume of the set $P \cap Q$ is not smaller than $\frac{V}{\text{md } H+1}$.

17. Let $H \subset S^{n-1}$ be a set which is non-one-sided and symmetric with respect to the origin. Let, furthermore, $M \subset \mathbf{R}^n$ be an H-convex set and $P_1, \cdots, P_s$ be H-convex half-spaces such that their interiors cover M. Prove that it is possible to choose among $P_1, \cdots, P_s$ no more than md $H+1$ half-spaces whose interiors cover M.

§24 Some remarks on connection between *d*-convexity and *H*-convexity

We now compare properties of d-convex sets and H-convex ones. Since each H-convex set is closed, our considerations are restricted to *closed* d-convex sets.

Let $\Re^n$ be a normed space in which (besides the norm) a scalar product is introduced. Thus $\Re^n$ can be considered as a Euclidean space with normal vectors of hyperplanes. Let $H(\Re^n)$ denote the set of all unit vectors (in the sense of the Euclidean metric) each of which is orthogonal to a d-convex hyperplane of $\Re^n$. With the help of the set $H = H(\Re^n)$, the H-convex half-spaces and the H-convex sets in $\Re^n$ can be studied. Theorem 10.8 shows that a closed half-space $P \subset \Re^n$ is H-convex if and only if it is d-convex. Hence each H-convex set in $\Re^n$ is d-convex. The converse assertion is, in general, false, i.e., a d-convex set (even a closed one) need not be H-convex.

In Example 9.3, there is no H-convex half-space, i.e., the set $H = H(\Re^3)$ is empty. Hence, besides $\Re^3$ there is no H-convex set. For instance, the segments parallel to spatial diagonals of $\sum$ are d-convex but not H-convex. An analogous situation is given in Example 11.6. Furthermore, in Example 10.4 (cf. also Example 11.4) the only H-convex planes are those parallel or orthogonal to L, i.e., the set $H = H(\Re^3)$ consists of the unit vectors lying in the plane L or on the line l. Lines (or segments), enclosing with L an angle α with $\frac{\pi}{3} \leq \alpha < \frac{\pi}{2}$, are d-convex but not H-convex.

In Example 11.11, the only H-convex planes are those parallel to the plane $x_2 = 0$ or to the plane $x_3 = 0$. Lines (or segments) parallel to the plane $x_3 = 0$, but not parallel to the x_1-axis, are d-convex but not H-convex. Also in Example 13.5, the existence of d-convex lines (or segments), which are not H-convex, is easily verified.

In the above examples, the closed d-convex sets not being H-convex are *one-dimensional*. The same can be shown for sets of higher dimensions. In view of Example 11.11, let $M \subset \Re^3$ be the set defined by

$$x_3 = 0; \quad -1 \leq x_1 + x_2 \leq 1,$$

i.e., M is a strip whose bounding lines are parallel to the angle bisectors of the second and fourth coordinate angles in the (x_1, x_2)-plane. The plane aff M is d-convex and the bounding lines of M are d-convex. Hence M is d-convex (cf. Theorems 10.8 and 11.12). But M is not H-convex. Indeed, the only H-convex half-spaces containing M are described by $x_3 \leq \lambda$, $\lambda \geq 0$, and $x_3 \geq \lambda'$, $\lambda' \leq 0$. The intersection of all these half-spaces is not M, but aff M.

These examples illustrate that there exist closed, d-convex sets which are not H-convex. But none of them is a *body*. For this case we have

Theorem 24.1: A closed, convex body $M \subset \Re^n$ is d-convex if and only if it is H-convex. $\square$

This assertion follows immediately from Theorem 10.10.

Theorem 24.1 shows that the family of all closed, d-convex *bodies* coincides with the family of all H-convex bodies (in the sense of the set $H = H(\Re^n)$). Consequently the number $\mathrm{him}^{(b)}\Re^n$ (i.e., the Helly dimension of the family of all closed, d-convex bodies) coincides with the Helly dimension of the family of all H-convex bodies, i.e., with md H (cf. Theorem 22.1). In other words, $\mathrm{him}^{(b)}\Re^n = \mathrm{md}\, H$ (where, as we have said above, $H = H(\Re^n)$ is the set of all unit vectors orthogonal to d-convex hyperplanes).

In view of the above examples, one might think that H-convexity is only a part of d-convexity. But this is not the case. Namely, not for every set $H \subset S^{n-1}$ one can find a norm such that $H = H(\Re^n)$. In fact, the definition of the set $H(\Re^n)$ implies that it is symmetric about the origin (if the vector $f \in S^{n-1}$ is orthogonal to a d-convex hyperplane, then also $-f$). Furthermore, Theorem 11.10 shows that $H(\Re^n)$ is *compact*. Therefore a set $H \subset S^{n-1}$, which is not symmetric about o or is not compact, cannot be represented in the form $H(\Re^n)$. Moreover, as the following example shows, there exist sets $H \subset S^{n-1}$ which satisfy these two conditions but are not representable in the form $H(\Re^n)$.

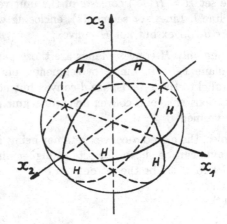

Fig. 109

EXAMPLE 24.2: In the space $\mathbf{R}^3$ with an orthonormal coordinate system (x_1, x_2, x_3) and unit sphere S^2, let L_1, L_2, L_3 denote the coordinate planes $x_1 = 0$, $x_2 = 0$ and $x_3 = 0$, respectively. We set $H = S^2 \cap (L_1 \cup L_2 \cup L_3)$. The set H (Fig. 109) is symmetric with respect to o and compact. But we will show that H is not representable in the form $H = H(\Re^3)$. Assume the contrary: there is a norm transforming $\mathbf{R}^3$ to a normed space $\Re^3$ with $H = H(\Re^3)$. The structure of H shows that any plane, containing a coordinate axis, is orthogonal to some vector $f \in H$, and therefore the plane is d-convex (by the assumption that $H = H(\Re^3)$). Since every line through o is the intersection of two suitable planes, each of which contains a coordinate axis, every line in $\Re^3$ is d-convex. But then, by Corollary 11.2, all boundary points of the unit ball $\sum \subset \Re^3$ are extreme points, and d-convexity in $\Re^3$ coincides with the linear convexity. Thus all planes are d-convex, i.e., $H(\Re^3) = S^2$. This contradicts the assumption, and therefore H is not representable in the form $H(\Re^3)$. $\square$

This example shows that in normed spaces the H-convexity is not a particular case of the d-convexity.

Thus also d-convexity does not lead to H-convexity. But it is possible to give a generalization of the notion "H-convexity" such that d-convexity is enclosed. We will only give the definition of this generalized H-convexity and renounce further investigations to this subject.

Let $\mathbf{R}^n$ be a vector space and H' be a family of subsets of $\mathbf{R}^n$ satisfying the following two conditions:

α)each set $M \in H'$ is either a flat or a closed half-flat (of arbitrary dimension) in $\mathbf{R}^n$,

β)if $M \in H'$, then also $(M + a) \in H'$, for arbitrary $a \in \mathbf{R}^n$.

Having such a family H', we can define H'-convex sets: a set $N \subset R^n$ is said to be H'-*convex* if it is representable as intersection of a family of sets from H'. We enter into the agreement that the whole space $\mathbf{R}^n$ is H'-convex.

On that base, we have

$$H' = \left(\bigcup_{r=0}^{n-1} F_r \right) \cup \left(\bigcup_{s=1}^{n} G_s \right),$$

where F_r denotes a (possibly empty) family of r-flats in $\mathbf{R}^n$ and G_s denotes a (possibly empty) family of s-dimensional half-flats in $\mathbf{R}^n$. In addition, F_r and G_s are invariant under translations. In general, there are no further restrictions with respect to H'. If all families F_r, G_s are empty, except for the family $G_n = H$, then the H'-convexity and the H-convexity coincide.

We will show now that H'-convexity is a generalization of d-convexity. More precisely, for arbitrary normed spaces $\Re^n$ one can choose a family H' of flats and half-flats (invariant under translations) such that a closed set $M \subset \mathbf{R}^n$ is d-convex if and only if it is H'-convex. Exactly H' will present the family of all flats and half-flats of $\Re^n$ which are d-convex. Thus, for showing the coincidence of d-convexity and H'-convexity (in the case of closed sets), the verification of the following assertion is sufficient.

Theorem 24.3: Each closed d-convex set $M \subset \Re^n$ can be represented in the form $\bigcap_{\alpha} P_\alpha$, where every set P_α is either a d-convex flat or a closed, d-convex half-flat. □

PROOF: Each closed, convex set M is the intersection of all its supporting cones (at points from M), and the supporting cones of a d-convex set are also d-convex (Theorem 10.5). Thus it is sufficient to prove the theorem only for the case when M is a d-convex cone. For showing this, we will use descending induction with respect to the dimension of the minimal face of the cone. If this minimal face is n-dimensional, then $M = \mathbf{R}^n$, and in this case the assertion is obvious. Assume that the theorem holds for all d-convex cones, the minimal faces of which are at least k-dimensional, with $k \in \{1, \cdots, n\}$. We choose a closed, d-convex cone M whose minimal face N is $(k-1)$-dimensional. For the case dim N = dim M (i.e., $M = N = $ aff M is a flat) and for dim N = dim $M - 1$ (i.e., M is a half-flat) the assertion of the theorem is also obvious. For dim $N \leq$ dim $M - 2$, we denote by K_x the supporting cone of M at a point $x \in M \setminus N$. By dim $N \leq$ dim $M - 2$, we have

$$M = \bigcap_{x \in M \setminus N} K_x. \tag{1}$$

In fact, if $a \notin M$ is a point from aff M, then the segment $[a, b]$ intersects rbd M for any point $b \in$ ri M. By the inequality dim $N \leq$ dim $M - 2$ and some motion of the point b we may assume that $[a, b] \cap N = \emptyset$, i.e., the point x, at which $[a, b]$ intersects rbd M, does not belong to N. But then obviously $a \notin K_x$. Hence the equality (1) holds. Besides, for $x \in M \setminus N$ each cone K_x is d-convex (Theorem 10.5) and its minimal face is at least k-dimensional. Hence the assertion of the theorem holds for each cone K_x, and by (1) the same holds for M. Therefore the theorem holds for d-convex cones whose minimal faces are $(k-1)$-dimensional, and the induction is verified. ∎

In conclusion, we consider two theorems which connect the dimensional invariants $\text{him}^{(\text{cl})}\mathfrak{R}^n$ and $\text{him}^{(b)}\mathfrak{R}^n$ with the functional md.

We remark that for any normed space $\mathfrak{R}^n$ an invariant md $H(\mathfrak{R}^n)$ can be defined, where $H(\mathfrak{R}^n)$ denotes the set of all unit normal vectors of d-convex hyperplanes in $\mathfrak{R}^n$. Shortly, we denote this invariant by md $\mathfrak{R}^n$. Using normal vectors, the following statement seems to be justified: For introducing md $\mathfrak{R}^n$ = md $H(\mathfrak{R}^n)$, besides the metric of the space $\mathfrak{R}^n$ itself a Euclidean metric is needed, giving the usual orthogonality and the sphere S^{n-1}. But one can define the invariant md $\mathfrak{R}^n$ also directly in terms of $\mathfrak{R}^n$, without this Euclidean metric. Namely, md $\mathfrak{R}^n$ is the largest of the positive integers m, for which there exist d-convex $(n-1)$-subspaces $L_1, \cdots, L_{m+1}$ in $\mathfrak{R}^n$ whose intersection $L = L_1 \cap \cdots \cap L_{m+1}$ is an $(n-m)$-subspace, where L is also the intersection of each m of the subspaces $L_1, \cdots, L_{m+1}$. (The case $n = 3$, $m = 2$ is shown in Fig. 110.) However, in the sequel we will use the first approach, by means of the additional Euclidean metric.

Theorem 24.4: Let $\mathfrak{R}^n$ be a normed space. If the set $H(\mathfrak{R}^n)$ is nonempty (i.e., in $\mathfrak{R}^n$ there exist d-convex hyperplanes), then

$$\text{him}^{(b)}\mathfrak{R}^n = \text{md } \mathfrak{R}^n. \quad \square$$

PROOF: Let $\mathfrak{V}_H$ denote the family of all H-convex sets in $\mathfrak{R}^n$ with $H = H(\mathfrak{R}^n)$ and $\mathfrak{V}_d$ denote the set of all closed, d-convex bodies in $\mathfrak{R}^n$. Since $\mathfrak{V}_d \subset \mathfrak{V}_H$ (cf. Theorem 10.10), we have him $\mathfrak{V}_d \leq$ him $\mathfrak{V}_H$, i.e., $\text{him}^{(b)}\mathfrak{R}^n \leq$

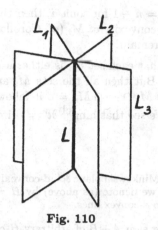

Fig. 110

him $\mathfrak{V}_H$. Furthermore, we have him $\mathfrak{V}_H = \mathrm{md}\,H = \mathrm{md}\,\mathfrak{R}^n$, since the non-empty set $H = H(\mathfrak{R}^n)$ is closed (see Theorem 11.10) and symmetric about the origin.

For showing the converse inequality, we observe two properties of $\mathfrak{V}_d$. First, each translate of a d-convex set is also d-convex, and second, for each vector $f \in H(\mathfrak{V}_d)$ one can find a set $V \in \mathfrak{V}_d$ such that V has an $(n-1)$-face with f as outer normal vector. In addition, we have $H(\mathfrak{V}_d) = H = H(\mathfrak{R}^n)$, see the proof of Theorem 22.1. Hence, by the observations above and the inequality him $\mathfrak{V}_d \geq \mathrm{md}\,H(\mathfrak{V}_d)$ following from them, we get $\mathrm{him}^{(b)}\mathfrak{R}^n \geq \mathrm{md}\,\mathfrak{R}^n$.   ∎

Theorem 24.5: Let the set $H(\mathfrak{R}^n)$ be nonempty. If $\mathrm{md}\,\mathfrak{R}^n < n$, then $\mathrm{him}^{(\mathrm{cl})}\mathfrak{R}^n < n$.   □

PROOF: Let $\mathrm{him}^{(\mathrm{cl})}\mathfrak{R}^n = n$. Then there exist closed, d-convex sets $M_o, M_1, \cdots, M_n$ such that each n of them have a nonempty intersection, but $M_o \cap M_1 \cap \cdots \cap M_n = \emptyset$. For every $i \in \{0, 1, \cdots, n\}$, we choose a point p_i lying in each of the sets $M_o, M_1, \cdots, M_n$ except for M_i, i.e., $p_i \in M_j$ for $i \neq j$ and $p_i \notin M_i$ with $i, j \in \{0, 1, \cdots, n\}$. If all the points $p_o, p_1, \cdots, p_n$ would lie in a hyperplane L, then (by Helly's theorem) the intersection of the convex sets $M_o \cap L, M_1 \cap L, \cdots, M_n \cap L$ would be nonempty, contradicting the equality $M_o \cap M_1 \cap \cdots \cap M_n = \emptyset$. Thus the points $p_o, p_1, \cdots, p_n$ form the vertex set of an n-simplex, leading to $\dim M_i \geq n-1$, $i = 0, 1, \cdots, n$.

Let $\dim M_i = n-1$ for $i = 0, 1, \cdots, n$. Denote by M_i' the intersection of all sets $M_o, M_1, \cdots, M_n$ except for M_i. Then $M_i' \cap \mathrm{aff}\,M_i = \emptyset$. Namely, if this intersection would be nonempty, we could assume that p_i is chosen from $M_i' \cap \mathrm{aff}\,M_i$ (p_i belongs to all sets M_j with $i \neq j$). Since, furthermore, for $i \neq j$ the relation $p_j \in M_i \subset \mathrm{aff}\,M_i$ holds, we would obtain that all the points $p_o, p_1, \cdots, p_n$ lie in the hyperplane $\mathrm{aff}\,M_i$, contradicting the fact that $\{p_o, p_1, \cdots, p_n\}$ is the vertex set of an n-simplex. Thus, $M_i' \cap \mathrm{aff}\,M_i = \emptyset$.

We see that if $\dim M_i = n - 1$ for some i, then the set M_i can be replaced by its affine hull; the d-convexity of M_i (cf. Corollary 10.6) and the equality $M_o \cap M_1 \cap \cdots M_n = \emptyset$ remain.

Hence we can assume that each set M_i is either a d-convex hyperplane or a closed, d-convex body. But then all the sets M_i are H-convex, where $H = H(\Re^n)$, and from $M_o \cap M_1 \cap \cdots \cap M_n = \emptyset$ it follows that him $\mathfrak{V}_H = n$, i.e., md $H = n$ (see §22). We see that $\text{him}^{(\text{cl})}\Re^n = n$ implies md $H = n$.   ∎

Exercises:

1. Prove that for every Minkowski plane $\Re^2$, d-convexity (for closed sets) coincides with H-convexity if we denote (as above) by $H = H(\Re^2)$ the set of all unit vectors orthogonal to d-convex lines.

2. Prove that the vector sum $A + B$ of arbitrary H-convex sets $A, B \subset \mathbf{R}^2$ is an H-convex set (for any $H \subset S^1$).

3. Prove that the vector sum $A + B$ of arbitrary closed, d-convex sets A, B in a Minkowski plane $\Re^2$ is a d-convex set.

4. Prove that for every Minkowski space $\Re^n$ and arbitrary sets $A, B \subset \Re^n$ the relation

$$\text{conv}_d(A + B) \supset \text{conv}_d A + \text{conv}_d B$$

holds.

5. Prove that, in general, the following assertion is false: "In every Minkowski space $\Re^n$ the vector sum $A + B$ of arbitrary d-convex sets $A, B \subset \Re^n$ is a d-convex set." Decide whether this statement is true in the Minkowski spaces considered in Example 10.4, in Example 12.6 and in Exercise 3 of §11.

6. Prove that in a Minkowski space $\Re^n$ the equality

$$\bigcap_{\varepsilon > 0} \text{conv}_d(U_\varepsilon(M)) = M$$

holds if and only if M is representable as an intersection of d-convex *bodies* (i.e., M is $H(\Re^n)$-convex).

7. Prove that if $M \subset \Re^n$ is representable as an intersection of d-convex *bodies*, then for every $\varepsilon < 0$ there exists a number $\delta > 0$ such that $\text{conv}_d U_\delta(M) \subset U_\varepsilon(M)$. Thus for closed sets $M \subset \Re^n$, which admit the indicated representation (and only for them), the operation conv_d possesses the upper semicontinuity property (cf. Theorem 20.4).

8. Is it possible to replace $\text{him}^{(\text{cl})}\Re^n$ in Theorem 24.5 by him $\Re^n$?

9. Show that in Example 14.8 the relation md $\Re^3 = 2$ holds. Consequently in this example him $\Re^3 >$ md $\Re^3$ (cf. Theorem 24.4).

10. Prove that in Example 14.9 the relation $\text{him}^{(\text{cl})}\Re^3 = 2$ holds.

11. Let $H \subset S^{n-1}$ be a set which is not one-sided, and $X \subset \mathbf{R}^n$ be a compact set. We say that a point $x \in X$ is H-visible (in X) from a point $x_o \in X$ if $\mathrm{conv}_H\{x, x_o\} \subset X$. If there exists a point $x_o \in X$ such that every point $x \in X$ is H-visible from x_o in X, then we say that X is H-star-shaped. Prove the following generalization of Krasnosel'ski's theorem: Let X be a (possibly nonconvex) polyhedral region in $\mathbf{R}^n$ and H be the set of unit outward normals for the facets of X. The set X is H-star-shaped if and only if every md $H + 1$ of its points are H-visible in X from a point of the set X.

12. Let $X \subset \Re^n$ be a polyhedral region all facets of which are situated in d-convex hyperplanes. Prove that X is d-star-shaped if and only if every md $\Re^n + 1$ of its points are d-visible in X from a point of the set X.

IV. The Szökefalvi-Nagy Problem

§25 The Theorem of Szökefalvi-Nagy and its generalization

Let $M \subset \mathbf{R}^n$ be a compact, convex body. We consider the family $T(M)$ of all its translates (Fig. 111) and denote by him M the Helly dimension of $T(M)$: him $M = $ him $T(M)$. This chapter is devoted to the *Helly-dimensional classification of compact, convex bodies*. In other words, we are going to consider the following problem: to give a geometrical description of the compact, convex bodies $M \subset \mathbf{R}^n$ satisfying him $M = r$, where $1 \leq r \leq n$. By reasons which will be mentioned in this section, this problem is said to be the *Szökefalvi-Nagy problem*.

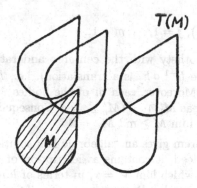

Fig. 111

First of all, we give an algebraic approach to this problem. It was obtained in 1976 [Bt 7].

Theorem 25.1: For every compact, convex body M the equality him $M = $ md M holds. $\square$

PROOF: We put $H = H(M)$. Then the body M is H-convex (Theorem 19.2). In other words, the family $T(M)$ is contained in the family C_H of the H-convex sets. Consequently,

$$\text{him } M = \text{him } T(M) \leq \text{him } C_H = \text{md } M.$$

To establish the opposite inequality, we put md $M = m$ and find $m+1$ regular boundary points $a_1, \cdots, a_{m+1}$ of M such that the unit outward normals $p_1, \cdots, p_{m+1}$ of the body M at these points are minimally dependent. Consider the half-spaces

$$P_i = \{x : \langle p_i, x \rangle \leq -1\}, \quad i = 1, \cdots, m+1.$$

Then each m of the half-spaces $P_1, \cdots, P_{m+1}$ have a common interior point, and the intersection $P_1 \cap \cdots \cap P_{m+1}$ is empty (cf. the beginning part of the proof of Theorem 22.1). Consequently there are points $b_1, \cdots, b_{m+1}$ such that $b_i \in \text{int } P_j$ for $i \neq j$. Let $M_j = (-p_j - a_j) + M$, $j = 1, \cdots, m+1$. Then M_j is a translate of M and $-p_j$ is a regular boundary point of M_j with the unit outward normal p_j (Fig. 112), i.e., P_j is the supporting half-space of M_j at $-p_j$. Since the point $-p_j \in \text{bd } M_j$ is regular, there exists a positive integer k (large enough) such that $b_i \in h_j(M_j)$ for $i \neq j$, where h_j is the homothety with the center $-p_j$ and ratio k. We may suppose that the number k is the same for $j = 1, \cdots, m+1$. It follows from the inclusion $b_i \in h_j(M_j)$, $i \neq j$, that every m of the bodies $h_1(M_1), \cdots, h_{m+1}(M_{m+1})$ have a common point, whereas the intersection of the $m+1$ bodies is empty (since $h_j(M_j) \subset P_j$). Finally, we put

$$M_j' = h^{-1}(h_j(M_j)), \, j = 1, \cdots, m+1,$$

where h is the homothety with the center o and ratio $\frac{1}{k}$. Then M_j' is a translate of M (since $h^{-1} \circ h_j$ is a translation), i.e., $M_j' \in T(M)$ for each $j \in \{1, \cdots, m+1\}$. Moreover, each m of the bodies $M_1', \cdots, M_{m+1}'$ have a common point, whereas $M_1' \cap \cdots \cap M_{m+1}' = \emptyset$. Consequently (Theorem 14.7), him $T(M) \geq m$, i.e., him $M \geq$ md M. ∎

The established theorem gives an "algebraic half-solution" of the Szökefalvi - Nagy problem. Indeed, it contains a description of the compact, convex bodies $M \subset \mathbf{R}^n$, for which him $M = r$, in terms of *linear algebra* (i.e., with the help of the functional md), whereas we are interested in a *geometrical* description.

We now account a result which was obtained by the well-known Hungarian mathematician B. Szökefalvi-Nagy in 1954 [Sz]. His proof was complicated, but now (on the basis of Theorem 25.1) we can give a more simple one. The result of Szökefalvi-Nagy is given by

Theorem 25.2: A compact, convex body $M \subset \mathbf{R}^n$ is 1-Helly-dimensional if and only if M is a parallelotope. In other words, him $M = 1$ for parallelotopes and only for them.

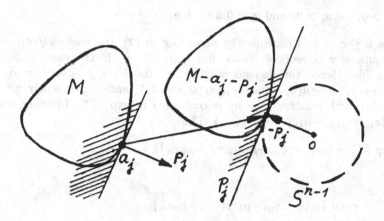

Fig. 112

PROOF: The part "if" is trivial, see Example 14.3. For proving the part "only if", let $M \subset \mathbf{R}^n$ be a compact, convex body with him $M = 1$. According to Theorem 25.1, we have md $M = 1$. We may suppose $o \in$ int M. By virtue of Theorem 18.1, we get the equality md (norm exp M^*) $= 1$, i.e., md exp $M^* = 1$.

Let $a \in$ exp M^*. Since $o \in$ int M^*, the vector system exp M^* is not one-sided. Hence, according to Theorem 16.7, there are vectors $b_1, \cdots, b_s \in$ exp M^* and positive numbers $\lambda_1, \cdots, \lambda_s$ such that $-a = \lambda_1 b_1 + \cdots + \lambda_s b_s$, i.e., we have the positive dependence

$$a + \lambda_1 b_1 + \cdots + \lambda_s b_s = o. \tag{1}$$

By virtue of Theorem 16.5, we may suppose that (1) is a minimal dependence. Since md exp $M^* = 1$, the case $s > 1$ is impossible, i.e., the dependence (1) has the form $a + \lambda_1 b_1 = o$ (with $\lambda_1 > 0$). Thus together with each vector $a \in$ exp M^* the system exp M^* contains an oppositely directed vector $b = -\mu a$, where $\mu = \frac{1}{\lambda_1} > 0$.

Since exp M^* is not one-sided, there are linearly independent vectors $e_1, \cdots, e_n \in$ exp M^*. Hence,

$$\text{exp } M^* \supset \{e_1, \cdots, e_n, -\mu_1 e_1, \cdots, -\mu_n e_n\}$$

with positive $\mu_1, \cdots, \mu_n$. Let, finally, $f \in$ exp M^*. Then

$$-f = \nu_1 e_1 + \cdots + \nu_n e_n. \tag{2}$$

We may suppose (replacing e_i by $-\mu_i e_i$ if necessary) that the coefficients in the right-hand side of (2) are nonnegative. Furthermore, without loss of generality we may suppose that

$$\nu_1 > 0, \cdots, \nu_q > 0, \text{ and } \nu_i = 0 \text{ for } i > q,$$

where $1 \leq q \leq n$. Consequently, according to (2), the vectors $f, e_1, \cdots, e_q$ are minimally dependent. Since md exp $M^* = 1$, the case $q > 1$ is impossible. Hence the dependence (2) has the form $f + \nu_1 e_1 = o$, i.e., $f = -\nu_1 e_1 = -\mu_1 e_1$ (since $-\mu_1 e_1 \in \exp M^*$, and consequently no vector $-\lambda e_1$ with positive $\lambda \neq \mu_1$ is contained in $\exp M^*$). In other words, $f \in \{e_1, \cdots, e_n, -\mu_1 e_1, \cdots, -\mu_n e_n\}$. Thus,

$$\exp M^* = \{e_1, \cdots, e_n, -\mu_1 e_1, \cdots, -\mu_n e_n\},$$

i.e.,

$$M^* = \operatorname{conv} \{e_1, \cdots, e_n, -\mu_1 e_1, \cdots, -\mu_n e_n\}.$$

This means that M^* is the polytope with the vertices $e_1, \cdots, e_n, -\mu_1 e_1, \cdots, -\mu_n e_n$. Consequently $M = M^{**}$ (its polar body) is a polytope with facets in the hyperplanes $(e_1)^*, \cdots, (e_n)^*, (-\mu_1 e_1)^*, \cdots, (-\mu_n e_n)^*$. In other words, M is the intersection of $2n$ half-spaces

$$\{x : \langle x, e_i \rangle \leq 1\}, \{x : \langle x, -\mu_i e_i \rangle \leq 1\}, i = 1, \cdots, n,$$

i.e., M is a parallelotope.   ∎

REMARK 25.3: From some general results obtained for Banach spaces by L. Nachbin [Na] and D.B. Goodner [Go] (in 1950), one may deduce the following statement: if a *centrally symmetric*, compact, convex body $M \subset \mathbf{R}^n$ possesses the property $\bigcap_1$ (i.e., him $M = 1$), then M is a parallelotope. Szökefalvi-Nagy's Theorem 24.2 *does not suppose* that M is centrally symmetric.

It should be noticed that the previous theorem shows the difference between the algebraic and geometrical solutions of the Szökefalvi-Nagy problem in a clear manner: in which cases do we have him $M = 1$? The algebraic "half-solution" (Theorem 25.1) answers: if and only if md $M = 1$. This does not cover the *geometrical* description of the bodies satisfying him $M = 1$. Theorem 25.2 gives a complete geometrical answer: if and only if M is a parallelotope.

Moreover, the proof of Theorem 25.2 indicates how it is possible to solve the Szökefalvi-Nagy problem for some r (i.e., to describe the bodies M with him $M = r$). A solution has to consist of two parts: (i) to find a (necessary and sufficient) condition under which a *vector system* H satisfies the relation md $H = r$; (ii) to describe the convex bodies M for which the vector system $\exp M^*$ satisfies the obtained condition.

Before we will give a generalization of the Szökefalvi-Nagy theorem, we formulate a general auxiliary proposition.

Theorem 25.4: Let $M \subset \mathbf{R}^n$ be a compact, convex body, decomposed into the direct vector sum

$$M = M_1 \oplus \cdots \oplus M_s.$$

Each compact, convex set M_q is a body in its affine hull aff M_q, *and therefore we can define its Helly dimension* him M_q *(in* aff M_q*) for $q = 1, \cdots, s$. In this situation, the following equality holds:*

him $M = \max($ him $M_1, \cdots,$ him $M_s)$. $\square$

PROOF: By virtue of Theorems 25.1 and 18.1, the equalities

him $M = $ md $M = $ md (norm exp (M^*)) $= $ md exp M^*

hold. Furthermore, according to Theorem 4.4, there exist a direct decomposition $\mathbf{R}^n = E_1' \oplus \cdots \oplus E_s'$ and convex sets M_q' with aff $M_q' = E_q'$, $q = 1, \cdots, s$, such that

$$M^* = (M_1 \oplus \cdots \oplus M_s)^* = M_1' \vee \cdots \vee M_s',$$

where $j_q(M_q') = M_q^* \subset E_q^*$. Consequently (cf. Exercise 3 in §4 and Exercises 11, 12 in §16) the equalities

$$
\begin{aligned}
\text{md } M \;\; &= \;\; \text{md exp } M^* = \text{md exp } (M_1' \vee \cdots \vee M_s') \\
&= \;\; \text{md } ((\text{exp } M_1') \cup \cdots \cup (\text{exp } M_s')) \\
&= \;\; \max (\text{md exp } M_1', \cdots, \text{md exp } M_s') \\
&= \;\; \max (\text{md exp } M_1^*, \cdots, \text{md exp } M_s^*) \\
&= \;\; \max (\text{him } M_1, \cdots, \text{him } M_s)
\end{aligned}
$$

hold.   ∎

Corollary 25.5: Under the conditions of Theorem 25.4, the inequality

him $M \leq \max($ dim $M_1, \cdots,$ dim $M_s)$

holds. $\square$

The following generaliziation of Szökefalvi-Nagy's result was obtained in 1976 [Bt 7], and it illustrates this way of solution. The problem is to give a list of *centrally symmetric*, compact, convex bodies M with him $M = 2$. Two theorems below realize the above parts (i), (ii) of the solution. $\square$

Theorem 25.6: Let $H \subset \mathbf{R}^n$ be a system of nonzero vectors which is symmetric with respect to the origin and not one-sided. If md $H \leq 2$, *then there exists a direct decomposition $\mathbf{R}^n = L_1 \oplus \cdots \oplus L_s$ such that* dim $L_i \leq 2$ *for each $i \in \{1, \cdots, s\}$ and $H \subset L_1 \cup \cdots \cup L_s$.* $\square$

PROOF: If md $H = 1$, then the theorem is true (see Exercise 1). Let now md $H = 2$. We choose three minimally dependent vectors $a, b, c \in H$ and denote the two-dimensional subspace lin (a, b, c) by L_1. We put $H_1 = H \cap L_1$, $H' = H \setminus H_1$. Let K' be the set of all vectors representable as linear combinations of vectors from H' with nonnegative coefficients. Then K' is a subspace of $\mathbf{R}^n$ (cf. Exercise 2).

We will prove that $L_1 \cap K' = \{o\}$. Indeed, assume $\dim(L_1 \cap K') > 0$, i.e., there exist vectors $g_1, \cdots, g_k \in H'$ and positive numbers $\lambda_1, \cdots, \lambda_k$ such that $\lambda_1 g_1 + \cdots + \lambda_k g_k$ is a nonzero vector belonging to L_1. Denote by q the *minimal* one of the integers k for which such a linear combination exists. Thus there are vectors $h_1, \cdots, h_q \in H'$ and positive numbers $\nu_1, \cdots, \nu_q$ such that $a = \nu_1 h_1 + \cdots + \nu_q h_q$ is a nonzero vector belonging to L_1. The vectors $(-a), h_1, \cdots, h_q$ are positively dependent:

$$(-a) + \nu_1 h_1 + \cdots + \nu_q h_q = o. \tag{3}$$

Since q is minimal, the vectors $h_1, \cdots, h_q$ are linearly independent (Theorem 16.4). The nonzero vector $-a \in L_1$ can be represented in the form $-a = \mu_1 e_1 + \mu_2 e_2$, where $e_1, e_2 \in L_1 \cap H$ are linearly independent and the numbers μ_1, μ_2 are positive (cf. Exercise 3). We obtain from (3) the positive dependence

$$\mu_1 e_1 + \mu_2 e_2 + \nu_1 h_1 + \cdots + \nu_q h_q = o,$$

which contains $q+2$ vectors $e_1, e_2, h_1, \cdots, h_q \in H$. Since, evidently, $q \geq 2$, the integer $q + 2$ is greater than 3, and consequently the vectors $e_1, e_2, h_1, \cdots, h_q$ cannot be minimally dependent (so far as md $H = 2$). Hence the vectors $e_1, e_2, h_1, \cdots, h_{q-1}$ are linearly dependent (Theorem 16.2), i.e., there exists a non-trivial linear dependence

$$\alpha_1 e_1 + \alpha_2 e_2 + \beta_1 h_1 + \cdots + \beta_{q-1} h_{q-1} = o.$$

Since the vector system H' is centrally symmetric, we may suppose (replacing h_i by $-h_i$, if necessary) that all the coefficients $\beta_1, \cdots, \beta_{q-1}$ are non-negative. Furthermore, since e_1 and e_2 are linearly independent, at least one of the coefficients $\beta_1, \cdots, \beta_{q-1}$ is not equal to zero. Hence, $\beta_1 h_1 + \cdots + \beta_{q-1} h_{q-1}$ is a nonzero vector (since $h_1, \cdots, h_{q-1}$ are linearly independent) which belongs to L_1, contradicting the minimality of the integer q. This contradiction shows that $L_1 \cap K' = \{o\}$.

It is clear that $\dim K' = n - 2$ (otherwise $\dim(L_1 \oplus K') < n$ and the vector system $H \subset L_1 \cup K'$ would be one-sided). Consequently $\mathbf{R}^n = L_1 \oplus K'$. Moreover, the centrally symmetric vector system $H' \subset K'$ is not one-sided in K' (cf. Exercise 4) and md $H' \leq 2$. Thus we can apply the above reasoning to the vector system $H' \subset K'$, etc. ∎

Theorem 25.7: Let $M \subset \mathbf{R}^n$ be a centrally symmetric, compact, convex body. The inequality him $M \leq 2$ *holds if and only if M is the direct vector sum of convex sets each of which has a dimension ≤ 2:*

$$M = M_1 \oplus \cdots \oplus M_s; \quad \dim M_i \leq 2, \quad i = 1, \cdots, s. \quad \square$$

PROOF: Let him $M \leq 2$. Then, by virtue of Theorem 25.1, md exp $(M^*) \leq 2$. Moreover, the vector system exp M^* is centrally symmetric (since M is). According to Theorem 25.6, there exists a decomposition $\mathbf{R}^n = L_1 \oplus \cdots \oplus L_s$ such that dim $L_i \leq 2$ for each $i \in \{1, \cdots, s\}$ and exp $M^* \subset L_1 \cup \cdots \cup L_s$. Consequently,

$$M^* = \text{conv cl exp } M^* = M_1' \vee \cdots \vee M_s',$$

where $M_i' = \text{conv cl } (L_i \cap \text{exp } M^*)$ is a convex body in the subspace L_i, i.e., dim $M_i' = $ dim $L_i \leq 2$, $i = 1, \cdots, s$. By virtue of Theorem 4.2, we conclude now that

$$M = M^{**} = M_1 \oplus \cdots \oplus M_s,$$

where dim $M_i = $ dim $M_i' \leq 2$ for $i = 1, \cdots, s$.

We obtain the part "if" as a direct consequence of Corollary 25.5, see also Exercise 5. ∎

Corollary 25.8: Let $M \subset \mathbf{R}^n$ be a centrally symmetric, compact, convex body. The equality him $M = 2$ holds if and only if M is distinct from a parallelotope and $M = M_1 \oplus \cdots \oplus M_s$, where dim $M_i \leq 2$ for $i = 1, \cdots, s$. □

We will finish this section by formulating some results (without proofs) which are connected with *lattice polytopes* and their corresponding *lattice Helly dimension.*

DEFINITION 25.9: Let Z^n be a lattice in $\mathbf{R}^n$, i.e., the set of all points $p_1 e_1 + \cdots + p_n e_n$, where $e_1, \cdots, e_n$ is a basis in $\mathbf{R}^n$ and the coefficients $p_1, \cdots, p_n$ are integers. A finite subset $P \subset Z^n$, satisfying dim conv $P = n$ and $P = Z^n \cap \text{conv } P$, is said to be an *n-dimensional lattice polytope*. In other words, P is an n-dimensional lattice polytope if it is not contained in a hyperplane of $\mathbf{R}^n$ and $P = Z^n \cap M$, where M is a compact, convex body. □

DEFINITION 25.10: Let $P \subset Z^n$ be an n-dimensional lattice polytope. Denote by $\Gamma(P)$ the family of all lattice n-polytopes in Z^n which are positively homothetic to P and by $\text{him}_l P$ the *Helly dimension* of this family: $\text{him}_l P = $ him $\Gamma(P)$. In other words, $\text{him}_l P$ is the smallest of the integers m for which $\Gamma(P)$ possesses the following property: for every collection of sets $P_1, \cdots, P_s \in \Gamma(P)$, $s \geq m + 2$, the relation $P_1 \cap \cdots \cap P_s \neq \emptyset$ holds if each $m + 1$ of the sets have a common point. □

We recall that a set $P_i \in \Gamma(P)$, which is not a single point, is *not convex* in the usual sense (it consists only of lattice points from $Z^n \subset \mathbf{R}^n$), and $P_1 \cap \cdots \cap P_s \neq \emptyset$ means that there exists a *lattice point* $x \in Z^n$ contained in each of the sets $P_1, \cdots, P_s$.

The following example shows that the Helly dimension with respect to lattices has properties quite different from those of the Helly dimension for compact, convex bodies.

EXAMPLE 25.11: Denote by P the two-dimensional lattice polytope consisting of the five points $o, \pm e_1, \pm e_2$ in the plane $\mathbf{R}^2$, where e_1, e_2 is the basis defining the lattice $Z^2 \subset \mathbf{R}^2$. It is easy to construct four lattice polytopes P_1, P_2, P_3, P_4 which are translates of P and satisfy $P_1 \cap P_2 \cap P_3 \cap P_4 = \emptyset$, although each three of them have a common point. (We notice that the convex hulls of P_1, P_2, P_3, P_4 have common points, but however these points are not lattice points.) Thus in this case $\mathrm{him}_l P = 3$, i.e., $\mathrm{him}_l P$ is *larger* than the dimension of the space $\mathbf{R}^2$ in which the lattice is considered.

If the basis e_1, e_2 is orthonormal, then the above two-dimensional lattice polytope P can be described as the intersection $Z^2 \cap S$, where S is a circle centered at a lattice point and having radius a little less than $\sqrt{2}$. Analogously, if $e_1, \cdots, e_n$ is an orthonormal basis of $\mathbf{R}^n$ and S is a ball centered at a lattice point whose radius r is a little less than $\sqrt{n}$ (more precisely, $\sqrt{n-1} < r < \sqrt{n}$), then the intersection $P = Z^n \cap S$ is an n-dimensional lattice polytope with $\mathrm{him}_l P = 2^n - 1$.   □

Looking at the above construction, the following result of J.-P. Doignon [Doi] seems to be natural.

Theorem 25.12: For every n-dimensional lattice polytope $P \subset Z^n$ the inequality

$$1 \leq \mathrm{him}_l P \leq 2^n - 1$$

holds.   □

DEFINITION 25.13: Let Z^n be a lattice in $\mathbf{R}^n$. Each n-parallelotope C in $\mathbf{R}^n$, whose vertices are the only points of C belonging to Z^n, is said to be a *cell* of Z^n.   □

DEFINITION 25.14: We say that an n-parallelotope P, whose vertices belong to Z^n, is *parallel to a cell* C of Z^n if conv P and conv C have correspondingly parallel facets.   □

The following result was obtained by A. Beutelspacher and K. Bezdek [B-B] (for the definition of *zonotopes*, we refer to chapter VII.)

Theorem 25.15: Let M be an n-dimensinal zonotope with the vertices belonging to $Z^n \subset \mathbf{R}^n$, $n \geq 2$. The Helly dimension of $P = M \cap Z^n$ with respect to the lattice Z^n is equal to 1 (i.e., $\mathrm{him}_l P = 1$) if and only if P is an n-parallelotope parallel to a cell of Z^n.   □

It is easy to construct a parallelogram M with the vertices from $Z^2 \subset \mathbf{R}^2$, such that $P = M \cap Z^2$ is not parallel to a cell of Z^2 and for which $\mathrm{him}_l P \neq 1$. More precisely (as it is shown in [B-B]), if P is a lattice polygon in Z^2 which is not a parallelogram parallel to a cell of Z^2, then $\mathrm{him}_l P \geq 2$. It is unknown whether for $n > 2$ the analogous assertion is true, i.e., whether Theorem 25.15 holds without the assumption $P = M \cap Z^n$, where M is a zonotope with the vertices in Z^n.

Exercises

1. Let $H \subset \mathbf{R}^n$ be a vector system which is not one-sided and satisfies the condition md $H = 1$. Prove that there exists a direct decomposition $\mathbf{R}^n = L_1 \oplus \cdots \oplus L_s$ such that each subspace L_i is one-dimensional, $i = 1, \cdots, s$, and $H \subset L_1 \cup \cdots \cup L_s$. Is this true without the requirement "H is not one-sided"?

2. Let $H \subset \mathbf{R}^n$ be a vector system symmetric with respect to the origin. Prove that the set K of the vectors, which are representable as linear combinations of vectors from H with nonnegative coefficients, is a subspace of $\mathbf{R}^n$.

3. Let $H \subset \mathbf{R}^n$ be a system of nonzero vectors symmetric with respect to the origin, and a, b, c be a minimally dependent triple. Prove that every nonzero vector $p \in \mathrm{lin}\,(a, b, c)$ is representable in the form $p = \lambda_1 e_1 + \lambda_2 e_2$, where the vectors $e_1, e_2 \in H \cap \mathrm{lin}\,(a, b, c)$ are linearly independent and the numbers λ_1, λ_2 are positive.

4. Let $\mathbf{R}^n = L_1 \oplus \cdots \oplus L_s$, and $H \subset L_1 \cup \cdots \cup L_s$ be a vector system. Prove that H is not one-sided if and only if each system $H_i = H \cap L_i$ is not one-sided in L_i, $i = 1, \cdots, s$.

5. Give a direct proof (as in the Example 14.3, without using polar bodies) that if $M = M_1 \oplus \cdots \oplus M_s$, then $\mathrm{him}\, M = \max\,(\mathrm{him}\, M_1, \cdots, \mathrm{him}\, M_s)$.

6. Let $M \subset \mathbf{R}^n$ be a compact, convex body, $a_1, \cdots, a_n$ be regular boundary points of the body, $\Gamma_1, \cdots, \Gamma_n$ be the corresponding supporting hyperplanes, and $p_1, \cdots, p_n$ be the outward unit normals of M at $a_1, \cdots, a_n$. Suppose that the vectors $p_1, \cdots, p_n$ are linearly independent. Prove that if a is the common point of $\Gamma_1, \cdots, \Gamma_n$, then for the body $N = \mathrm{conv}\,(M \cup \{a\})$, all the cases $\mathrm{him}\, N < \mathrm{him}\, M$, $\mathrm{him}\, N = \mathrm{him}\, M$, $\mathrm{him}\, N > \mathrm{him}\, M$ are possible.

7. a) Let $M \subset \mathbf{R}^n$ be a compact, convex body. Let, furthermore, $P_1, \cdots, P_{m+1}$ be supporting half-spaces of M at regular boundary points $a_1, \cdots, a_{m+1}$. Suppose that there exists an $(n - m)$-dimensional subspace N that is contained in a translate of each hyperplane $\mathrm{bd}\, P_1, \cdots, \mathrm{bd}\, P_{m+1}$ and, moreover, the orthogonal projection of $P_1 \cap \cdots \cap P_{m+1}$ into the orthogonal complement of N is an m-dimensional simplex. Prove that $\mathrm{him}\, M \geq m$.

   b) Decide whether the converse assertion is true (i.e., if $\mathrm{him}\, M \geq m$, then there exist supporting half-spaces $P_1, \cdots, P_{m+1}$ as above).

   c) Prove that $\mathrm{him}\, M$ is the maximal one of the integers m for which there exist the half-spaces as above.

8. Prove that if a compact, convex body $M \subset \mathbf{R}^n$ is *smooth* (i.e., each of its boundary points is regular), then $\mathrm{him}\, M = n$.

9. Let $M = M_1 \vee I$, where M_1 is an $(n-1)$-dimensional simplex in $\mathbf{R}^n$ and I is a segment such that $o \in \mathrm{ri}\, M_1 \cap \mathrm{ri}\, I$. Prove that $\mathrm{him}\, M$ is equal to either $n - 1$ or n. Can you describe a condition under which $\mathrm{him}\, M = n - 1$?

10. Prove that if $M \subset \mathbf{R}^3$ is the join of a two-dimensional compact, convex set and a segment, then $\mathrm{him}\, M$ is equal to either 2 or 3. Can you generalize this for a join of an $(n-1)$-dimensional compact, convex set and a segment?

11. Show by an example that the requirement "M is centrally symmetric" in Theorem 25.5 is essential. In other words, give an example of a compact, convex body M (not centrally symmetric) with him $M = 2$, dim $M > 2$, which is indecomposable.

12. Let e_1, e_2, e_3, e_4 be an orthonormal basis of $\mathbf{R}^4$ and H be a set consisting of the 12 vectors

$$\frac{1}{\sqrt{3}}(\pm e_1, \pm e_2 \pm e_3), \; \frac{1}{\sqrt{2}}(\pm e_1 \pm e_4)$$

(all combinations of the signs). Show that H is symmetric with respect to the origin and non-one-sided. Show that md $H = 3$ and the system H is not splittable.

13. In the space $\mathbf{R}^4$ with coordinate system (x_1, x_2, x_3, x_4), let M be the polytope described by the inequalities

$$|x_1 \pm x_2 \pm x_3| \leq \sqrt{3}, \; |x_1 \pm x_4| \leq \sqrt{2}$$

(all combinations of the signs). Show that him $M = 3$ and M is indecomposable. This example shows that it is impossible to replace the number 2 in Theorem 25.5 by 3 or a larger integer.

14. Let $M \subset \mathbf{R}^n$ be a compact, convex body. Denote by $T(M)$ the family of all its translates and by $Y(M)$ the family which consists of the translates of M and the bodies each of which is homothetic to M with positive ratio. Prove that him $T(M) = $ him $Y(M)$. This explains why in the sequel we consider only the family $T(M)$ instead of $Y(M)$.

15. Let $M_1, \cdots, M_{m+1} \subset \mathbf{R}^n$ be compact, convex sets such that every m of them have a common point, whereas $M_1 \cap \cdots \cap M_{m+1} = \emptyset$. Prove that there exist half-spaces $P_1, \cdots, P_{m+1} \subset \mathbf{R}^n$ such that $M_i \subset P_i$ for $i = 1, \cdots, m+1$ and $P_1 \cap \cdots \cap P_{m+1} = \emptyset$.

16. Let $M_1, \cdots, M_{m+1}$ be sets as in the previous exercise, and $N_1, \cdots, N_{m+1} \subset \mathbf{R}^n$ be compact sets. Prove that there are translates $N_1', \cdots, N_{m+1}'$ of $N_1, \cdots, N_{m+1}$ such that every m of the sets

$$M_1 + N_1', \cdots, M_{m+1} + N_{m+1}' \tag{4}$$

have a common point, whereas the intersection of the sets (4) is empty.

17. Let $Q_1, Q_2 \subset \mathbf{R}^n$ be compact, convex sets. Prove that him $(Q_1 + Q_2) \geq$ him Q_1. Generalize this for the case when the vector sum of sets $Q_1, \cdots, Q_s$ is considered.

18. Show by an example that, in general, the inequality in the previous exercise is strict. More detailed, for every $r = 1, \cdots, n-1$ there are compact sets $Q_1, Q_2 \subset \mathbf{R}^n$ such that him $Q_1 \leq r$, him $Q_2 \leq r$ and him $(Q_1 + Q_2) > r$.

19. Let $M \subset \mathbf{R}^4$ be a compact, convex body symmetric with respect to the origin. Prove that the inequality him $M \leq 2$ holds if and only if there exist two-dimensional convex sets M_1, M_2 symmetric with respect to the origin such that $\mathbf{R}^4 = \text{aff } M_1 \oplus \text{aff } M_2$ and $M^* = \text{conv } (M_1 \cup M_2)$.

§26 Description of vector systems with md $H = 2$ that are not one-sided

The main purpose of the following sections is to state and to prove a theorem [Bt 13] that contains a complete list of compact, convex bodies $M \subset \mathbf{R}^n$ (not assumed to be centrally symmetric) satisfying him $M = 2$. For the first reading it is sufficient to look at the statements (§26 and the first half of §29) and to omit the proofs (§27, §28, and the end of §29).

First we consider the part (i) indicated in the previous section, i.e., we give a description of vector systems $M \subset \mathbf{R}^n$ that are not one-sided and satisfy md $H = 2$. Then we carry out the part (ii), i.e., we obtain a list of the compact, convex bodies M in $\mathbf{R}^n$ with md $M = 2$.

A vector system $H \subset \mathbf{R}^n$ is named a 2-*system* if it is not one-sided and md $H = 2$. Certainly, it is sufficient to describe only *nonsplittable* 2-systems, i.e., such that there is no direct decomposition $\mathbf{R}^n = L_1 \oplus \cdots \oplus L_s$, $s > 1$, with $\dim L_i > 0$, $i = 1, \cdots, s$, and $H \subset L_1 \cup \cdots \cup L_s$ (cf. Exercise 1).

A vector $e \in H$ is said to be *particular* if it cannot be represented as a positive linear combination of vectors from $H \setminus \{e\}$.

Theorem 26.1: Let $H \subset S^{n-1}$ be a nonsplittable 2-system with $n \geq 3$. If H contains no particular vector, then $n = 4$, and in a basis e_1, e_2, e_3, e_4 of $\mathbf{R}^4$, the system H coincides with

$$H^{(o)} = \text{norm } \{e_1, e_2, e_3, e_4, -e_1 - e_2, e_1 - e_3, -e_2 - e_4, -e_3 - e_4,$$
$$e_1 + e_2 + e_3 + e_4\}. \quad \square$$

This theorem will be proved in §27.

Theorem 26.2: Let $H \subset S^{n-1}$ be a nonsplittable 2-system that contains at least one particular vector, $n \geq 3$. Then there exist an integer k, $1 \leq k \leq n$, and linearly independent particular vectors $e_1, \cdots, e_k$ from H such that the following assertions are true. For $k = n$,

$$H \subset \text{norm } \left(\{e_1, \cdots, e_n\} \cup \left(\bigcup_{i<j\leq n} F_{i,j} \right) \right), \tag{1}$$

where $F_{i,j}$ is the angle

$$F_{i,j} = \{v : v = x_i e_i + x_j e_j;\ x_i \leq 0,\ x_j \leq 0\}$$

(see Fig. 113). Furthermore, let $k < n$. Then there exist vectors $e_{k+1}, \cdots, e_n$ and a map

$$\varphi : \{k+1, \cdots, n\} \to \{1, \cdots, k\}$$

such that $e_1, \cdots, e_n$ is a basis in $\mathbf{R}^n$ (not necessarily orthonormal) and, denoting by $\Pi_{l,i}$ the closed half-plane $\{v : v = x_l e_l + x_i e_i; \ x_i \leq 0\}$ (Fig. 114), the following inclusions hold:

$$H \subset \mathrm{norm}\left(\{e_1\} \cup \left(\bigcup_{l>1} \Pi_{l,1}\right)\right) \text{ for } k = 1, \tag{2}$$

$$H \subset \mathrm{norm}\left(\{e_1, e_2\} \cup F_{1,2} \cup \left(\bigcup_{l>2} \Pi_{l,\varphi(l)}\right)\right)$$

for $k = 2$ where the map φ is onto, $\tag{3}$

$$H \subset \mathrm{norm}\left(\{e_1, \cdots, e_k\} \cup \left(\bigcup_{i<j\leq k} F_{i,j}\right) \cup \left(\bigcup_{l>k} \Pi_{l,\varphi(l)}\right)\right)$$

for $2 < k < n$. $\tag{4}$

Besides, the right-hand sides in the inclusions (1)–(4) are maximal 2-systems contained in S^{n-1}, i.e., any additional vector makes md larger than 2. □

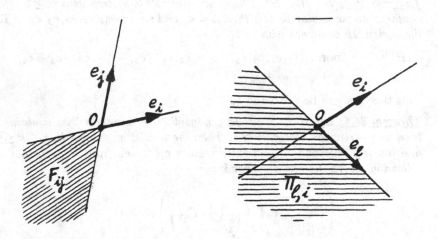

Fig. 113   Fig. 114

This theorem will be proved in §28.

Theorems 26.1 and 26.2 were proved in [Bt 13]. They contain a complete list of the nonsplittable 2-systems. This means that for any 2-system $H \subset \mathbf{R}^n$ there exists a direct decomposition $\mathbf{R}^n = L_1 \oplus \cdots \oplus L_s$ such that $H \subset L_1 \cup \cdots \cup L_s$ and each system $H \cap L_i$ has a form described either in Theorem 26.1 or in Theorem 26.2, $i = 1, \cdots, s$.

EXAMPLE 26.3: It follows from Theorem 26.1 that every 2-system $H \subset \mathbf{R}^3$ has to contain at least one particular vector (since $\mathbf{R}^3$ cannot contain the system $H^{(o)}$). Now it follows from Theorem 26.2 that there are only the following types of *nonsplittable* 2-systems $H \subset \mathbf{R}^3$:

$$H \subset \text{ norm } (\{e_1\} \cup \Pi_{2,1} \cup \Pi_{3,1}), \tag{5}$$

$$H \subset \text{ norm } (\{e_1, e_2, e_3\} \cup F_{1,2} \cup F_{1,3} \cup F_{2,3}). \tag{6}$$

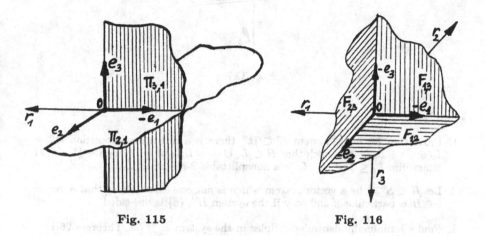

Fig. 115   Fig. 116

The case (5) is obtained when $k = 1$ (cf. (2) in Theorem 26.2), and (6) is obtained for $k = 3$ (cf. (1) in Theorem 26.2). We note that the case $k = 2$ is nonrealizable, because there is no mapping of the set $\{3\}$ onto $\{1,2\}$, see (3) in Theorem 26.2. The set in the right-hand side of (5) is contained in $r_1 \cup \Pi_{2,1} \cup \Pi_{3,1}$ (Fig. 115), where the half-planes $\Pi_{2,1}$ and $\Pi_{3,1}$ with bounding lines through the origin have the negative x_1-semiaxis in their relative interiors and r_1 is the positive x_1-semiaxis. Furthermore, the set on the right-hand side of (6) is contained in $r_1 \cup r_2 \cup r_3 \cup F_{1,2} \cup F_{1,3} \cup F_{2,3}$ (Fig. 116): the boundary $F_{1,2} \cup F_{1,3} \cup F_{2,3}$ of the coordinate octant $x_1 \leq 0, x_2 \leq 0, x_3 \leq 0$ is combined with the positive x_1-, x_2-, x_3-semiaxes. Finally, if a 2-system is *splittable*, then

$$H \subset \text{ norm } (P_{1,2} \cup l_3), \tag{7}$$

where $P_{1,2}$ is the (x_1, x_2)-plane and l_3 is the x_3-axis (Fig. 117). Thus for every 2-system $H \subset S^2$ there exists a basis e_1, e_2, e_3 (not necessarily orthonormal) such that one of the inclusions (5), (6), (7) holds.

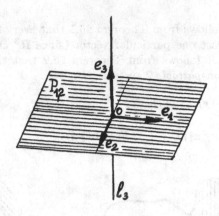

Fig. 117

Exercises

1. Prove that for any 2-system $H \subset \mathbf{R}^n$ there is a direct decomposition $\mathbf{R}^n = L_1 \oplus \cdots \oplus L_s \ (s \geq 1)$ such that $H \subset L_1 \cup \cdots \cup L_s$ and for every $i \in \{1, \cdots, s\}$ either dim $L_i \leq 2$ or $H \cap L_i$ is a nonsplittable 2-system in L_i.

2. Let $H \subset S^{n-1}$ be a vector system which is not one-sided. Prove that a vector $e \in H$ is particular if and only if the system $H \setminus \{e\}$ is one-sided.

3. Find all minimally dependent triples in the system $H^{(o)}$ (cf. Theorem 26.1).

4. Prove that the vector system $H^{(o)}$ does not contain any minimally dependent vectors $a_1, \cdots, a_{m+1}$ with $m = 3$ or 4.

5. Prove (applying a permutation of the indices $1, \cdots, n$) that the function φ in Theorem 26.2 may be supposed to satisfy the following conditions: (i) φ is nondecreasing; (ii) for the cardinalities of the inverse images $\varphi^{-1}(i)$, the inequalities

$$|\varphi^{-1}(1)| \geq \cdots \geq |\varphi^{-1}(k)|$$

hold.

6. Prove that under the conditions (i) and (ii), indicated in the previous exercise, the function φ is uniquely defined for each case (1)–(4).

7. Prove that if a vector system $H \subset \mathbf{R}^3$ is contained in $r_1 \cup \Pi_{2,1} \cup \Pi_{3,1}$ (Fig. 115), then md $H \leq 2$. In what case does a vector system $H \subset r_1 \cup \Pi_{2,1} \cup \Pi_{3,1}$ satisfy md $H = 1$?

8. Prove that if a vector system $H \subset \mathbf{R}^3$ is contained in $r_1 \cup r_2 \cup r_3 \cup F_{1,2} \cup F_{1,3} \cup F_{2,3}$ (Fig. 116), then md $H \leq 2$.

9. In what case does a 2-system $H \subset S^2$ satisfy both the inclusions (5), (6)?

10. Give an example of a polytope $M \subset \mathbf{R}^3$ with md $M = 2$ for which the vector system $H = H(M)$ satisfies the inclusion (5).

11. Give an example of a polytope $M \subset \mathbf{R}^3$ with md $M = 2$ for which the vector system $H = H(M)$ satisfies the inclusion (6).

12. List (with the help of Theorems 26.1 and 27.1) the cases which are possible for a 2-system H in $\mathbf{R}^4$.

13. Let $H \subset S^{n-1}$ be a 2-system. Let, furthermore, $L \subset \mathbf{R}^n$ be a subspace such that $0 < \dim L < n$ and no vector $v \in L$ is particular with respect to the vector system $(H \cap L) \cup \{v\}$. Prove that H is splittable.

14. Let $H \subset S^{n-1}$ be a nonsplittable 2-system and $a_1, a_2, a_3 \in H$ be minimally dependent vectors. Prove that at least one of them is particular with respect to the vector system $H \cap \text{lin} \, (a_o, a_1, a_2)$.

§27 The 2-systems without particular vectors

In this section, we prove Theorem 26.1. The proofs of some auxiliary propositions are omitted (see the exercises at the end of this section or a more detailed account in [Bt 13]).

Thus let $H \subset S^{n-1}$ be a nonsplittable 2-system without particular vectors, $n \geq 3$. Let, furthermore, $a_1, a_2, a_3 \in H$ be three minimally dependent vectors. There exist vectors $b_1, b_2, b_3, c_1, c_2, c_3$ in H which do not belong to $L = \text{lin} \, (a_1, a_2, a_3)$ such that the triples (a_1, b_1, c_1), (a_2, b_2, c_2), (a_3, b_3, c_3) are minimally dependent (cf. Exercise 8). Then the spaces $L_1 = \text{lin} \, (a_1, a_2, a_3, b_1, c_1)$, $L_2 = \text{lin} \, (a_1, a_2, a_3, b_2, c_2)$, $L_3 = \text{lin} \, (a_1, a_2, a_3, b_3, c_3)$ are three-dimensional. First we prove that L_1, L_2, L_3 are pairwise distinct (and consequently $n \geq 4$).

Suppose, conversely, that two of these subspaces coincide, say $L_2 = L_3$. Then the vectors a_2, a_3, b_2 form a basis in the space $L_2 = L_3$. Without loss of generality we may suppose (changing the roles of the vectors b_2, c_2, if necessary) that b_2, b_3 lie in one open half-space with boundary L (in the three-dimensional space $L_2 = L_3$), and c_2, c_3 lie in the other open half-space. Since a_2, b_2, c_2 are minimally dependent,

$$b_2 = -\alpha_2 a_2 - \gamma_2 c_2 \; (\alpha_2 > 0, \gamma_2 > 0). \tag{1}$$

Analogously,

$$b_3 = -\alpha_3 a_3 - \gamma_3 c_3 \; (\alpha_3 > 0, \gamma_3 > 0). \tag{2}$$

Furthermore, since b_2, b_3 lie in one open half-space with boundary L (in the three-dimensional space $L_2 = L_3$), we have

$$\lambda b_3 = b_2 + \mu a_2 + \nu a_3 \; (\lambda > 0). \tag{3}$$

Without loss of generality we may suppose (changing the roles of the vectors b_2, b_3, if necessary) that $\mu \geq 0$. From (2) and (3), we obtain the equation

$$\lambda\gamma_3 c_3 + b_2 + \mu a_2 + (\nu + \lambda\alpha_3)a_3 = o \quad (\lambda > 0, \mu \geq 0, \gamma_3 > 0). \tag{4}$$

Besides, we have the dependence

$$p_1 a_1 + p_2 a_2 + a_3 = o \quad (p_1 > 0, p_2 > 0), \tag{5}$$

since a_1, a_2, a_3 are minimally dependent.

If $\nu + \lambda\alpha_3$ is negative (i.e., $\nu + \lambda\alpha_3 = -\delta$, with $\delta > 0$), then, from (4) and (5), we obtain

$$\lambda\gamma_3 c_3 + b_2 + \delta p_1 a_1 + (\delta p_2 + \mu)a_2 = o.$$

This is a minimal dependence (since the coefficients are positive and a_1, a_2, b_2 are linearly independent), contradicting the equality md $H = 2$. Hence $\nu + \lambda\alpha_3 \geq 0$. Furthermore, if both the coefficients μ, $\nu + \lambda\alpha_3$ in (4) are positive, then (4) is a minimal dependence, contradicting the equality md $H = 2$. Consequently both the coefficients μ and $\nu + \lambda\alpha_3$ are nonnegative and at least one of them is equal to 0. However, the case $\mu = \nu + \lambda\alpha_3 = 0$ is impossible. Indeed, from (4) we would have the equality $\lambda\gamma_3 c_3 + b_2 = o$ for this case, and consequently $c_3 + b_2 = o$ (since $\| c_3 \| = \| b_2 \| = 1$, i.e., $-b_2 = c_3 \in H$), contradicting the result of Exercise 7 (applied to the minimally dependent triple (a_2, b_2, c_2)).

Thus the following two cases are possible:

(a) $\lambda\gamma_3 c_3 + b_2 + \mu a_2 = o \quad (\lambda\gamma_3 > 0, \mu > 0)$;

(b) $\lambda\gamma_3 c_3 + b_2 + (\nu + \lambda\alpha_3)a_3 = o \quad (\lambda\gamma_3 > 0, \nu + \lambda\alpha_3 > 0)$.

In the case (a), the vectors c_3, b_2, a_2 are minimally dependent and we have $c_3 = c_2$ (cf. Exercise 7). In the case (b), the vectors c_3, b_2, a_3 are minimally dependent and we have $b_2 = b_3$ (see Exercise 7). Thus, without loss of generality we may suppose (changing, if necessary, the roles of the pairs b_2, b_3 and c_2, c_3) that $b_2 = b_3$, i.e., we have three distinct vectors $b_2 = b_3, c_2, c_3$.

Now consider the third three-dimensional space L_1. Since a_1, b_1, c_1 are minimally dependent, we have

$$b_1 = -\alpha_1 a_1 - \gamma_1 c_1 \quad (\alpha_1 > 0, \gamma_1 > 0). \tag{6}$$

Consequently, by virtue of (1), (2), and (5),

$$p_1\alpha_2\alpha_3(b_1 + \gamma_1 c_1) + p_2\alpha_1\alpha_3(b_2 + \gamma_2 c_2) + \alpha_1\alpha_2(b_3 + \gamma_3 c_3)$$

$$= -\alpha_1\alpha_2\alpha_3(p_1 a_1 + p_2 a_2 + a_3) = o \tag{7}$$

This is a positive dependence of the vectors b_1, b_2, c_1, c_2, c_3 (since $b_2 = b_3$). If now $b_1 \notin L_2$, then the vectors b_2, c_2, c_3, b_1 are linearly independent and, according to (7), the vectors b_1, c_1, b_2, c_2, c_3 are minimally dependent, contradicting the equality md $H = 2$. Hence $b_1 \in L_2$, and this means $L_1 = L_2 = L_3$.

Thus from the equality $L_2 = L_3$ we have deduced that among b_2, b_3, c_2, c_3 there are only three distinct vectors; say $b_2 = b_3$. Furthermore, from the equality $L_1 = L_2$ we analogously obtain that among b_1, b_2, c_1, c_2 there are only three distinct vectors. So, there are two possibilities:

(i) $b_1 = b_2$, $c_1 \neq c_2$; (ii) $c_1 = c_2$, $b_1 \neq b_2$.

In the case (ii) we have $b_2 = b_3$, $c_2 \neq c_3$ and $c_1 = c_2$, $b_1 \neq b_2$. This means that $b_1 \neq b_3$, $c_1 \neq c_3$, contradicting the equality $L_1 = L_3$. Consequently the case (i) must hold, i.e., $b_1 = b_2 = b_3$, and c_1, c_2, c_3 are pairwise distinct. By virtue of (7), we have in this case

$$p_1\alpha_2\alpha_3(b_1 + \gamma_1 c_1) + p_2\alpha_1\alpha_3(b_1 + \gamma_2 c_2) + \alpha_1\alpha_2(b_1 + \gamma_3 c_3) = o,$$

i.e., the vectors b_1, c_1, c_2, c_3 are positively dependent. Besides, the vectors b_1, c_1, c_2 are linearly independent. Consequently b_1, c_1, c_2, c_3 are minimally dependent, contradicting the equality md $H = 2$. The obtained contradiction shows that the spaces L_1, L_2, L_3 are pairwise different.

Denote by Q the four-dimensional space lin $(a_1, a_2, a_3, b_2, b_3, c_2, c_3)$. The vectors b_2, b_3, c_2, c_3 form its basis. If $b_1 \notin Q$, then the vectors b_2, b_3, c_2, c_3, b_1 are linearly independent. Consequently, according to (7), the vectors b_1, b_2, b_3, c_1, c_2, c_3 are minimally dependent, contradicting the equality md $H = 2$. This shows that $b_1 \in Q$ and hence $c_1 \in Q$, i.e., the vectors $a_1, b_1, c_1, a_2, b_2, c_2$, a_3, b_3, c_3 are contained in the four-dimensional space Q.

Consider now the minimally dependent triplets (a_2, b_2, c_2) and (a_3, b_3, c_3). We have $b_3 \notin$ lin $(a_2, a_3, b_2) = L_2$ (since $L_2 \neq L_3$). Consequently a_2, a_3, b_2, b_3 is a basis in Q. The vector $-b_1$ has a representation $-b_1 = f_2 + f_3$, where $f_2 \in$ lin (a_2, b_2), $f_3 \in$ lin (a_3, b_3). It is clear that $f_2 \neq o, f_3 \neq o$, since $b_1 \notin$ lin $(a_2, b_2, c_2) \subset L_2$, $b_1 \notin$ lin $(a_3, b_3, c_3) \subset L_3$.

Admit that f_2 coincides with none of the vectors a_2, b_2, c_2 (up to a positive coefficient). Then f_2 is a positive linear combination of two of the vectors a_2, b_2, c_2, say $f_2 = \beta b_2 + \gamma c_2$, where $\beta > 0, \gamma > 0$. Hence,

$$b_1 + \beta b_2 + \gamma c_2 = -f_3 \in \text{lin } (a_3, b_3, c_3).$$

Furthermore, $f_3 \neq o$ is a positive linear combination of one or two of the vectors a_3, b_3, c_3. This means that we obtain four or five minimally dependent vectors (i.e., b_1, b_2, c_2 and one or two of the vectors a_3, b_3, c_3), contradicting the equality md $H = 2$. This contradiction shows that f_2 must coincide (up to a positive coefficient) with one of the vectors a_2, b_2, c_2. Similarly, f_3 must coincide (up to a positive coefficient) with one of the vectors a_3, b_3, c_3. The same reasoning shows that $-c_1 = g_2 + g_3$, where g_2 coincides (up to a positive coefficient) with one of the vectors a_2, b_2, c_2 and g_3 coincides (up to a positive coefficient) with one of the vectors a_3, b_3, c_3.

Admit $f_2 = ta_2$, $t > 0$. Then we have three possibilities: either $f_3 = ua_3$, $u > 0$, or $f_3 = vb_3$, $v > 0$, or $f_3 = wc_3$, $w > 0$. In the case $f_3 = ua_3$, $u > 0$, we obtain $b_1 = -f_2 - f_3 \in \text{lin}\,(a_2, a_3) = \text{lin}\,(a_1, a_2, a_3)$, which is impossible by virtue of the result of Exercise 7 (recall that the nine vectors $a_1, a_2, a_3, b_1, b_2, b_3, c_1, c_2, c_3$ are pairwise different). Furthermore, in the case $f_3 = vb_3$, $v > 0$, we conclude that b_1, a_2, b_3 are linearly dependent and consequently $b_3 \in \text{lin}\,(b_1, a_2) \subset L_1$, i.e., $L_3 = L_1$, contradicting what was previously proved. Finally, in the case $f_3 = wc_3$, $w > 0$, the vectors b_1, a_2, c_3 are linearly dependent and consequently $c_3 \in \text{lin}\,(b_1, a_2) \subset L_1$, i.e., again $L_3 = L_1$, which is impossible. Thus the case $f_2 = ta_2, t > 0$, does not hold.

In other words, f_2 is equal to either b_2 or c_2 (up to a positive coefficient). Similarly, f_3 is equal to either b_3 or c_3 (up to a positive coefficient). The same reasoning shows that g_2 is equal to either b_2 or c_2 and g_3 is equal to either b_3 or c_3 (up to positive coefficients).

We remark now that (by virtue of the pairwise distinctness of the spaces L_1, L_2, L_3) we may interchange the symbols b_2 and c_2; similarly, we may interchange the symbols b_3 and c_3. Therefore, we may suppose that $f_2 = \lambda_2 b_2$, $f_3 = \lambda_3 b_3$ ($\lambda_2 > 0, \lambda_3 > 0$), i.e., the three vectors $b_1 = -f_2 - f_3, b_2, b_3$ are minimally dependent.

Furthermore, if $g_2 = sb_2$, $s > 0$, then the vectors $b_1 = -\lambda_2 b_2 - \lambda_3 b_3$ and $c_1 = -g_2 - g_3$ are contained in the three-dimensional space $L' = \text{lin}\,(b_2, b_3, c_3)$. Consequently $a_1 \in L'$ (since a_1, b_1, c_1 are minimally dependent) and $a_3 \in L'$ (since a_3, b_3, c_3 are minimally dependent), i.e., $L' = L_1 = L_3$, contradicting what was previously proved. This contradiction shows that $g_2 = \mu_2 c_2$, $\mu_2 > 0$. Similarly, $g_3 = \mu_3 c_3$, $\mu_3 > 0$. This means that the three vectors $c_1 = -g_2 - g_3, c_2, c_3$ are minimally dependent.

We may sum up. The vectors a_2, b_2, a_3, b_3 form a basis in Q. Moreover, each triplet

$$(a_1, a_2, a_3),\ (b_1, b_2, b_3),\ (c_1, c_2, c_3),\ (a_1, b_1, c_1),\ (a_2, b_2, c_2),\ (a_3, b_3, c_3)$$

is minimally dependent, i.e., we have positive dependences:

$$a_1 + \lambda_2 a_2 + \lambda_3 a_3 = o, \quad b_1 + \mu_2 b_2 + \mu_3 b_3 = o, \quad pa_1 + p'b_1 + c_1 = o, \quad (8)$$

$$c_1 + \nu_2 c_2 + \nu_3 c_3 = o, \quad qa_2 + q'b_2 + c_2 = o, \quad \tau a_3 + \tau'b_3 + c_3 = o. \qquad (9)$$

We now introduce a basis e_1, e_2, e_3, e_4 in Q by the equalities

$$e_1 = p\lambda_2 a_2, \quad e_2 = p\lambda_3 a_3, \quad e_3 = p'\mu_2 b_2, \quad e_4 = p'\mu_3 b_3. \qquad (10)$$

Then, by virtue of (8),

$$a_1 = -\frac{1}{p}(e_1 + e_2), \quad b_1 = -\frac{1}{p'}(e_3 + e_4), \quad c_1 = e_1 + e_2 + e_3 + e_4. \tag{11}$$

Furthermore, according to (9) and (10), we have

$$\begin{aligned}
c_1 &= \nu_2(qa_2 + q'b_2) + \nu_3(\tau a_3 + \tau' b_3) \\
&= \frac{\nu_2 q}{p\lambda_2}e_1 + \frac{\nu_2 q'}{p'\mu_2}e_3 + \frac{\nu_3 \tau}{p\lambda_3}e_2 + \frac{\nu_3 \tau'}{p'\mu_3}e_4.
\end{aligned}$$

Consequently,

$$\frac{\nu_2 q}{p\lambda_2} = 1, \quad \frac{\nu_2 q'}{p'\mu_2} = 1, \quad \frac{\nu_3 \tau}{p\lambda_3} = 1, \quad \frac{\nu_3 \tau'}{p'\mu_3} = 1.$$

According to these relations, the last two equalities in (9) take the form

$$\frac{p\lambda_2}{\nu_2}a_2 + \frac{p'\mu_2}{\nu_2}b_2 + c_2 = 0, \quad \frac{p\lambda_3}{\nu_3}a_3 + \frac{p'\mu_3}{\nu_3}b_3 + c_3 = 0,$$

and consequently,

$$c_2 = -\frac{1}{\nu_2}(e_1 + e_3), \quad c_3 = -\frac{1}{\nu_3}(e_2 + e_4). \tag{12}$$

By virtue of the relations (10)–(12) (and taking into account that $\| a_i \| = \| b_i \| = \| c_i \| = 1$, since $H \subset S^{n-1}$) we obtain that

$$\begin{aligned}
\{a_1, &a_2, a_3, b_1, b_2, b_3, c_1, c_2, c_3\} \\
&= \text{norm } \{-e_1 - e_2, e_1, e_2, -e_3 - e_4, e_3, e_4, e_1 + e_2 + e_3 + e_4, \\
&\quad -e_1 - e_3, -e_2 - e_4\} = H^{(o)}
\end{aligned}$$

(see the statement of Theorem 26.1). Consequently $H \supset H^{(o)}$ (because $a_i, b_i, c_i \in H$).

It remains to prove that $n = 4$ and H coincides with $H^{(o)}$. For this purpose, we remark first that the system $H^{(o)}$ is not one-sided in the space Q (even the vector system $e_1, e_2, -e_1 - e_2, e_3, e_4, -e_3 - e_4$ contained in Q is not one-sided). Furthermore, each vector $v \in H \cap Q$ is nonparticular with respect to the vector system $H \cap Q$. Indeed, if $v \notin H^{(o)}$, then v is nonparticular, since $(H \cap Q)\backslash\{v\} \supset H^{(o)}$ is not one-sided. If $v \in H^{(o)}$, then also v is nonparticular, since

$$e_1 = (e_1 + e_2 + e_3 + e_4) + e_4 + (-e_2 - e_4) + (-e_3 - e_4)$$

(analogously for e_2, e_3, e_4),

$$-e_1 - e_2 = (-e_1 - e_3) + (-e_2 - e_4) + e_3 + e_4$$

(similarly for $-e_1 - e_3, -e_2 - e_4, -e_3 - e_4$), and $e_1 + e_2 + e_3 + e_4$ is, evidently, nonparticular, too. Now the result of Exercise 10 shows that Q is not a proper subspace of $\mathbf{R}^n$ (otherwise the system H would be decomposable), i.e., $Q = \mathbf{R}^n$, and consequently $n = 4$.

It remains to prove that the 2-system $H \subset \mathbf{R}^4$ coincides with $H^{(o)}$. Indeed, let $v \in H$. If $v \in \text{lin } (e_1, e_3)$, then v coincides with one of the vectors $e_1, e_3, -e_1 - e_3$ (Exercise 7), i.e., $v \in H^{(o)}$. Similarly, if $v \in \text{lin } (e_2, e_4)$, then $v \in H^{(o)}$.

Let, finally, $v \in H \setminus \text{lin } (e_1, e_3) \setminus \text{lin } (e_2, e_4)$. Then $v = v^* + v^{**}$, where $v^* \in \text{lin } (e_1, e_3)$, $v^{**} \in \text{lin } (e_2, e_4)$ and $v^* \neq o$, $v^{**} \neq o$. The vector $-v^*$ is representable as a positive linear combination of one or two of the vectors $e_1, e_3, -e_1 - e_3$. If $-v^*$ is expressed by *two* vectors, say

$$-v^* = \lambda e_1 + \mu(-e_1 - e_3); \quad \lambda > 0, \mu > 0,$$

then we have $v + \lambda e_1 + \mu(-e_1 - e_3) - v^{**} = o$, and (representing $-v^{**}$ by a positive linear combination of one or two of the vectors $e_2, e_4, -e_2 - e_4$) we obtain a minimal dependence between four or five vectors, contradicting the equality md $H = 2$. So, $-v^*$ coincides with one of the vectors $e_1, e_3, -e_1 - e_3$ (up to a positive coefficient), and similarly $-v^{**}$ coincides with one of the vectors $e_2, e_4, -e_2 - e_4$ (up to a positive coefficient). Thus we have the following cases (α, β are positive):

$$v = -\alpha e_1 - \beta e_2, \quad v = -\alpha e_3 - \beta e_4, \quad v = \alpha(e_1 + e_3) + \beta(e_2 + e_4); \quad (13)$$

$$v = -\alpha e_1 - \beta e_4, \quad v = -\alpha e_3 - \beta e_2; \quad (14)$$

$$v = \alpha(e_1 + e_3) - \beta e_2, \quad v = \alpha(e_1 + e_3) - \beta e_4; \quad (15)$$

$$v = \alpha(e_2 + e_4) - \beta e_1, \quad v = \alpha(e_2 + e_4) - \beta e_3. \quad (16)$$

In the first case of (13) we have

$$v + \alpha(e_1 + e_2 + e_3 + e_4) + (\alpha - \beta)(-e_2 - e_4) + \alpha(-e_3 - e_4)$$
$$+(\alpha - \beta)e_4 = o \text{ (for } \alpha > \beta),$$

$$v + \beta(e_1 + e_2 + e_3 + e_4) + (\beta - \alpha)(-e_1 - e_3) + \beta(-e_3 - e_4)$$
$$+(\beta - \alpha)e_3 = o \text{ (for } \alpha < \beta).$$

Hence if $\alpha \neq \beta$, then md $(H \cup \{-\alpha e_1 - \beta e_2\}) \geq 4$, and this is impossible. Consequently the vector $v = -\alpha e_1 - \beta e_2$ belongs to H only if $\alpha = \beta$, i.e., $v = \text{norm } (-e_1 - e_2) \in H^{(o)}$. The last two cases of (13) may be considered analogously.

If the equalities (14), (15) or (16) hold, then there exist no positive numbers α, β for which $v \in H$. For example, if $v = -\alpha e_1 - \beta e_4$, then

$$v + \alpha(e_1 + e_2 + e_3 + e_4) + \alpha(-e_2 - e_4) + \alpha(-e_3 - e_4) + (\alpha + \beta)e_4 = o;$$

if $v = \alpha(e_1 + e_3) - \beta e_2$, then

$$v + \alpha(-e_2 - e_3) + \alpha(-e_1 - e_4) + \alpha e_4 + (\alpha + \beta)e_2 = o,$$

contradicting the equality md $H = 2$. Thus, if $v \in H$, then $v \in H^{(o)}$, i.e., $H \subset H^{(o)}$. ∎

Exercises

1. Let $H \subset S^{n-1}$ be a 2-system, $n \geq 3$. Let, furthermore, a_1, a_2, a_3 be minimally dependent vectors, and $L = \mathrm{lin}\,(a_1, a_2, a_3)$. Denote by K the convex cone with the apex o consisting of the nonnegative linear combinations of the vectors from $H \setminus L$. Prove that if $v \in K \cap L$ is a unit vector not contained in int K (Fig. 118), then v coincides with one of the vectors $-a_1, -a_2, -a_3$.

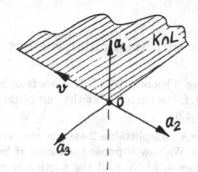

Fig. 118

2. In the notation of Exercise 1, prove that v is representable in the form $v = \lambda_1 b_1 + \lambda_2 b_2$, where $b_1, b_2 \in H \setminus L$ and the numbers λ_1, λ_2 are positive.

3. In the notation of Exercise 1, prove that if $u \in K \cap L$ is a unit vector such that $-u \in H$, then u coincides with one of the vectors $-a_1, -a_2, -a_3$.

4. In the notation of Exercise 1, prove that if a unit vector $w \in L$ satisfies $w + \lambda_1 b_1 + \lambda_2 b_2 = o$ (where $b_1, b_2 \in H \setminus L$ and the numbers λ_1, λ_2 are positive), then w is not representable as a positive linear combination of two vectors from $H \cap L$ distinct from w.

5. Let $H \subset S^{n-1}$, $n > 2$, be a nonsplittable 2-system, $a_1, a_2, a_3 \in H$ be minimally dependent, nonparticular vectors, and $L = \mathrm{lin}\,(a_1, a_2, a_3)$. Denote by K the set of all nonnegative linear combinations of vectors from $H \setminus L$. Prove that from the following complete list of logical possibilities the cases (a), (b), (c) are not realizable:

   (a) $K \cap L = \{o\}$ and $o \notin \mathrm{ri}\, K$;

   (b) $K \cap L = \{o\}$ and $o \in \mathrm{ri}\, K$;

   (c) $\dim (K \cap L) = 1$;

   (d) $\dim (K \cap L) = 2$.

6. In the notation of Exercise 5, prove that $L \subset K$.

7. In the notation of Exercise 5, prove that the intersection $H \cap L$ contains only the vectors a_1, a_2, a_3.

8. In the notation of Exercise 5, prove that there exist vectors $b_1, b_2 \in H$ not belonging to L such that the vectors b_1, b_2, a_3 are minimally dependent.

9. Let $H \subset S^{n-1}$ be a 2-system. Let, furthermore, $L \subset \mathbf{R}^n$ be a proper subspace such that the system $H \cap L$ is not one-sided in L and no vector from $H \cap L$ is particular for the system $H \cap L$. Denote by K the set of the nonnegative linear combinations of vectors from $H \setminus L$. Prove that $K \cap L = \{o\}$.

10. In the notation of Exercise 9, prove that the system H is splittable.

§28 The 2-system with particular vectors

In this section we prove Theorem 26.2. The proofs of some auxiliary propositions are omitted (cf. the exercises at the end of this section or a more detailed account in [Bt 13]).

Thus let $H \subset S^{n-1}$ be a nonsplittable 2-system that contains at least one particular vector, $n \geq 3$. We may suppose (replacing H by a larger 2-system, if necessary; cf. Exercises 4, 5) that all the particular vectors $e_1, \cdots, e_k$ of the system H are linearly independent and the vectors $-e_1, \cdots, -e_k$ belong to H. Denote by L the subspace $\mathrm{lin}\,(e_1, \cdots, e_k)$ and by N its orthogonal complement, i.e., $\mathbf{R}^n = L \oplus N$. Furthermore, denote by $\pi : \mathbf{R}^n \to N$ the orthogonal projection (hence L is the kernel of π).

First, we consider the case when $1 \leq k < n$, i.e., $1 \leq \dim N < n$. Then $H^* = \mathrm{norm}\,\pi(H \setminus L)$ is a vector system in N that is not one-sided and satifies the condition md $H^* = 1$ (Exercise 14). Hence

$$H^* = \{e_{k+1}^*, \cdots, e_n^* - e_{k+1}^*, \cdots, -e_n^*\},$$

where $e_{k+1}^*, \cdots, e_n^*$ is a basis in N (cf. the proof of Theorem 25.2).

Denote by G_l the subspace $\mathrm{lin}\,(L \cup \pi^{-1}(e_l^*))$, $l = k+1, \cdots, n$. Each subspace G_l has the dimension $k+1$ and the inclusion $H \subset G_{k+1} \cup \cdots \cup G_n$ is true. Since e_i is a particular vector, there exists a hyperplane $Q_i \subset \mathbf{R}^n$ through the origin such that $H \setminus \{e_i\} \subset P_i$, $e_i \notin P_i$, where P_i is a closed half-space with the bounding hyperplane Q_i, $i = 1, \cdots, k$. Since $e_j \in H$ and $-e_j \in H$, we have $e_j \in Q_i$ for $i \neq j$. The intersection $Q_i \cap G_l$ is a hyperplane in the space G_l such that $e_j \in Q_i \cap G_l$ for $i \neq j$ and $e_i \notin Q_i \cap G_l$. Hence the intersection $Q_1 \cap \cdots \cap Q_k \cap G_l$ is a line which is not contained in L. We denote by e_l a unit vector of this line. Then $e_1, \cdots, e_k, e_l$ is a basis of the subspace G_l, and consequently $e_1, \cdots, e_k, e_{k+1}, \cdots, e_n$ is a basis in $\mathbf{R}^n$.

Since $H \setminus \{e_i\} \subset P_i \, (i = 1, \cdots, k)$, each vector $v \in (H \cap G_l) \setminus \{e_i\}$ has the form

$$v = \alpha_1 c_1 + \cdots + \alpha_k c_k + \beta e_l \; (\alpha_i \leq 0).$$

Consequently for each vector $v \in (H \cap G_l) \setminus L$ we have

$$-v = x_1 e_1 + \cdots + x_k e_k + x_l e_l \; (x_i \geq 0 \text{ for all } i = 1, \cdots, k). \tag{1}$$

Now we denote by $S_l^{(+)}, S_l^{(-)}$ the half-spaces of all vectors

$$x_1 e_1 + \cdots + x_k e_k + x_{k+1} e_{k+1} + \cdots + x_n e_n$$

for which $x_l \geq 0$ and $x_l \leq 0$, respectively, and by Q_l their intersection, i.e., $Q_l = S_l^{(+)} \cap S_l^{(-)}$ is the $(n-1)$-dimensional subspace spanned by all the vectors $e_1, \cdots, e_n$ except for e_l (l is one of the indices $k+1, \cdots, n$). Then $G_l \cap S_l^{(+)}$ and $G_l \cap S_l^{(-)}$ are two closed half-spaces in G_l with the common boundary hyperplane $L \subset G_l$. It is easily shown that each of the sets $H_l^{(+)} = H \cap G_l \cap \operatorname{int} S_l^{(+)} = H \cap \operatorname{int} S_l^{(+)}$, and $H_l^{(-)} = H \cap G_l \cap \operatorname{int} S_l^{(-)} = H \cap \operatorname{int} S_l^{(-)}$ contains at least two elements. Indeed, if $H \cap \operatorname{int} S_l^{(+)}$ contains no element of H, then the system H is one-sided (contradicting the condition of the theorem), and if $H \cap \operatorname{int} S_l^{(+)}$ contains only one element e of H, then e is a particular element of H (contradicting $e \notin L$). Hence $H \cap \operatorname{int} S_l^{(+)}$ contains at least two elements of H. Similarly, $H \cap \operatorname{int} S_l^{(-)}$ contains at least two elements of H.

Let now $v \in H_l^{(+)}, v' \in H_l^{(-)}$, i.e., $-\alpha v = x_1 e_1 + \cdots + x_k e_k + e_l$, $-\alpha' v' = x_1' e_1 + \cdots + x_k' e_k - e_l$, where $\alpha > 0, \alpha' > 0$ and the coefficients $x_1, \cdots, x_k$, $x_1', \cdots, x_k'$ are nonnegative. Then

$$\alpha v + \alpha' v' + (x_1 + x_1') e_1 + \cdots + (x_k + x_k') e_k = o,$$

and hence at most one of the sums $x_1 + x_1', \cdots, x_k + x_k'$ is positive (since $v, v', e_1, \cdots, e_k \in H$ and md $H = 2$). It follows that each vector $v \in H_l^{(+)}$ has the form $-\alpha v = x_i e_i + e_l$, where i is one of the indices $1, \cdots, k$ and $\alpha > 0, x_i \geq 0$. Moreover, there is at least one element $v \in H_l^{(+)}$ for which one of the coefficients $x_1, \cdots, x_k$ is positive, say $-\alpha v = x_i e_i + e_l$, $x_i > o$ (since there are at least two elements of the set $H_l^{(+)}$, and if $x_1 = \cdots = x_k = o$ for one of them, then we may take another element). Similarly, each vector $v' \in H_l^{(-)}$ has the form $-\alpha' v' = x_j' e_j - e_l$; $x_j' \geq 0, \alpha' > 0$, and there is an element $v' \in H_l^{(-)}$ for which $x_j' > 0$. Finally, we conclude that $-\alpha v = x_i e_i + e_l$, $-\alpha' v' = x_j' e_j - e_l$, where the index $i = j$ is the *same* for all the elements v, v'. We denote this index by $\varphi(l)$. So we have $-\gamma w = x_{\varphi(l)} e_{\varphi(l)} \pm e_l$ for each $w \in (H \cap G_l) \setminus L$, where $\varphi(l)$ is one of the numbers $1, \cdots, k$ and the coefficients

$\gamma > 0, x_{\varphi(l)} \geq 0$ depend on w. This means that $H \setminus L \subset$ norm $\left(\bigcup_{l>k} \Pi_{l,\varphi(l)} \right)$ (see the statement of Theorem 26.2). More detailed,

$$H \setminus L \subset \text{ norm } \bigcup_{l>1} \Pi_{l,1} \text{ for } k = 1,$$

$$H \setminus L \subset \text{ norm } \bigcup_{l>k} \Pi_{l,\varphi(l)} \text{ for } 2 \leq k \leq n-1,$$

$$H \setminus L = \emptyset \text{ for } k = n$$

(see (1)–(4) in Theorem 26.2).

It remains to investigate the part $H \cap L$ of the system H. We investigate this point in the general case, that is, for dim $N = 0, 1, \cdots, n-1$ (or, what is the same, for $k = \dim L = 1, \cdots, n$). Since $H \setminus \{e_i\} \subset P_i$ for $i = 1, \cdots, k$, then

$$(H \cap L) \quad \setminus \quad \{e_1, \cdots, e_k\} \subset P_1 \cap \cdots \cap P_k \cap L$$
$$= \{v : -v = x_1 e_1 + \cdots + x_k e_k; x_1 \geq 0, \cdots, x_k \geq 0\}.$$

This means that

$$H \cap L \subset \text{ norm } \{e_1, -e_1\} \text{ for } k = 1,$$

$$H \cap L \subset \text{ norm } \left(\{e_1, \cdots, e_k\} \cup \left(\bigcup_{i<j\leq k} F_{i,j} \right) \right) \text{ for } 2 \leq k \leq n.$$

Summarizing the obtained results, we have for the set $H = (H \cap L) \cup (H \setminus L)$:

$$H \subset \text{ norm } \left(\{e_i\} \cup \left(\bigcup_{l>1} \Pi_{l,1} \right) \right) \text{ for } k = 1,$$

$$H \subset \text{ norm } \left(\{e_1, \cdots, e_k\} \cup \left(\bigcup_{i<j\leq k} F_{i,j} \right) \cup \left(\bigcup_{l>k} \Pi_{l,\varphi(l)} \right) \right)$$
$$\text{for } 2 \leq k \leq n-1,$$

$$H \subset \text{ norm } \left(\{e_1, \cdots, e_n\} \cup \left(\bigcup_{i<j\leq n} F_{ij} \right) \right) \text{ for } k = n.$$

Furthermore, we notice that in the case $k = 2$, if φ is not onto, say $\varphi(l) = 1$ for $l = 3, \cdots, n$, then

$$H \subset \text{norm}\left(\{e_1, e_2\} \cup F_{1,2} \cup \left(\bigcup_{l>2} \Pi_{l,1}\right)\right) \subset \text{norm}\left(\{e_1\} \cup \left(\bigcup_{l>1} \Pi_{l,1}\right)\right)$$

(since $\{e_2\} \cup F_{1,2} \subset \Pi_{2,1}$). This means that for $k = 2$ it is sufficient to consider the case when φ is onto (see (3) in Theorem 26.2).

It remains to prove that md $H' = 2$ for every vector system H' indicated on the right-hand side of the inclusions (1)–(4) in Theorem 26.2.

Let $a_o, \cdots, a_m \in H'$ be minimally dependent vectors. First we remark that for every $j = 1, \cdots, n$ the intersection $H' \cap Q_j$ is a vector system which is contained in a vector system of the same type as (1)–(4), but situated in $(n-1)$-dimensional space Q_j (we write "is contained" instead of "coincides" because, for example, in the case $k = 1$ the intersection $H' \cap Q_1$ has the form $\{e_2, \cdots, e_n, -e_2, \cdots, -e_n\}$, i.e., md $(H' \cap Q_1) < 2$). Thus, by an obvious induction, we conclude that if $\{a_o, \cdots, a_m\} \subset Q_j$, then md $\{a_o, \cdots, a_m\} \leq 2$ (i.e., $m \leq 2$).

Furthermore, in the case $k < n$ we may suppose that each intersection $\{a_o, \cdots, a_m\} \cap \text{int } S_n^{(+)}$, $\{a_o, \cdots, a_m\} \cap \text{int } S_n^{(-)}$ is nonempty (otherwise we have $\{a_o, \cdots, a_m\} \subset Q_n$ and consequently $m \leq 2$). Without loss of generality we may suppose that $a_o \in \text{int } S_n^{(+)}$, $a_1 \in \text{int } S_n^{(-)}$ (and hence $a_o, a_1 \in \Pi_{n,\varphi(n)}$). Moreover, $e_{\varphi(n)} \in \{a_2, \cdots, a_m\}$ (otherwise $\{a_o, \cdots, a_m\} \subset P_{\varphi(n)}$ and consequently $\{a_o, \cdots, a_m\} \subset Q_{\varphi(n)}$, i.e., $m \leq 2$). However, in this case the vectors $a_o, a_1, e_{\varphi(n)}$ are linearly dependent and consequently $m \leq 2$, too.

Finally, in the case $k = n$ there exist indices i, j such that $\{a_o, \cdots, a_m\} \cap F_{i,j} \neq \emptyset$ (otherwise $\{a_o, \cdots, a_m\} \subset \{e_1, \cdots, e_n\}$, contradicting the minimal dependence of the vectors $a_o, \cdots, a_m$). Without loss of generality we may suppose that $\{a_o, \cdots, a_m\} \cap F_{1,2} \neq \emptyset$ and $a_o \in F_{1,2}$. Then $e_1 \in \{a_o, \cdots, a_m\}$ (otherwise $\{a_o, \cdots, a_m\} \subset Q_1$ and $m \leq 2$); similarly, $e_2 \in \{a_o, \cdots, a_m\}$. However, in this case the vectors a_o, e_1, e_2 are linearly dependent and, consequently, $m \leq 2$.

The maximality of the vector systems indicated on the right-hand sides of the relations (1)–(4) is evident, since none of these systems is contained in any other. ∎

Exercises

1. Let $H \subset S^{n-1}$ be a 2-system and e be a particular vector of H. Prove that $H \cup \{-e\}$ is a 2-system, too.

2. Let $H \subset S^{n-1}$ be a 2-system that contains at least one particular vector. Prove that the set $H^\triangle$ of all its particular vectors is finite.

3. Let $H \subset H'$ be 2-systems contained in S^{n-1}. Prove that the set of the particular vectors of H' is contained in the set $H^\triangle$ of the particular vectors of the system H.

4. Let $H \subset S^{n-1}$ be a 2-system. Prove that in S^{n-1} there exists a 2-system $H_1 \supset H$ such that, together with every of its particular vector e, the system H_1 contains the vector $-e$, too.

5. Let $H_1 \subset S^{n-1}$ be a 2-system such that, together with every particular vector e of it, the system H_1 contains the vector $-e$. Prove that the particular vectors of H_1 are linearly independent.

6. Show by an example that for every $n \geq 2$ there exists a 2-system $H \subset S^{n-1}$ such that its particular vectors are linearly dependent.

7. Let $H \subset S^{n-1}$ be a nonsplittable 2-system and e be a particular vector of H. Prove that there exist vectors $b_1, b_2 \in H$ such that e, b_1, b_2 are minimally dependent.

8. Let $H \subset S^{n-1}$ be a vector system, $L \subset \mathbf{R}^n$ be a proper subspace, and $\pi : \mathbf{R}^n \to N$ be the orthogonal projection onto the orthogonal complement of L. Prove that if $H \cap L$ is not one-sided in L and norm $\pi(H \setminus L)$ is not one-sided in N, then H is not one-sided.

9. In the notation of Exercise 8, prove that if $H \cap L$ is not one-sided in L, then md norm $\pi(H \setminus L) \leq$ md H.

10. Let $H \subset S^{n-1}$ be a 2-system such that the set $H^\triangle$ of all its particular vectors is nonempty, and together with every $e \in H^\triangle$ the system H contains the opposite vector $-e$. Let, furthermore, $L \subset \mathbf{R}^n$ be a proper subspace such that the vector system $H \cap L$ is not one-sided in L. Denote by $\pi : \mathbf{R}^n \to N$ the orthogonal projection onto the orthogonal complement N of the subspace L. Prove that if md norm $\pi(H \setminus L) = 2$ and $a_1^*, a_2^*, a_3^* \in$ norm $\pi(H \setminus L)$ are minimally dependent, then their representatives $a_1, a_2, a_3 \in H$ are uniquely defined and are minimally dependent (with the same coefficients as a_1^*, a_2^*, a_3^*).

11. In the notation of Exercise 10, prove that if the vector system norm $\pi(H \setminus L)$ is not splittable and md norm $\pi(H \setminus L) = 2$, then the vector system norm $\pi(H \setminus L)$ contains no particular vector.

12. In the notation of Exercise 10, prove that if dim $N = 4$ and the vector system norm $\pi(H \setminus L)$ is isomorph to $H^{(o)}$ (cf. Theorem 26.1), then the system H is splittable.

13. In the notation of Exercise 10, prove that if dim $N = 2$ and the vector system norm $\pi(H \setminus L)$ is nonsplittable, then the system H is splittable.

14. Let $H \subset S^{n-1}$ be a nonsplittable 2-system, $n \geq 3$. Assume that the set $H^\triangle = \{e_1, \cdots, e_k\}$ of all its particular vectors is nonempty and linearly independent. Assume, furthermore, that $k < n$ and denote by $\pi : \mathbf{R}^n \to N$ the orthogonal projection onto the orthogonal complement of the subspace $L = \operatorname{lin} H^\triangle$. Prove that md norm $\pi(H \setminus L) = 1$.

§29 The compact, convex bodies with md $M = 2$

We now can prove a theorem that contains the complete list of the compact, convex, indecomposable bodies $M \subset \mathbf{R}^n$ with md $M = 2$. In the following theorem [Bt 13], we preserve the notations of Theorem 26.2 (see also Exercises 5, 6 in §26).

Theorem 29.1: Let $M \subset \mathbf{R}^n$ be a compact, convex body, $n \geq 3$. Assume that $o \in \text{int } M$ and M is indecomposable. The inequality md $M \leq 2$ *holds in and only in the following cases:*

(A) *There exist a basis $e_1, \cdots, e_n$ in $\mathbf{R}^n$ and closed, two-dimensional, convex sets B_{ij} ($i < j \leq n$) such that*

   (i) $B_{ij} \subset \text{lin } (e_i, e_j)$ *and* ext $B_{ij} \subset \{e_i, e_j\} \cup F_{i,j}$,

   (ii) $M^* = \text{conv } (\bigcup_{i<j\leq n} B_{ij})$.

(B) *There exist a basis $e_1, \cdots, e_n$ in $\mathbf{R}^n$ and closed, two-dimensional, convex sets $C_2, \cdots, C_n$ such that*

   (i) $C_l \subset \text{lin } (e_1, e_l)$ *and* ext $C_l \subset \{e_1\} \cup \Pi_{l,1}$ *for $l = 2, \cdots, n$,*

   (ii) $M^* = \text{conv } (C_2 \cup \cdots \cup C_l)$.

(C) *There exist a basis $e_1, \cdots, e_n$ in $\mathbf{R}^n$, an integer k ($2 \leq k < n$), a map $\varphi : \{k+1, \cdots, n\} \to \{1, \cdots, k\}$ (that is onto in the case $k = 2$), and closed, two-dimensional, convex sets B_{ij}, C_l ($i < j \leq k; l > k$) such that*

   (i) $B_{ij} \subset \text{lin } (e_i, e_j)$ *and* ext $B_{ij} \subset \{e_i, e_j\} \cup F_{i,j}$,

   (ii) $C_l \subset \text{lin } (e_l, e_{\varphi(l)})$ *and* ext $C_l \subset \{e_{\varphi(l)}\} \cup \Pi_{l,\varphi(l)}$,

   (iii) $M^* = \text{conv } ((\bigcup_{i<j\leq k} B_{ij}) \cup (C_{k+1} \cup \cdots \cup C_l))$.

(D) $n = 4$ *and there exists a basis e_1, e_2, e_3, e_4 in $\mathbf{R}^4$ such that $H(M) = H^{(o)}$ (see Theorem 26.1).* $\square$

We give the proof of this theorem in the second part of the section.

Theorem 29.1 gives a list of compact, convex bodies $M \subset \mathbf{R}^n$ with md $M = 2$ in terms of their polar bodies. We are going now to give a direct description of these bodies. First, we introduce a class of bodies which are named *stacks*. Let $W \subset \mathbf{R}^n$ be a parallelotope and Π be an $(n-1)$-dimensional face of W. A compact convex body M satisfying the inclusions $\Pi \subset M \subset W$ is said to be a *stack* with the basis Π if for each hyperplane L, which is parallel to Π and passes through an interior point of M, the intersection $L \cap M$ is an $(n-1)$-dimensional parallelotope whose faces are correspondingly parallel to the faces of Π.

We remark that the second basis of a stack M (i.e., the intersection of M with the supporting hyperplane parallel to Π) is a parallelotope that may have any of the dimensions $0, 1, \cdots, n-1$. Three-dimensional stacks are shown in the Fig. 119.

There is another description of stacks. Let W, Π be as above, and $[q, a_1], \cdots, [q, a_n]$ be edges of W with the common vertex $q \in \Pi$, where $[q, a_i] \subset \Pi$ for $i = 2, \cdots, n-1$. For each index $i = 2, \cdots, n$ denote by Γ_i the parallelogram

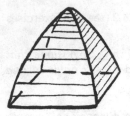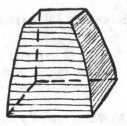

Fig. 119

with two sides $[q, a_1]$, $[q, a_i]$. Furthermore, choose a two-dimensional convex figure Φ_i such that $[q, a_i] \subset \Phi_i \subset \Gamma_i$. Finally, denote by G_i the "beam" $\Phi_i \oplus L_i$, where L_i is the $(n-2)$-dimensional flat that contains all the edges $[q, a_2], \cdots, [q, a_n]$ except for $[q, a_i]$. Then $M = G_2 \cap \cdots \cap G_n$ is a stack, and every stack may be represented in such a form.

Theorem 29.2: A compact, convex body $M \subset \mathbf{R}^n$ is a stack if and only if it satisfies the condition (B) in Theorem 29.1. $\square$

A proof is given in the second part of the section.

We now describe another class of bodies with md $M = 2$; these bodies are called *outcuts*. Let $W \subset \mathbf{R}^n$ be an n-parallelotope and $[q, a_1], \cdots, [q, a_n]$ be its edges with the common vertex q. For each pair i, j (where $1 \le i < j \le n$) we denote by Γ_{ij} the parallelogram spanned by $[q, a_i]$ and $[q, a_j]$; this is a two-dimensional face of H. Furthermore, we take a convex figure $\Phi_{ij} \subset \Gamma_{ij}$ which contains the points q, a_i, a_j and denote by E_{ij} the "beam" $\Phi_{ij} \oplus L_{ij}$, where L_{ij} is the $(n-2)$-subspace spanned by all the vectors $a_1 - q, \cdots, a_n - q$ except for $a_i - q$ and $a_j - q$. Then the intersection $\bigcap\limits_{i<j} E_{ij}$ of these "beams" is said to be an *outcut* with the edges $[q, a_1], \cdots, [q, a_n]$.

There is another description of outcuts. Consider a point $d \in W$ which does not belong to any facet of W containing q and take n arcs $A_1, \cdots, A_n \subset W$, where A_i connects the points d and a_i. The arcs must be taken in such a way that for every $i < j$, denoting by $\pi_{ij} : \mathbf{R}^n \to \operatorname{aff}\{q, a_i, a_j\}$ the projection parallel to L_{ij}, the curve $\pi_{ij}(A_i \cup A_j) \cup [q, a_i] \cup [q, a_j]$ bounds a convex figure $\Phi_{ij} \subset \operatorname{aff}(q, a_i, a_j)$ (Fig. 120). Then

$$\operatorname{conv}\left(\left(\bigcup_{ij} \Phi_{ij} \right) \cup \left(\bigcup_{i} A_i \right) \right)$$

is an outcut, and every outcut may be represented in such a form.

A three-dimensional outcut is shown in Fig. 121.

Theorem 29.3: A compact, convex body $M \subset \mathbf{R}^n$ is an outcut if and only if it satisfies the condition (A) in Theorem 29.1. $\square$

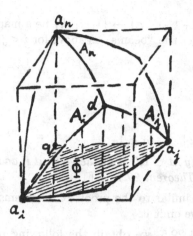

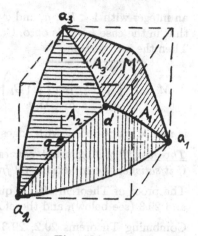

Fig. 120   Fig. 121

A proof is given at the end of the section.

Corollary 29.4: _A three-dimensional compact, convex body_ $M \subset \mathbf{R}^3$, _which is not a parallelotope, satisfies the condition_ md $M = 2$ _if and only if it is either a stack (Fig. 119), or an outcut (Fig. 121), or a direct vector sum of two convex sets (one-dimensional and two-dimensional)._   □

Indeed, as we have seen in Example 26.3, for $n = 3$ only the possibilities (5), (6), (7) are realizable. The possibility (5) corresponds to the case (B) in Theorem 29.1; consequently M is a stack (Theorem 29.2). The possibility (6) of Example 26.3 corresponds to the case (1) in Theorem 26.2 (for $n = 3, k = 3$), i.e., to the case (A) in Theorem 29.1; consequently M is an outcut (Theorem 29.3). Finally, the possibility (7) in Example 26.3 corresponds to the case when M is the direct vector sum of one-dimensional and two-dimensional convex sets.   ■

Corollary 29.4 (first published in [B-Ch 1]) gives a complete geometrical solution of the Szökefalvi-Nagy problem for three-dimensional bodies. It is contained in the following theorem which can be easily derived from Corollary 29.4.

Theorem 29.5: _A compact, convex body_ $M \subset \mathbf{R}^3$ _satisfies the condition_ him $M = 1$ _if and only if_ M _is a parallelotope. A compact, convex body_ $M \subset \mathbf{R}^3$, _that is not a parallelotope, satisfies the condition_ him $M = 2$ _if and only if_ M _is either a stack, or an outcut, or a direct sum. A compact, convex body_ $M \subset \mathbf{R}^3$ _satisfies the condition_ him $M = 3$ _if and only if_ M _is neither a stack, nor an outcut, nor a direct vector sum._   □

We now describe a third class of bodies M (with md $M = 2$) which can be named the class of "stack-outcuts". Let $W \subset \mathbf{R}^n$ be a parallelotope and $[q, b_1], \cdots, [q, b_n]$ be its edges with a common vertex q. Furthermore, let k be an integer with $1 < k < n$, and $\varphi : \{k+1, \cdots, n\} \to \{1, \cdots, k\}$ be a mapping that in the case $k = 2$ is onto. Consider the "beams" as above for $i < j \le k$. Then the intersection

$$M = \left(\bigcap_{i<j\le k} E_{ij} \right) \cap \left(\bigcap_{l>k} G_l \right)$$

is named *stack-outcut* with edges $[q, a_i], \cdots, [q, a_n]$.

Theorem 29.6: A compact, convex body $M \subset \mathbf{R}^n$ is a stack-outcut if and only if it satisfies the condition (C) from Theorem 29.1.   □

The proof of Theorem 29.6 is quite similar to the proofs of Theorems 29.2 and 29.3 (see below), and therefore we omit it.

Combining Theorems 29.2, 29.3 and 29.6, we obtain the following result, which generalizes Theorem 29.5.

Theorem 29.7: A compact, convex, indecomposable body $M \subset \mathbf{R}^n$, $n \ge 3$, satisfies the condition md $M = 2$ if and only if M is either a stack, or an outcut, or a stack-outcut, or, finally, $n = 4$ and M is a four-dimensional polytope with $H(M) = H^{(o)}$.   □

We now give the proofs of the above theorems.

PROOF OF THEOREM 29.1: Let md $M = 2$, i.e., md norm ext $M^* = 2$ (Theorems 18.1 and 2.6). The vector system $H^* = $ norm ext M^* is not splittable. Indeed, if H^* were splittable, then the system $H(M) \subset H^*$ would be splittable, i.e., the body M would be decomposable (Corollary 4.3), contradicting the assumption.

If now H^* contains no particular vector, then $n = 4$ and $H^* = H^{(o)}$ (Theorem 26.1). This means that H^* is finite, and consequently $H(M) \subset H^*$ is also finite, i.e., $H(M) = H^* = $ cl $H(M)$. Thus in this case $H(M) = H^{(o)}$, in concordance with the case (D) of Theorem 29.1.

Let now $H^* = $ norm ext M^* contain at least one particular vector. Then, according to Theorem 26.2, one of the inclusios (1)–(4) holds for the system H^*. Consider, for example, the case (2), i.e., H^* contains only one particular vector e_1 and

$$H^* \subset \{e_1\} \cup \text{norm} \left(\bigcup_{l>1} \Pi_{l,1} \right) \tag{1}$$

(for a basis $e_1, \cdots, e_n$ of $\mathbf{R}^n$). Thus $H^* \subset \bigcup_{l>1} \text{lin} \, (e_1, e_l)$, and therefore

$$\text{ext} \, M^* \subset \bigcup_{l>1} \text{lin} \, (e_1, e_l).$$

According to Exercise 6 in §2, we conclude that

$$M^* \subset \text{conv} \left(\bigcup_{l>1} (M^* \cap \text{lin} \, (e_1, e_l)) \right). \tag{2}$$

The converse inclusion is evident, i.e., in (2) equality holds. Moreover, it follows from (1) that all the points of the set ext M^*, except for e_1, are situated in the half-space Π which does not contain e_1 and is defined by the hyperplane Γ spanned by $e_2, \cdots, e_n$. Consequently,

$$M^* = \text{conv ext } M^* = \text{conv} \, (\{e_1\} \cup (M^* \cap \Pi)).$$

In particular,

$$M^* \cap \text{lin} \, (e_1, e_l) \subset \text{conv} \, (\{e_1\} \cup (M^* \cap \Pi \cap \text{lin} \, (e_1, e_l))$$
$$= \text{conv} \, (\{e_1\} \cup (M^* \cap \Pi_{l,1})).$$

Hence the extreme points of $M^* \cap \text{lin} \, (e_1, e_l)$, except for e_1, are situated in the half-space $\Pi_{l,1}$, i.e.,

$$\text{ext} \, (M^* \cap \text{lin} \, (e_1, e_l)) \subset \{e_1\} \cup \Pi_{l,1}$$

for every $l = 2, \cdots, n$. Thus, denoting by C_l the set $M^* \cap \text{lin} \, (e_1, e_l)$, we conclude, according to (2) above, that the case (B) of Theorem 29.1 is realized.

In detail, we have considered the case of the inclusion (2) in Theorem 26.2. Analogous reasonings show that the inclusion (1) leads us to the case (A), while the inclusions (3) and (4) lead us to the case (C). This means that if md $M = 2$, then one of the cases (A), (B), (C), (D) is realized.

Conversely, let a compact, convex body $M \subset \mathbf{R}^n$ (which is indecomposable and has the origin in its interior) realizes one of the cases (A), (B), (C), (D) for $n \geq 3$. We show that md $M = 2$. Indeed, in the case (D) this is evident, since md $H^{(o)} = 2$.

Furthermore, in the cases (A), (B), (C) the vector system norm ext M^* satisfies one of the inclusions (1)–(4). Consequently md $H(M) \leq$ md (norm ext $(M^*)) \leq 2$, according to the last sentence of Theorem 26.2. ∎

PROOF OF THEOREM 29.2: Let M be a stack (with the origin in its interior). Denote by e_1 the unit outward normal of the face $\Pi \subset M$ and by $e_2, \cdots, e_n$ the unit outward normals of the other facets of W (with a common vertex $q \in \Pi$). Let $a \notin \Pi$ be a regular boundary point of M. We denote by P the supporting half-space of M at the point a, by Γ its bounding hyperplane, and by v the unit outward normal of P. At first, consider the case when Γ is not parallel to Π. Let L be the hyperplane which contains a and is parallel to Π. Then the intersection $L \cap M$ is an $(n-1)$-parallelotope whose faces are correspondingly parallel to the faces of Π. Since a is a regular boundary point of M, there is an $(n-2)$-dimensional face $\triangle$ of the parallelotope $L \cap M$ such that $a \in \text{ri} \, \triangle$. Hence there exists an index $l \in \{2, \cdots, n\}$ such that the vectors e_1 and e_l are orthogonal to aff $\triangle$. The vector v is also orthogonal to aff $\triangle$ (since $\triangle \subset \Gamma$), and therefore v, e_1, e_l are linearly dependent, i.e., $v = x_1 e_1 + x_l e_l$. Moreover, the coefficient x_1 is nonpositive, since the half-space P contains the basis Π (in the Fig. 122 we see the projection of M into the two-dimensional plane $\mathrm{lin}\,(e_1, e_l)$ parallel to aff $\triangle$). This means that $v \in \Pi_{l,1}$. In the omitted case, when Γ is parallel to Π, the inclusion $v \in \Pi_{l,1}$ holds, too (since in this case $v = -e_1$). Thus,

$$H(M) \subset \mathrm{norm}\,\left(\{e_1\} \cup \left(\bigcup_{l>1} \Pi_{l,1}\right)\right),$$

i.e., for the vector system $H = H(M)$ the inclusion (2) of Theorem 26.2 holds. This means that the body M satisfies the condition (B) of Theorem 29.1.

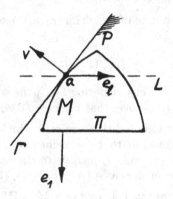

Fig. 122

Conversely, suppose that a compact, convex body $M \subset \mathbf{R}^n$ realizes the case (B) of Theorem 29.1. Then

$$H(M) \subset \mathrm{norm}\,\mathrm{ext}\,M^* \subset \{e_1\} \cup \left(\bigcup_{l>1} \Pi_{l,1}\right).$$

We put

$$D_l = H(M) \cap (\{e_1\} \cup \Pi_{l,1}),\ l = 2, \cdots, n, \tag{3}$$

i.e., $H(M) = D_2 \cup \cdots \cup D_l$. For every unit vector v, denote by $P(v)$ the supporting half-space of the body M with the outward normal v. Then

$$M = \bigcap_{v \in H(M)} P(v) = \left(\bigcap_{v \in D_2} P(v)\right) \cap \cdots \cap \left(\bigcap_{v \in D_n} P(v)\right) = G_2 \cap \cdots \cap G_n,$$

where $G_2, \cdots, G_n$ are "beams" defined by the equalities

$$G_2 = \bigcap_{v \in D_2} P(v), \cdots, G_n = \bigcap_{v \in D_n} P(v).$$

According to (3), $G_l = L_l \oplus M_l$, where L_l is the orthogonal complement of the plane lin (e_1, e_l) and $M_l = G_l \cap \mathrm{lin}\,(e_1, e_l)$, see Fig. 123.   ■

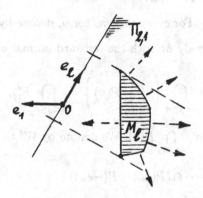

Fig. 123

PROOF OF THEOREM 29.3: Let M be an outcut containing the origin in its interior. Denote by e_i the unit outward normal of the $(n-1)$-dimensional face of W which contains all the points $q, a_1, \cdots, a_n$, except for a_i. Then e_i is orthogonal to $a_k - q$ for $i \neq k$. Let $a \notin \bigcup_{i<j} \Phi_{ij}$ be a regular boundary point of M. Denote by P the supporting half-space of M at the point a, by Γ its bounding hyperplane, and by v the unit outward normal of P. Since $M = \bigcap_{i<j} E_{ij}$, the boundary of M is contained in $\bigcup_{i<j} \mathrm{bd}\, E_{ij}$; hence there are indices i, j such that

$$a \in \mathrm{bd}\, E_{ij} = \mathrm{bd}\, (\Phi_{ij} \oplus L_{ij}) = (\mathrm{rbd}\, \Phi_{ij}) \oplus L_{ij}.$$

Consequently v is orthogonal to the $(n-2)$-flat L_{ij}. The vectors e_i, e_j are also orthogonal to L_{ij}, and therefore $v = x_i e_i + x_j e_j$. Moreover, both the coefficients x_i, x_j are nonpositive, since $a \notin \bigcup_{i<j} \Phi_{ij}$. This means that $v \in F_{i,j}$ (cf. the statement of Theorem 26.2). Thus,

$$H(M) \subset \mathrm{norm}\,\left(\{e_1, \cdots, e_n\} \cup \left(\bigcup_{i<j\leq n} F_{i,j} \right) \right), \tag{4}$$

i.e., for the vector system $H = H(M)$ the inclusion (1) from Theorem 26.2 holds. This means that the body M satisfies the condition (A) of Theorem 29.1.

Conversely, assume that a compact, convex body $M \subset \mathbf{R}^n$ realizes the case (A) of Theorem 29.1., i.e., the above inclusion (4) holds. We put

$$A_{ij} = H(M) \cap (\{e_i, e_j\} \cup F_{i,j}),$$

i.e., $H(M) = \bigcup\limits_{i<j\leq n} A_{ij}$. For every unit vector v, denote by $P(v)$ the supporting half-space of the body M with the outward normal v. Then

$$M = \bigcap_{v \in H(M)} P(v) = \bigcap_{i<j\leq n} \left(\bigcap_{v \in A_{ij}} P(v) \right) = \bigcap_{i<j\leq n} E'_{ij},$$

where E'_{ij} is the "beam" $\bigcap\limits_{v \in A_{ij}} P(v)$. We denote by W' the parallelotope

$$(P(e_1) \cap P(-e_1)) \cap \cdots \cap (P(e_n) \cap P(-e_n)),$$

by q' its vertex bd $P(e_1) \cap \cdots \cap$ bd $P(e_n)$, and by $[q', a'_i]$ the edge of W' that is not contained in bd $P(e_i)$, $i = 1, \cdots, n$. By the inclusion (4), the points $q', a'_1, \cdots, a'_n$ belong to M. Furthermore, the "beam" E'_{ij} has the form $E'_{ij} = \Phi'_{ij} \oplus L'_{ij}$, where L'_{ij} is the $(n-2)$-subspace spanned by all the vectors $a'_1 - q', \cdots, a'_n - q'$ except for $a'_i - q'$, $a'_j - q'$, and $\Phi'_{ij} \subset W$ is a two-dimensional convex set containing the points q', a'_i, a'_j. This means that M is an outcut.

■

Exercises

1. Give a list of the compact, convex bodies with md $M = 2$ in $\mathbf{R}^4$.

2. Prove that the two definitions of stacks given above are equivalent.

3. Let $W \subset \mathbf{R}^n$ be a parallelotope, Π be one of its facets, and $P \subset \mathbf{R}^n$ be a closed half-space such that $\Pi \subset$ int P and bd P contains an $(n-2)$-face of W. Prove that $W \cap P$ is a stack. Consider the case $n = 3$.

4. Let $M \subset \mathbf{R}^n$ be the direct vector sum of a triangle and an $(n-2)$-parallelotope. Prove that M is a stack.

5. The polytope $M \subset \mathbf{R}^3$ with the vertices $(\pm 1, \pm 1, 0)$, $(\pm 2, \pm 2, 1)$, $(0, 0, 3)$ has a square basis (situated in the plane $x_3 = 0$) and, moreover, each plane parallel to the basis and passing through an interior point of M intersects M by a parallelogram, whose sides are parallel to the sides of the basis. Explain why M is not a stack. Prove that md $M = 3$.

6. Let $Q \subset \mathbf{R}^3$ be the cube defined in a cartesian coordinate system with orthonormal basis by the inequalities $0 \leq x_i \leq 1, i = 1, 2, 3$. Denote by P the pyramid with the vertex $(\frac{1}{2}, \frac{1}{2}, 2)$ whose basis coincides with the face of Q situated in the plane $x_3 = 1$. Prove that the union $Q \cup P$ is a stack whose second basis is a point.

7. Let $M \subset \mathbf{R}^3$ be a compact, convex body such that a parallelogram P is contained in its boundary. Prove that M is a stack with the basis P if and only if the following conditions are satisfied: (i) if Γ is a supporting plane of M at a regular boundary point, then Γ is parallel to at least one of the sides of P; (ii) if Γ_1, Γ_2 are parallel supporting planes of M, then at least one of them has a common point with P.

8. Prove that the two definitions of outcuts given above are equivalent.

9. There are outcuts which are convex polytopes. Prove that the convex polytope with the vertices

$$(0,0,0), (2,0,0), (0,2,0), (0,0,2), (1,1,1)$$

is an example (Fig. 124).

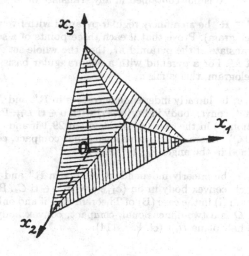

Fig. 124

10. Prove that the convex polytope with the vertices

$$(0,0,0), (4,0,0), (0,4,0), (0,0,4), (3,3,3), (0,3,3), (3,0,3), (3,3,0)$$

is an outcut (Fig. 125), but not a stack.

11. Formulate a condition under which a compact, convex body $M \subset \mathbf{R}^n$ is, at the same time, a stack and an outcut.

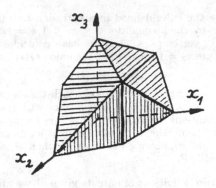

Fig. 125

12. Let $M \subset \mathbf{R}^3$ be a compact, convex body. Prove that if M is neither a stack, nor an outcut, nor a direct vector sum, then there exists a set $X = \{b_1, b_2, b_3, b_4\}$ such that every three of its points are contained in a translate of M, but the whole set X is not contained in any translate of M.

13. Let $M \subset \mathbf{R}^3$ be an affinely regular pyramid with five vertices (i.e., its basis is a parallelogram). Prove that if each three points of a set $X \subset \mathbf{R}^3$ are contained in a translate of the pyramid M, then the whole set X is contained in a translate of M. For a pyramid with a quadrangular basis, which is distinct from a parallelogram, this is false.

14. Let e_i, e_j be linearly independent vectors in $\mathbf{R}^n$ and B_{ij} be a two-dimensional, compact, convex body in lin (e_i, e_j) with $o \in \mathrm{ri}\ B_{ij}$. Prove that B_{ij} satisfies the condition (i) in the case (A) of Theorem 29.1 if and only if $B_{ij} = \mathrm{conv}\ (\{e_i\} \cup \{e_j\} \cup Q)$, where Q is a two-dimensional compact, convex body in lin (e_i, e_j) contained in the angle $F_{i,j}$ (cf. Fig. 113).

15. Let e_1, e_l be linearly independent vectors in $\mathbf{R}^n$ and C_l be a two-dimensional, compact, convex body in lin (e_1, e_l) with $o \in \mathrm{ri}\ C_l$. Prove that C_l satisfies the condition (i) in the case (B) of Theorem 29.1 if and only if $C_l = \mathrm{conv}\ (\{e_1\} \cup Q)$, where Q is a two-dimensional, compact, convex body in lin (e_1, e_l) contained in the half-plane $\Pi_{l,1}$ (cf. Fig. 114).

§30 Centrally symmetric bodies

In this section we will throw light upon the results obtained by the Hungarian mathematician J. Kincses [Ki]. He considers only *centrally symmetric*, compact, convex bodies in $\mathbf{R}^n$ and obtains in this case fine achievements. Namely, Kincses gives necessary and sufficient conditions under which him $M = 3$ or him $M = 4$ (the case him $M = 2$ was considered in Theorem 25.7). In other words, he describes the 3-Helly dimensional and the 4-Helly dimensional centrally symmetric, compact, convex bodies in $\mathbf{R}^n$.

According to Theorem 25.4, we only need to list the *indecomposable* bodies with him $M = r$, to solve the Szökefalvi-Nagy Problem for a class of compact, convex bodies (i.e., in the frames of this section, for the class of all centrally symmetric bodies).

We now formulate the first Theorem obtained by Kincses.

<u>*Theorem 30.1:*</u> *Let $M \subset \mathbf{R}^n$ be an indecomposable, compact, convex body centered at the origin, $n \geq 4$. The equality him $M = 3$ holds if and only if there exists a basis $e_1, \cdots, e_n$ in $\mathbf{R}^n$ and two-dimensional, compact, convex sets $K_1, \cdots, K_{n-1}$, which are centred at the origin such that $K_i \subset \mathrm{lin}\,(e_1, e_{i+1})$ for $i = 1, \cdots, n - 1$, and*

$$M^* = \mathrm{conv}\,(K_1 \cup \cdots \cup K_{n-1}). \quad \Box$$

The proof will be sketched in Exercises 1–8.

We remark that every centrally symmetric, *three-dimensional*, indecomposable, compact, convex body M satisfies the condition him $M = 3$ (by virtue of Theorem 25.7). Thus the complete list of centrally symmetric, compact, convex bodies M with him $M = 3$ is obtained in the following way. Let

$$M = M_1 \oplus \cdots \oplus M_q,$$

where each convex set $M_1, \cdots, M_q$ is indecomposable. The equality him $M = 3$ holds if and only if each of the centrally symmetric, compact, convex sets $M_1, \cdots, M_q$ either has a dimension ≤ 3, or satisfies the condition indicated in Kincses's Theorem 30.1 and, moreover, at least one of the sets $M_1, \cdots, M_q$ either is three-dimensional, or satisfies the condition of Theorem 30.1. Below we give an illustrative example to this Theorem of Kincses (see Example 30.8).

We now formulate the second theorem of Kincses which gives, in a similar way, a complete list of centrally symmetric, compact, convex bodies with him $M = 4$.

<u>*Theorem 30.2:*</u> *Let $M \subset \mathbf{R}^n$ be an indecomposable, compact, convex body centred at the origin, $n \geq 5$. The inequality him $M \leq 4$ holds in and only in the following two cases:*

(A) *There exist a basis $e_1, \cdots, e_n$ in $\mathbf{R}^n$ and centrally symmetric (with respect to the origin) compact, convex sets $K_1, \cdots, K_{n-2}$ such that K_1 is a three-dimensional set contained in $\mathrm{lin}\,(e_1, e_2, e_3)$, the set K_i is two-dimensional and is contained in $\mathrm{lin}\,(e_1, e_{i+2})$ for $i = 2, \cdots, n - 2$, and the polar body M^* has the form*

$$M^* = \mathrm{conv}\,(K_1 \cup \cdots \cup K_{n-2}).$$

(B) *There exist a basis $e_o, \cdots, e_{n-1}$ in $\mathbf{R}^n$, a decomposition of the index set $\{2, \cdots, n-1\}$ into two nonempty disjoint sets J_o, J_1, and two-dimensional, compact, convex sets $K_1, \cdots, K_{n-1}$ centred at the origin such that K_1 is contained in* $\operatorname{lin}(e_o, e_1)$, *the set K_i is contained in* $\operatorname{lin}(e_o, e_i)$ *for $i \in J_o$, the set K_i is contained in* $\operatorname{lin}(e_1, e_i)$ *for $i \in J_1$, and*

$$M^* = \operatorname{conv}(K_1 \cup \cdots \cup K_{n-1}). \qquad \square$$

The proof will be sketched in Exercises 9–18.

We now consider the graph technique of Kincses which was kindly explained by him during our talks on the problem. He uses this technique not for proofs, but in order to foresee the statement of a result. Let G be a graph with vertices $e_1, \cdots, e_{n+1}$. The same symbols $e_1, \cdots, e_{n+1}$ will denote an orthogonal basis in $\mathbf{R}^{n+1}$ such that $|e_i| = \frac{1}{\sqrt{2}}$ for $i = 1, \cdots, n+1$. Consider the set H of the unit vectors $\pm(e_i - e_j)$ taken for the cases when e_i and e_j are joined by an edge in G. Then the convex polytope $\operatorname{conv} H$ is centrally symmetric (with respect to the origin) and situated in the n-dimensional space $\mathbf{R}^n \subset \mathbf{R}^{n+1}$ defined by the equation $x_1 + \cdots + x_{n+1} = 0$.

Proposition 30.3: If the graph G is connected, then the polytope $\operatorname{conv} H$ is n-dimensional, i.e., it is a body in $\mathbf{R}^n$. $\quad \square$

PROOF: Denote by L the affine hull of the polytope $\operatorname{conv} H$. Since G is connected, there is a chain of edges in G that connects the vertices e_1 and e_{n+1}, say the edges $[e_1, e_i], [e_i, e_j], [e_j, e_{n+1}]$ are belonging to G (Fig. 126). Then the vectors $e_1 - e_i, e_i - e_j, e_j - e_{n+1}$ are contained in L, and consequently their sum $e_1 - e_{n+1}$ is contained in L. The same reasoning shows that each of the vectors $e_2 - e_{n+1}, \cdots, e_n - e_{n+1}$ belongs to L. Since the vectors $e_1 - e_{n+1}, \cdots, e_n - e_{n+1}$ are linearly independent, it follows that $\dim L \geq n$. The converse inequality is obvious, since $L \subset \mathbf{R}^n$. Thus $L = \mathbf{R}^n$. ∎

Proposition 30.4: If the graph G is connected, then the polar body $M = (\operatorname{conv} H)^* \subset \mathbf{R}^n$ is a centrally symmetric, convex polytope which satisfies the condition $H(M) = H$. $\quad \square$

PROOF: For each vertex $e_i - e_j$ of the polytope $\operatorname{conv} H$ the corresponding $(n-1)$-dimensional face of M is $\triangle_{ij} \cap M$, where $\triangle_{ij} = \{y \in \mathbf{R}^n : \langle e_i - e_j, y \rangle = 1\}$. The unit outward normal of this face is equal to $e_i - e_j$. This means that $H \subset H(M)$. Conversely, each $(n-1)$-dimensional face of M has the above form $\triangle_{ij} \cap M$. ∎

Proposition 30.5: Assume that the graph G is 2-connected, i.e., it is connected and cannot be represented as the union of two subgraphs with a unique common vertex (as in Fig. 126). Then the polar body $M = (\operatorname{conv} H)^*$ is indecomposable. $\quad \square$

PROOF: Assume that M is decomposable. Then there are proper subspaces $L_1, L_2 \subset \mathbf{R}^n$ such that $\mathbf{R}^n = L_1 \oplus L_2$ and $H \subset L_1 \cup L_2$. Denote by G_k the union of all edges $[e_i, e_j]$ of the graph G such that $\pm(e_i - e_j) \in L_k$, $k = 1, 2$.

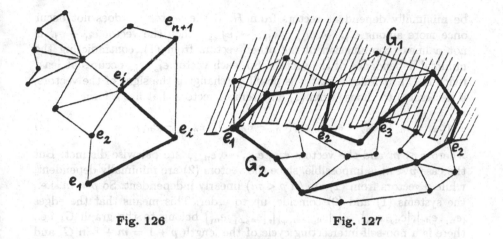

Fig. 126   Fig. 127

Then G_1 and G_2 are subgraphs of G and, moreover, $G_1 \cup G_2 = G$. Since G is 2-connected, there is a non-self-intersecting cycle in G that has edges in G_1 and in G_2. Let $e_{i_1}, e_{i_2}, \cdots, e_{i_{2q}}$ be vertices of this cycle such that the edges between e_{i_1} and e_{i_2} belong to G_1, the edges between e_{i_2} and e_{i_3} belong to G_2, the edges between e_{i_3} and e_{i_4} belong again to $G_1, \cdots$, and finally, the edges between $e_{i_{2q}}$ and e_{i_1} belong to G_2 (Fig. 127). Then $e_{i_1} - e_{i_2} \in L_1$ (since the edges of the cycle between e_{i_1} and e_{i_2} belong to G_1, i.e., the corresponding vectors belong to L_1). Similarly, $e_{i_2} - e_{i_3} \in L_2$, $e_{i_3} - e_{i_4} \in L_1, \cdots, e_{i_{2q}} - e_{i_1} \in L_2$. Consequently the left-hand side of the equality

$$(e_{i_1} - e_{i_2}) + (e_{i_3} - e_{i_4}) + \cdots + (e_{i_{2q-1}} - e_{i_{2q}})$$
$$= -(e_{i_2} - e_{i_3}) - \cdots - (e_{i_{2q}} - e_{i_1})$$

belongs to L_1, while the right-hand side belongs to L_2. This means that L_1 and L_2 have a nonzero common vector, contradicting the decomposition $\mathbf{R}^n = L_1 \oplus L_2$. ∎

Proposition 30.6: If the graph G is connected, then md M *is equal to the maximal length of a non-self-intersecting cycle decreased by 1.*   □

PROOF: Let $k + 1$ be the maximal length of a non-self-intersecting cycle contained in G and $m = $ md M. Let, furthermore, $(e_{i_1}, \cdots, e_{i_{k+1}})$ be a non-self-intersecting cycle in G, i.e., the edges $[e_{i_1}, e_{i_2}], \cdots, [e_{i_k}, e_{i_{k+1}}], [e_{i_{k+1}}, e_{i_1}]$ belong to G. Then

$$(e_{i_1} - e_{i_2}) + \cdots + (e_{i_k} - e_{i_{k+1}}) + (e_{i_{k+1}} - e_{i_1}) = o$$

is a minimal dependence between $k + 1$ vectors from $H = H(M)$, i.e., $m \geq k$. Conversely, let

$$(e_{i_1} - e_{j_1}), (e_{i_2} - e_{j_2}), \cdots, (e_{i_m} - e_{j_m}), (e_{i_{m+1}} - e_{j_{m+1}}) \qquad (1)$$

be minimally dependent vectors from H. If the vector e_{i_1} does not occur once more among $e_{i_2}, \cdots, e_{i_{m+1}}, e_{j_1}, \cdots, e_{j_{m+1}}$, then the vector $e_{i_1} - e_{j_1}$ is not a linear combination of the other vectors from (1), contradicting the minimal dependence of the vectors. So, each vector $e_{i_\alpha}, e_{j_\alpha}$ occures at least twice in (1). It follows that it is possible (changing the signs of the vectors, if necessary) to choose from (1) a system of vectors that has a form

$$(e_{n_1} - e_{n_2}), (e_{n_2} - e_{n_3}), \cdots, (e_{n_p} - e_{n_{p+1}}), (e_{n_{p+1}} - e_{n_1}), \tag{2}$$

where $p \leq m$ and the vectors $e_{n_1}, e_{n_2}, \cdots, e_{n_{p+1}}$ are pairwise distinct. But the case $p < m$ is impossible, since the vectors (2) are minimally dependent, while p vectors from (1) are (if $p < m$) linearly independent. So $p = m$, i.e., the systems (1) and (2) coincide, up to order. This means that the edges $[e_{n_1}, e_{n_2}], [e_{n_2}, e_{n_3}], \cdots, [e_{n_p}, e_{n_{p+1}}], [e_{n_{p+1}}, e_{n_1}]$ belong to the graph G, i.e., there is a non-self-intersecting cycle of the length $p + 1 = m + 1$ in G, and hence $k \geq m$. ∎

The Propositions 30.3–30.6 form the basis of the graph technique by Kincses. We now consider examples of application of this technique.

EXAMPLE 30.7: The graph G in the Fig. 128 is Hamiltonian, i.e., there exists in G a non-selfintersecting cycle passing through all the vertices. This means, according to Proposition 30.6, that the corresponding centrally symmetric, convex polytope $M = (\text{conv } H)^*$ is *Helly-maximal*, i.e., him $M = \dim M = n$. The set H consists of the vectors

$$\pm(e_1 - e_{n+1}), \cdots, \pm(e_n - e_{n+1}), \pm(e_1 - e_2), \pm(e_2 - e_3), \cdots,$$
$$\pm(e_{n-1} - e_n), \pm(e_n - e_1).$$

In the case $n = 3$ the polytope $M = (\text{conv } H)^*$ is the regular icosahedron. □

EXAMPLE 30.8: The maximal length of a non-self-intersecting cycle in the graph G drawn in the Fig. 129 is, evidently, equal to 4. Consequently the

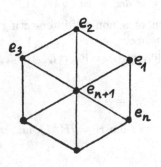

Fig. 128

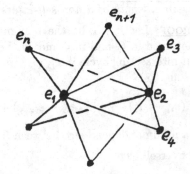

Fig. 129

corresponding polytope $M \subset \mathbf{R}^n$ is indecomposable, centrally symmetric, and md $M = 3$. If we denote the vector $\frac{1}{2}(e_1 - e_2)$ by b_1 and the vector $e_{i+1} - \frac{e_1 + e_2}{2}$ by b_i, then we obtain that $H = H(M)$ consists of the vectors $\pm b_1 \pm b_i (i = 2, \cdots, n)$.   □

This example (cf. Exercises 12 and 13 in §25) illustrates the first theorem of Kincses. Indeed,

$$M^* = \mathrm{conv}\,(K_1 \cup \cdots \cup K_{n-1}),$$

where $K_i = \mathrm{conv}\,(\pm b_1 \pm b_{i+1}) \subset \mathrm{lin}\,(e_1, e_{i+1}), i = 1, \cdots, n-1$. On the other hand, this example illustrates Theorem 25.7, i.e., it shows that for centrally symmetric, compact, convex bodies with him $M > 2$ the assertion analogous to Theorem 25.7 is false.   □

Kincses's Theorems 30.1 and 30.2 show that the complete solution of the Szökefalvi-Nagy Problem obtained in §29 for the space $\mathbf{R}^3$ (Theorem 29.5) may be extended (only in the case of centrally symmetric bodies) to the spaces $\mathbf{R}^4$ and $\mathbf{R}^5$. Indeed, for example, in $\mathbf{R}^5$ we have a complete list of centrally symmetric, compact, convex bodies with him $M = 1$ (Theorem 25.2), with him $M = 2$ (Corollary 25.8), with him $M = 3$ (Theorem 30.1) and with him $M = 4$ (Theorem 30.2). For all other centrally symmetric bodies in $\mathbf{R}^5$ we have him $M = 5$. But in $\mathbf{R}^6$ we do not have a complete picture, since we cannot distinguish the cases him $M = 5$ and him $M = 6$. In this connection it would be interesting to describe the centrally symmetric, compact, convex bodies $M \subset \mathbf{R}^n$ with him $M = n - 1$. The following example illustrates this problem.

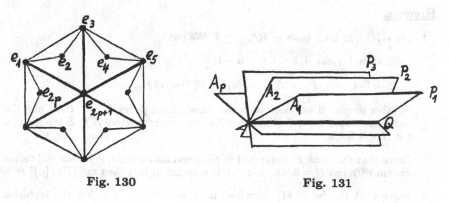

Fig. 130   Fig. 131

EXAMPLE 30.9: Consider the graph G shown in the Fig. 130. It differs from the graph in the Fig. 128 by supplementary vertices, each of which is joined with two former ones. The maximal length of a non-selfintersecting cycle in this graph G is equal to $2p = n$, i.e., according to Proposition 30.6, the corresponding indecomposable, centrally symmetric polytope $M = (\mathrm{conv}\,H)^*$

has Helly dimension him $M = \text{md } M = n - 1$. If we denote the sub-space lin $(e_{2p+1} - e_1, e_{2p+1} - e_3, \cdots, e_{2p+1} - e_{2p-1})$ by S, the subspaces lin $(e_2 - e_1)$, lin $(e_4 - e_3), \cdots$, lin $(e_{2p} - e_{2p-1})$ by $A_1, \cdots, A_p$, and the subspaces $S \oplus A_1, \cdots, S \oplus A_p$ by $P_1, \cdots, P_p$, then we obtain the following visual picture (Fig. 131). There is a direct decomposition $\mathbf{R}^n = S \oplus A_1 \oplus \cdots \oplus A_p$. Further-more, the subspaces $P_1, \cdots, P_p$ are $(p+1)$-dimensional "pages" of the "book" $B = P_1 \cup \cdots \cup P_p$ with the p-dimensional "spine" S. So, the vector system H is contained in the "book" B.   □

In the theorems of Kincses we have books of another type. In Theorem 30.1 the pages $P_i = \text{lin } (e_1, e_{i+1})$ are two-dimensional and the spine $S = \text{lin } (e_1)$ is one-dimensional. The theorem affirms that $H(M)$ is contained in the book $B = P_1 \cup \cdots \cup P_{n-1}$.

In the case (A) of Theorem 30.2 the book B has $n-2$ two-dimensional pages $P_i = \text{lin } (e_1, e_{i+2})$ and one three-dimensional page $P_o = \text{lin } (e_1, e_2, e_3)$, while the spine of this book $B = P_o \cup P_1 \cup \cdots \cup P_{n-2}$ is the line $S = \text{lin } (e_1)$. In this case the vector system H is contained in the book B, too.

Finally, in the case (B) of Theorem 30.2 we have two books B_o, B_1. The first book B_o has the spine $S_o = \text{lin } (e_o)$ and two-dimensional pages $P_i^{(o)} = \text{lin } (e_o, e_i)$, $i \in J_o$, whereas the second book B_1 has the spine $S_1 = \text{lin } (e_1)$ and two-dimensional pages $P_i^{(1)} = \text{lin } (e_1, e_i)$, $i \in J_1$. Moreover, there is the "shelf" $F = \text{lin } (e_o, e_1)$ that contains both the spines S_o, S_1. In this case the vector system H is contained in the "library" $L = F \cup B_o \cup B_1$.

These "library" arguments will be considered in Problem 11 of Chapter VIII.

Exercises

1. Let $e_1, \cdots, e_n$ be a basis in $\mathbf{R}^n$, $n \geq 4$. We put

   $$P_i = \text{lin } (e_1, e_{i+1}), i = 1, \cdots, n - 1;$$

   $$B = P_1 \cup \cdots \cup P_{n-1}, S = \text{lin } (e_1) = P_1 \cap \cdots \cap P_{n-1}.$$

   In other words, B is a *book* with two-dimensional *pages* $P_1, \cdots, P_{n-1}$ and the one-dimensional *spine* S. Prove that for every vector system $H \subset B$ the relation md $H \leq 3$ holds.

2. Prove that the book B, described in the previous exercise, is a *maximal* vector system with md $B = 3$, i.e., if $e \notin B$ is a vector in $\mathbf{R}^n$, then md $(B \cup \{e\}) > 3$.

3. Prove that the body M, described in Theorem 30.1, satisfies the inclusion $H(M) \subset B$, where B is a book as in Exercise 1. This means that md $M \leq 3$ (and the inequality md $M < 3$ is excluded by virtue of Theorem 25.7), i.e., the part "if" in Theorem 30.1 holds.

4. Let $H \subset \mathbf{R}^n$, $n \geq 4$, be a system of nonzero vectors which is nonsplittable, non-one-sided, symmetric with respect to the origin, and satisfying md $H = 3$. Let, furthermore, $a_o, a_1, a_2, a_3 \in H$ be minimally dependent vectors. We put $L = \text{lin } (a_o, a_1, a_2, a_3)$ and $K = \text{lin } (H \setminus L)$. Prove that dim $(L \cap K) \geq 1$.

5. In the notation of Exercise 4, let $b_o, b_1, \cdots, b_l \in H$ be minimally dependent vectors such that $b_o \in L$; $b_1, \cdots, b_l \in H \setminus L$. Prove that $l = 2$ and (up to an enumeration of the vectors a_o, a_1, a_2, a_3) the relation $b_o = \lambda_o a_o + \lambda_1 a_1 = -\lambda_2 a_2 - \lambda_3 a_3$ holds, where $\lambda_o a_o + \lambda_1 a_1 + \lambda_2 a_2 + \lambda_3 a_3 = o$ is the minimal dependence between a_o, a_1, a_2, a_3. In other words, the vectors $a_o, a_1, a_2, a_3, b_o, b_1, b_2$ are contained in the book $P_1 \cup P_2 \cup P_3$, where $P_1 = \mathrm{lin}\ (b_o, a_o, a_1)$, $P_2 = \mathrm{lin}\ (b_o, a_2, a_3)$, $P_3 = \mathrm{lin}\ (b_o, b_1, b_2)$ and $S = P_1 \cap P_2 \cap P_3 = \mathrm{lin}\ (b_o)$ is the spine of the book.

6. In the notation of Exercise 5, we put $L' = \mathrm{lin}\ (P_1 \cup P_2 \cup P_3)$, $K' = \mathrm{lin}\ (H \setminus L')$. Prove that $\dim\ (L' \cap K') \geq 1$. Prove, in addition, that if $b'_o \in L'$, $b'_1, \cdots, b'_l \in H \setminus L$ are minimally dependent vectors, then $l = 2$ and $b'_o \in S$. In other words, putting $P_4 = \mathrm{lin}\ (b'_o, b'_1, b'_2)$, we obtain a book $P_1 \cup P_2 \cup P_3 \cup P_4$ with two-dimensional pages and one-dimensional spine S. Moreover, every vector $e \in H$ contained in $\mathrm{lin}\ (P_1 \cup P_2 \cup P_3 \cup P_4)$ belongs to the book $P_1 \cup P_2 \cup P_3 \cup P_4$.

7. Let $H \subset \mathbf{R}^n$, $n \geq 4$, be a vector system which is non-one-sided, nonsplittable, symmetric with respect to the origin, and satisfying md $H = 3$. Prove by induction (generalizing the reasoning of Exercise 6) that H is contained in a book described in Exercise 1.

8. Using Exercise 7, prove that the part "only if" in Kincses's Theorem 30.1 holds.

9. Let $e_1, \cdots, e_n$ be a basis in $\mathbf{R}^n$, $n \geq 5$. We put

$$P_1 = \mathrm{lin}\ (e_1, e_2, e_3);\ P_i = \mathrm{lin}\ (e_1, e_{i+2})\ \text{for}\ i = 2, \cdots, n - 2;$$

$$B = P_1 \cup \cdots \cup P_{n-2},\ S = \mathrm{lin}\ (e_1) = P_1 \cap \cdots \cap P_{n-2}.$$

In other words, B is a book in $\mathbf{R}^n$ that has a three-dimensional page P_1, two-dimensional pages $P_2, \cdots, P_{n-2}$, and a one-dimensional spine S. Prove that for every vector system $H \subset B$ the relation md $H \leq 4$ holds. Moreover, B is a maximal vector system in $\mathbf{R}^n$ with md $B = 4$, i.e., if $e \notin B$ is a vector in $\mathbf{R}^n$ then md $(B \cup \{e\}) > 4$.

10. Prove that the body M described in the case (A) of Kincses' Theorem 30.2 satisfies the inclusion $H(M) \subset B$, where B is a book as in the previous exercise. This means that md $M = 4$, i.e., the part "if" in the case (A) of Theorem 30.2 holds.

11. Let $e_o, \cdots, e_{n-1}$ be a basis in $\mathbf{R}^n$, $n \geq 5$, and J_o, J_1 be a partition of the set $\{2, \cdots, n - 1\}$ as in Theorem 30.2. We put

$$P_i = \mathrm{lin}\ (e_o, e_i)\ \text{for}\ i \in J_o,\ P_i = \mathrm{lin}\ (e_1, e_i)\ \text{for}\ i \in J_1;$$

$$S_o = \mathrm{lin}\ (e_o),\ S_1 = \mathrm{lin}\ (e_1),\ F = \mathrm{lin}\ (e_o, e_1);$$

$$B_o = \bigcup P_i\ \text{over}\ i \in J_o,\ B_1 = \bigcup P_i\ \text{over}\ i \in J_1;$$

$$L = B_o \cup B_1 \cup F.$$

So we have two books B_o, B_1 with two-dimensional pages and one-dimensional spines $S_o, S_1 \subset F$, and L is a library with two books B_o, B_1 and the shelf F. Prove that for every vector system $H \subset L$ the relation md $H \leq 4$ holds. Moreover, L is a maximal vector system in $\mathbf{R}^n$ with md $L = 4$, i.e., if $e \notin L$ is a vector in $\mathbf{R}^n$, then md $(L \cup \{e\}) > 4$.

12. Prove that the body M described in the case (B) of the Kincses theorem 30.2 satisfies the inclusion $H(M) \subset L$, where L is the library as in the previous exercise. This means that md $M \leq 4$, i.e., the part "if" in the case (B) of Theorem 30.2 holds.

13. Generalize Exercise 4 for the case when md $H = m$ and $a_o, \cdots, a_m$ are minimally dependent vectors $(m < n)$.

14. Let a_o, a_1, a_2, a_3, a_4 be minimally dependent vectors in $\mathbf{R}^n$. Prove that, for every nonzero vector $b_o \in L = \text{lin}\,(a_o, a_1, a_2, a_3, a_4)$, it is possible to choose three or four vectors among $\pm a_o, \pm a_1, \pm a_2, \pm a_3, \pm a_4$ such that b_o and the chosen vectors are minimally dependent.

15. Let $H \subset \mathbf{R}^n$, $n \geq 5$, be a vector system which is symmetric with respect to the origin and satisfies md $H = 4$. Let, furthermore, $a_o, a_1, a_2, a_3, a_4 \in H$ be minimally dependent vectors. Let, finally, $b_o \in L = \text{lin}\,(a_o, \cdots, a_4)$, $b_1, \cdots, b_l \in H \setminus L$ be also minimally dependent vectors. Prove that $l = 2$ and (up to an enumeration of the vectors $a_o, \cdots, a_4$) the relation

$$\pm b_o = \lambda_o a_o + \lambda_1 a_1 = -\lambda_2 a_2 - \lambda_3 a_3 - \lambda_4 a_4$$

holds, where $\lambda_o a_o + \lambda_1 a_1 + \lambda_2 a_2 + \lambda_3 a_3 + \lambda_4 a_4 = o$ is the minimal dependence between a_o, a_1, a_2, a_3, a_4.

16. Let $H \subset \mathbf{R}^n$, $n \geq 5$, be a system of nonzero vectors that is symmetric with respect to the origin, nonsplittable, non-one-sided, and satisfies the condition md $H = 4$. Let, furthermore, $a_o, a_1, a_2, a_3, a_4 \in H$ be minimally dependent vectors. We put $L = \text{lin}\,(a_o, \cdots, a_4)$ and $K = \text{lin}\,(H \setminus L)$. In the case dim $(L \cap K) = 1$, conduct an induction (similar to the reasoning in the Exercises 6 and 7) which shows that H is contained in a book (described in Exercise 9). This means that for dim $(L \cap K) = 1$ the part "only if" in the case (A) of Theorem 30.2 holds.

17. In the notation of Exercise 15, we put

$$P_o = \text{lin}\,(b_o, b_1, b_2), \quad L' = L + P_o, \quad K' = \text{lin}\,(H \setminus L').$$

Prove that if $b_o' \in L'$, $b_1', \cdots, b_l' \in K'$ are minimally dependent vectors, then either $b_o' \in \text{lin}\,(b_o)$, or dim $(L' \cap K') = 2$ and (up to an enumeration of the vectors a_2, a_3, a_4) the relation

$$\pm b_o' = \lambda_2 a_2 + \lambda_3 a_3 = -\lambda_o a_o - \lambda_1 a_1 - \lambda_4 a_4$$

holds.

18. In the notation of Exercise 16 and for the case dim $(L \cap K) = 2$, conduct an induction (using the reasoning of Exercise 17) which shows that H is contained in a library L described in Exercise 11. This means that for dim $(L \cap K) = 2$ the part "only if" in the case (B) of Theorem 30.2 holds. Explain, why the case dim $(L \cap K) \geq 3$ is impossible.

19. Let $Q \subset \mathbf{R}^n$ be an $(n-1)$-dimensional cross-polytope with the center at the origin and I be a segment centered at the origin and not lying in aff Q. We put $M = Q + I$. Prove (with the help of the results of J. Kincses or without them) that him $(M^*) = 3$.

20. Let $e_1, \cdots, e_n$ be a basis in $\mathbf{R}^n$ and q be an integer such that $3 \le q \le n-1$. Denote by M the convex polytope with the vertices

$$\pm e_1 \pm e_i, \, i = 3, \cdots, q; \quad \pm e_2 + e_j, \, j = q+1, \cdots, n$$

(all combinations of signs). What is the Helly dimension of the polar polytope M^*?

21. Let $e_1, \cdots, e_n$ be a basis in $\mathbf{R}^n$. Denote by M the convex polytope with the vertices

$$\pm e_1 \pm e_2 \pm e_3; \quad \pm e_1 \pm e_k, \, k = 4, \cdots, n$$

(all combinations of signs). What is the Helly dimension of the polar polytope M^*?

22. Can you obtain 2-connected graphs for which corresponding Kincses polytopes coincide with M, as in the Exercises 19, 20, and 21?

23. Let H be a set and $\mathfrak{M}$ be a family of its nonempty subsets. The system $(H, \mathfrak{M})$ is said to be a *matroid* (cf. [Ai]) if the following axioms are satisfied:

(i) if $C \subset \mathfrak{M}$, then no proper subset of C belongs to $\mathfrak{M}$;

(ii) if $C_1, C_2 \in \mathfrak{M}$ and $p \in C_1 \cap C_2, q \in C_1 \setminus C_2$, then there exists a set $D \in \mathfrak{M}$ with $q \in D \subset C_1 \cup C_2 \setminus \{p\}$;

(iii) there exists an integer n such that, if a nonempty subset $A \subset H$ does not contain any $C \in \mathfrak{M}$, then the cardinality of A is not larger than n.

Let now $H \subset S^{n-1}$ be a nonempty vector system symmetric with respect to the origin and $\mathfrak{M}$ be the set of all nonempty subsystems $C = \{a_o, \cdots, a_m\} \subset H$ such that there exist numbers $\lambda_o = \pm 1, \cdots, \lambda_m = \pm 1$, for which $\lambda_o a_o, \cdots, \lambda_m a_m$ are minimally dependent. Prove that $(H, \mathfrak{M})$ is a matroid. In the language of matroids, J. Kincses gave an account of his results. For details the reader is referred to the original paper [Ki].

24. Let $H \subset S^{n-1}$ be a nonsplittable, centrally symmetric vector system. Prove that H does not contain particular vectors.

V. Borsuk's partition problem

If a bounded set $F \subset \mathbf{R}^n$ consists of at least two points, then there exists a finite partition $F = F_1 \cup \cdots \cup F_k$ such that for every $i = 1, \cdots, k$ the diameter diam $F_i = \sup \{\| x - y \|: x, y \in F_i\}$ of the part F_i is *smaller* than diam F. The least positive integer k for which such a partition exists is said to be the *Borsuk number* of F, since K. Borsuk considered this question for two-dimensional sets and for the n-dimensional ball $B \subset \mathbf{R}^n$. One motivation for these investigations was given by the famous theorem of Borsuk and Ulam, referring to continuous mappings of the n-sphere into $\mathbf{R}^n$.

The question for (upper bounds on) the Borsuk number of an arbitrary bounded set $F \subset \mathbf{R}^n$ is one of the most famous and longstanding problems of combinatorial geometry.

There are various surveys on Borsuk's partition problem. An excellent overview until 1963 was given by B. Grünbaum [Grü 3]. In addition the reader is referred to [B-G 1,2], [Dol], [C-F-G; A27, D14, D15], [B-So], and [Schm; §2.5]. For the special class of bodies of constant width the surveys [C-G; §9] and [H-M; §5] are worth mentioning.

§31 Formulation of the problem and a survey of results

It is impossible to dissect a circle of diameter h into two parts each of which has a diameter smaller than h. However, using a dissection into three parts, this is possible (Fig. 132). A parallelogram of diameter h can be dissected into two parts of smaller diameter (Fig. 133).

Thus we have a motivation for the following question which was first formulated in [Br 2]: Let $F \subset \mathbf{R}^n$ be a set of diameter h (hence F is bounded). What is the smallest integer k such that F is the union of k sets each of which has a diameter strictly less than h? Denote this smallest number k by $a(F)$. The integer $a(F)$ is said to be the *Borsuk number of F*. For a circle F we have $a(F) = 3$, and for a parallelogram F the equality $a(F) = 2$ holds. For the plane, Borsuk's question is answered by

Theorem 31.1: For any bounded set $F \subset \mathbf{R}^2$, the inequality $a(F) \leq 3$ holds.
□

Fig. 132   Fig. 133

An elegant proof of this theorem was published in [Br 2]. It is known (cf. [Pa] and also [Y-B, p. 45]) that each set $F \subset \mathbf{R}^2$ of diameter h can be covered by a regular hexagon V having h as distance of its parallel sides, see Fig. 134. (This is a consequence of Theorem 32.1 below.) The hexagon can be dissected into three parts each of which has a diameter smaller than h, cf. Fig. 135. Hence in such a manner also $F \subset V$ is dissected into three parts of smaller diameters. This yields Theorem 31.1. ■

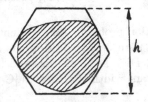
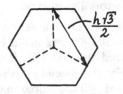

Fig. 134   Fig. 135

From this proof we can easily deduce that each set $F \subset \mathbf{R}^2$ of diameter h can be dissected into three parts whose diameters are not larger that $\frac{\sqrt{3}}{2}h \approx 0,866h$. This estimate is the best, since already a circle of diameter h cannot be dissected into three parts each of which has a diameter smaller than $\frac{\sqrt{3}}{2}h$.

Theorem 31.2: A ball $E \subset \mathbf{R}^n$ cannot be partitioned into n parts of smaller diameters. □

This theorem was proved by K. Borsuk [Br 1] in 1932, but L.A. Lyusternik and L.G. Schnirel'man obtained this result (in a modified formulation) already in 1930, cf. [L-S].

Theorem 31.2 says that for a ball $E \subset \mathbf{R}^n$ the inequality $a(E) \geq n+1$ holds. It is easy to show that $a(E) = n+1$, i.e., the ball can be dissected into $n+1$ parts of smaller diameters. For example, if $T_o, T_1, \cdots, T_n$ are the facets of a regular simplex inscribed into E and K_i is the cone that is the union of all

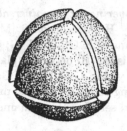

Fig. 136	Fig. 137

rays emanating from the center of E through points from T_i $(i = 0, 1, \cdots, n)$, then each of the parts $E \cap K_o, E \cap K_1, \cdots, E \cap K_n$ of E has a smaller diameter, see Fig. 136. Another partition is shown in Fig. 137.

Hence in the inequality $a(F) \leq 3$ for bounded sets $F \subset \mathbf{R}^2$ (Theorem 31.1) we have *equality* if F is a circle. This means that

$$\max_F a(F) = 3,$$

where F runs over all bounded sets $F \subset \mathbf{R}^2$, and for a circle this maximal value is achieved. One might assume that an analogous situation holds for $n > 2$. This yields the famous Borsuk conjecture, which was first formulated in [Br 2]: *each bounded set $F \subset \mathbf{R}^n$ can be dissected into $n + 1$ subsets each of which has a diameter smaller than F* (i.e., $a(F) \leq n + 1$, and therefore $\max_F a(F) = n + 1$ for every n). Actually, K. Borsuk was careful enough to formulate this not as a conjecture, but merely as a question. We will see that, for sufficiently large n, counterexamples exist.

For $n = 3$, the Borsuk conjecture was verified by J. Perkal and H.G. Eggleston using rather complicated arguments, see [Pe] and [Eg 1]. Thus we have

Theorem 31.3: For an arbitrary bounded set $F \subset \mathbf{R}^3$, the inequality $a(F) \leq 4$ holds.   □

Simpler proofs (which are similar to that for $n = 2$ given above) were indepentently obtained by A. Heppes and P. Révész [H-R], B. Grünbaum [Grü 2], and A. Heppes [He]. Their approach is based on the following result of D. Gale [Ga]: every three-dimensional body of diameter 1 can be embedded into a regular octahedron whose opposite facets have the distance equal to 1. With this starting point, one can find a body $V \subset \mathbf{R}^3$ with the following properties:

(A) an arbitrary set $F \subset \mathbf{R}^3$ of diameter 1 can be embedded into V (by using a suitable motion),

(B) the body V can be dissected into four parts, each having a diameter smaller than 1.

Each set satisfying (A) from above is called a *covering set* or *universal cover*. Surveys referring to covering sets are given by H. Meschkowski [Mes], B. Weissbach [We 2], and H. Croft, K.J. Falconer and R.K Guy [C-F-G; §D 16]. Obviously, if such a covering set additionally satisfies (B) or, in the plane, the property described after Theorem 31.1, then it yields a solution of Borsuk's problem. In particular, the existence of a covering set $V \subset \mathbf{R}^3$ with property (B) would imply that $a(F) \leq 4$ for any bounded set $F \subset \mathbf{R}^3$ of diameter 1.

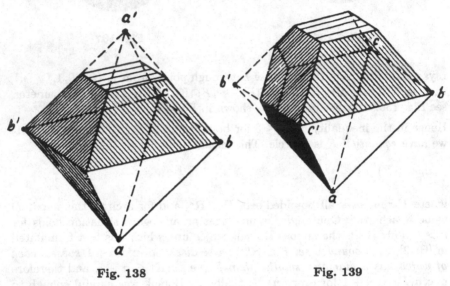

Fig. 138   Fig. 139

In [Grü 2], B. Grünbaum modified the result of D. Gale in the following manner: any set of diameter 1 is contained in a regular octahedron of minimal width 1, with three of its vertices cut off by planes orthogonal to the diagonals at a distance $\frac{1}{2}$ from the center, see Figs. 138–140. More precisely, for every diagonal of the octahedron one cutting-off plane has to be used. Doing this for the first diagonal (Fig. 138), it is easy to see that the property (A) remains, and the polyhedra after the other two truncations (see the Figures 139 and 140) satisfy the same condition. So it remains to show that property (B) is satisfied for the final polyhedron V shown in Fig. 140. A suitable dissection of this polyhedron into four parts of a diameter smaller than 1 is presented in Fig. 141, where

$$\| a_1 - l_1 \| = \| a_2 - l_1' \| = \| c_1 - l_3 \| = \cdots = \frac{15\sqrt{3} - 10}{46\sqrt{2}};$$

$$\| k_1 - y_1 \| = \| k_3 - y_3 \| = \| k_2 - y_2 \| = \frac{1231\sqrt{3} - 1986}{1518\sqrt{2}},$$

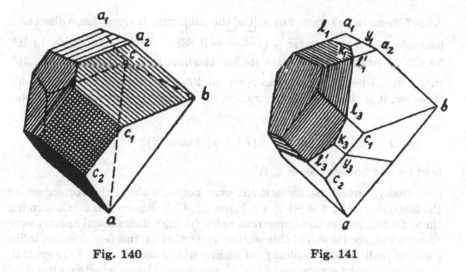

Fig. 140   Fig. 141

and the center of the former octahedron is the only common point of the four polyhedral parts. A calculation shows that each of these parts (cf. Fig. 142) has a diameter given by

$$\sqrt{\frac{6129030 - 937419\sqrt{3}}{1518\sqrt{2}}} \approx 0,9887.$$

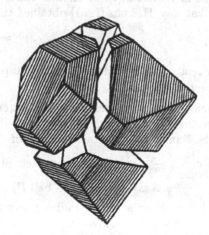

Fig. 142

The partition presented by A. Héppés [He] yields the estimate $0,9977\cdots$, which is greater than Grünbaum's one. At present, no smaller value than

$0,9887$ seems to be known, but in [Ga] the conjecture is given that a dissection into four pieces of diameter $\leq \sqrt{\frac{3+\sqrt{3}}{6}} = 0,888\cdots$ is always possible. In $\mathbf{R}^n$ for $n > 3$, the problem of Borsuk (to find the number $\max_F a(F)$ when $F \subset \mathbf{R}^n$ runs over all bounded sets) was open till 1993 (see the end of the section). However, it is clear that the restriction to compact, convex sets is sufficient, since

$$\text{diam } F = \text{diam cl conv } F, \quad a(F) \leq a(\text{cl conv } F)$$

hold for any bounded set $F \subset \mathbf{R}^n$.

Repeatedly, several mathematicians were not sure about the correctness of the inequality $a(F) \leq n+1$ if $n > 3$, see e.g. C.A. Rogers [Ro 2]. Confirming these doubts, meanwhile counterexamples (in high-dimensional spaces) were constructed, see the end of this section. Nevertheless, this fact does not influence the undiminished topicality of finding upper bounds on $a(F)$ in general, or of the exact determination of Borsuk numbers of bounded sets with special geometrical properties. Now we shall present a collection of results of the first type.

If a set $D \subset \mathbf{R}^n$ has the property that each bounded set F of diameter 1 can be covered by D (as an example was used by B. Grünbaum for $n = 3$), then, also in n-space, D is said to be a *covering set*. It is easy to show that a cube of edge-length 1 has this property. It is also obvious that a cube of edge-length 1 can be covered by m^n congruent cubes of edge-length $\frac{1}{m}$, where m has to be large enough if the diameters of these smaller cubes are demanded to be smaller than 1. On that way, H. Lenz [Le 1] obtained the estimate

$$a(F) \leq [\sqrt{n} + 1]^n.$$

Replacing the congruent cubes by suitable parallelotopes, one can improve this bound, cf. [Br 3] and [La 1].

By lattice arrangements of such parallelotopes, no better bound than 2^n can be obtained. It was already confirmed by R. Knast [Kn] that this upper bound on $a(F)$ holds. Namely, the famous theorem of H. Jung [J] says that a ball B_1 of radius $r = \left(\frac{n}{2n+2}\right)^{\frac{1}{2}}$ is a covering set. Since the limit of the increasing sequence $\sqrt{\frac{n}{2n+2}}$ is given by $\frac{1}{2}\sqrt{2}$, a ball B_2 of radius $r^* = \frac{1}{2}\sqrt{2}$ is a covering set, too. By splitting a ball of radius r into 2^n congruent pieces, the diameter of each of these pieces is given by $\sqrt{2}r$, yielding the upper bound 2^n mentioned above.

A slight modification of this ball partition gives a better bound: By a suitable hyperplane, we split a layer from that ball which is smaller than a semiball. The remaining part is splitted into 2^{n-1} parts by $n-1$ hyperplanes through the center of the former ball, where all n hyperplanes are pairwise orthogonal. This partition yields

$$a(F) \leq 2^{n-1} + 1,$$

see [La 1].

There are other possibilities to obtain better bounds by using the balls B_1 and B_2 from above. One of them shall be described in a more detailed manner. In the survey [Grü 3; p. 275], the estimate

$$a(F) < \sqrt{\frac{1}{3}(n+1)^3(2+\sqrt{2})^{\frac{n-1}{2}}}$$

of L. Danzer is given. But the paper cited by B. Grünbaum (see, e.g., also [B-G 1]) was never published. Moreover, another contribution of L. Danzer (cf. [Da 1]), which is cited in view of this result at several places, does not contain this estimate. But Danzer's results on intersection families of balls given there imply that bound, if a slight modification is accomplished. Namely, one has to multiply this bound by a factor $q \geq \left(\frac{3\pi}{2\sqrt{2}}\right)^{\frac{1}{2}} > 1,8$, which perhaps was lost by the transmission of that result. This estimate

$$a(F) \leq q\sqrt{\frac{(n+2)}{3}}(2+\sqrt{2})^{\frac{n-1}{2}} \tag{1}$$

can be derived in the following manner: A cap of the ball B_2, having the spherical radius $\frac{\pi}{4}$, has the Euclidean diameter $\sqrt{2}r^*$. If the set bd B_2 is covered by k caps (each having the spherical radius $\frac{\pi}{4}$), then the corresponding sectors yield a covering of B_2, and each of these sectors has itself the Euclidean diameter $\sqrt{2}r^*$. Since the radius of B_1 is smaller than that of B_2 for each n, the ball B_1 can be covered by k congruent sectors whose diameter is smaller than $\sqrt{2}r^*$. Thus the representation

$$a(F) \leq \overline{N}_n\left(\frac{\pi}{4}\right) \tag{2}$$

is obtained, where $\overline{N}_n(\alpha)$ denotes the smallest positive integer k such that a covering of an $(n-1)$-sphere by k caps of spherical radius α is possible. L. Danzer gave an estimate for $0 < \alpha \leq \frac{\pi}{2}$, namely:

$$\overline{N}_n(\alpha) \leq \sqrt{\frac{\pi}{(2n-3)\cos\alpha}} \cdot \frac{n^2-1}{n+3}(3+n\cdot\cos\alpha)\cdot(\sqrt{2}\sin\frac{\alpha}{2})^{1-n}. \tag{3}$$

For $\alpha = \frac{\pi}{4}$, this yields

$$\overline{N}_n\left(\frac{\pi}{4}\right) \leq \sqrt{q(n)}(2+\sqrt{2})^{\frac{n-1}{2}},$$

where $q(n)$ is a rational function of n with

$$\lim_{n\to\infty}\frac{q(n)}{n^3}=\frac{\pi}{2\sqrt{2}},\ q(n)<\frac{\pi}{2\sqrt{2}}(n+2)^3.$$

For getting an upper bound on $a(F)$ by means of (2) and (3), one cannot omit the factor q in (1), since otherwise a contradiction to the growth of $q(n)$ would be obtained.

Since C.A. Rogers [Ro 3] derived a better upper bound on $\overline{N}_n(\alpha)$ already in 1963, it is understandable that L.Danzer did not take up this subject in a later contribution. On the other hand, it is surprising that no further authors used the natural approach (2) to bounds on $a(F)$.

One can compare the given upper bounds on $a(F)$ by the asymptotic behaviour of $a(F)^{\frac{1}{n}}$ for $n\to\infty$. The estimates from [Kn] and [La 1] yield $a(F)^{\frac{1}{n}}\le 2(1+o(1))$, and that of L.Danzer gives $a(F)^{\frac{1}{n}}\le(2+\sqrt{2})^{\frac{1}{2}}(1+o(1))$. By the estimate from [Ro 3] for $\overline{N}_n(\alpha)$, $a(F)^{\frac{1}{2}}\le\sqrt{2}(1+o(1))$ is obtained, and the best known upper bounds on $a(F)$ up to now can be described by $a(F)^{\frac{1}{n}}\le\sqrt{\frac{3}{2}}(1+o(1))$.

Using the illumination method for bodies of constant width (see the next chapter), O. Schramm [Schr] obtained the general upper bound $a(F)\le 5n^{3/2}(e+ln\,n)(\frac{3}{2})^{d/2}$, which was reproved by J. Bourgain and J. Lindenstrauss [B-L]. It can be formulated also in the following form: for every ε and sufficiently large n, the inequality $a(F)\le\left(\sqrt{\frac{3}{2}}+\varepsilon\right)^n$ holds.

The following assertions refer to bounded sets $F\subset\mathbf{R}^n$ which, by their special geometrical properties, satisfy the inequality $a(F)\le n+1$. First, we mention a result of H. Lenz [Le 1]: for all bodies of constant width in $\mathbf{R}^n$ the lower bound $n+1\le a(F)$ holds, and if F is a smooth convex body which is not of constant width, then even $a(F)\le n$. C.A. Rogers [Ro 1] proved that any n-dimensional body of constant width with the symmetry group of the regular n-simplex can be covered by $n+1$ sets of smaller diameter, if there are no symmetry requirements on the covering sets, leading to the Borsuk number $n+1$ for bodies invariant under the symmetry group of the regular n-simplex.

A.S. Riesling [Ri] showed that $a(F)\le n+1$ holds for $F\subset\mathbf{R}^n$ a centrally symmetric, compact, convex body. To see this, it suffices to take a regular n-simplex, whose centroid coincides with the symmetry center of F, and to consider the partition of F by means of this simplex as it is described in the proof of Theorem 35.11. In particular, if F is a centrally symmetric polytope, then even $a(F)=2$ holds (since there exist hyperplanes through the symmetry center of F which do not meet a vertex of it). Thus the "Euclidean part" of Riesling's results is more or less a triviality. His main merit is the extension of the bound above to bodies in 3-spaces of constant curvature.

H. Hadwiger [Ha 1,2] verified the relation $a(F) \leq n + 1$ if $F \subset \mathbf{R}^n$ is a compact, convex body with smooth boundary. He used the known Gaussian mapping of bd F onto the unit sphere S^{n-1} by means of the outer normal vector of each $x \in$ bd F. Namely, if there is a covering of S^{n-1} by $n + 1$ closed sets, none of which contains an antipodal pair of points from S^{n-1}, then the corresponding original closed sets $F_i \subset$ bd F yield a covering of bd F, each set F_i having a smaller diameter than F, $i = 1, \cdots, n + 1$. It is remarkable that in [Ha 1] the relation $a(F) \leq n+1$ was "verified" even for all bounded sets $F \subset \mathbf{R}^n$. For getting this false statement, Hadwiger erroneously assumed that a diametrical chord of a spherical simplex necessarily connects two vertices (as in the Euclidean situation). However, a correction was given in [Ha 2].

A sharpening of the result from [Ha 2] was obtained by H. Lenz [Le 1]: Let $F \subset \mathbf{R}^n$ be a compact, convex body with smooth boundary and π_s be the orthogonal projection onto an s-dimensional subspace of $\mathbf{R}^n$ ($s \geq 1$) such that the diameter of $\pi_s(F)$ is smaller than that of F. Then $a(F) \leq n+1-s$. This statement implies that $a(F) = n + 1$ holds if F is a smooth body of constant width. Related extensions were investigated by [A-K] and [De 2]. B. Dekster [De 1,3] gave estimates on the diameters of pieces in a Borsuk partition if again special assumptions on the boundary points (such as restrictions to the diameters of their spherical images) of a compact, convex body $F \subset \mathbf{R}^n$ with $a(F) \leq n + 1$ are made, see also [De 2, Ha 3,4, We 1]. He also showed that if such a body has a supporting cylinder, *each* supporting line of which contains a regular boundary point of F, then $a(F) \leq n+1$ [De 4]. In [Bt 1], the bound $a(F) \leq n + 1$ was extended to all compact, convex bodies having at most n *corners* (i.e., boundary points with nonunique supporting hyperplane, see also the next chapter).

Furthermore, D. Kolodzieczyk [Ko 1] proved $a(F) \leq n + 1$ for F a compact, convex body of revolution, and a proof of B.V. Dekster [De 5] shows that this statement can be extended to spaces of constant curvature. Here the approach is based on the fact that each diametrical chord has to meet the axis of revolution. Moreover, in [Ko 1] the number $a(F) \leq n + 1 - s$ was verified, if each open diametrical chord of a compact, convex body meets the same s-dimensional flat, $0 \leq s \leq n - 1$.

A.D. Milka [Mi] investigated the Borsuk problem on convex hypersurfaces in Euclidean n-space. He considered the Riemannian metric on such hypersurfaces, induced by the ambient space, and asked whether this hypersurface can be partitioned into $n + 1$ pieces of smaller diameters in the metric of the hypersurface; for $n = 3$ the answer is affirmative, and for $n \geq 4$ at least the hypersurfaces of revolution were shown to have the Borsuk number $n + 1$.

On this basis, one might ask the following: which classes of convex bodies $F \subset \mathbf{R}^n$ are even characterized by special Borsuk numbers smaller than $n + 1$?

For example, Theorem 31.1 gives only the *upper bound* on $a(F)$ for bounded figures $F \subset \mathbf{R}^2$. Since obviously $a(F) > 1$, we have the (more precise) statement: *for any bounded plane figure F the value of $a(F)$ is equal to 2 or 3*. Which bounded figures F are characterized by $a(F) = 2$? This problem was answered in [Bt 2]. To formulate the corresponding theorem, we give here the notion of *bodies of constant width*. Let $F \subset \mathbf{R}^n$ be a compact, convex body and Γ be an arbitrary $(n-1)$-subspace of $\mathbf{R}^n$. The distance of the two supporting hyperplanes of F, which are parallel to Γ, is said to be the *width* of F at Γ. If the width of F is constant for all Γ, then F is called a *body of constant width*.

For example, let P be a regular, convex polygon with odd vertex number and diameter h; if two neighboring vertices of P are connected by a circular arc, whose center is the (unique) opposite vertex, then a figure F of constant width h is obtained, see Fig. 143. Figures of such a type are called *Reuleaux polygons*, since the simplest such "polygon", obtained when P is an equilateral triangle, was considered already by F. Reuleaux [Re]. This construction can be successfully extended to nonregular polygons P with odd vertex number if each vertex belongs to two diagonals of length h, cf. Fig. 144.

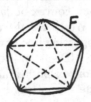

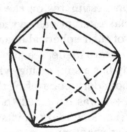

Fig. 143   Fig. 144

There exist many interesting results about bodies of constant width in Euclidean and finite-dimensional normed spaces. The interested reader is referred to the surveys [C-G] and [H-M]. Here we will give only a few of their basic properties. If $F \subset \mathbf{R}^n$ is a body of constant width h, we clearly have diam $F = h$. Each boundary point of F is contained in at least one diametrical chord of F (i.e., a chord of F having the length h). If $[a, b]$ is such a diametrical chord of F, then F is contained in the strip between its parallel supporting hyperplanes which are orthogonal to $[a, b]$ and contain a and b, respectively.

As the forthcoming Theorem 32.1 says, each figure $F \subset \mathbf{R}^2$ of diameter h is contained in a figure Φ of constant width h. In several cases, this *completion process* is unique. (An example is given by the equilateral triangle of side length h, for which the only containing figure of constant width h is the Reuleaux triangle, see Fig. 145.) To show that in general this process is not unique, let F be a figure obtained from a circle K by cutting off less than a half-circle. Then F has the same diameter h as the circle K, and obviously K is a figure of constant width h containing F. Another completion process of F is shown in Fig. 146, where the equalities

$$\| a - x \| = \| x - b \| = \| a - p \| = \| b - q \| = h$$

hold and arcs of radius h are drawn.

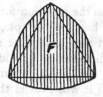

Fig. 145   **Fig. 146**

The announced result from [Bt 2] (see also Exercise 19) is given by the following

Theorem 31.4: Let $F \subset \mathbf{R}^2$ be a figure of diameter h. The equality $a(F) = 3$ holds if and only if F can be completed to a figure of constant width h in a unique way. □

We remark that the analogous statement for three-dimensional bodies is not correct. For example, if the body $F \subset \mathbf{R}^3$ is a regular tetrahedron of edge length h, then obviously $a(F) = 4$. On the other hand, there is not a unique way to complete the tetrahedron to a body of constant width h (cf. [Y-B, pp. 103-104] and Problem 13 in the final chapter). This means that for three-dimensional bodies F of diameter h the equality $a(F) = 4$ depends on conditions which are sharper than the uniqueness of the completion process described above. (This uniqueness was also investigated in higher-dimensional Euclidean and Minkowski spaces, see [Sal] and [Groe].)

For further results on covering sets and the Borsuk problem, connected with bodies of constant width, the reader should consult papers of P.L. Bowers [Bw], V.L. Dol'nikov [Dol], V.V. Makeev [Mak], O. Stefani [St] and B. Weissbach [We 3,4], more references can be found in the latter paper and in the surveys [C-G, p. 79] and [H-M, §5]. Moreover, we have $a(F) = 2$ if and only if a convex figure $F \subset \mathbf{R}^2$ has a non-diametrical chord U such that the relative interior of every diametrical chord of F intersects U. On this basis, in [Ko 3] it is shown that F has a unique completion if and only if for each non-diametrical chord U of F there exists a diametrical chord V of F whose relative interior is disjoint from U.

The exact determination of the Borsuk number $a(F)$ can be considered for bounded subsets F of an arbitrary metric (in particular, normed) space. For the special case of a Minkowski plane whose unit ball is a parallelogram, this problem was solved by B. Grünbaum [Grü 1] (cf. also Theorem 33.4 below), and for arbitrary Minkowski planes it was investigated in [B-SV]. In particular, the above formulated Theorem 31.4 can be derived from the results of that paper. And also the following section is mainly devoted to results from the paper [B-SV], which are also based on investigations of H.G. Eggleston [Eg 4].

A natural generalization of the Borsuk problem is the question for the minimal number $a_n(k)$ of pieces of F if each of them has, for example, a k-times less diameter than F. So H. Lenz [Le 2] showed that any plane set of diameter ≤ 1 can be split into 4 subsets, each of diameter $\leq \frac{1}{2}$, and that 7 is the smallest possible integer (i.e., $a_2(2) = 7$, see also [B-V]). From Borsuk's paper [Br 3] it follows that $a_2(3) \leq 20$. In [D-L] it was proved that in fact $a_2(3) = 14$. Furthermore, from [Br 3] the estimate $a_3(2) \leq 48$ can be taken. Showing even $a_3(2) \leq 31$, M. Lassak [La 2] improved this result. Moreover, L. Danzer [Da 1] and M. Lassak [La 2] showed that for a ball $F \subset \mathbf{R}^3$ the bounds $12 \leq a_3(2) \leq 20$ hold. A discussion of further related questions is given in [C-F-G, D 14], see also [St].

We continue with a strongly related extension of the Borsuk problem. If $F \subset \mathbf{R}^n$ is a bounded set of positive diameter, then one can consider a sequence $D_1(F), D_2(F), \cdots$ of real numbers, where $D_k(F) = \delta$ means the following: F can be covered by k sets whose diameters are not greater than δ and F cannot be covered by k such sets if, in each case, their diameters are strictly smaller than δ. (Obviously, $D_1(F)$ coincides with the usual diameter of F.) This definition of $D_k(F)$ is due to H. Lenz [Le 2], and for sets satisfying certain (geometric) conditions the question for bounds on the numbers $D_k(F)$ makes sense. For example, one can set

$$\alpha_n(k) = \sup\{D_k(F) : F \subset \mathbf{R}^n, d(F) = \delta\},$$

where $d(F)$ denotes the usual diameter of F. Asking for the smallest positive integer k satisfying $\alpha_n(k) < \delta$, one has a direct connection with the Borsuk problem, see also [Grü 3]. Seemingly, except for results in the plane nearly no estimates on $D_k(F)$, $k \geq 2$, by means of other functionals are known.

If $B^n \subset \mathbf{R}^n$ denotes an n-dimensional ball of diameter $d(B^n) = \delta$, then the Theorem of Lyusternik, Shnirel'man and Borsuk implies

$$D_1(B^n) = D_2(B^n) = \cdots = D_n(B^n) = \delta,$$

whereas $D_{n+1}(B^n) < \delta$ (since there are partitions of B^n into $n+1$ pieces of diameters smaller than δ). Clearly, we have $D_2(B^1) = \frac{1}{2}\delta$. An upper bound for $D_{n+1}(B^n)$, n arbitrary, is obtained by dissecting bd B^n into $n+1$ congruent spherical simplices. A corresponding lower bound on $D_{n+1}(B^n)$, $n \geq 2$, was obtained already by H. Hadwiger in 1954 (cf. [Ha 4]), namely

$$D_{n+1}(B^n) \geq \left(\sqrt{\frac{1}{2}} + \frac{1}{2}\sqrt{\frac{n-1}{2n}} \right) \delta.$$

For $n = 2$ and $n = 3$, this bound is sharp, but for larger values of n it can be improved. For $n \to \infty$, the given lower bound converges to $\left(\sqrt{2 + \sqrt{2}} \right) \frac{\delta}{2}$, whereas the upper bound described above attains the value δ. More recently, D.G. Larman and N.K. Tamvakis [L-T] derived a lower bound which converges to δ, too. In other words, one has

$$D_{n+1}(B^n) = \delta(1 + o(1)), \ n \to \infty.$$

D.G. Larman [L] has suggested to investigate related problems for 2-distance sets (i.e., if x_1, x_2 are from such a set, then $\| x_1 - x_2 \| = 1$ or $\| x_1 - x_2 \| = c$, for some $c < 1$). For such 2-distance sets the reader should consult Problem 27 in chapter VIII.

Now we will be concerned with the (recently obtained) counterexamples to the inequality $a(F) \leq n + 1$ for arbitrary bounded sets $F \subset \mathbf{R}^n$. The first breakthrough in this direction was presented by J. Kahn and G. Kalai [K-K]. They showed that $a(F) \geq (1,1)^{\sqrt{n}}$, i.e., for some high-dimensional space a counterexample can be constructed, and, (more surprisingly) $a(F)$ is even growing exponentially if $n \to \infty$. A more recent contribution was given by N. Alon [Ni], showing the existence of counterexamples already in $\mathbf{R}^n$ for $n = 946$. In the following, we will give a detailed account on Alon's approach.

Let p be an odd prime number. We put $n = 4p$ and consider two Euclidean spaces $\mathbf{R}^n, \mathbf{R}^\triangle$ of dimensions n and n^2, respectively. Coordinates in $\mathbf{R}^n$ we denote by $x_1, \cdots, x_n$ and coordinates in $\mathbf{R}^\triangle$ by x_{ij}, where $i, j = 1, \cdots, n$, independently. In both the spaces we consider scalar products: if $x, y \in \mathbf{R}^n$, then $\langle x, y \rangle = x_1 y_1 + \cdots + x_n y_n$, and if $v, w \in \mathbf{R}^\triangle$, then $\langle v, w \rangle = \sum_{i,j=1}^{n} v_{ij} w_{ij}$.

Furthermore, by $\mathfrak{F}$ we denote the set of all vectors $x = (x_1, \cdots, x_n) \in \mathbf{R}^n$ with $x_1 = 1$ and $x_i = \pm 1$ for all $i = 2, \cdots, n$ with an even number of the coordinates equal to -1. For each vector $x = (x_1, \cdots, x_n) \in \mathbf{R}^n$ denote by $x * x$ its tensorial square, i.e., the vector from $\mathbf{R}^\triangle$ with coordinates $x_{ij} = x_i x_j; i, j = 1, \cdots, n$. We consider the set $\mathfrak{S}$ of all the vectors $x * x$ taken for $x \in \mathfrak{F}$. Then $\mathfrak{S}$ is a finite set of vectors in $\mathbf{R}^\triangle$, i.e., conv $\mathfrak{S} \subset \mathbf{R}^\triangle$ is a convex polytope. N. Alon [Ni] proves that the diameter of this polytope is equal to $n\sqrt{2}$, and for p large enough it is impossible to divide conv $\mathfrak{S}$ into dim (conv $\mathfrak{S}$) + 1 pieces of smaller diameters. This gives a counterexample for the Borsuk problem. For realizing this idea, N. Alon uses several lemmas.

Lemma A: *The scalar product* $\langle a, c \rangle$ *of two vectors* $a, c \in \mathfrak{F}$ *is an integer which is divisible by 4. If* $a \neq c$ *and* a *is not orthogonal to* c, *then the integer* $\langle a, c \rangle$ *is not divisible by* p.   □

PROOF: The vector $a \in \mathfrak{F}$ can be represented in the form $a = e - 2b$, where $e = (1, \cdots, 1)$ and b is a vector with coordinates $x_i = 0$ or $x_i = 1$ and an even number of coordinates equal to 1. Analogously, $c = e - 2d$. Consequently,

$$\langle a, c \rangle = \langle e, e \rangle - 2\langle e, d \rangle - 2\langle e, b \rangle + 4\langle b, d \rangle.$$

Since $\langle e, e \rangle = n = 4p$, $\langle e, b \rangle = 2k$, $\langle e, d \rangle = 2l$, the integer $\langle a, c \rangle$ is divisible by 4. Moreover, since

$$|\langle a, c \rangle| = |x_1 y_1 + \cdots + x_n y_n| \leq n = 4p,$$

the integer $\langle a, c \rangle \neq 0$ is divisible by p only in the case when $x_1 y_1 = \cdots = x_n y_n = \pm 1$, i.e., $a = \pm c$ (as far as $|x_i| = |y_i| = 1$). But this is possible only in the case when $a = c$ (since $x_1 = 1$ for all $a \in \mathfrak{F}$, i.e., $\mathfrak{F}$ does not contain two vectors symmetric with respect to the origin).   ■

Lemma B: *For every* $a \in \mathfrak{F}$ *we define a polynomial*

$$P_a(x) = \prod_{i=1}^{p-1} ((\langle a, x \rangle - i)), \tag{4}$$

where x *is a vector with integer coordinates* $x_1, \cdots, x_n$ *and the values of* $P_a(x)$ *are considered in the field* F_p *of the order* p. *Then*

(i) $P_a(b) \equiv 0 \pmod{p}$ *for every distinct, nonorthogonal* $a, b \in \mathfrak{F}$;

(ii) $P_a(a) \not\equiv 0 \pmod{p}$ *for every* $a \in \mathfrak{F}$.   □

PROOF: If a is not equal to b and is not orthogonal to b, then $\langle a, b \rangle$ is an integer which is divisible by 4, but not divisible by p, i.e., $\langle a, b \rangle = 4i$, where i is one of the numbers $1, \cdots, p - 1$. Consequently $P_a(b) \equiv 0 \pmod{p}$, according to (4). Furthermore, $P_a(a) = \prod_{i=1}^{p-1} (\langle a, a \rangle - i) = \prod_{i=1}^{p-1} (4p - i)$. Since each multiplier $4p - i$ is distinct from $0 \pmod{p}$, then $P_a(a) \not\equiv 0 \pmod{p}$.   ■

To formulate the next lemma, we introduce some additional notions. The polynomial $P_a(x)$ has the degree $p - 1$. Introducing additional relations $x_i^2 = 1$ between the variables $x_1, \cdots, x_n$, we obtain from $P_a(x)$ (in its standard representation) a multilinear polynomial $\overline{P}_a(x)$, i.e. a polynomial at each summand of which every variable x_i is contained in the degree 0 or 1.

Lemma C: *For every vector* $x = b \in \mathfrak{F}$ *the polynomials* $P_a(x)$ *and* $\overline{P}_a(x)$ *take the same value:* $P_a(b) = \overline{P}_a(b)$.   □

PROOF: This assertion is an immediate consequence of the fact that $x_i = \pm 1$, i.e., $x_i^2 = 1$ for each $b = (x_1, \cdots, x_n) \in \mathfrak{F}$.   ■

Lemma D: _If a subset_ $\mathfrak{G} \subset \mathfrak{F}$ _contains no two orthogonal vectors, then all the multilinear polynomials_ $\overline{P}_a(x)$, $a \in \mathfrak{G}$, _are linearly independent (over the field_ F_p). $\square$

PROOF: Suppose

$$\sum_{a \in \mathfrak{G}} \lambda_a \overline{P}_a(x) \equiv 0. \tag{5}$$

By substituting $x = b$, we obtain (by virtue of Lemmas B and C) that $\lambda_b \equiv 0$ in (5). This completes the proof. ■

Lemma E: _If a subset_ $\mathfrak{G} \subset \mathfrak{F}$ _contains no two orthogonal vectors, then the cardinality_ $|\mathfrak{G}|$ _is not greater than_

$$\sum_{i=0}^{p-1} \binom{n}{i}. \quad \square \tag{6}$$

PROOF: Since the polynomials $\overline{P}_a(x), a \in \mathfrak{G}$, are linearly independent (Lemma D), the number of these polynomials (i.e., $|\mathfrak{G}|$) is not greater than the dimension of the space of all multilinear polynomials of degrees $\leq p-1$ on n variables $x_1, \cdots, x_n$. But by $\binom{n-1}{0} = 1$, there is only one linearly independent polynomial of degree 0 and, furthermore, there are only $\binom{n-1}{1}$ linearly independent, homogeneous polynomials of degree 1 (namely $x_1, \cdots, x_n$), only $\binom{n-1}{2}$ linearly independent, homogenous polynomials of degree 2 (namely $x_i x_j$, $i \neq j$), etc. ■

Lemma F: _For every_ $a, b \in \mathfrak{F}$ _the vectors_ $a * a, b * b \in \mathbf{R}^{\triangle}$ _have a nonnegative scalar product._ $\square$

PROOF: We have

$$\langle a * a, b * b \rangle = \sum_{i,j=1}^{n} (a_i a_j)(b_i b_j) = \sum_{i,j=1}^{n} (a_i b_i)(a_j b_j) = \left(\sum_{i=1}^{n} a_i b_i \right)^2 \geq 0. \quad ■$$

Lemma G: _The set_ conv $\mathfrak{G}$ _has the diameter_ $n\sqrt{2}$, _and the number of vertices for each of its subsets with smaller diameter cannot be greater than the number given in (6)._ $\square$

PROOF: Since

$$\| a * a \| = \sqrt{\langle a * a, a * a \rangle} = \sqrt{\sum_{i,j=1}^{n} (\pm 1)^2} = \sqrt{n^2} = n,$$

we have

$$|\langle a * a - b * b, a * a - b * b \rangle| = |\langle a, a \rangle|^2 + |\langle b, b \rangle|^2 - 2\langle a * a, b * b \rangle \leq 2n^2.$$

Taking two orthogonal vectors $a * a$ and $b * b$, we obtain $\| a * a - b * b \| = n\sqrt{2}$. Finally, if a set $A \subset$ conv $\mathfrak{G}$ has a diameter smaller than $n\sqrt{2}$, then it cannot contain two orthogonal vectors $a * a, b * b$. By virtue of Lemma E, this means that the number of vertices of conv $\mathfrak{S}$ contained in A is not greater than that given in (6) (we remark that $a * a$ and $b * b$ are orthogonal if and only if the vectors a, b are orthgonal). ∎

We now claim the proof of the main theorem of N. Alon.

Theorem 31.5: If

$$\frac{2^{n-1}}{\sum_{i=1}^{p-1} \binom{n}{i}} > \binom{n}{2} + 1, \tag{7}$$

then the polytope conv $\mathfrak{S}$ *cannot be divided into* dim conv $\mathfrak{S} + 1$ *parts of smaller diameters.* □

PROOF: The number of vertices of conv $\mathfrak{S}$ (i.e., the number of points $x * x$ with $x = (1, x_2, \cdots, x_n)$ for $x_i = \pm 1$; $i = 2, \cdots, n$) is equal to 2^{n-1}. The dimension of conv $\mathfrak{S}$ is equal to $\binom{n}{2}$, since conv $\mathfrak{S}$ is situated in the space defined in $\mathbf{R}^\triangle$ by equalities $x_{ij} = x_{ji}$, $x_{ii} = 1$. As far as the cardinality of any vertex set of $A \subset$ conv $\mathfrak{S}$, which has a smaller diameter, is not greater than the number given in (6), we have the assertion. ∎

It remains to notice that for $p = 11$, i.e., for $n = 44$ and $\binom{n}{2} = 946$, the inequality (7) holds. Thus, in the space of dimension 946 there is a convex polytope $M = $ conv $\mathfrak{S}$ whose Borsuk number is greater than dim $M + 1$.

Exercises:

1. Prove that for a convex regular polygon $M \subset \mathbf{R}^2$ with $k \geq 3$ vertices the Borsuk number $a(M)$ is equal to 2 or 3, depending on the parity of k.

2. Let $M \subset \mathbf{R}^2$ be a convex polygon with $2k + 1$ vertices (for an integer $k \geq 1$) and diameter d. We say that M is a *dia-polygon* if for every its vertex a and a successive numeration $a_o = a, a_1, \cdots, a_{2k}$ of the vertices, we have the relations $\| a_o - a_k \| = \| a_o - a_{k+1} \| = d$ (cf. Fig. 147). Prove that for every dia-polygon $M \subset \mathbf{R}^n$ the equality $a(M) = 3$ holds.

3. Prove that if a polygon $M \subset \mathbf{R}^2$ satisfies $a(M) = 3$, then it is possible to choose (for a positive integer k) $2k + 1$ vertices of M which are the vertices of a dia-polygon having the same diameter as M.

4. A compact, convex body $D \subset \mathbf{R}^n$ is said to be a *covering set* (with respect to the diameter d) if for every set $X \subset \mathbf{R}^n$ with diam $X \leq d$ there exists a body which is *congruent* to D and contains X. Prove that there exists a *minimal covering set*, i.e., a compact, convex body $D \subset \mathbf{R}^n$ which is a covering set with respect to diameter d, whereas no compact, convex body $D' \subset D$ not coinciding with D is a covering set (with respect to d). Prove that in the plane $\mathbf{R}^2$ there is a minimal covering set (with respect to d) which is contained in the regular hexagon whose opposite sides have the distance d. Borsuk's idea was to find a covering set $D \subset \mathbf{R}^n$ (with respect to d) which can be partitioned into $n + 1$ pieces each of which has a diameter less than d. (Borsuk's proof of Theorem 31.1 and Grünbaum's proof of Theorem 31.3 follow this idea.)

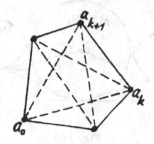

Fig. 147

5. Prove that for every $k \in \{2, \cdots, n+1\}$ there is a set $X \subset \mathbf{R}^n$ with $a(X) = k$.

6. Let $M \subset \mathbf{R}^2$ be a dia-polygon (cf. Exercise 2) of diameter d, see Fig. 147. Prove that there exists a unique figure of constant width d, which contains M. These figures are said to be Reuleaux polygons, cf. Figs. 143–145.

7. Prove that any two diametrical chords of a figure $M \subset \mathbf{R}^2$ of constant width have a common point (Fig. 148).

Fig. 148

8. Prove that the perimeter of every figure of constant width d is equal to πd (Theorem of Barbier, see [Y-B]).

9. Prove that if $M_1, M_2 \subset \mathbf{R}^2$ are figures of constant width d, which are symmetric to each other with respect to the origin, then $M_1 + M_2$ is a ball of radius d centered at the origin (Fig. 149). Generalize this to n-dimensional bodies of constant width.

10. Prove that if $M \subset \mathbf{R}^2$ is a figure of constant width d, then its inscribed circle and its circumscribed circle have the same center, and the sum of their radii is equal to d.

11. Prove that if a sequence $M_1, M_2, \cdots$ of figures of constant width d in $\mathbf{R}^2$ converges to a compact, convex figure M, then M is a figure of constant width d.

12. Prove that every figure of constant width d is the limit of a convergent sequence of Reuleaux polygons of constant width d.

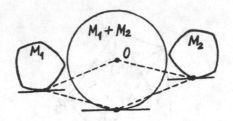

Fig. 149

13. Let $X \subset \mathbf{R}^2$ be a figure with a diameter not exceeding h. The intersection of all disks of radius h which contain X is said to be the h-hull of X and denoted by $X^{(h)}$, see Fig. 150. Prove that diam $X =$ diam $X^{(h)}$. Prove, moreover, that if X is a figure of constant width h, then $X^{(h)} = X$.

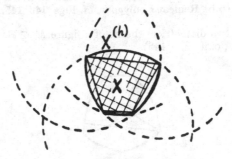

Fig. 150

14. Let $X \subset \mathbf{R}^2$ be a figure of diameter h. Prove that if the width of $X^{(h)}$, in at least one direction, is less than h, then $a(X) < 3$.

15. Prove that if X is a figure of diameter h and $X^{(h)}$ is a figure of constant width h, then at least one endpoint of every diametrical chord of $X^{(h)}$ belongs to X.

16. Prove that if $X \subset \mathbf{R}^2$ is a figure of diameter h and $a(X) = 2$, then $a(X^{(h)}) = 2$.

17. Prove that if $M \subset \mathbf{R}^2$ is a figure of constant width, then $a(M) = 3$.

18. Give a proof of Theorem 31.4.

19. Let $X \subset \mathbf{R}^2$ be a compact, convex figure of diameter h. A chord $[a, b]$ of X is called a *diametrical* one if $\| a - b \| = h$. Deduce from Theorem 31.4 the following theorem which is due to Kolodziejczyk: A compact, convex figure $X \subset \mathbf{R}^2$ satisfies the condition $a(X) = 2$ if and only if there exists a nondiametrical chord $[c, d]$ of X such that the endpoints of each diametrical chord are situated at different sides of the line passing through c and d (Fig. 151).

20. Let a compact, convex body $M \subset \mathbf{R}^n$ be centrally symmetric and not a ball. Prove that $a(M) \leq n$.

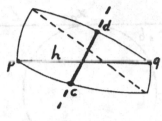

Fig. 151

§32 Bodies of constant width in Euclidean and normed spaces

For the following theorem, we refer to [Pa], [Eg 3], and [C-G].

Theorem 32.1: Each set $F \subset \mathbf{R}^n$ of diameter h can be covered by a body of the same constant width h.   □

PROOF: We will consider all convex sets of diameter h which contain F (such a set is also conv F). Since every set M of such a type is contained in a ball cl $U_h(x)$ (where $x \in F$) and hence M is contained in a cube of edge-length $2h$, the volume of M cannot be larger than $(2h)^n$. Let v denote the actual upper bound on the volumes of bodies with diameter h containing F. Then a sequence of convex bodies can be constructed such that each of them contains F, has diameter h, and v is the limit of their volumes. Since one can pass on to a suitable subsequence (if necessary), such a sequence can be assumed to be convergent. Its limit is a convex body $\Phi \supset F$ of diameter h and volume v. We will prove that Φ is a body of constant width h.

Let $\Phi^{(h)}$ denote the intersection of all closed balls of radius h which contain Φ in each case (such balls exist since, if $x \in \Phi$, then cl $U_h(x) \supset \Phi$). We will show that diam $\Phi^{(h)} = h$. Let $a, b \in \Phi^{(h)}$ and x be an arbitrary point from Φ. Then cl $U_h(x) \supset \Phi$, i.e., cl $U_h(x)$ is one of the balls whose intersection is $\Phi^{(h)}$. Hence cl $U_h(x) \supset \Phi^{(h)}$. In particular, $a \in$ cl $U_h(x)$, i.e., $\| a - x \| \leq h$. Since this holds for any point $x \in \Phi$, we have $\Phi \subset$ cl $U_h(a)$. This means that cl $U_h(a)$ is one of the balls whose intersection is $\Phi^{(h)}$, and therefore $\Phi^{(h)} \subset$ cl $U_h(a)$. In particular, $b \in$ cl $U_h(a)$, i.e., $\| a - b \| \leq h$ holds for any points $a, b \in \Phi^{(h)}$, which implies diam $\Phi^{(h)} \leq h$. Since $\Phi^{(h)} \supset \Phi$, the converse inequality is trivial.

Thus $\Phi^{(h)}$ is a body of diameter h containing F, i.e., its volume cannot be larger than v. Hence (since $\Phi \subset \Phi^{(h)}$ and Φ has the volume v) the bodies Φ and $\Phi^{(h)}$ coincide.

Let now Π be a supporting half-space of $\Phi = \Phi^{(h)}$ and $\Gamma = $ bd Π. For $x_o \in \Gamma \cap \Phi$, denote by K the ball of radius h with $x_o \in K \subset \Pi$ and by a the center of K.

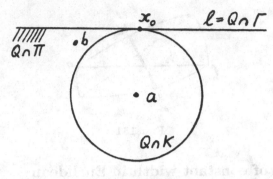

Fig. 152

We show that $\Phi \subset K$. Admit the contrary, i.e., there exists a point $b \in \Phi$ with $b \notin K$. Since $x_o, b \in \Phi$, we have $\| x_o - b \| \le h$. Furthermore, let Q be the two-dimensional plane spanned by the points x_o, a, b (Fig. 152). By S denote the circumference through x_o and b with radius h, whose center is situated on the same side of the line $l = Q \cap \Gamma$ as a, see Fig. 153. Since S and $Q \cap \mathrm{bd}\, K$ do not coincide, the intersection $S \cap l$ consists of two points. Hence the smaller arc A of S with endpoints x_o and b intersects l, i.e., there is a point $x \in A$ with $x \notin \Pi \cap Q$. Any circle C of radius $\le h$ containing x_o and b contains also the whole arc A (Fig. 154), since $S \cap \mathrm{bd}\, C$ consists of two points outside of A

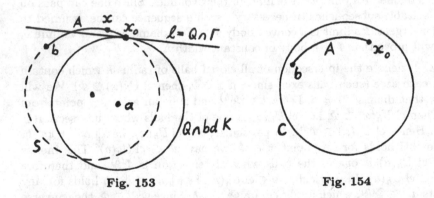

Fig. 153   Fig. 154

(Fig. 155). Let now K_1 be a ball of radius h with $\Phi \subset K_1$. Then $x_o, b \in K_1$. Since the circle $Q \cap K_1$ has a radius $\le h$ and satisfies $x_o, b \in Q \cap K_1$, we have also $A \subset Q \cap K_1$; in particular, $x \in Q \cap K_1$. Thus each ball K_1 of radius h with $\Phi \subset K_1$ contains the point x, i.e., $x \in \Phi^{(h)} = \Phi$. But this is impossible, since $x \notin \Phi$. Hence $\Phi \subset K$ is proved.

The inclusion $\Phi \subset K$ implies that also $\Phi \cup \{a\}$ has the diameter h (since $\| y - a \| \le h$, for arbitrary $y \in K$). Therefore conv $(\Phi \cup \{a\})$ has the diameter

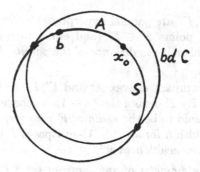

Fig. 155

h. If a would not belong to the convex set Φ, then conv $(\Phi \cup \{a\})$ would have a larger volume than Φ, a contradiction. Hence $a \in \Phi$. Consequently the distance of the parallel supporting hyperplanes Γ, Γ' of Φ cannot be smaller than h. By diam $\Phi = h$, this distance cannot be greater than h, too. Thus the distance of an arbitrary pair of parallel supporting hyperplanes of Φ is equal to h, i.e., Φ is a body of constant width h. ∎

After Theorem 32.1, we obtain the following theorem of I. Pál [Pa]: each set $F \subset \mathbf{R}^2$ of diameter h can be covered by a regular hexagon whose three pairs of parallel sides have the distance h, in each case. In fact, let $F \subset \mathbf{R}^2$ have diameter h, and Φ be a figure of constant width h that contains F (Theorem 32.1). We choose an arbitrary ray l with starting point o and draw six supporting lines of Φ, two of which are parallel to l, such that the lines form a hexagon M with angles equal to $\frac{2\pi}{3}$ circumscribed to Φ. The distance of opposite sides of the hexagon is equal to h. The hexagon M is uniquely determined by l, and a rotation of l changes the shape of M continuously. Following its contour, we enumerate the sides of M and denote by f the difference between the sum of lengths of the first, third and fifth sides and the sum of lengths of the second, fourth and sixth ones. Clearly, this difference continuously depends on the direction of l. Since the even and odd sides of M change under the rotation about the angle $\frac{\pi}{3}$, the sign of f changes. Hence there is a position of l with $f = 0$, i.e., M is a regular hexagon. ∎

Now we will give a generalization of Theorem 32.1, in view of normed spaces. Let $\Re^n$ be an n-dimensional normed space. By

$$d(a, \Gamma) = \min_{x \in \Gamma} d(a, x)$$

we denote the *distance of the point* $a \in \Re^n$ *to the hyperplane* $\Gamma \subset \Re^n$. It is obvious that for $r < d(a, \Gamma)$ the closed ball cl $U_r(a) = \{x : d(a, x) \le r\}$ has an empty intersection with Γ, and this intersection is nonempty if $r \ge d(a, \Gamma)$. For $r = d(a, \Gamma)$, this intersection is a face of cl $U_r(a)$, i.e., Γ is a supporting hyperplane of the ball.

Furthermore, if Γ', Γ'' are parallel hyperplanes in $\Re^n$, the number $d(x', \Gamma'')$ is constant for all points $x' \in \Gamma'$ (and equal to the number $d(x'', \Gamma')$ for $x'' \in \Gamma''$); this number is the *distance of the parallel hyperplanes* Γ', Γ'', and it is denoted by $d(\Gamma', \Gamma')$.

Let $F \subset \Re^n$ be a compact, convex set and Γ', Γ'' be two parallel supporting hyperplanes of F. By Γ denote the $(n-1)$-subspace parallel to Γ' and Γ''. Then $d(\Gamma', \Gamma'')$ is said to be the *width of F with respect to the subspace* Γ. If F has the same width h for all $(n-1)$-subspaces $\Gamma \subset \Re^n$, then F is said to be a *body of constant width* h *in* $\Re^n$.

Theorem 32.2: The diameter of any compact set $F \subset \Re^2$ is the largest width of F.   □

PROOF: Let the points $a, b \in F$ satisfy the condition $d(a, b) = \| a - b \| = $ diam F. We consider the ball cl $U_r(x_o)$ with $x_o = \frac{a+b}{2}$ and $r = \frac{d(a,b)}{2}$. The points a, b are diametrical boundary points of the ball. Hence there are two parallel supporting hyperplanes Γ', Γ'' of cl $U_r(x_o)$ containing a and b, respectively. Their distance $d(\Gamma', \Gamma'')$ is equal to the diameter of the ball, i.e., we have $d(a, b) = $ diam F. By $a, b \in F$, the width of F with respect to the corresponding direction cannot be smaller than diam F. Moreover, the maximal width of F cannot be smaller than diam F.

On the other hand, for two parallel supporting hyperplanes Π', Π'' of F with $x' \in F \cap \Pi'$, $x'' \in F \cap \Pi''$ we have $d(\Pi', \Pi'') \le d(x', x'') \le$ diam F. Hence every width of F is not larger than diam F and the maximal width of F cannot be larger than diam F. ∎

Following Eggleston [Eg 4], we say that a set $F \subset \Re^n$ is *complete* if there is no set $F' \supset F$ with $F' \ne F$ and diam $F' = $ diam F. In other words, for a complete set F the properties $F \subset F'$ and $F \ne F'$ imply that there is some $(n-1)$-subspace Γ such that the width of F' with respect to Γ is larger than diam F (Theorem 32.2). If $\mathbf{R}^n$ is the Euclidean n-space, then each complete set is a body of constant width and vice versa (this was established in the proof of Theorem 31.1). In Minkowski spaces this is, in general, not true. More precisely, the notions "complete sets" and "bodies of constant width" are equivalent only in two-dimensional normed spaces; for $n \ge 3$ only one implication holds: a body of constant width is complete. For discussions of these and related questions we refer to the surveys [C-G], §4, and [H-M], §5.4. By the following example, we will demonstrate that these notions differ from each other for $n = 3$.

EXAMPLE 32.3: Let $\Re^3$ be the join of its subspaces $\Re^1$ and $\Re^2$, where $\Re^2$ has a regular hexagon $\sum_1$ as its unit ball. Thus the unit ball $\sum$ of $\Re^3$ is a double pyramid having this hexagon as its basis, Fig. 156. Following a circuit around $\sum_1$, we denote by a, b, c the midpoints of each second side of the hexagon (Fig. 157) and determine

$$U_1 = \sum - \frac{2}{3}a, \ U_2 = \sum - \frac{2}{3}b, \ U_3 = \sum - \frac{2}{3}c;$$

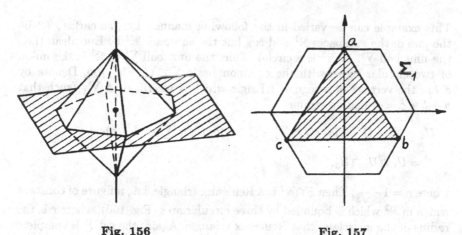

Fig. 156   Fig. 157

$$F = U_1 \cap U_2 \cap U_3.$$

Then F is a complete set of diameter 1 in $\Re^3$. Indeed, the set F is a double pyramid over the regular triangle $T \subset \Re^2$ with the vertices $\frac{2}{3}(a + b)$, $\frac{2}{3}(a + c)$, $\frac{2}{3}(b+c)$, see Fig. 158. For $x \notin F$, at least one of the conditions $x \notin U_1$, $x \notin U_2$, $x \notin U_3$ is satisfied, say $x \notin U_1$. Since $\frac{2}{3}(b + c) \in F$ is the center of the ball U_1 (of radius 1), we have $d(x, \frac{2}{3}(b + c)) > 1$, and therefore the diameter of $F \cup \{x\}$ is greater than 1. This means that F is a complete set.

At the same time, F is not a body of constant width. In fact, the width of F with respect to $\Gamma = \Re^2 = \operatorname{aff} T$ is smaller than 1. (If we denote by p, q the boundary points of the ball $\sum$ lying in $\Re^1$, then $\frac{1}{3}p$ and $\frac{1}{3}q$ are the apices of the double pyramid, and therefore the width of F regarding Γ is equal to $\frac{2}{3}$.)

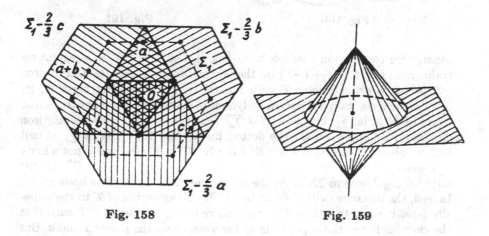

Fig. 158   Fig. 159

This example can be varied in the following manner. Let, as earlier, $\Re^3$ be the join of the subspaces $\Re^1$ and $\Re^2$, but the subspace $\Re^2$ be Euclidean (i.e., the unit ball $\sum_1$ of $\Re^2$ is a circle). Thus the unit ball $\sum \subset \Re^3$ is the union of two circular cones with the common basis $\sum_1$, cf. Fig. 159. Denote by a, b, c the vertices of a regular triangle which is inscribed to $\sum_1$ (such that $a + b + c = o$) and determine

$$U_1 = \sum - (1 - r)a, \ U_2 = \sum - (1 - r)b, \ U_3 = \sum - (1 - r)c;$$

$$F = U_1 \cap U_2 \cap U_3,$$

where $r = 1 - \frac{1}{\sqrt{3}}$. Then $F \cap \Re^2$ is a Reuleaux triangle, i.e., a figure of constant width in $\Re^2$ which is bounded by three circular arcs (Fig. 160), where r is the radius of the incircle of that Reuleaux triangle. Again, the set F is complete and a simple calculation shows that the width of F with respect to $\Gamma = \Re^2$ is equal to $2r = \frac{2(\sqrt{3}-1)}{\sqrt{3}} < 1$. $\square$

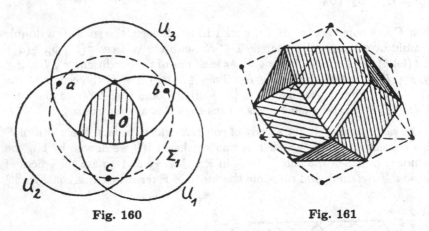

Fig. 160   Fig. 161

Finally, we consider an example from [Eg 4]. Let $T \subset \mathbf{R}^3$ be a regular tetrahedron and $M = T + (-T)$ be the vector sum of T and the tetrahedron $-T$, which is symmetric to T with repsect to the origin. Furthermore, let $\sum \supset M$ be a centrally symmetric polyhedron containing the eight triangular facets of M in bd $\sum$, but with $M \neq \sum$. (For example, $\sum$ is the octahedron circumscribed about M, see the dotted line in Fig. 161). Taking $\sum$ as unit ball, we obtain the normed space $\Re^3$ from $\mathbf{R}^3$. The tetrahedron is not a body of constant width in $\Re^3$, since $M = T + (-T)$ is not a ball in $\Re^3$ (cf. the forthcoming Theorem 32.5). At the same time, T is a complete body in $\Re^3$. Indeed, the distance (with respect to $\Re^3$) from any vertex of T to the opposite facet is equal to 1. If $x \notin T$, then there exists a vertex $a \in T$ such that the carrying flat of the opposite facet for a separates the points x and a. But then $d(a, x) > 1$, i.e., diam $(T \cup \{x\}) > $ diam $T = 1$. $\square$

Lemma 32.4: *If a complete set $M \subset \Re^n$ of diameter $2h$ is symmetric with respect to a point $a \in \Re^n$, then $M = \mathrm{cl}\, U_h(a)$.*   □

PROOF: Since two arbitrary parallel supporting hyperplanes Γ', Γ'' of M are symmetric with respect to a, we have

$$d(a, \Gamma') = d(a, \Gamma'') = \frac{1}{2} d(\Gamma', \Gamma'') \le \frac{1}{2} \mathrm{diam}\, M = h.$$

Thus the distance of an arbitrary supporting hyperplane of M from a is not larger than h. Therefore $M \subset \mathrm{cl}\, U_h(a)$ and, by the completeness, $M = \mathrm{cl}\, U_h(a)$.   ∎

The following statement is taken from [Eg 4].

Theorem 32.5: *Let $F \subset \Re^n$ be a compact, convex set and F' be a set which is symmetric to F with respect to a point. Then F is a body of constant width h in $\Re^n$ if and only if $F + F'$ is a ball of radius h in $\Re^n$.*   □

PROOF: Since the width is invariant under the group of translations, we may assume that $F' = -F$. Then $F + F'$ is symmetric about the origin and the width of the body $F + F'$ with respect to any $(n-1)$-subspace $\Gamma \subset \Re^n$ is equal to the doubled width of F regarding the subspace. If F is a body of constant width h in $\Re^n$, then $F + F'$ is a body of constant width $2h$ symmetric with respect to the origin, i.e., $F + F' = \mathrm{cl}\, U_h(o)$ (cf. Lemma 32.4).

Conversely, assume $F + F' = \mathrm{cl}\, U_h(o)$. Since the ball is a body of constant width $2h$, the same holds for $F + F'$, i.e., F is of constant width h in $\Re^n$.   ∎

Corollary 32.6: *Let F be a convex body in the n-dimensional vector space (without norm) and h be a positive number. Then there is (exactly one) norm in the n-space for which F is a body of constant width h.* □

In fact, for this norm the body $F + F'$ (with $F = -F'$) must be a closed ball of radius $2h$ centered at o. Hence the unit ball $\sum$ of the corresponding normed space $\Re^n$ is $\frac{1}{2h}(F + F')$.

Theorem 32.7: *Each set $F \subset \Re^n$ of diameter h is contained in a complete body of diameter h.*   □

PROOF: Consider all the convex sets of diameter h which contain F (such a set is conv F). The union of an arbitrary increasing sequence of such sets is again a convex set of diameter h. The Lemma of Zorn implies that there is at least one *maximal* convex body of diameter h which contains F. Because of maximality, the body is complete.   ∎

Example 32.3 shows that in the theorem above the notion "complete set" cannot be replaced by "body of constant width". Since in Euclidean space both notions coincide, Theorem 32.1 gives a stronger result than Theorem 32.7.

Also the next theorem is taken from [Eg 4].

Theorem 32.8: _A set_ $M \subset \Re^n$ _of diameter_ h _is complete if and only if it coincides with_

$$N = \bigcap_{x \in M} \operatorname{cl} U_h(x). \quad \square$$

PROOF: Since diam $M = h$, we have $M \subset \operatorname{cl} U_h(x)$ for any point $x \in M$. In other words, $M \subset N$. Therefore, the statement to be proved can be reformulated: the set M of diameter h is complete if and only if $M \supset N$.

Let M be a complete set. We choose an arbitrary point $y_o \in N$. Then $y_o \in \operatorname{cl} U_h(x)$ holds for any point $x \in M$, i.e., $d(y_o, x) \leq h$. Therefore the set $M \cup \{y_o\}$ has the diameter h, too. The completeness of M implies that $y_o \in M$ and $N \subset M$.

Let now M be incomplete, i.e., there exists a point $z_o \in \Re^n$ such that diam $(M \cup \{z_o\}) = h$ and $z_o \notin M$. The relation diam $(M \cup \{z_o\}) = h$ implies that $z_o \in \operatorname{cl} U_h(x)$ for arbitrary $x \in M$, i.e., $z_o \in N$. Hence the set N is not contained in M (since $z_o \notin M$). ∎

Theorem 32.9: _Let_ $M \subset \Re^n$ _be a complete set of diameter_ h. _The set_ M _is a body of constant width if and only if for any supporting hyperplane_ Γ _of_ M _there exists a point_ $x_o \in \Re^n$ _such that_ $M \subset \operatorname{cl} U_h(x_o)$ _and_ Γ _is a supporting hyperplane of the ball_ $\operatorname{cl} U_h(x_o)$. $\quad \square$

PROOF: Assume that for any supporting hyperplane there exists such a point x_o. Then by $M \subset \operatorname{cl} U_h(x_o)$ we have diam $(M \cup \{x_o\}) = h$ and $x_o \in M$. At the same time, $d(x_o, \Gamma) = h$ (since Γ is a supporting hyperplane of $\operatorname{cl} U_h(x_o)$). Hence the width of M with respect to Γ is not smaller than h; by diam $M = h$ it is also not greater than h. Thus for any hyperplane the corresponding width of M is equal to h, i.e., M is a body of constant width h.

Admit now that, for a supporting hyperplane Γ of M, the required point x_o does not exist. This means that (if we denote by Γ' the hyperplane parallel to Γ at the distance h from Γ and on the same side of Γ as M) we have $M \cap \Gamma' = \emptyset$. Consequently the width of M with respect to Γ is _smaller_ than h, i.e., M is not a body of constant width. ∎

Lemma 32.10: _In_ $\Re^2$, _let_ a, b _be points from the boundary of_ $\operatorname{cl} U_h(x)$ _satisfying_ $d(a, b) < 2h$. _Denote by_ P _the closed half-plane regarding the line_ (a, b), _which does not contain the point_ x, _and by_ p _the arc from the boundary of_ $\operatorname{cl} U_h(x)$ _which lies in_ P. _Then any circle_ $\operatorname{cl} U_h(z)$, _containing_ a _and_ b, _contains_ p, _too._ $\quad \square$

PROOF: We have $x \notin (a, b)$, since otherwise $d(a, b) = 2h$ would hold. Hence the half-plane P is uniquely determined. Let t denote the translation from x to z and $a' = t(a)$, $b' = t(b)$, see Fig. 162. The arc $p' = t(p)$ belongs to the strip Q between the lines (a, a'), (b, b'), since a' and b' are boundary points of $\operatorname{cl} U_h(z)$. Hence also the arc p (obtained from p' by t^{-1}) belongs to Q. This means that $p \subset P \cap Q$, yielding the relation $p \subset \operatorname{cl} U_h(z)$. ∎

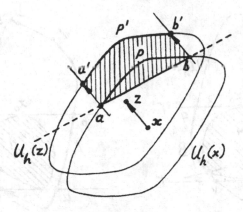

Fig. 162

The next theorem is again taken from [Eg 4].

Theorem 32.11: In an arbitrary two-dimensional normed space $\Re^2$, each complete set M of diameter h is a figure of constant width h.   □

PROOF: Let Γ be any supporting line of M. By Π denote the closed half-plane with respect to Γ satisfying $M \subset \Pi$. For the supporting line $\Gamma' \parallel \Gamma$ of M, we choose a point $x_o \in \Gamma' \cap M$. Clearly, $M \subset \operatorname{cl} U_h(x_o)$, and we will show that $d(\Gamma, \Gamma') = h$.

Admit that $d(\Gamma, \Gamma') < h$. Then there exists a boundary point y of $\operatorname{cl} U_h(x_o)$ outside the half-plane Π, see Fig. 163. The intersection points of the boundary of $\operatorname{cl} U_h(x_o)$ with the lines Γ and Γ' are denoted by u, v and p, q, respectively. The boundary part of M between Γ and Γ' consists of two arcs, denoted by A and B.

First, we will assume that the sets $M \cap A$ and $M \cap B$ are nonempty and choose arbitrary points $a \in M \cap A$, $b \in M \cap B$. Since $d(p,q) = 2h$, $d(a,b) \leq h$, at least one of the points a, b is different from p and q. Hence the line (a,b) does not contain x_o. If P is the half-plane with respect to (a,b) with $x_o \notin P$, then $u, v, y \in P$. By Lemma 32.10, an arbitrary circle covering M (and hence a and b) contains also the intersection of P and the boundary of $\operatorname{cl} U_h(x_o)$. By Theorem 32.8, this intersection is contained in M. In particular, we have $y \in M$, contradicting that Γ supports M.

Thus at least one of the sets $M \cap A$, $M \cap B$ has to be empty. Without loss of generality we set $M \cap A = \emptyset$, see Fig. 164. The diametrical points v', y' with respect to v and y (i.e., $v' = 2x_o - v, y' = 2x_o - y$) lie outside the strip between Γ and Γ'. By $M \cap A = \emptyset$, the whole arc with endpoints y and v', which contains A, does not contain points from M. If s_λ denotes the translation with the vector $\lambda(v - y)$, then, for $\lambda > 0$ small enough, the circle $s_\lambda(\operatorname{cl} U_h(x_o))$ will contain the figure M. The center z of this circle belongs to M, since $\operatorname{diam}(M \cup \{z\}) = h$. But this center $z = x_o + \lambda(v - y)$ is situated

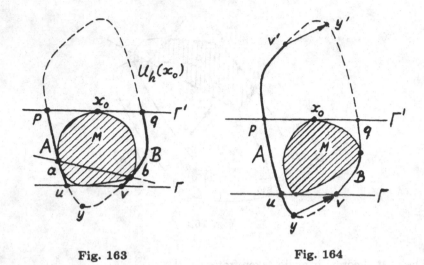

Fig. 163   Fig. 164

outside of the strip between Γ and Γ', and therefore it does not belong to M. This contradiction shows that the inequality $d(\Gamma, \Gamma') < h$ cannot be satisfied. Hence $d(\Gamma, \Gamma') = h$. This means that M is a figure of constant width h. ∎

Theorem 32.12: In each two-dimensional normed space $\Re^2$ there is at least one possibility to complete an arbitrary figure of diameter h to a figure of constant width h. □

This statement is an immediate consequence of the Theorems 32.7 and 32.11.

Finally we will consider generalizations of complete sets and bodies of constant width which are connected with the concepts of d-convexity and H-convexity. We assume that, besides the corresponding norm of the space $\Re^n$, a Euclidean metric is given, which allows the use of the scalar product and of the unit sphere S^{n-1}. Furthermore, we assume that a closed set $H \subset S^{n-1}$, symmetric about o, is given. By Theorem 32.2, the following definition is justified: Let $F \subset \Re^n$ be a compact set and h_Γ denote its width regarding an H-convex $(n-1)$-subspace $\Gamma \subset \Re^n$; the number

$$\mathrm{diam}_H F = \max_\Gamma h_\Gamma$$

is said to be the *H-diameter* of the set F. (Here the maximum is taken over all H-convex $(n-1)$-subspaces.) For $H = \emptyset$, we will say that $\mathrm{diam}_H F = 0$, for any set $F \subset \Re^n$. If $H = S^{n-1}$, then the H-diameter and the usual diameter $\mathrm{diam}\, F$ coincide. For $H = H(\Re^n)$, the H-diameter of F is called the *d-diameter* and denoted by $\mathrm{diam}_d F$. In other words,

$$\mathrm{diam}_d F = \max_\Gamma h_\Gamma,$$

where the maximum is taken over all d-convex $(n-1)$-subspaces $\Gamma \subset \Re^n$.

A set $F \subset \Re^n$ is said to be H-*complete* if there is no set $F' \supset F$ with $F' \neq F$ and $\text{diam}_H F' = \text{diam}_H F$. Furthermore, a convex set $F \subset \Re^n$ is called a *body of constant H-width h* if the width of F with respect to each H-convex $(n-1)$-subspace Γ is equal to h. For $H = S^{n-1}$, we obtain the usual definition of bodies of constant width in $\Re^n$. For $H = H(\Re^n)$, a body of constant H-width is also said to be a *body of constant d-width*. In other words, F is a body of constant d-width h in $\Re^n$ if the width of F with respect to any d-convex $(n-1)$-subspace is equal to h. Each body of constant H-width h has h as its H-diameter, and also each body of constant d-width h has h as its d-diameter.

Obviously, if $H' \supset H$ with $H \neq \emptyset$, each body of constant H'-width in $\Re^n$ is also of constant H-width in $\Re^n$. Since the unit ball $\sum \subset \Re^n$ is a body of constant width 2 in $\Re^n$, this set $\sum$ is also of constant H-width 2 in $\Re^n$, for any nonmepty set $H \subset S^{n-1}$ (which is closed and symmetric about the origin).

EXAMPLE 32.13: In the space $\mathbf{R}^2$ with a fixed coordinate system (x_1, x_2), let the unit ball $\sum$ of the corresponding normed space $\Re^2$ be described by $|x_1| + |x_2| \leq 1$. The only d-convex 1-subspaces of $\Re^2$ are the coordinate axes. The unit ball $\sum$ is of constant d-width 2, but is not d-complete. Namely, the square F', given by $|x_1| \leq 1$, $|x_2| \leq 1$, contains $\sum$, but in spite of $\sum \neq F'$ the width of F' with respect to each d-convex 1-subspace equals just 2, like for the unit ball $\sum$. Thus $\text{diam}_d F' = \text{diam}_d \sum = 2$. This example shows that for $H \neq S^{n-1}$ there exist bodies of constant H-width in $\Re^n$ which are not H-complete. $\square$

Theorem 32.14: Let $\sum$ be the unit ball of a normed space $\Re^n$. Then $\text{conv}_H \sum$ is an H-complete body of constant H-width 2. $\square$

PROOF: Let Γ be an H-convex $(n-1)$-subspace. Denote by Γ', Γ'' the supporting hyperplanes of $\sum$ parallel to Γ, and by P', P'' the corresponding closed half-spaces containing $\sum$, in each case. Since the half-spaces P', P'' are H-convex, also the strip $P' \cap P''$ is an H-convex set. By $\sum \subset P' \cap P''$ and hence $\text{conv}_H \sum \subset P' \cap P''$ it follows that the width of $\text{conv}_H \sum$ with respect to Γ cannot be greater than 2 (since $d(\Gamma', \Gamma'') = 2$).

Obviously, this width cannot be smaller than 2, since $\sum$ has the width 2. Thus $\text{conv}_H \sum$ is a body of constant H-width 2. We will show that $\text{conv}_H \sum$ is also H-complete. Let $F' \supset \text{conv}_H \sum$ denote a set with $F' \neq \text{conv}_H \sum$. Then there exists a point $a \in F'$ not belonging to $\text{conv}_H \sum$, and therefore an H-convex half-space $\Pi \supset \text{conv}_H \sum$ is constructable such that the bounding hyperplane L' of Π supports $\sum$ with $a \notin \Pi$. Let L'' denote the second supporting hyperplane of $\sum$ parallel to L', and L be the parallel $(n-1)$-subspace. This subspace L is H-convex. Since a lies outside of the closed strip between L' and L'', we have $d(a, L'') > d(L', L'')$, i.e., $d(a, L'') > 2$. Hence for the hyperplane L_1 containing a, which is parallel to L, we obtain

$d(L_1, L'') = d(a, L'') > 2$. Thus the width of F' with respect to L is larger than 2. ∎

Theorem 32.14 shows that, considering H-complete sets and bodies of constant H-width, the role of the sets $\mathrm{conv}_H U_h(x)$ is analogous to that of the balls cl $U_h(x)$ in the proofs of previous statements in this section. It is easily checked that the above theorems remain true for H-complete sets and bodies of constant H-width (together with the proofs, if we replace cl $U_h(x)$ by $\mathrm{conv}_H U_h(x)$ and if all supporting hyperplanes under consideration are assumed to be H-convex). For example, Theorem 32.8 takes the following form: *let $M \subset \Re^n$ be an H-complete set of H-diameter h; then*

$$M = \bigcap_{x_o \in M} \mathrm{conv}_H U_h(x_o).$$

The details are left to the reader.

Exercises:

1. Let M be a square and d be the length of its diagonal. Find the Reuleaux polygon of diameter d containing M. What is the minimal number of its vertices?

2. Let $M \subset \mathbf{R}^n$ be a set of diameter h and $M^{(h)}$ be the intersection of all (closed) balls of radius h which contain M. Prove that if $M^{(h)}$ is not a body of constant width h, then a body of constant width h containing M is not unique.

3. Let $M \subset \mathbf{R}^n$ be a set of diameter h and Q be a body of constant width h containing M. Prove that the body $M^{(h)}$ (see the previous exercise) is contained in Q.

4. Let $M \subset \mathbf{R}^n$ be a set and h be a positive number. Prove that if diam $M < h$, then diam $M^{(h)} < h$.

5. In the proof of Pál's theorem, the assertion was used that the hexagon Φ continuously depends on the direction of the ray l. Give a confirmation of this statement.

6. Prove that for every set $X \subset \mathbf{R}^3$ with diam $X = d$ there exists a regular octahedron $M \supset X$ with the distance d between its opposite facets. This result was obtained by D. Gale [Ga], and it was used in Grünbaum's construction (cf. Figs. 138–142).

7. Let $\sum \subset \mathbf{R}^2$ be a regular hexagon with the succesive vertices $a_1, \cdots, a_6$ and $\Re^2$ be the Minkowski plane with unit ball $\sum$ (Fig. 157). Denote by M the triangle whose vertices are the midpoints of the segments $[a_1, a_2], [a_3, a_4], [a_5, a_6]$. Prove that M is a figure of constant width with respect to $\Re^2$.

8. Let $\Re^n$ be a Minkowski space and $M \subset \Re^n$ be a body of constant width h. Prove that M is *complete*, i.e., for every set $X \supset M$ with $X \neq M$ the inequality diam $X >$ diam M holds.

9. Describe the figures of constant width in the Minkowski plane $\Re^2$ considered in Example 9.1.

10. Let $\Re^n$ be a Minkowski space and $M \subset \Re^n$ be a set of diameter h. Denote by $M^{(h)}$ the intersection of all balls of radius h containing M. Prove that if $Q \subset \Re^n$ is a body of constant width h containing M, then $Q \supset M^{(h)}$.

11. Prove the H-generalization of Theorem 32.8 (cf. the last sentences of the section).

§33 Borsuk's problem in normed spaces

Let $\Re^n$ be a normed space with unit shpere $\sum$. We will investigate the following problem, which was stated by B. Grünbaum [Grü 1] as a generalization of Borsuk's one (cf. §31): *Let $F \subset \Re^n$ be a set of diameter h. What is the smallest positive integer k such that F can be represented as the union of k sets each of which has a diameter smaller than h?* This smallest integer will be denoted by $a_\Sigma(F)$.

EXAMPLE 33.1: In the Euclidean plane, a parallelogram can be dissected into two parts of smaller diameter (Fig. 165). But if this parallelogram is the unit ball $\sum$ of a normed space $\Re^2$, then the diameters of $\sum$ and of its parts, as in Fig. 165, are equal to 2. Since the pairwise distance of the four vertices of $\sum$ is equal to 2, we obtain $a_\Sigma(\sum) = 4$ (Fig. 166). Thus in this case the number $a_\Sigma(\sum)$ is larger than the Borsuk number $a(\sum) = 2$. But if F is a circle inscribed to a square $\sum$ (which, again, will denote the unit ball of a Minkowski plane $\Re^2$), then $a_\Sigma(F) = 2$, i.e., in this case the number $a_\Sigma(F)$ is smaller than the Borsuk number $a(F) = 3$, see Fig. 167.

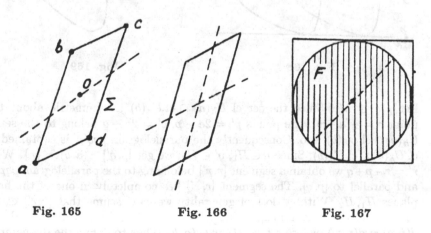

Fig. 165   Fig. 166   Fig. 167

Lemma 33.2: Let $F \subset \Re^2$ be a set of diameter h. If the way to complete F to a figure of constant width h is not unique, then $a_\Sigma(F) = 2$. □

PROOF: Let A, B be two different figures of constant width h which contain F, and $a \in A$ be a point not contained in B. By Theorem 32.8, there exists a point $b \in B$ such that $a \notin \mathrm{cl}\, U_h(b)$, i.e., $d(a, b) > h$. By $a \in A, b \in B$ we obtain $B \subset \mathrm{cl}\, U_h(b)$ and $A \subset \mathrm{cl}\, U_h(a)$. Hence $F \subset A \cap B \subset \mathrm{cl}\, U_h(a) \cap \mathrm{cl}\, U_h(b)$.

Denote by Π_1, Π_2 the closed half-planes with respect to the line (a, b) and by l_1, l_2 the lines parallel to (a, b) at the distance h, in each case (with $l_1 \subset \Pi_1$, $l_2 \subset \Pi_2$). The lines l_1, l_2 are supporting lines of the closed circles $\mathrm{cl}\, U_h(a)$ and $\mathrm{cl}\, U_h(b)$, respectively (Fig. 168). We will show that the diameter of each of the sets

$$M_1 = \mathrm{cl}\, U_h(a) \cap \mathrm{cl}\, U_h(b) \cap \Pi_1,$$

$$M_2 = \mathrm{cl}\, U_h(a) \cap \mathrm{cl}\, U_h(b) \cap \Pi_2$$

cannot be larger than h.

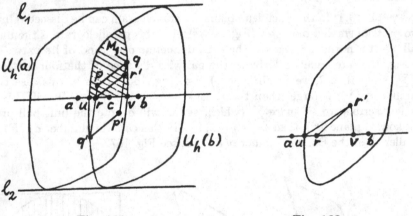

Fig. 168   Fig. 169

Let $p, q \in M_1$. Since the set $\mathrm{cl}\, U_h(a) \cap \mathrm{cl}\, U_h(b)$ is symmetric about the point $c = \frac{1}{2}(a + b)$, the points $p' = 2c - p$, $q' = 2c - q$ belong to the set $\mathrm{cl}\, U_h(a) \cap \mathrm{cl}\, U_h(b)$, too. Consequently the parallelogram $pqq'p'$ is contained in $\mathrm{cl}\, U_h(a) \cap \mathrm{cl}\, U_h(b)$. Since $p \in \Pi_1$, $q' \in \Pi_2$, we get $[p, q'] \cap (a, b) = \{r\}$. With $r' = r - p + q$ we obtain a segment $[r, r']$ belonging to the parallelogram $pqp'q'$ and parallel to $[p, q]$. The segment $[r, r']$ lies completely in one of the half-planes Π_1, Π_2. Without loss of generality we can assume that $[r, r'] \subset \Pi_1$ (otherwise, the mirror image with respect to c can be considered). Hence $d(p, q) = d(r, r')$, where $r, r' \in M_1$, $r \in (a, b)$. Thus to define the diameter of a set M_1, it suffices to take segments like $[r, r']$ into consideration.

Let the segment $[u, v]$ be the intersection of $\mathrm{cl}\, U_h(a) \cap \mathrm{cl}\, U_h(b)$ and (a, b). If $r' \in M_1$ and l denotes the larger one of the numbers $d(u, r')$ and $d(v, r')$, then the ball $\mathrm{cl}\, U_l(r')$ contains the points u and v (Fig. 169) and also the whole segment $[u, v]$. Hence we have $d(r, r') \leq l$ for any point $r \in [u, v]$, i.e., at least one of the inequalities $d(r, r') \leq d(u, r')$, $d(r, r') \leq d(v, r')$ is satisfied for any point $r \in [u, v]$. Thus to define the diameter of a set M_1, it is sufficient to consider the distances $d(u, r')$, $d(v, r')$ for $r' \in M_1$.

We will assume that a lies *left-side* regarding b, and u lies *left-side* with respect to v. (We remark that because of $d(a, b) > h$, the points r, u, v are interior points of the segment $[a, b]$.) Let now t denote the translation which takes b to a and cl $U_h(b)$ onto cl $U_h(a)$. Hence for any point $r' \in M \subset$ cl $U_h(a) \cap$ cl $U_h(b)$ the inclusion $t(r') \in$ cl $U_h(a)$ holds. Therefore $[r', t(r')] \subset$ cl $U_h(a)$. If $r' \notin l_1$ holds (i.e., $[r', t(r')]$ lies between the lines (a, b) and l_1), then the segment (possibly except for the endpoints) is situated in the interior of $U_h(a)$. This implies that the fourth vertex $w = a - v + r'$ of the parallelogram $avr'w$ (Fig. 170) lies in cl $U_h(a)$, i.e., $d(v, r') = d(a, w) < h$. Analogously, $d(u, r') < h$. Thus, if $r' \notin l_1$ (for $r' \in M$), then the inequalities $d(u, r') < h$, $d(v, r') < h$ hold.

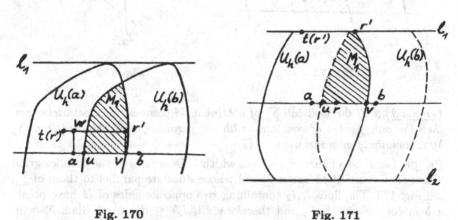

Fig. 170   **Fig. 171**

Finally, let the point $r' \in l_1$ be contained in M_1. In this case $r' \in$ bd $U_h(a) \cap$ bd $U_h(b)$, i.e., both the points $r', t(r')$ lie in the boundary of $U_h(a)$. Since there is no point of this circle above l_1, the whole segment $[r', t(r')]$ is contained in the boundary of $U_h(a)$ (Fig. 171). For any point $r \in [a, b]$, this yields the equality $d(r, r') = h$ (in particular, this holds for any point $r \in [u, v]$).

Comparing the above conclusions, we obtain that diam $M_1 \leq h$, where the equality $d(r, r') = h$ for points $r, r' \in M_1$ is only satisfied if one of them lies on (a, b) and the other one on l_1.

Reverting to the set F, we see that cl $F \subset$ cl $U_h(a) \cap$ cl $U_h(b)$ implies the inclusion cl $F \subset M_1 \cup M_2$. If then cl $F \cap l_1 = \emptyset$ and cl $F \cap l_2 = \emptyset$, the above conclusions imply that each of the intersections cl $F \cap M_1$, cl $F \cap M_2$ has a diameter smaller than h, i.e., F can be dissected into two parts of smaller diameter. But if, for example, cl $F \cap l_1 \neq \emptyset$ holds, then $F \subset M_1$ (under the line (a, b), there is no point from F, since the distance of (a, b) and l_1 equals h). If M_1 is partitioned into two parts by a line l' between l_1 and (a, b) (cf. Fig. 172), then these parts of M_1 (and hence of $F \subset M_1$) have a smaller diameter.   ∎

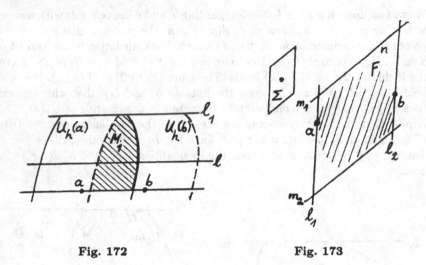

Fig. 172   Fig. 173

Lemma 33.3: If the unit ball $\sum$ of a Minkowski plane $\Re^2$ is a parallelogram, then the only figures of constant width are parallelograms obtained from $\sum$ by a homothety or translation.   □

PROOF: Let F be a figure of constant width h. Denote by Π the parallelogram which is circumscribed about F and whose sides are parallel to those of $\sum$, see Fig. 173. The lines l_1, l_2 containing two opposite sides of Π have points a, b in common with F, and therefore $d(l_1, l_2) \leq d(a, b) \leq \operatorname{diam} F = h$. Analogously, for the two other lines through opposite sides of Π the inequality $d(m_1, m_2) \leq h$ holds. Hence diam $\Pi \leq h$, and the completeness of $F \subset \Pi$ implies $F = \Pi$. The distances $d(l_1, l_2)$, $d(m_1, m_2)$ have to be equal to h, since otherwise an enlargement of Π without increasing of diameter would be possible, see Fig. 174. Therefore $F = \Pi$ is obtainable from $\sum$ by means of a homothety or translation.   ∎

Theorem 33.4: Let $\Re^2$ be a Minkowski plane whose unit ball $\sum$ is a parallelogram. For a compact set $F \subset \Re^2$ of diameter h the equality $a_\Sigma(F) = 2$ holds if and only if F does not contain three points which are the vertices of an equilateral triangle (in the metric of $\Re^2$) having the side-length h. The equality $a_\Sigma(F) = 4$ holds if and only if conv F is a parallelogram obtained from $\sum$ by a homothety or translation. In the remaining cases, $a_\Sigma(F) = 3$ holds.   □

PROOF: Let $a_\Sigma(F) > 2$. Then, by Lemma 33.2, there is a unique way of completing F to a figure Π of constant width h. According to Lemma 33.3,

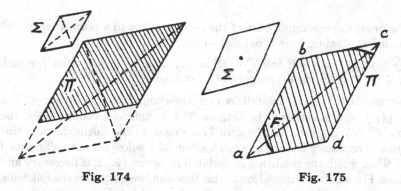

Fig. 174   Fig. 175

this figure Π is a parallelogram $abcd$, obtained from $\sum$ by a homothety or translation. If none of the points a, c belongs to F, then (since F is closed) there are neighbourhoods of a and c not containing points of F. This yields $a_\Sigma(F) = 2$ (Fig. 175), a contradiction. Therefore, at least one of the points a, c has to lie in F, say a. Analogously, one of the points b, d belongs to F, say b. Furthermore, let p be a common point of F and the side $[c, d]$ of the parallelogram Π (since Π is the circumscribed parallelogram). The triangle with the vertices $a, b, p \in F$ is the desired one (since $d(a, b) = d(a, p) = d(b, p) = h$).

If, conversely, $a, b, p \in F$ are the vertices of an equilateral triangle with side-length h, then, under a partition into three different pieces of smaller diameter, no piece can contain two of these points. Hence $a_\Sigma(F) > 2$.

Furthermore, for $a_\Sigma(F) > 2$ the figure $\Pi \supset F$ can be dissected into four pieces of smaller diameter (Fig. 166) and therefore $a_\Sigma(F) \leq 4$. If then one of the points a, b, c, d does not lie in F, it follows that $a_\Sigma(F) \leq 3$ (Fig. 176). Hence the equality $a_\Sigma(F) = 4$ holds only in the case conv $F = \Pi$. ∎

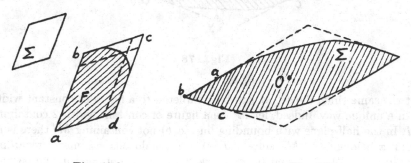

Fig. 176   Fig. 177

The theorem above gives a complete answer regarding $a_\Sigma(F)$ if $\sum$ is a parallelogram. And the following theorem refers to the case when $\sum$ is not a parallelogram. It will be proved in the next chapter (see Theorem 35.7). As this theorem, the remaining part of the section refers to a normed space $\Re^2$ with unit ball distinct from a parallelogram.

Theorem 33.5: If the unit ball $\sum \subset \Re^2$ is not a parallelogram, then for each bounded set $F \subset \Re^2$ the inequality $a_\Sigma(F) \leq 3$ holds. $\square$

In other words, if $\sum$ is not a parallelogram, then one of the equalities $a_\Sigma(F) = 2$ or $a_\Sigma(F) = 3$ is satisfied. In Lemma 33.2 a sufficient condition for the case $a_\Sigma(F) = 2$ is given. As Theorem 31.4 shows, in the Euclidean case this condition is necessary and sufficient. Theorem 33.7 below refers to all normed spaces $\Re^2$ in which the condition formulated in Lemma 33.2, is necessary and sufficient. For a better formulation of this theorem, we introduce the following notion: we will say that a normed space $\Re^2$ with unit ball $\sum$ is an _angled space_ if there are three noncollinear points $a, b, c \in$ bd $\sum$ such that $[a, b] \subset$ bd $\sum$ and $[b, c] \subset$ bd $\sum$ (Fig. 177).

Theorem 33.6: If the normed space $\Re^2$ is angled, then there exists a figure M of diameter h which can be completed to a figure of constant width h in a unique way and satisfies $a_\Sigma(M) = 2$. $\square$

PROOF: Let $a, b, c \in$ bd $\sum$ be three non-collinear points with $[a, b] \subset$ bd $\sum$, $[b, c] \subset$ bd $\sum$ and $a' = -a$, $b' = -b$, $c' = -c$. Denote by M the intersection of $\sum$ and the strip bounded by the lines (a, c) and (a', c'). It is easy to see that the line (b, b') dissects M into two parts of diameter < 2 (Fig. 178). Since diam $M = 2$ (by $d(a, a') = 2$), the relation $a_\Sigma(M) = 2$ holds.

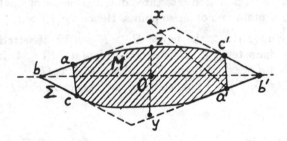

Fig. 178

At the same time, the set M can be completed to a figure of constant width 2 in a unique way. Indeed, let M' be a figure of constant width 2 containing M. In the half-plane with bounding line (a, b) not containing M, there is no point x belonging to M', since $d(a', x) > 2$. Analogous arguments regarding the lines $(a', b'), (b, c)$, and (b', c') show that M' is contained in the parallelogram spanned by these four lines. Furthermore, if $y \notin \sum$ belongs to the parallelogram, then we have $d(y, z) > 2$, $z \in M$, for a point $z \in$ bd $\sum$ satisfying $o \in [y, z]$. Hence $M' \subset \sum$, and by the completeness of M' we have $M' = \sum$. Therefore $\sum$ is the only figure of constant width 2 which contains

M. For obtaining an analogous figure of diameter h, it is sufficient to use a homothety. ■

The following theorem is taken from [B-SV].

Theorem 33.7: Let the normed space $\Re^2$ be not angled. For a figure $F \subset \Re^2$ of diameter h the equality $a_\Sigma(F) = 3$ holds if and only if F can be completed in a unique way to a figure of constant width h. □

For proving the theorem, we compile some lemmas which are also interesting for themselves. We remark that Theorem 31.4 (which was formulated above without proof) is an immediate consequence of Theorem 33.7.

Lemma 33.8: Let $F \subset \Re^n$ be a bounded set. Then $a_\Sigma(F) = a_\Sigma(\operatorname{conv} F)$. □

PROOF: Let $F = M_1 \cup \cdots \cup M_m$, where diam $F = h$ and diam $M_i < h$ ($i = 1, \cdots, m$). We write $F_k = \operatorname{cl} \operatorname{conv} (M_1 \cup \cdots \cup M_k)$, $k = 1, \cdots, m$, and we will show (by induction over k) that F_k can be represented as union of k sets whose diameters are smaller than h. For $k = 1$, this is obvious. Assume that the assertion is true for an integer k, i.e., $F_k = A_1 \cup \cdots \cup A_k$ with diam $A_k < h$, $i = 1, \cdots, k$. Choose an $\varepsilon > 0$ such that the inequalities diam $U_\varepsilon(A_i) < h$ ($i = 1, \cdots, k$), diam $M_{k+1} < h - \varepsilon$ are satisfied. We put

$$B_1 = \operatorname{cl} U_\varepsilon(A_1) \cap F_{k+1}, \quad \cdots, \quad B_k = \operatorname{cl} U_\varepsilon(A_k) \cap F_{k+1},$$

$$B_{k+1} = F_{k+1} \setminus (U_\varepsilon(A_1) \cup \cdots \cup U_\varepsilon(A_k)).$$

It is clear that $F_{k+1} = B_1 \cup \cdots \cup B_{k+1}$ and, furthermore, diam $B_i \leq$ diam $U_\varepsilon(A_i) < h$, for $i = 1, \cdots, k$. Finally, also the inequality diam $B_{k+1} < h$ holds. In fact, we have $F_{k+1} = \operatorname{cl} \operatorname{conv} (F_k \cup M_{k+1})$, and hence for arbitrary points $a, b \in F_{k+1}$ the inclusions $a \in [a', a'']$, $b \in [b', b'']$ with $a', b' \in F_k$ and $a'', b'' \in \operatorname{cl} \operatorname{conv} M_{k+1}$ hold. If, in addition, $a, b \in B_{k+1}$, then $d(a, a') \geq \varepsilon$, $d(b, b') \geq \varepsilon$. It follows that the points

$$a^* = a + \frac{\varepsilon}{h}(a - a''), \quad b^* = b + \frac{\varepsilon}{h}(b - b'')$$

lie in the set F_{k+1}. Since $a - b = \frac{h}{h+\varepsilon}(a^* - b^*) + \frac{\varepsilon}{h+\varepsilon}(a'' - b'')$, we obtain

$$\| a - b \| \leq \frac{h}{h + \varepsilon} \| a^* - b^* \| + \frac{\varepsilon}{h + \varepsilon} \| a'' - b'' \|$$

$$\leq \frac{h^2}{h + \varepsilon} + \frac{\varepsilon}{h + \varepsilon}(h - \varepsilon) = h - \frac{\varepsilon^2}{h + \varepsilon}.$$

Therefore diam $B_{k+1} \leq h - \frac{\varepsilon^2}{h+\varepsilon} < h$, i.e., F_{k+1} is representable as union of $k + 1$ sets whose diameters are smaller than h.

For $k = m$ we obtain that the set $F_m = \operatorname{cl} \operatorname{conv} F$ is representable as union of m sets whose diameters are smaller than h. Hence $a_\Sigma(\operatorname{cl} \operatorname{conv} F) \leq a_\Sigma(F)$. The converse inequality is obvious. ■

Also for the next theorem, we refer to [B-SV].

Theorem 33.9 Let $\Re^2$ be a two-dimensional normed space. For a closed figure $F \subset \Re^2$ of diameter h the inequality $a_\Sigma(F) > 2$ holds if and only if the following two conditions are satisfied:

(1) There is a unique way to complete F to a figure Φ of constant width h.

(2) For any two parallel supporting lines of Φ, at least one has a nonempty intersection with F. □

PROOF: At first, we will show that, if a closed figure $F \subset \Re^2$ satisfies (1) and (2), the inequality $a_\Sigma(F) > 2$ holds.

Assume $a_\Sigma(F) = 2$, i.e., $F = Q_1 \cup Q_2$, where Q_1, Q_2 are figures of diameter smaller than h. We can assume that F is convex (cf. Lemma 33.8).

For each figure M of diameter $\leq h$, denote by $M^{(h)}$ the intersection of all circles cl $U_h(x)$ which contain M. We show that $F^{(h)}$ coincides with the (uniquely determined) figure Φ of constant width h which contains F. By Theorem 32.8

$$\Phi = \bigcap_{x \in \Phi} \text{cl } U_h(x) \supset F^{(h)}$$

(because each circle cl $U_h(x)$, $x \in \Phi$, contains Φ and also F). Hence $\Phi \supset F^{(h)}$.

We show the converse. For doing this, assume that there exists a point $x_o \in \Phi$ not contained in $F^{(h)}$. This means that there is a circle cl $U_h(y_o)$ containing F but not x_o. The inclusion $F \subset$ cl $U_h(y_o)$ implies diam $(F \cup \{y_o\}) = h$ and the existence of a figure Φ' of constant width h satisfying $\Phi' \supset F \cup \{y_o\}$. By $d(x_o, y_o) > h$, we have $x_o \notin \Phi'$. Therefore $\Phi \neq \Phi'$, contradicting (1). This contradiction shows that $\Phi = F^{(h)}$.

Now we show that each of the figures $Q_1^{(h)}, Q_2^{(h)}$ has a diameter smaller than h. Let diam $Q_1 = \lambda < h$. We choose an arbitrary figure Ψ_1 of constant width λ with $Q_1 \subset \Psi_1$. Then $\Psi_1 = \bigcap_{x \in \Psi_1}$ cl $U_\lambda(x)$. But each circle cl $U_\lambda(x)$ is the intersection of all circles cl $U_h(y)$ of radius h containing it. Therefore Ψ_1 is the intersection of a set of circles cl $U_h(y)$ each of which contains Ψ_1 (and hence also Q_1). It follows that the intersection of all circles cl $U_h(y)$, which contain Q_1, is contained in Ψ_1, i.e., $Q_1^{(h)} \subset \Psi_1$. Therefore diam $Q_1^{(h)} = \lambda < h$, and analogously diam $Q_2^{(h)} = $ diam $Q_2 < h$ is derived.

Furthermore, we verify the inclusion

$$\text{bd } F^{(h)} \subset Q_1^{(h)} \cup Q_2^{(h)}.$$

Let $x \in$ bd $F^{(h)}$. We construct a supporting line Γ of $F^{(h)}$ through the point x. According to Theorem 32.9, there is a point $y \in \Re^2$ such that $F^{(h)} \subset$

cl $U_h(y)$ and Γ is a supporting line of cl $U_h(y)$. From $F^{(h)} \subset$ cl $U_h(y)$ it follows that diam $(F^{(h)} \cup \{y\}) = h$, and therefore $y \in F^{(h)}$. Let Γ' denote the line which is parallel to Γ and contains y. Since $d(\Gamma, \Gamma') = h$, the line Γ' supports $F^{(h)}$. For the sake of convenience, we say that Γ, Γ' are "horizontals". Let $[u, v]$ denote the segment $\Gamma \cap F^{(h)}$ and $[w, z]$ the segment $\Gamma' \cap F^{(h)}$. It is possible that $u = v$ and $w = z$. Clearly, $x \in [u, v]$ and $y \in [w, z]$.

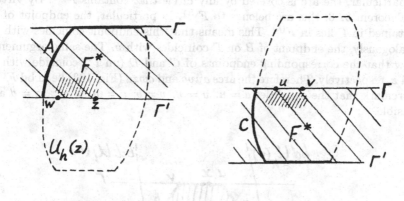

Fig. 179   Fig. 180

For $z \in F^{(h)}$, the inclusion $F^{(h)} \subset$ cl $U_h(z)$ holds. The boundary part of cl $U_h(z)$ between Γ and Γ' consists of two open arcs. The arc on the left (together with its two endpoints) will be denoted by A (Fig. 179), the left arc from $U_h(v)$ by C (Fig. 180). The correspondingly right arcs from the boundaries of $U_h(w), U_h(u)$ are denoted by B and D, respectively (Figs. 181 and 182).

It is easy to see that the intersection of each of the arcs A, B, C, D with F is nonempty. In fact, for $A \cap F = \emptyset$ the circle cl $U_h(z)$ could be translated to the right in such a manner that the translate completely contains F and

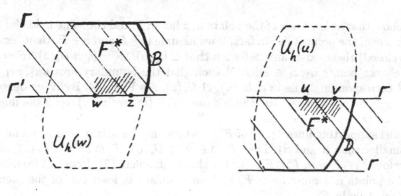

Fig. 181   Fig. 182

hence $F^{(h)}$. But then the center z' of the translate would lie in $F^{(h)}$ (since diam $F^{(h)} \cup \{z'\} = h$) with $z' \in F^{(h)} \cap \Gamma'$, a contradiction to $z' \notin [w, z]$. This shows that $A \cap F \ne \emptyset$, and analogously $B \cap F \ne \emptyset$, $C \cap F \ne \emptyset$, $D \cap F \ne \emptyset$.

We choose arbitrary points $a \in A \cap F$, $b \in B \cap F$, $c \in C \cap F$, $d \in D \cap F$. Since $a \in F \subset F^{(h)}$, $x \in F^{(h)}$, the arc from the boundary of cl $U_h(z)$ between a and x lies in any circle cl $U_h(\xi)$ containing a and x (according to Lemma 32.10). In particular, the arc is covered by any circle that contains $F^{(h)}$. By virtue of Theorem 32.8, this arc belongs to $F^{(h)}$. In particular, the endpoint of A contained in Γ lies in $F^{(h)}$. This means that this endpoint coincides with u. Analogously, the endpoint of B on Γ coincides with v. The same arguments show that the corresponding endpoints of C and D (on Γ') coincide with w and z, respectively. Therefore the arcs $auvb$ and $cwzd$ (Fig. 183) lie in bd $F^{(h)}$; we remark that the equalities $a = u$, $u = v$, $v = b$; $c = w$, $w = z$, $z = d$ are possible.

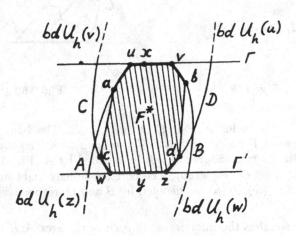

Fig. 183

We show that at least one of the points u, z lies in F and the same holds for at least one of the points v, w. In fact, if we assume $u \notin F$ and $z \notin F$, then (since F is closed) there exists an $\varepsilon > 0$ such that cl $U_\varepsilon(u) \cap F = \emptyset$, cl $U_\varepsilon(z) \cap F = \emptyset$. We choose points $u_1, z_1 \in$ bd $F^{(h)}$ such that the boundary arcs uu_1, zz_1 of $F^{(h)}$ correspondingly lie in cl $U_\varepsilon(u)$, cl $U_\varepsilon(z)$, see Fig. 184. By α we denote the angle between $[u, u_1]$ and the left part of Γ (regarding u), by β the angle between $[z, z_1]$ and the right part of Γ' (regarding z). Finally, we introduce parallel supporting lines Γ_1, Γ_1' of $F^{(h)}$, which enclose with Γ and Γ' an angle γ (smaller than α and β). Then $\Gamma_1 \cap F^{(h)} \subset U_\varepsilon(u)$, $\Gamma_1' \cap F^{(h)} \subset U_\varepsilon(z)$, and therefore $\Gamma_1 \cap F = \emptyset$, $\Gamma_1' \cap F = \emptyset$. But this contradicts (2). Hence at least one of the points u, z must lie in F, and analogously at least one of the points v, w must lie in F.

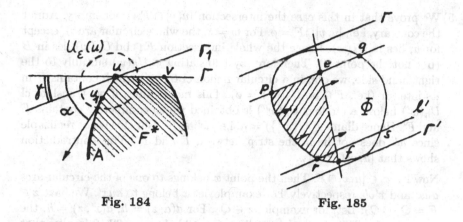

Fig. 184   Fig. 185

Now the following five cases (excluding each other) can be represented:

1) $z \notin F$, $w \notin F$;

2) $z \notin F$, $w \in F$;

3) $z \in F$, $w \in F$;

4) $z \in F$, $w \in F$, $z = w$;

5) $z \in F$, $w \in F$, $z \neq w$.

We consider these cases separately:

1) In this case the inclusions $u \in F$, $v \in F$ hold. Hence, by convexity of F, we have $x \in [u, v] \subset F$, i.e. $x \in F = Q_1 \cup Q_2 \subset Q_1^{(h)} \cup Q_2^{(h)}$.

2) In this case $u \in F$, i.e., the whole triple b, u, w lies in $F = Q_1 \cup Q_2$. For example, let $w \in Q_2$. For $d(w, u) = d(w, b) = h$, the points b, u lie in Q_1 (since diam $Q_2 < h$). Furthermore, the circular arc uvb belongs to the boundary of cl $U_h(w)$. According to Lemma 32.10, each circle cl $U_h(\xi)$ containing Q_1 (i.e., also the points b, u) contains the whole arc uvb. In particular, we get $x \in$ cl $U_h(\xi)$ (for any circle cl $U_h(\xi)$ containing Q_1) and therefore $x \in Q_1^{(h)}$. Hence in this case $x \in Q_1^{(h)} \cup Q_2^{(h)}$.

3) This case is analogous to the previous one.

4) The points a, b, w belong to $F = Q_1 \cup Q_2$. For example, let $w \in Q_2$. Since the equalities $d(w, a) = d(z, a) = h$ and $d(w, b) = h$ hold, we have $a, b \in Q_1$. Furthermore, the circular arc auv belongs to the boundary of cl $U_h(z)$, the circular arc uvb to that of cl $U_h(w)$, and therefore (by $w = z$) the whole circular arc $auvb$ lies in the boundary of cl $U_h(w)$. As in the second case, we conclude that the whole circular arc $auvb$ (and particularly the point x) lies in $Q_1^{(h)}$. Hence $x \in Q_1^{(h)} \cup Q_2^{(h)}$.

5) We prove that in this case the intersection $[u,v] \cap F$ is nonempty. Admit the contrary, i.e., $[u,v] \cap F = \emptyset$. For $w \neq z$, the whole circular arc A, except for u, lies in $U_h(w)$. Hence the whole intersection $F \cap \operatorname{bd} U_h(w)$ lies in B (but not the point v). Therefore, by translating $\operatorname{cl} U_h(w)$ suitably to the right-hand side, we obtain a circular region $\operatorname{cl} U_h(w')$ which contains F in its interior (i.e., $F \cap \operatorname{bd} U_h(w') = \emptyset$). This means that, if we translate $\operatorname{cl} U_h(w')$ below, a circle $\operatorname{cl} U_h(w'')$ is obtained which completely contains F and F'. Hence $\operatorname{diam}(F' \cup \{w''\}) = h$, i.e., $w'' \in F'$. But this is not realizable since w'' does not lie in the strip between Γ and Γ'. This contradiction shows that $[u,v] \cap F \neq \emptyset$.

Now let $x' \in [u,v] \cap F$. Then the point x belongs to one of the circular arcs aux' and $x'vb$, respectively. For example, let x belong to aux'. We have $z \in F = Q_1 \cup Q_2$. Let, for example, $z \in Q_2$. For $d(a,z) = h$, $d(x',z) = h$, the points $a, x' \in F$ lie in Q_1 (since $\operatorname{diam} Q_2 < h$). Lemma 32.10 implies that the whole arc aux' lies in $Q^{(h)}$ (cf. the discussion of the second case) and, in particular, $x \in Q_1^{(h)} \subset Q_1^{(h)} \cup Q_2^{(h)}$. Hence, in any case, $x \in Q_1^{(h)} \cup Q_2^{(h)}$, i.e., the inclusion $\operatorname{bd} F \subset Q_1^{(h)} \cup Q_2^{(h)}$ is proved.

Since each of the closed sets $Q_1^{(h)} \cap \operatorname{bd} F^{(h)}$, $Q_2^{(h)} \cap \operatorname{bd} F^{(h)}$ has a diameter smaller than h, none of them can coincide with $\operatorname{bd} F^{(h)}$. The union of the sets is $\operatorname{bd} F^{(h)}$. Since $\operatorname{bd} F^{(h)}$ is a connected set, the intersection of the sets cannot be empty, i.e., there is a point $\xi \in Q_1^{(h)} \cap Q_2^{(h)} \cap \operatorname{bd} F^{(h)}$.

Since $F^{(h)} = \Phi$ is a figure of constant width h in $\Re^2$, there exists a point $\eta \in \operatorname{bd} F^{(h)}$ such that $d(\xi,\eta) = h$. From $\xi \in Q_1^{(h)}$, $d(\xi,\eta) = h$, and $\operatorname{diam} Q_1^{(h)} < h$ it follows that $\eta \notin Q_1^{(h)}$. Analogously, we have $\eta \notin Q_2^{(h)}$. Therefore $\eta \notin Q_1^{(h)} \cup Q_2^{(h)}$, $\eta \in \operatorname{bd} F^{(h)}$, contradicting the inclusion $\operatorname{bd} F^{(h)} \subset Q_1^{(h)} \cup Q_2^{(h)}$. This contradiction shows that one cannot have the equality $a_\Sigma(F) = 2$. Thus the conditions (1) and (2) of the theorem are sufficient for the inequality $a_\Sigma(F) > 2$.

For the necessity we have to show that $a_\Sigma(F) = 2$ holds if one of the conditions is not satisfied. In fact, according to Lemma 33.2 we have $a_\Sigma(F) = 2$ if (1) is not satisfied. Assume now that (1) holds, but not (2). In other words, the closed figure F of diameter h can be completed in a unique way to a figure Φ of constant width h, and in addition Φ has two parallel supporting lines Γ, Γ' with $\Gamma \cap F = \emptyset$, $\Gamma' \cap F = \emptyset$. We introduce two supporting lines L, L' of F which are parallel to Γ, Γ' (Fig. 185). They intersect Φ in two nondegenerate chords $[p,q]$, $[r,s]$. Let e,f denote interior points of the chords. Then the line (e,f) dissects F in such a way that the two parts have a diameter smaller than h (by a suitable translation along Γ, each part can be brought into the interior of the figure Φ). Therefore $a_\Sigma(F) = 2$. ∎

Now we are ready for the

<u>PROOF OF THEOREM 33.7:</u> Since the space $\Re^2$ is not angled, the unit ball $\sum$ is not a parallelogram, and therefore the inequality $a_\Sigma(F) \leq 3$ holds for each bounded figure F (Theorem 33.5). Hence, by Lemma 33.2, it suffices to show that $a_\Sigma(F) > 2$ if F (of diameter h) can be completed to a figure of constant width h in a unique way. By Theorem 33.9, it remains to show that, if a closed figure F of diameter h can be completed in a unique way to a figure Φ of constant width h, then (in the case of a nonangled space) the condition (2) from Theorem 33.9 is satisfied. For showing this, we use the notation $A, B, C, D, a, b, c, u, v, w, z$ as in the proof of Theorem 33.9.

Assume that each of the parallel supporting lines Γ, Γ' of $\Phi = F^{(h)}$ has empty intersection with F (Fig. 183). Then $a \neq u$, $b \neq v$, $c \neq w$, $d \neq z$. We show that in this case the arcs A, B, C, D contain segments whose corresponding endpoints are u, v, w, z. Assume that A and D do not contain such segments. Let l be the line through the midpoint of $[u, z]$ which intersects A and D and whose angle α with (u, z) is smaller than the angles azu and duz (Fig. 186). The line $l_1 \parallel l$ containing z and the line $l_2 \parallel l$ containing u are considered. Let u', z' denote the intersection points of l with A and D, respectively. The line l_1 intersects A in a point u'' and the line (u, u') in a point u^*. The line l_2 intersects D in a point z'', and the line (z, z') in a point z^*. Since the part of A between u and u'' is not a segment, the points u, u', u'' are not collinear and therefore $u'' \neq u$. Analogously, $z'' \neq z$. Then we have

$$d(o, u') = \frac{1}{2}d(z, u^*) > \frac{1}{2}d(z, u'') = \frac{1}{2}h,$$

$$d(o, z') = \frac{1}{2}d(u, z^*) > \frac{1}{2}d(u, z'') = \frac{1}{2}h,$$

and hence

$$d(u', z) = d(o, u') + d(o, z') > h.$$

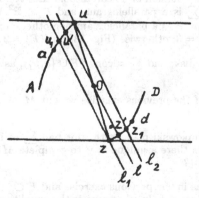

Fig. 186

But this contradicts $u', z' \in \Phi = F^{(h)}$, since Φ has the diameter h. This contradiction shows that u, z are endpoints of segments in A and D, and the same holds for v and w regarding B and C.

These facts imply that (in both the cases $w = z$ and $w \neq z$) the boundary of cl $U_h(z)$ contains two segments with u as common endpoint, and hence the corresponding normed space $\Re^2$ is angled, a contradiction. Therefore at least one of the intersections $F \cap \Gamma$, $F \cap \Gamma'$ is nonempty. Since this holds for arbitrary pairs Γ, Γ' of parallel supporting lines of F', the condition (2) is satisfied. ∎

Finally, we consider a generalization of the problem of Borsuk and Grünbaum. Let $\Re^n$ be a normed space with unit ball $\sum$. Introduce additionally a Euclidean metric in $\Re^n$ (which allows the consideration of a unit sphere $S^{n-1} \subset \mathbf{R}^n$ and a scalar product). We choose a closed, nonempty set $H \subset S^{n-1}$, which is symmetric about o. Let $F \subset \mathbf{R}^n$ have the H-diameter h and k be a positive integer such that F can be represented as union of k sets each of which has an H-diameter smaller than h. Denote by $a_{\sum,H}(F)$ the smallest k with respect to F, H and $\sum$. The problem of investigating the number $a_{\sum,H}(F)$ includes that of investigating $a_{\sum}(F)$, discussed in this chapter. Namely, for $H = S^{n-1}$ we get the problem of Borsuk and Grünbaum: in this case $a_{\sum,H}(F) = a_{\sum}(F)$.

Exercises:

1. Prove the following theorem obtained by B. Grünbaum in 1957: Let $\Re^2$ be a Minkowski plane with the unit ball $\sum$. For any bounded figure $F \subset \Re^2$ the inequality $a_{\sum}(F) \leq 4$ holds. Equality is attained if and only if $\sum$ is a parallelogram and conv F is a parallelogram which is homothetic or congruent to $\sum$.

2. Let $\Re^2$ be the Minkowski plane whose unit ball is a parallelogram. Describe all equilateral triangles in $\Re^2$ (i.e., the triangles abc satisfying $\| a-b \| = \| b-c \| = \| c-a \|$).

3. Prove the following version of Theorem 33.4: Let $\Re^2$ be the Minkowski plane whose unit ball $\sum$ is a parallelogram, and $F \subset \Re^2$ be a compact figure of diameter d. If conv F is a parallelogram homothetic or congruent to $\sum$ (Fig. 187), then $a_{\sum}(F) = 3$; otherwise (Fig. 188) $a_{\sum}(F) = 2$.

4. Let $F \subset \Re^2$ be a disk, and $\sum = $ conv $(F \cup \{a, b\})$, as in Fig. 189. Prove that $a_{\sum}(F) = 2$.

5. In the notation of the previous exercise we put $M = $ conv $(F \cup \{a\})$. What is the number $a_{\sum}(M)$?

6. Let $\Re^2$ be the Minkowski plane with unit ball $\sum$ as in Exercise 4 and $M = $ conv $(F \cup \{a\})$. Is there a unique way to complete M to a figure of constant width $h = $ diam M?

7. Let $\Re^2$ be given as in the previous exercise and $F \subset \sum$ be a closed set containing both the points a, b. In what case is the equality $a_{\sum}(F) = 2$ attained?

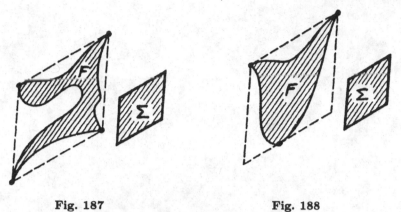

Fig. 187   Fig. 188

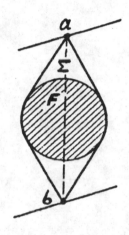

Fig. 189

8. In the Minkowski plane $\Re^2$ introduced in Exercise 6, let Π be the half-plane with $a, b \in$ bd Π and $Q = \sum \cap \Pi$. Find $a_\Sigma(Q)$.

9. Let $\Re^2$ be the Minkowski plane whose unit ball is a regular hexagon and M be the triangle as in Fig. 157. Find $a_\Sigma(Q)$

10. Let $\Re^2$ be a Minkowski plane whose unit ball is not a parallelogram and $M \subset \Re^2$ be a figure of constant width. Prove that $a_\Sigma(M) = 3$.

VI. Homothetic covering and illumination

In the first three sections of this chapter, we will investigate four affine invariant problems referring to convex bodies in $\mathbf{R}^n$. It is shown that these problems are equivalent for compact, convex bodies, whereas they differ from each other in the unbounded case. Among these four problems, the central one is the question for the minimal number of smaller homothets of a convex body $M \subset \mathbf{R}^n$ which are sufficient to cover M. In addition, the problem of illuminating of the boundary bd M by the smallest number of directions is discussed. A lot of partial results regarding both the problems are known, but for $n \geq 3$ the general solutions are still unknown. We give a survey on the contributions up to the recent state.

The last two sections refer to inner illumination of convex bodies, also leading to an equivalent covering problem. Here we show once more some relations to basic notions of other chapters in this book, such as Helly-dimension, d-convexity and H-convexity.

§34 The main problem and a survey of results

Let K be a closed, convex body in $\mathbf{R}^n$. The problem of Gohberg, Markus and Hadwiger [G-M, Ha 1] is to find the smallest one of positive integers m such that there are smaller homothetical copies $K_1, \cdots, K_m$ of K which are sufficient to *cover* K, i.e., $K \subset K_1 \cup \cdots \cup K_m$. We denote this smallest integer by $b(K)$. If a body K cannot be covered by a finite number of smaller homothetical copies, we set $b(K) = \infty$.

We notice that no restriction on the ratio k is put, except $0 < k < 1$. In other words, k may by arbitrarily close to 1.

If, for example, $P \subset \mathbf{R}^2$ is a parallelogram, then $b(P) = 4$. Indeed, no smaller copy of P (i.e., its image under a homothety with a positive ratio $k < 1$) can contain *two* vertices of P. This means that every vertex of P requires a separate smaller copy, and therefore $b(P) \geq 4$. On the other hand, it is sufficient to have four "copies" (Fig. 190). However, if $M \subset \mathbf{R}^2$ is a compact, convex, two-dimensional figure distinct from a parallelogram, then $b(M) = 3$ (cf. [G-M, Bt 1, B-SP 3,4] and Fig. 191 for a disk).

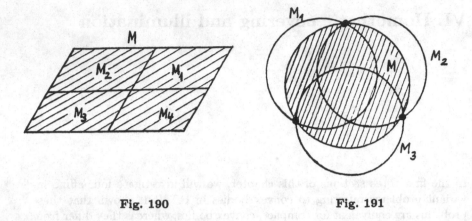

Fig. 190   Fig. 191

A similar (and, as we will see, even equivalent) problem was posed by Levi [Lev]: what is the smallest positive integer m such that there are translates $K_1, \cdots, K_m$ of the convex body $K \subset \mathbf{R}^n$ with $K \subset \operatorname{int} K_1 \cup \cdots \cup \operatorname{int} K_m$? This smallest integer is denoted by $b'(K)$.

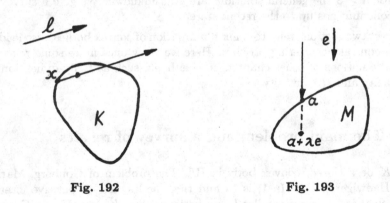

Fig. 192   Fig. 193

We now turn to the illumination problem. We say that a point $x \in \operatorname{bd} K$ is *illuminated* by an (oriented) direction l in $\mathbf{R}^n$ if a ray, emanating from x and having the direction l, meets an interior point of K (Fig. 192). In other words, a boundary point a of the body M is said to be *illuminated* by l if $a + \lambda e \in \operatorname{int} M$ for sufficiently small $\lambda > 0$, where e is the unit vector of the direction l (Fig. 193). We say that the directions determined by nonzero vectors $e_1, \cdots, e_k$ *illuminate* the boundary of the body M if every point $a \in \operatorname{bd} M$ is illuminated by at least one of these directions (Fig. 194). By $c(M)$ we denote the smallest one of the integers k such that there exist k nonzero vectors, whose directions illuminate the boundary of M. (This problem was considered in [Bt 1].) If $P \subset \mathbf{R}^2$ is a parallelogram, then $c(P) = 4$. Indeed, no direction can illuminate *two* vertices of P, i.e., every vertex of P requires a

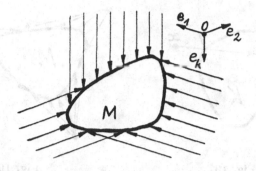

Fig. 194

separate direction, and therefore $c(P) \geq 4$. On the other hand, it is possible to illuminate bd P by four directions (Fig. 195). If $M \subset \mathbf{R}^2$ is a disk, then $c(M) = 3$ (Fig. 196).

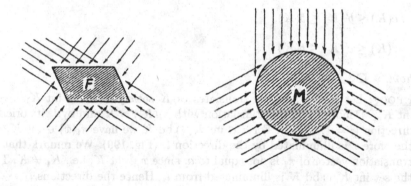

Fig. 195   **Fig. 196**

Finally, we consider the illumination problem for $K \subset \mathbf{R}^n$ by means of point sources *(central illumination)*. Let $x \in \mathbf{R}^n \setminus K$ be a "light source". A point $y \in \mathrm{bd}\, K$ is said to be illuminated by x (Fig. 197) if the ray $[x, y)$ and the interior of K have common points outside the segment $[x, y]$. Furthermore, a set $N \subset \mathrm{bd}\, K$ is said to be illuminated by a set $M \subset \mathbf{R}^n \setminus K$ if each point of N is illuminated by at least one point from M. The problem is to find the smallest positive integer m such that there is a set $M = \{x_1, \cdots, x_m\} \subset \mathbf{R}^n \setminus K$ that illuminates the whole boundary of K (Fig. 198). This smallest integer is denoted by $c'(K)$.

As $b(K)$, also the other quantities $b'(K)$, $c(K)$, $c'(K)$ can be finite or infinite.

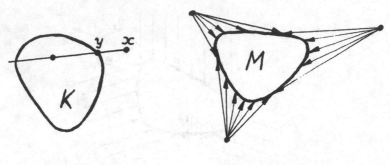

Fig. 197   Fig. 198

Summarizing the aforesaid, we see that $b(M)$ and $c(M)$ take the same values for the parallelogram and for the disk, in each case. As the following theorems [Bt 1, B-SP 3,4] show, this fact is valid in the general case.

Theorem 34.1: For any closed, convex body $K \subset \mathbf{R}^n$, not identical with $\mathbf{R}^n$, the relations

$$c(K) \leq b'(K) \leq b(K), \tag{1}$$

$$c(K) \leq c'(K) \leq b(K) \tag{2}$$

hold.   □

<u>PROOF:</u> Let $K_1, \cdots, K_m$ be translates of K satisfying $K \subset \operatorname{int} K_1 \cup \cdots \cup \operatorname{int} K_m$. Denote by π_i the translation with $\pi_i(K_i) = K$ and by l_i its oriented direction ($i = 1, \cdots, m$). For $x \in \operatorname{int} K_i \cap \operatorname{bd} K$ we have $\pi_i(x) \in \operatorname{int} K$, and therefore x is illuminated by the direction l_i (Fig. 199). We remark that the translation vector of π_i is not equal to o, since $x \notin \operatorname{int} K$; i.e., $K_i \neq K$. Thus the set $\operatorname{int} K_i \cap \operatorname{bd} K$ is illuminated from l_i. Hence the directions $l_1, \cdots, l_m$ illuminate the whole boundary $\operatorname{bd} K = \bigcup_{i=1}^{m} (\operatorname{int} K_i \cap \operatorname{bd} K)$. Therefore $c(K) \leq m$. Thus, if $b'(K) < \infty$, then $c(K) \leq b'(K)$.

We will show now the second inequality in (1). Let $K_1, \cdots, K_m$ be smaller homothetical copies of K with $K \subset K_1 \cup \cdots \cup K_m$ and h_i denote the homothety (with ratio $k_i < 1$) which transforms K onto K_i. For a point $a_i \in \operatorname{int} K_i$ (Fig. 200), let h'_i be the homothety with the center a_i and ratio $\frac{1}{k_i}$. Obviously, $\pi_i = h'_i \circ h_i$ is a translation, and therefore $h'_i(K_i) = h'_i(h_i(K)) = \pi_i(K)$ is a translate of K. Furthermore, by $\frac{1}{k_i} > 1$ and $a_i \in \operatorname{int} K_i$ the inclusion $K_i \subset \operatorname{int} h'_i(K_i)$ holds. Thus $K \subset K_1 \cup \cdots \cup K_m \subset \operatorname{int} h'_1(K_1) \cup \cdots \cup \operatorname{int} h'_m(K_m)$, i.e., $K \subset \operatorname{int} \pi_1(K) \cup \cdots \cup \pi_m(K)$. This implies $b'(K) \leq m$, and therefore $b'(K) \leq b(K)$ holds, showing the complete relation (1).

Turning to the first inequality in (2), let $y_1, \cdots, y_m$ from $\mathbf{R}^n \setminus K$ be light sources which illuminate the whole boundary $\operatorname{bd} K$ of K. Let $a \in \operatorname{int} K$ and l_i be the direction of the ray $[y_i, a)$, $i = 1, \cdots, m$. Assume that $x \in \operatorname{bd} K$ is

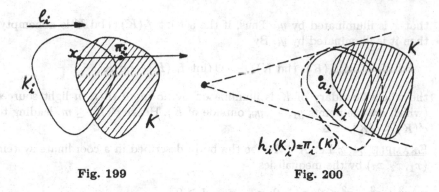

Fig. 199   Fig. 200

illuminated by y_i. We choose a point $z \in \operatorname{int} K$ such that $x \in]y_i, z[$ and denote by u the point from the segment $[a, z]$ such that the vectors $u - x$ and $a - y_i$ are parallel (Fig. 201). Then u is from the interior of K (since $a, z \in \operatorname{int} K$). Since l_i is the direction of the ray $[x, u)$, the point x is illuminated by l_i. Hence each point $x \in \operatorname{bd} K$ is illuminated by one of the directions $l_1, \cdots, l_m$, and therefore $c(K) \leq m$ holds. This shows the inequality $c(K) \leq c'(K)$.

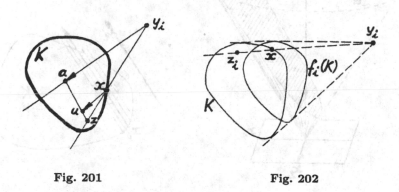

Fig. 201   Fig. 202

Finally, we show the second inequality in (2). Let $K_1, \cdots, K_m$ be smaller homothetical copies of K with $K \subset K_1 \cup \cdots \cup K_m$. Let h_i denote the homothety (with ratio $k_i < 1$) which transforms K onto K_i. We choose a point $a_i \in \operatorname{int} K_i$ and denote by g_i the homothety with center a_i and ratio λ_i, where $1 < \lambda_i < \frac{1}{k_i}$. Then $f_i = g_i \circ h_i$ is a homothety with $k_i \lambda_i$ as its ratio; the center of f_i we denote by y_i. Since $\lambda_i > 1$ and $a_i \in \operatorname{int} K_i$, we have $K_i \subset \operatorname{int} g_i(K_i) = \operatorname{int} f_i(K)$ and hence $K \subset \operatorname{int} f_1(K) \cup \cdots \cup \operatorname{int} f_m(K)$. Let $x \in \operatorname{int} f_i(K) \cap \operatorname{bd} K$. Then $y_i \notin K$. Indeed, since $k_i \lambda_i < 1$, for $y_i \in K$ we would have $f_i(K) \subset K$, $\operatorname{int} f_i(K) \subset \operatorname{int} K$ and hence $\operatorname{int} f_i(K) \cap \operatorname{bd} K = \emptyset$, contradicting $x \in \operatorname{int} f_i(K) \cap \operatorname{bd} K$. We will prove that x is illuminated by y_i. Since $x \in \operatorname{int} f_i(K)$, we have $x = f_i(z_i)$, where $z_i \in \operatorname{int} K$. The point z_i lies on $[y_i, x)$ but not on the segment $[y_i, x]$, since $k_i \lambda_i < 1$ (Fig. 202). This shows that x is illuminated by y_i. Thus, if the set int $f_i(K) \cap \operatorname{bd} K$ is nonempty, then it is illuminated by y_i. By

$$\operatorname{bd} K = (\operatorname{int} f_1(K) \cap \operatorname{bd} K) \cup \cdots \cup (\operatorname{int} f_m(K) \cap \operatorname{bd} K),$$

the whole boundary of K is illuminated by no more than m light sources (which are the points $y_1, \cdots, y_m$ outside of K). Hence $c'(K) \le m$, leading to $c'(K) \le b(K)$. ∎

EXAMPLE 34.2: Let $K \subset \mathbf{R}^3$ be the body described in a coordinate system (x_1, x_2, x_3) by the inequalities

$$x_1^2 + x_2^2 - x_3^2 \le 0; \ x_3 \ge 0; \ x_1 - x_3 + 1 \ge 0.$$

In other words, $K = Q \cap P$, where Q is an unbounded circular cone (one pole) and P is a half-space which contains the apex $(0,0,0)$ of Q and whose bounding plane $\Gamma = \operatorname{bd} P$ intersects $\operatorname{bd} Q$ by a parabola L (Fig. 203).

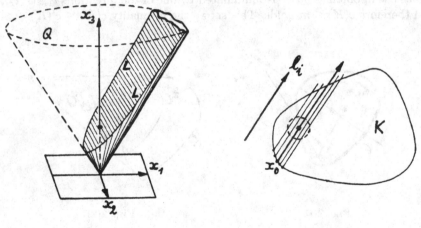

Fig. 203   Fig. 204

We will show now that $c(K) = \infty$. Indeed, each point x of the parabola $L = \Gamma \cap \operatorname{bd} Q$ is an extreme point of K, where the angle between Γ and the tangent plane of Q at x tends to zero as x tends to infinity. Hence each direction illuminates only a *bounded* arc of the parabola L in $\operatorname{bd} K$. This easily may be confirmed by a calculation. Let $x \in L$, i.e., the coordinates x_1, x_2, x_3 of x satisfy the equalities $x_1^2 + x_2^2 - x_3^2 = 0$, $x_1 - x_3 + 1 = 0$. For $x_2 = \xi$, we have

$$x_1 = \frac{\xi^2 - 1}{2}, \ x_2 = \xi, \ x_3 = \frac{\xi^2 + 1}{2},$$

i.e.

$$x = \left(\frac{\xi^2 - 1}{2}, \, \xi, \, \frac{\xi^2 + 1}{2} \right).$$

Let l be a direction defined by a vector $p = (\alpha, \beta, \gamma)$. If x is illuminated from l, then for a suitable $t > 0$ the point

$$x + tp = \left(\frac{\xi^2 - 1}{2} + t\alpha, \, \xi + t\beta, \, \frac{\xi^2 + 1}{2} + t\gamma \right) \tag{3}$$

lies in int K. Therefore the coordinates of this point satisfy

$$x_1^2 + x_2^2 - x_3^2 < 0; \quad x_1 - x_3 + 1 > 0; \quad x_1 - x_3 < 0 \tag{4}$$

(the latter inequality holds, since K lies in the half-space $x_1 - x_3 \leq 0$). Substituting the coordinates of the point $x + tp$ in these inequalities (cf (3)), we obtain

$$(\alpha - \gamma)\xi^2 + 2\beta\xi + (t\beta^2 + t\alpha^2 - t\gamma^2 - \alpha - \gamma) < 0; \tag{5}$$

$$\alpha - \gamma > 0; \quad t\alpha - t\gamma - 1 < 0. \tag{6}$$

If $|\alpha| \geq |\gamma|$, then the last bracket in (5) is not smaller than $-\alpha - \gamma$. If $|\alpha| < |\gamma|$, then by the first inequality in (6) also $\gamma < 0$ and $\alpha + \gamma < 0$. Therefore the last bracket in (5) (which is also representable by $t\beta^2 + (\alpha + \gamma)(t\alpha - t\gamma - 1)$) has to be positive, cf. (6). Thus, in each case, the last bracket from (5) is not smaller than a number δ which only depends on α and γ. Therefore by (5)

$$(\alpha - \gamma)\xi^2 + 2\beta\xi + \delta < 0$$

holds, i.e., $|\xi| < \lambda$, where λ is a constant depending on α, β, γ. We see that only those points $x \in L$ are illuminated by l, for which $|\xi| < \lambda$. This holds for a bounded arc. Hence to illuminate all points of L (and, moreover, the whole boundary of K), infinitely many directions are needed, showing $c(K) = \infty$. By the inequalities (1), (2) we have also $b'(K) = \infty, c'(K) = \infty$ and $b(K) = \infty$. $\square$

In this example we have a face $M = K \cap \Gamma$ of K such that for an arbitrary $\varepsilon > 0$ a point $x \in \text{rbd } M$ with the following property can be found: x is contained in two supporting hyperplanes of K whose angle (containing K) has a value $< \varepsilon$. It is easy to see that if this condition is satisfied, then $c(K) = \infty$ (but this condition is not necessary for the validity of $c(K) = \infty$). Here we refer to Exercise 8.

Finally we remark that the occurence of the value ∞ for $c(K)$ is based on the assumption that K is unbounded. If K is bounded (and hence compact), then $b(K) < \infty$, and by (1), (2) the same is true for $b'(K), c(K)$ and $c'(K)$.

The following theorem was obtained in [Bt 1].

Theorem 34.3: For any compact, convex body $K \subset \mathbf{R}^n$ the relations

$$b(K) = b'(K) = c(K) = c'(K)$$

hold.   □

PROOF: By (1) and (2), we have to show that $b(K) \leq c(K)$. Let $l_1, \cdots, l_m$ be directions which illuminate the whole boundary of K. Let, furthermore, Δ_i denote the set of all points $x \in$ bd K which are illuminated by l_i, $i = 1, \cdots, m$. If $x_o \in$ bd K is illuminated from l_i, i.e. $x_o \in \Delta_i$, then each point $x \in$ bd K sufficiently near to x_o belongs to Δ_i, too (Fig. 204). In other words, $\Delta_1, \cdots, \Delta_m$ are *open* sets in bd K. And since the directions $l_1, \cdots, l_m$ illuminate the whole boundary of K, we have $\Delta_1 \cup \cdots \cup \Delta_m =$ bd K.

We will show that there exist open sets $V_1, \cdots, V_m$ in bd K which satisfy the relations

$$\text{cl } V_1 \subset \Delta_1, \cdots, \text{cl } V_m \subset \Delta_m; \quad V_1 \cup \cdots \cup V_m = \text{bd } K. \tag{7}$$

For showing this, we use induction over k. Assume that, for an integer $k \in \{1, \cdots, m\}$, there are sets $V_1, \cdots, V_{k-1}$ satisfying cl $V_1 \subset \Delta_1, \cdots,$ cl $V_{k-1} \subset \Delta_{k-1}$ and $V_1 \cup \cdots \cup V_{k-1} \cup \Delta_k \cup \cdots \cup \Delta_m =$ bd K (the case $k = 1$ is trivial). We will construct the set V_k. For this we consider the sets

$$F_k = \text{bd } K \setminus (V_1 \cup \cdots \cup V_{k-1} \cup \Delta_{k+1} \cup \cdots \cup \Delta_m),$$

$$H_k = \text{bd } K \setminus \Delta_k.$$

Then F_k, H_k are closed, disjoint sets. Let h_k denote the distance of the sets: $h_k = \min \| x - y \|$, where the minimum is taken over $x \in F_k$, $y \in H_k$. We choose a positive number $\varepsilon < h_k$ and set $V_k = U_\varepsilon(F_k) \cap$ bd K. The set V_k is open in bd K and has the required property: cl $V_k \cap H_k = \emptyset$, and therefore

$$\text{cl } V_k \subset \Delta_k; V_1 \cup \cdots \cup V_k \cup \Delta_{k+1} \cup \cdots \cup \Delta_m = \text{bd } K.$$

This induction shows the existence of sets $V_1, \cdots, V_m$ which satisfy the condition (7).

For a point $x \in \Delta_i$, denote by $l_i(x)$ the ray emanating from x with the direction l_i. Since x is illuminated by l_i, the ray $l_i(x)$ meets int K and intersects bd K in a point $a_i(x) \neq x$. Thus the segment $[x, a_i(x)]$ lies in K, with its endpoints belonging to bd K and its interior points belonging to int K. Let $f_i(x)$ be the length of $[x, a_i(x)]$. Consequently $f_i(x)$ is a positive function on Δ_i which is (by general properties of convex sets) continuous. Since cl $V_i \subset \Delta_i$ is compact, the function $f_i(x)$ has a positive minimum on cl V_i, and hence there exists some $q_i > 0$ such that $f_i(x) > q_i$ for all $x \in$ cl V_i. In other words, for any point $x \in$ cl V_i the length of $[x, a_i(x)]$ is larger than q_i. This means that the translation π_i with direction l_i and length q_i transfers cl V_i onto int K, i.e., π_i (cl V_i) $\subset$ int K. Thus π_i^{-1}(int K) $\supset$ cl V_i for $i = 1, \cdots, m$.

Let now y_o be an aribtrary interior point of K. The positive numbers q_i from the previous construction can be assumed to be sufficiently small such that $y_o \in \pi_i^{-1}$(int K) for all $i \in \{1, \cdots, m\}$. Since the closed set cl $V_i \cup \{y_o\}$ is contained in π_i^{-1}(int K) = int $\pi_i^{-1}(K)$, there exists a homothety g_i with a positive ratio $k_i < 1$ such that cl $V_i \cup \{y_o\} \subset g_i(\pi_i^{-1}(K))$, $i = 1, \cdots, m$.

Thus $g_i \circ \pi_i^{-1}$ is a homothety with a positive ratio $k_i < 1$, i.e., the body $K_i = g_i(\pi_i^{-1}(K))$ is a smaller homothetical copy of K.

Finally, we prove that $K \subset K_1 \cup \cdots \cup K_m$. For $z \in K$ we choose a boundary point x of K such that $z \in [x, y_o]$. Furthermore, we choose some $i \in \{1, \cdots, m\}$ such that $x \in$ cl V_i (note that $V_1 \cup \cdots \cup V_m =$ bd K). Since the points y_o and $x \in$ cl V_1 lie in the convex set $K_i = g_i(\pi_i^{-1}(K))$, the inclusion $[x, y_o] \subset K_i$ is obtained, and therefore $z \in K_i \subset K_1 \cup \cdots \cup K_m$. Hence $K \subset K_1 \cup \cdots \cup K_m$, i.e., $b(K) \leq m$ and $b(K) \leq c(K)$. ∎

The proved theorem says that on the family of compact, convex bodies in $\mathbf{R}^n$ the functions $b(K), b'(K), c(K)$ and $c'(K)$ coincide. The following examples will demonstrate that for unbounded convex bodies these four functions differ from each other.

EXAMPLE 34.4: Consider an unbounded convex figure $K \subset \mathbf{R}^2$ with two parallel asymptotes, having no point in common with K. For example, in a coordinate system (x_1, x_2) the set K can be given by the inequality

$$x_2 \geq \frac{1}{\sqrt{x_1(1 - x_1)}}$$

(see Fig. 205). The width of K with respect to the x_2-axis is 1. Hence the corresponding width of a smaller homothetical copy K_1 is < 1, and K_1 cannot cover K, i.e., $b(K) \geq 2$. It is easy to show that $b(K) = 2$. Moreover, $b'(K) = 1$, since by a translation π of K (with a translation vector $(0, \alpha), \alpha < 0$) a figure $K_1 = \pi(K)$ with int $K_1 \supset K$ is obtainable. Hence $b'(K) < b(K)$, and by (1) also $c(K) < b(K)$, i.e., $c(K) = 1$. □

EXAMPLE 34.5: In the space $\mathbf{R}^3$ with a coordinate system (x_1, x_2, x_3), let K be a convex body defined by the inequalities

$$0 \leq x_1 < 1; \ 0 \leq x_2 \leq 1; \ x_3(2 - x_1 - x_2) \geq 1,$$

see Fig. 206. We will show that $b(K) = 4$. Since K is contained in the both way infinite cylinder which is described by $0 \leq x_1 \leq 1, 0 \leq x_2 \leq 1$, its widths with respect to the (x_1, x_3)-plane and with respect to the (x_2, x_3)-plane cannot be greater than 1. More precisely, these widths are equal to 1, since for x_3 increasing the body asymptotically approaches the boundary of the infinite cylinder. Hence the copy K' of K, obtained by a homothety with

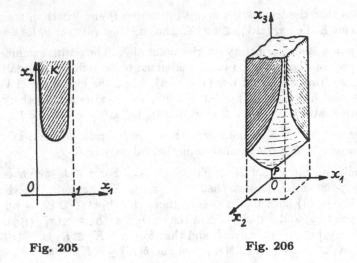

Fig. 205   **Fig. 206**

ratio $k > 0$, has the width k regarding the (x_1, x_3)-plane and (x_2, x_3)-plane, respectively.

Let now K_1, K_2, K_3 be smaller homothetical copies of K with k denoting the largest occuring ratio. Then the width of each of the bodies K_1, K_2, K_3 with respect to the (x_1, x_3)-plane cannot exceed k, and the same holds for the width regarding the (x_2, x_3)-plane. We choose a number λ with $k < \lambda < 1$ and consider the points

$$a = (0, 0, \frac{1}{2 - 2\lambda}), \quad b = (\lambda, 0, \frac{1}{2 - 2\lambda}),$$

$$c = (0, \lambda, \frac{1}{2 - 2\lambda}, \quad d = (\lambda, \lambda, \frac{1}{2 - 2\lambda}).$$

These points (obviously lying in K) are the vertices of a square whose sides are parallel to the axes x_1 and x_2, and whose side-length is $\lambda > k$. Hence no two of these vertices can belong to one of the bodies K_1, K_2, K_3, and therefore at least one of them is not contained in $K_1 \cup K_2 \cup K_3$. In other words, the body K is not contained in $K_1 \cup K_2 \cup K_3$. Thus $b(K) \geq 4$. It is easy to see that $b(K) = 4$, i.e., K can be covered by four smaller homothetical copies.

We show now that $c'(K) = 3$. In fact, consider the light sources

$$y_1 = (-1, -1, -1); \quad y_2 = (-1, 4, 0); \quad y_3 = (4, -1, 0),$$

which lie outside of K. The point y_1 illuminates all points from bd K which lie in the coordinate planes (x_1, x_3) and (x_2, x_3), except for the points in the surface $x_3(2 - x_1 - x_2) = 1$ or in the planes $x_1 = 1$, $x_2 = 1$. (We remark that the corner point $p = (0, 0, \frac{1}{2})$ is illuminated by y_1, too.) All further points from bd K are illuminated by y_2 and y_3; those which satisfy $x_2 \neq 1$ by y_3 and those which satisfy $x_1 \neq 1$ by y_2. Thus $c'(K) \leq 3$. Moreover, since no two of the boundary points $(0, 0, \frac{1}{2})$, $(1, 0, 1)$, $(0, 1, 1)$ can be illuminated by one light source, $c'(K) = 3$ holds. This yields $c'(K) < b(K)$, i.e., the function $b(K)$ differs not only from $b'(K)$ and $c(K)$ (cf. Example 34.4), but also from the function $c'(K)$.   □

EXAMPLE 34.6: Now we consider the relations between the remaining functions $b'(K)$, $c'(K)$ and $c(K)$. In Example 34.4, we had $c(K) = 1$. At the same time, the figure K shown in Fig. 205 satisfies $c'(K) = 2$. Indeed, let $y \in \mathbf{R}^2 \backslash K$ be an arbitrary light source and (ξ, η) be the coordinates of y. Then at least one of the inequalities $\xi > 0$, $\xi < 1$ is satisfied. Consider the case $\xi > 0$ (Fig. 207). We choose a point $a \in \text{int } K$ whose x_1-coordinate is smaller than ξ. Then the ray $[y, a)$ intersects bd K in a point x which is not illuminated from y. In other words, no light source $y \in \mathbf{R}^2 \backslash K$ can illuminate the whole boundary of K. This yields $c'(K) > 1$. By using the points $(0, 0)$ and $(1, 0)$ as light sources, the equality $c'(K) = 2$ is clarified. Thus $c(K) < c'(K)$, i.e., the functions $c(K)$ and $c'(K)$ differ from each other.   □

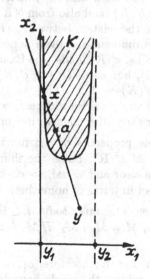

Fig. 207

EXAMPLE 34.7: Consider once more the body K from Example 34.5. We obtained $c'(K) = 3$ and therefore, by (2), $c(K) \leq 3$. Since no two of the points $(0, 0, \frac{1}{2})$, $(1, 0, 1)$, $(0, 1, 1)$ can be illuminated by one direction, this could be restricted to $c(K) = 3$. We will show now that for this body K the equality $b'(K) = 4$ holds. In fact, assume $b'(K) \leq 3$, i.e., there exist translates K_1, K_2, K_3 of K satisfying $K \subset \text{int } K_1 \cup \text{int } K_2 \cup \text{int } K_3$. For two arbitrary points $a, b \in \text{int } K_1$, the difference of the first coordinates has to be smaller than 1, and the same holds for the difference of the second coordinates. The analogous assertion holds for $a, b \in \text{int } K_2$ and $a, b \in \text{int } K_3$. Hence no two of the points $a = (0, 0, \frac{1}{2}), b = (1, 0, 1), c = (0, 1, 1)$ (which are contained in K) can lie in one of the sets int K_1, int K_2, int K_3. For example, consider the case $a \in \text{int } K_1$, $b \in \text{int } K_2$, $c \in \text{int } K_3$. From $a \in \text{int } K_1$ it follows that K_1 is contained in the strip $-\alpha \leq x_1 \leq 1 - \alpha$ for $0 < \alpha < 1$. Furthermore, $b \in \text{int } K_2$ implies that K_2 lies in the strip $-\beta \leq x_2 \leq 1 - \beta, 0 < \beta < 1$, and $c \in \text{int } K_3$ yields the condition $-\gamma \leq x_1 \leq 1 - \gamma$, with $0 < \gamma < 1$, for all points from K_3. Let ε be a positive number which is smaller than each of the numbers α, β, γ. Then the point $p = (1 - \varepsilon, 1 - \varepsilon, \frac{1}{2\varepsilon}) \in K$ is not contained in any of the three strips introduced above, and therefore $p \notin K_1 \cup K_2 \cup K_3$. Hence the body K is not contained in $K_1 \cup K_2 \cup K_3$, and this contradicts the assumption that $K \subset \text{int } K_1 \cup \text{int } K_2 \cup \text{int } K_3$. Thus $b'(K) \geq 4$ is obtained and $b'(K) > 4$ is easily excluded. Consequently $b'(K) = 4$, i.e., there exist four bodies K_1, K_2, K_3, K_4 which are translates of K and satisfy the inclusion $K \subset \text{int } K_1 \cup \text{int } K_2 \cup \text{int } K_3 \cup \text{int } K_4$. Hence also the functions $c(K)$ and $b'(K)$ differ from each other, with $c(K) < b'(K)$. $\square$

The Examples 34.6 and 34.7 show that the function $c(K)$ differs from each of the functions $c'(K)$ and $b'(K)$ (and also from $b(K)$, see Example 34.4). Thus it remains to investigate the relation between $b'(K)$ and $c'(K)$. It is obvious that these functions are different and even noncomparable, i.e., there exist bodies K for which $b'(K) < c'(K)$ holds (see Examples 34.4 and 34.6, where $b'(K) = 1 < c'(K) = 2$), but also $b'(K) > c'(K)$ is possible (cf. Examples 34.7 and 34.5, where $b'(K) = 4 > c'(K) = 3$). Hence all four functions $b(K), b'(K), c(K), c'(K)$ are pairwise different, where $b'(K)$ and $c'(K)$ are noncomparable, and for each other pair an inequality holds, cf. (1) an (2).

Now we will prove some properties of the functional $b(M) = c(M)$ for a compact, convex body $M \subset \mathbf{R}^n$. Since the illumination problem is (from our point of view) more clear and visual, we conduct reasonings in terms of illumination (rather than in terms of homothetic covering).

Theorem 34.8: Let a compact, convex body $M \subset \mathbf{R}^n$ be the vector sum of two convex sets M_1, M_2, i.e., $M = M_1 + M_2$. If M_1 is a body, then $c(M) \leq c(M_1)$. $\square$

PROOF: Let $c(M_1) = h$. Let, furthermore, $e_1, \cdots, e_h$ be vectors in $\mathbf{R}^n$ such that their directions illuminate the whole boundary of M_1. For every boundary point x of the body M there exist points $x_1 \in M_1, x_2 \in M_2$ such that $x = x_1 + x_2$. Moreover, x_1, x_2 are *boundary* points of the sets M_1, M_2, respectively (otherwise, x could not be a boundary point of M). Let i be an index such that the point $x_1 \in \text{bd } M_1$ is illuminated by the direction e_i, i.e., $y_1 = x_1 + \lambda e_i \in \text{int } M_1, \lambda > 0$. We put $y = y_1 + x_2$. Then $y \in \text{int } M$, since $y_1 \in \text{int } M_1$. Furthermore,

$$y - x = (y_1 + x_2) - (x_1 + x_2) = y_1 - x_1 = \lambda e_i,$$

and hence the point $x \in \text{bd } M$ is illuminated by the direction e_i. Thus each point $x \in \text{bd } M$ is illuminated by at least one of the directions $e_1, \cdots, e_h$, i.e., $c(M) \leq h = c(M_1)$. ∎

Theorem 34.9: *The function $c(M)$, defined on the family of all compact, convex bodies in $\mathbf{R}^n$, possesses the upper semicontinuity property. More detailed, for every compact, convex body $M \subset \mathbf{R}^n$ there exists a real number $\delta > 0$ such that if $\varrho(M, N) < \delta$, then $c(N) \leq c(M)$. In other words, if a sequence $M_1, M_2, \cdots$ of compact, convex bodies converges to a compact, convex body M and $c(M_k) = r$ for all $k = 1, 2, \cdots$, then $c(M) \geq r$.* □

PROOF: Let $c(M) = p$ and $e_1, \cdots, e_p$ be vectors whose directions illuminate the whole boundary of the body M. Then there exist compact sets $F_1, \cdots, F_p$ such that $F_1 \cup \cdots \cup F_p = \text{bd } M$ and every point $x \in F_i$ is illuminated by the direction e_i (cf. the proof of Theorem 34.1). There exists a number $\lambda > 0$ such that $t_i(F_i) \subset \text{int } M$ for every $i = 1, \cdots, p$, where t_i is the translation defined by the vector λe_i. Denote by $V \subset \mathbf{R}^n$ an open set satisfying the conditions $\text{cl } V \subset \text{int } M$ and $V \supset t_1(F_1) \cup \cdots \cup t_p(F_p)$. Furthermore, for every $i = 1, \cdots, p$ we choose an open set W_i such that $W_i \supset F_i$ and $t_i(W_i) \subset V$. Then $W_1 \cup \cdots \cup W_p \supset \text{bd } M$.

Let now $\delta > 0$ be a number that possesses the following property: If $\varrho(M, N) < \delta$, then $\text{cl } V \subset \text{int } N$ and $\text{bd } N \subset W_1 \cup \cdots \cup W_p$. We will show that if $\varrho(M, N) < \delta$, then $c(N) \leq c(M)$. Indeed, let $x \in \text{bd } N$. Then $x \in W_i$ for an index i. Hence $x + \lambda e_i = t_i(x) \in t_i(W_i) \subset V \subset \text{int } N$, i.e., the point x is illuminated by the direction e_i. Thus every point $x \in \text{bd } N$ is illuminated by at least one of the directions $e_1, \cdots, e_p$, i.e., $c(N) \leq p = c(M)$. ∎

EXAMPLE 34.10: Let $M \subset \mathbf{R}^n$ be an n-dimensional parallelotope and M_k be its $\frac{1}{k}$-neighbourhood, $k = 1, 2 \cdots$ (Fig. 208). Then $\lim_{k \to \infty} M_k = M$, $c(M) = 2^n$ and $c(M_k) = n + 1$, cf. Theorem 35.2 below. This means that the inequality $c(M) \leq \lim_{k \to \infty} c(M_k)$ does not hold in this case, i.e., there is no corresponding lower semicontinuity property. □

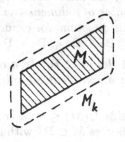

Fig. 208

EXAMPLE 34.11: Let $M \subset \mathbf{R}^n$ be a compact, convex body and $\{l_1, \cdots, l_s\}$ be a system of directions which illuminates the whole boundary of M. The system is said to be *primitive* if it has no proper illuminating subsystem. By $c_{\max}(M)$ denote the *largest* of the integers s such that there exists a primitive illuminating system $\{l_1, \cdots, l_s\}$. If for any positive integer m there is a primitive illuminating system consisting of more than m directions, then $c_{\max}(M) = \infty$. For example, if M is a convex polytope, then $c_{\max}(M)$ is finite (not greater than the number of the vertices). The existence of compact, convex bodies with $c_{\max}(M) = \infty$ was noticed recently by V. Soltan [SV 6]. Namely, let $M \subset \mathbf{R}^3$ be a rotational cone with the apex a and circular basis C. Choose m points $b_1, \cdots, b_m$ which divide the circumference rbd C into m equal arcs. Furthermore, for an $\varepsilon > 0$ and $i \in \{1, \cdots, m\}$ denote by $l_i(\varepsilon)$ the direction determined by the vector $a - b_i + \varepsilon(c - b_i)$, where c is the center of C. It is easily shown that for a choice of $\varepsilon > 0$ the directions $l_1(\varepsilon), \cdots, l_m(\varepsilon)$ illuminate rbd C (with a certain neighbourhood in bd M) and, moreover, for any $i = 1, \cdots, m$, the point b_i is illuminated only by the direction $l_i(\varepsilon)$. Adding the direction l_o determined by the vector $c - a$, we obtain $m + 1$ directions which form a *primitive* system illuminating bd M. Thus (since the positive integer $m \geq 3$ is arbitrary), we have $c_{\max}(M) = \infty$. $\square$

In practice, light sources are not zero-dimensional. For example, one can consider a "fluorescent line" instead of a point source (which, in fact, is also no more than a mathematical model of a "real situation", e.g. with luminiscent tubes). Moreover, we may replace a line by a flat. So we have a motivation for the following

DEFINITION 34.12: Let $M \subset \mathbf{R}^n$ be a compact, convex body and L be a flat that has no common point with M. A boundary point x of M is said to be *illuminated* by the flat L, if there are points $p \in L$, $q \in \text{int } M$ such that x is an interior point of the segment $[p, q]$. In other words, a point $p \in L$ illuminates the point x. $\square$

Now, if r is an integer satisfying the condition $0 \leq r \leq n-1$, we may consider the problem of *illuminating the whole boundary of a compact, convex body* $M \subset \mathbf{R}^n$ *by a minimal number of r-dimensional flats*. This minimal number is denoted by $I_r(M)$. In this direction, for example the following two results are known (see [Be 1, Be 2, Be 3, Be 4, Be 5, B-K-M]).

Theorem 34.13: For $n \geq 3$, let $M \subset \mathbf{R}^n$ be a convex n-polytope with affine symmetry, i.e., there exists a non-identical affine mapping of M onto itself. Then $I_{n-3}(M) \leq 8$ and $I_{n-2}(M) = 2$. $\square$

For $n = 3$ this theorem yields $I_o(M) \leq 8$ for convex 3-polytopes with affine symmetry, i.e., for convex bodies $M \subset \mathbf{R}^3$ with finitely many extreme points it generalizes the result obtained by M. Lassak [La 3] (see also [SP-SV 3]) for centrally symmetric bodies in $\mathbf{R}^3$. The next theorem refers to *zonoids* which are introduced in chapter 7.

Theorem 34.14: If $n = 2^k - 1$ for an integer k, then each n-dimensional zonoid $Z \subset \mathbf{R}^n$ can be illuminated by 2^{n-k} lines. In particular, every zonoid $Z \subset \mathbf{R}^3$ can be illuminated by two lines. $\square$

Finishing this section, we will give the announced survey on known results about the functionals $b(K)$, $b'(K)$, $c(K)$, $c'(K)$. Further overviews in this direction are contained in [B-G 1], [B-SP 3], [Mar 2], §D 17 in [C-F-G], [Be 3], and [Schm].

The first contribution was given by F.W. Levi [Lev] in 1955. He showed that for an arbitrary compact, convex figure $K \subset \mathbf{R}^2$, which is not a parallelogram, $b'(K) = 3$ holds. Inspired by Levi's result, in 1957 H. Hadwiger [Ha 6] published the famous covering conjecture $b(K) \leq 2^n$ for any compact, convex body $K \subset \mathbf{R}^n$ with equality precisely if K is an n-parallelotope. Independently, in 1960 I.Z. Gohberg and A.S. Markus [G-M] showed $b(K) = 3$ for non-parallelograms in the plane, posing the same conjecture $b(K) \leq 2^n$ for n-dimensional compact, convex bodies K. From the historical point of view, the following remarks seem to be interesting. Namely, the functional $b(K)$ was introduced by I.Z. Gohberg already in 1956. His motivation came from functional analysis. M. Krein was interested in the definition of the dimension of a Banach space in terms of covering the unit ball by translates of smaller balls. He invited I.Z. Gohberg to work on this question for infinite-dimensional Banach spaces, continuing results of M.G. Krein, M.A. Krasnosel'ski, and D.P. Milman [K-K-M]. After finishing the corresponding paper [G-K], Gohberg tried to understand the intriguing finite-dimensional meaning of assertions needed in [G-K], and he was led to the following problem: What is the smallest number of balls of radius < 1 in a Minkowski space $\Re^n$ such that the unit ball of $\Re^n$ can be covered by these smaller balls? In affine-geometric terms, this yields the covering problem, restricted to the centrally symmetric case. I.Z. Gohberg gave this question to his student A.S. Markus, who soonly proposed a (complicated) solution for the planar case. After a simplification of this solution by I.Z. Gohberg, in 1957 both authors sent a manuscript (with the complete solution for the planar case and with the posed conjecture for the n-dimensional situation) to I. Yaglom, for publishing it in the journal "Math. Education" in Moscow. But this journal did not appear regularly, and its periodical issues were stopped at this time. So, the authors had no other possibility than to publish their contribution in 1960 (cf. [G-M]). Thus, in our opinion, it would be better to speak about the _Gohberg-Markus-Hadwiger covering conjecture_ as well as the _Gohberg-Markus-Hadwiger number_ $b(K)$, whereas $b'(K)$ should be called the _Levi number_.

Furthermore, in 1960 the functional $c(K)$ was introduced, and it was shown that for an arbitrary compact, convex body K in $\mathbf{R}^n$ the equality $b(K) = b'(K) = c(K)$ holds, see [Bt 1]. In the same year, H. Hadwiger [Ha 7] introduced the functional $c'(K)$ and posed the conjecture $c'(K) \leq 2^n$, for any compact, convex body $K \subset \mathbf{R}^n$. It is remarkable that [Ha 7] does not contain any hint about the equivalence to the Gohberg-Markus-Hadwiger problem, which easily follows from the considerations in [Bt 1].

The lower bound $n + 1$ on the four functionals above with respect to $\mathbf{R}^n$ was verified in [Bt 1], and another proof of K.A. Post was published by H. Hadwiger [Ha 9].

M. Lassak [La 4] proved that for a compact convex body $K \subset \mathbf{R}^3$ the relation $b(K) \leq 20$ holds. He showed this by covering the boundary of K with the convex hulls of smaller homothetical copies of certain subsets of bd K. In addition, he gave the following upper bounds for compact, convex bodies $K \subset \mathbf{R}^n$ (cf. [La 7]):

$$b(K) \leq (n+1)^n - (n-1)^n,$$

$$b(K) \leq (n+1)n^{n-1} - (n-1)(n-2)^{n-1}.$$

For compact, convex bodies $K \subset \mathbf{R}^n$ with a center of symmetry, the following results are known: $b(K) \leq 8$ for $K \subset \mathbf{R}^3$, obtained by M. Lassak [La 3] and, independently, for 3-polytopes in [SP-SV 3]; $b(K) \leq 2^n$ ($n \log n + n \log \log n + 5n$) for $K \subset \mathbf{R}^n$, derived by C.A. Rogers (cf. [Grü 3], p. 284, and [B-G 1], §11); $b(K) \leq (n+1)^n$, obtained by A.J. Levin and J.I. Petunin (cf. [B-G 1], §11). It should be noticed that the bound of Rogers can be extended to all compact, convex bodies, namely, by multiplying it with the volume quotient $\frac{V(K+(-K))}{V(K)}$, which itself is satisfying the inequalities $2^n \leq \frac{V(K+(-K))}{V(K)} \leq \binom{2n}{n}$, see [R-S]. At this place, we repeatedly mention the result of K. Bezdek [Be 1] that for a convex 3-polytope P, whose affine symmetry group consists of the identity and at least one other affinity, the relation $b(P) \leq 8$ holds. This is a generalization of the corresponding assertion about convex 3-polytopes having central symmetry (see [La 3] and [SP-SV 3]).

The next class of convex bodies, for which the Gohberg-Markus-Hadwiger conjecture is even completely verified, is that of *belt bodies* ([Bt 14], [B-B 2], [B-M 1]) together with its subclasses (such as *zonotopes* and *zonoids*), see chapter VII and, in particular, §42. For a convex n-polytope $Z \subset \mathbf{R}^n$ all whose r-faces with $2 \leq r \leq n$ are centrally symmetric (i.e., for Z a zonotope), the estimate $b(Z) \leq 2^n$ was verified in [B-SP 3], Theorem 27.12. This was extended to the larger class of *belt polytopes* (i.e., to convex n-polytopes having a segment summand parallel to each of their edges) in [Mar 2], even with the stronger estimate $3 \cdot 2^{n-2}$ instead of 2^n. Zonoids are known to be the limits of convergent sequences of zonotopes with respect to the Hausdorff metric. Using approximate continuity instead of a direct limit passing from zonotopes to zonoids, in [B-SP 5,6] the inequality $b(Z) \leq 3 \cdot 2^{n-2}$ (for any zonoid Z distinct from a parallelotope) was derived, see also chapter VII. Finally, the same upper bound was verified for the more general class of belt bodies, which is dense in the set of all compact, convex bodies, see [Bt 14], [B-B 2], [B-M 1] and §41.

Furthermore, one should mention results on $b(K)$ obtained by M. Lassak, O. Schramm and B. Weissbach for a *body of constant width*. The first contribution was presented by M. Lassak [La 4]: For any body $K \subset R^3$ of constant width $w > 0$ the relation $b(K) \leq 6$ holds. The proof is based on an elementary property of a body K of constant width in 3-space: if two planes support $K \subset \mathbf{R}^3$ at the same boundary point, then the angle between them has at least 120 degrees. On that basis, two subcases have to be considered: (i) a regular tetrahedron T of side-length w is contained in K, and (ii) if there is no such a tetrahedron. In the first case, every six directions parallel to three pairwise orthogonal lines (none of which is parallel to some side of T) illuminate bd K, i.e., they yield $c(K) = b(K) \leq 6$. In the second case the proof is indirect, where the contradiction is obtained by means of the property of supporting planes mentioned above. It should be noticed that the same estimate ($b(K) \leq 6$ for K a three-dimensional body of constant width) can be simply derived from more general results of B. Weissbach. Namely, in [We 4] (see also [We 3]) the following theorem is given: If K is an n-dimensional body of constant width and $\overline{N}_n(\alpha)$ denotes the smallest positive integer m, for which there exists a covering of the unit sphere S^{n-1} with m congruent caps of spherical radius α, then $b(K) \leq \overline{N}_n(\frac{\pi}{4})$. Using estimates on $\overline{N}_n(\alpha)$ from [Ro 3], one gets

$$b(K)^{1/n} \leq \sqrt{2}(1 + o(1)), n \to \infty.$$

With a more complicated method, O. Schramm [Schr] obtained the estimate

$$b(K) \leq 5n\sqrt{n}(4 + \log n)\left(\sqrt{\frac{3}{2}}\right)^n,$$

leading also to the bound

$$b(K)^{1/n} \leq \sqrt{\frac{3}{2}}(1 + o(1)), n \to \infty,$$

again with a body K of constant width in n-space. For small values of n, the estimate of Weissbach is better, whereas in higher dimensions the bound of Schramm is much stronger. (For consequences of the given estimates on $\overline{N}_n(\alpha)$ and $b(K)$ with respect to the Borsuk problem, the reader is referred to §31.)

Here we also mention the confirmation of the Gohberg-Markus-Hadwiger conjecture for polars of cyclic polytopes if $n \in \{3, 4, 5\}$, see [Be-Bi]. Recently, these authors extended the result to arbitrary n (oral communication).

We continue this report with results about convex bodies satisfying *special boundary conditions*. The sharp assertion $b(K) = n + 1$ for compact, convex bodies $K \subset \mathbf{R}^n$ with *smooth boundary* was already mentioned by F.W. Levi [Lev], but the first complete proof was given in [Bt 1]. This equality even holds if K has no more than n non-regular boundary points (*corners*), cf. [Bt 1] and

§35. Moreover, for $n = 3$ even no more than four corners guarantee $b(K) = 4$, cf. [Ch] and §35. On the other hand, there exist compact, convex bodies $K \subset \mathbf{R}^n$, $n \geq 4$, with $n + 1$ corners and $b(K) = n + 2$, and in 3-space with five corners and $b(K) = 5$, see also the survey [Be 3], §4. Moreover, one can show that for any sufficiently large n there exists a constant c $(1 < c < 2)$ and a compact, convex body $K \subset \mathbf{R}^n$ with finitely many corners and $b(K) \geq c^n$, see [E-F]. B. Weissbach [We 1] and B.V. Dekster [De 2] obtained the result $b(K) = n + 1$ for bodies with arbitrarily many non-regular boundary points which, however, have to be "good-natured", see again §35. Recently, B.V. Dekster [De 6] proved that $b(K) = n + 1$ holds if $K \subset \mathbf{R}^n$ has at least one shadow-boundary (with respect to some parallel illumination) consisting only of regular boundary points.

The same result can be obtained if $K \subset \mathbf{R}^n$ is sufficiently symmetric. Namely, $b(K) = n + 1$ holds if there is a translate of K whose symmetry group G is generated by reflections at $(n - 1)$-subspaces, where (besides the origin and the whole n-space) no fixed subspace regarding all mappings from G exists, cf. [Schr].

For *unbounded* convex bodies $K \subset \mathbf{R}^n$ the equality $b(K) = c(K)$ no longer holds. This was first confirmed by V. Vizitei [V]. More precisely, he showed $c(K) \leq b(K) = b'(K) = c'(K)$ for any closed, convex figure K in the plane. The generalization to higher dimensions was clarified in [SP 2,3], see also §36.

Equivalent versions of the covering and illumination problem considered here were published by N.A. Bobylev [Bo] and K. Bezdek [Be 2,3], further surveys and papers containing related informations are: [B-G 1], paragraph D 17 in [C-F-G], [Mar 2], [Fu 2] and [Schm].

Modifications or sharpenings of the covering problem were considered in [Bel], [Fu 2], [La 5], [L-V] and [Kr 1,2], see also the surveys [La 6] and D 17 in [C-F-G].

Extensions of the illumination problem, namely by using r-dimensional linear light sources, $0 < r < d$, instead of points (at infinity), were considered by K. Bezdek and others, see [Be 1,2,3,4,5], [B-K-M] as well as the Theorems 34.13 and 34.14.

Finally, we mention an isometrical invariant introduced by B. Weissbach [We 5]: let $c^*(K)$ denote the smallest number of directions illuminating the boundaries of all congruent copies of a convex body $K \subset \mathbf{R}^n$. For a body of constant width, an upper bound on $c^*(K)$ is given in [We 5].

Exercises:

1. Let $M \subset \mathbf{R}^2$ be a convex polygon which is not a parallelogram. Prove that there exist three sides of M such that the lines containing them form a circumscribed triangle of M, see Fig. 209. With the help of this assertion, prove that $b(M) = c(M) = 3$.

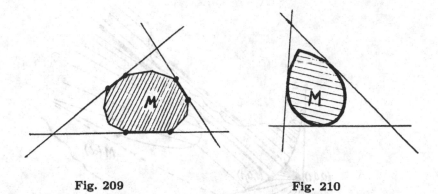

Fig. 209   Fig. 210

2. Let $M \subset \mathbf{R}^2$ be a two-dimensional convex figure distinct from a parallelogram. Prove that there exist three regular boundary points of M such that the corresponding supporting lines form a circumscribed triangle of M (Fig. 210). With the help of this assertion, prove that $b(M) = c(M) = 3$.

3. Prove that for every unbounded, closed, convex figure $M \subset \mathbf{R}^2$ the inequality $c(M) \leq 2$ holds.

4. Prove that if the inscribed cone of an unbounded, closed, convex figure $M \subset \mathbf{R}^2$ is two-dimensional, then $c(M) = 1$. Generalize this to $\mathbf{R}^n$.

5. Let an unbounded, closed, convex figure $M \subset \mathbf{R}^2$ have a one-dimensional inscribed cone. Formulate the conditions under which $c(M) = 1$ and $c(M) = 2$.

6. Prove that for the convex figure M, defined in $\mathbf{R}^2$ by the inequality $x_2 \geq x_1^2$, the equality $b(M) = \infty$ holds.

7. Prove that for every unbounded, closed, convex figure $M \subset \mathbf{R}^2$ one of the equalities $b(M) = 1$, $b(M) = 2$, $b(M) = \infty$ holds. Can you formulate conditions under which each of the indicated equalities holds?

8. Let $K \subset \mathbf{R}^3$ be the body defined by the inequalities
$$x_2^2 + x_1(x_3 - 1) \leq 0,$$
$$x_2 x_3 \geq 1,$$
$$x_1 \geq 0, \quad x_2 \geq 0,$$
see Fig. 211. Prove that $c(K) = \infty$. Furthermore, prove that there exists a rotational convex cone $C \subset \mathbf{R}^3$ such that for every boundary point $a \in \mathrm{bd}\, K$ the supporting cone $\mathrm{supcone}_a K$ contains a cone with the apex a, which is congruent to C. This example was given in [SP 9], cf. the end of Example 34.2.

9. Show by an example that on the family of all unbounded, closed, convex bodies in $\mathbf{R}^n (n \geq 2)$ the functional $c(M)$ does not possess the upper semicontinuity property (cf. Theorem 34.9). Generalize Theorem 34.8 for unbounded, closed, convex bodies in $\mathbf{R}^n$. Can you investigate this situation with respect to the functionals $b(M), b'(M), c'(M)$?

10. Show by an example that on the family of all unbounded, closed, convex bodies in $\mathbf{R}^n (n \geq 2)$ the functional $c(M)$ does not possess the lower semicontinuity property.

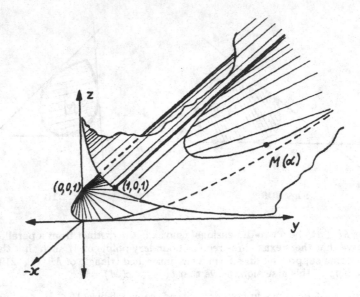

Fig. 211

11. Let $e_o, \cdots, e_n$ be minimally dependent vectors in $\mathbf{R}^n$. Prove that every regular boundary point of any convex body $M \subset \mathbf{R}^n$ is illuminated by at least one of the directions determined by the vectors $e_o, \cdots, e_n$.

12. Decide whether the requirement "M_1 is a body" in the assertion $c(M_1 + M_2) \leq c(M_1)$ is essential?

13. Let $M \subset \mathbf{R}^n$ be a smooth, compact, convex body and $l_1, \cdots, l_s$ be directions determined by vectors $e_1, \cdots, e_s$. Prove that $l_1, \cdots, l_s$ illuminate the whole boundary of M if and only if the vector system $\{e_1, \cdots, e_s\}$ is not one-sided.

14. Let $M \subset \mathbf{R}^n$ be a smooth, compact, convex body. Prove that any primitive illuminating system for M contains no more than $2n$ directions. This theorem is proved by B. Grünbaum [Grü 5].

15. Let $M \subset \mathbf{R}^n$ be a compact, convex body that has no more than q nonregular boundary points. Prove that any primitive illuminating system for M contains no more than $2n + q - 2$ directions. Can you improve this estimate?

16. Prove that for any compact, convex body $M \subset \mathbf{R}^n$ every its primitive illuminating system contains only finite number of directions.

17. Prove that if directions $l_1, \cdots, l_s$ illuminate the whole boundary of a compact, convex body $M \subset \mathbf{R}^n$ and $s = c(M)$, then the system $\{l_1, \cdots, l_s\}$ is primitive.

18. Let $\{l_1, \cdots, l_s\}$ be a primitive illuminating system for a compact, convex body $M \subset \mathbf{R}^n$. Prove that there exist points $a_1, \cdots, a_s \in \mathrm{bd}\, M$ such that each of the points is illuminated by only one of the directions.

19. Using the ideas of Example 34.11, prove that for any integer $m \geq 4$ there exists a convex polytope $M \subset \mathbf{R}^3$ with m vertices such that it has a primitive illuminating system consisting of m directions (i.e., each vertex of M is illuminated by only one direction of the system). Generalize this to $\mathbf{R}^n$.

20. Prove that for any integer $m \geq 4$ there exists a compact, convex body $M \subset \mathbf{R}^3$ with only m nonregular boundary points such that M has a primitive illuminating system consisting of m directions. Generalize this to $\mathbf{R}^n$.

21. Prove that the three-dimensional cube $M \subset \mathbf{R}^3$ can be illuminated by two lines, i.e., there exist two lines $l_1, l_2 \subset \mathbf{R}^3 \setminus M$ such that each point $x \in$ bd M is illuminated by a point $a \in l_1 \cup l_2$. (This is a special case of a result of K. Bezdek on the illumination of compact, convex bodies in $\mathbf{R}^n$ by r-dimensional flats, see the end of this section.)

§35 The hypothesis of Gohberg-Markus-Hadwiger

The assertions 35.1 up to 35.5 were originally proved in [Bt 1], but for Theorem 35.1 one should also look at [Ha 9], where a proof of K.A. Post is presented.

Theorem 35.1: For an arbitrary compact, convex body $K \subset \mathbf{R}^n$ the inequality $c(K) \geq n + 1$ holds.   □

PROOF: Let $l_1, \cdots, l_n$ be arbitrary directions. We will prove that these n directions cannot illuminate the whole boundary of any body $K \subset \mathbf{R}^n$. Let Γ denote a hyperplane which contains the directions $l_1, \cdots, l_{n-1}$ (Fig. 212). Furthermore, let Γ', Γ'' be the two supporting hyperplanes of K parallel to Γ and Π', Π'' be the closed half-spaces with respect to Γ', Γ'' which do not contain K, in each case. We choose points $x' \in \Gamma' \cap K$, $x'' \in \Gamma'' \cap K$ and denote by l', l'' the rays of the direction l_n with the starting points x' and x'', respectively. Then at least one of the inclusions $l' \subset \Pi'$, $l'' \subset \Pi''$ must hold. (If the ray l' is not contained in Π', then it is, except for x', contained in the open half-space $\mathbf{R}^n \setminus \Pi'$, and therefore the parallel half-line l'' has to be contained in Π''.) Thus we may assume $l' \subset \Pi'$. Then x' is not illuminated by any of the directions $l_1, \cdots, l_n$, since all the half-lines of directions $l_1, \cdots, l_n$ emanating from x' belong to Π' (thus they do not meet interior points of K, by int $K \subset \mathbf{R}^n \setminus \Pi'$). Hence n directions are not sufficient for illuminating bd K, i.e., $c(K) \geq n + 1$. ∎

Theorem 35.2: For a convex body $F \subset \mathbf{R}^n$ with regular boundary, the equality $c(F) = n + 1$ holds.   □

PROOF: Let $T \subset \mathbf{R}^n$ be an arbitrary n-dimensional simplex and a be an interior point of T. Furthermore, we consider the half-lines $[b_1, a), \cdots, [b_{n+1}, a)$ with $\{b_1, \cdots, b_{n+1}\}$ as vertex set of T (Fig. 213). These half-lines present $n+1$ directions $l_1, \cdots, l_{n+1}$ which illuminate the whole boundary of F.

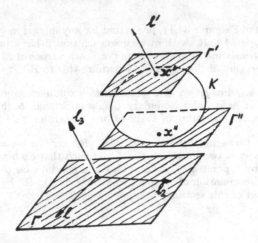

Fig. 212

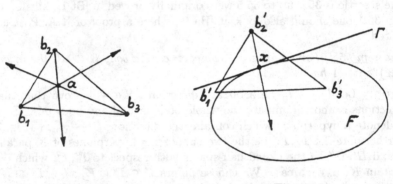

Fig. 213   Fig. 214

Indeed, let x be an arbitrary boundary point of F and Γ be the supporting hyperplane of F which contains x. We translate the simplex T such that a and x coincide. The vertices of the translated simplex T' are denoted by $b'_1, \cdots, b'_{n+1}$. Since Γ contains the interior point x of T', this simplex has points on both sides of Γ (Fig. 214). Let b'_i be a vertex of T' lying on that side of Γ which does not contain F. The half-line $[b'_i, x)$ has the direction l_i and contains interior points of F. (This holds because the closed half-space Π with bounding hyperplane Γ is the supporting cone of F at the regular boundary point x.) Therefore x is illuminated by l_i. Thus the directions $l_1, \cdots, l_{n+1}$ illuminate the whole boundary of F. This yields $c(F) \leq n+1$, and Theorem 35.1 implies $c(F) = n+1$. ∎

Corollary 35.3: Let T be an arbitrary n-dimensional simplex in $\mathbf{R}^n$ and $l_1, \cdots, l_{n+1}$ be directions from the vertices of T to a point $a \in \operatorname{int} T$. Then *each regular boundary point of any convex body $F \subset \mathbf{R}^n$ is illuminated by at least one of the directions $l_1, \cdots, l_{n+1}$.*   □

Exactly this was established in the proof of Theorem 35.2.

Theorem 35.4: If a convex body $F \subset \mathbf{R}^n$ has no more than n nonregular boundary points, then $c(F) = n + 1$.   □

PROOF: Let $x_1, \cdots, x_k$ be all nonregular boundary points of $F \subset \mathbf{R}^n$, where $k \leq n$. We choose arbitrary directions $l_1, \cdots, l_k$ illuminating the points $x_1, \cdots, x_k$ correspondingly. Since for any small perturbation of the direction l_i the point x_i remains to be illuminated by l_i, we may assume that the directions $l_1, \cdots, l_k$ are independent (i.e., they are not parallel to a $(k-1)$-flat). We choose points $a, b_1, \cdots, b_k \in \mathbf{R}^n$ such that the direction of the vector $a - b_i$ is given by l_i, $i = 1, \cdots, k$. Furthermore, we complete this set by $b_{k+1}, \cdots, b_n$ in such a manner that the vectors $a - b_1, \cdots, a - b_n$ are linearly independent, and we add $b_{n+1} = a - (b_1 - a) - \cdots - (b_n - a)$. Thus we obtain by $b_1, \cdots, b_{n+1}$ the vertex set of an n-simplex with a as interior point. Hence, the directions $l_1, \cdots, l_{n+1}$ of the vectors $a - b_1, \cdots, a - b_{n+1}$ illuminate all regular boundary points of F (Corollary 35.3). By the choice of the directions $l_1, \cdots, l_k$, the points $x_1, \cdots, x_k$ are also illuminated. Hence $c(F) \leq n + 1$, and by Theorem 35.1 we have $c(F) = n + 1$.   ■

Corollary 35.5: For any compact, convex body $K \subset \mathbf{R}^n$ the inequality $a(K) \leq b(K) = c(K)$ holds. If the body K is regular (or has no more than n nonregular boundary points), then $a(K) \leq n+1$. Therefore for any n-dimensional convex body with regular boundary (or with at most n nonregular boundary points) the hypothesis of Borsuk is true.   □

In fact, let $K \subset K_1 \cup \cdots \cup K_m$, where each of the sets $K_1, \cdots, K_m$ is a smaller homothetical copy of K with positive ratio. Then diam $K_i <$ diam K holds for each $i \in \{1, \cdots, m\}$, and therefore $a(K) \leq m$. From this it follows that $a(K) \leq b(K)$. The remaining assertions are clear by the Theorems 35.3, 35.2, and 35.4.

Lemma 35.6: An arbitrary bounded, convex figure $F \subset \mathbf{R}^2$, which is not a parallelogram, has regular boundary points x_1, x_2, x_3 such that the supporting lines of F through them form a triangle circumscribed about F.   □

PROOF: Let $a \in$ bd F be an arbitrary regular boundary point (Theorem 2.7). Let l denote the supporting line of F through a, and l' the parallel supporting line of F (Fig. 215). Let, furthermore, b be an arbitrary regular boundary point of F not lying in the lines l, l' (Theorem 2.7). Then the supporting line m of F through b is not parallel to l, i.e., $l \cap m$ is nonempty. Let now m' be the supporting line of F which is parallel to m. If the point $\{p\} = l' \cap m'$ is not from F, then the circumscribed triangle is obtained with the help of the supporting line through a regular boundary point $c \in$ bd F, where c is situated between the points $x \in l' \cap F$ and $y \in m' \cap F$, which are the nearest to p. Thus there remains the case $p \in F$. At first we assume that $\{q\} = l' \cap m$ is not from F. Let u, v be the nearest points to q from $l' \cap F$ and $m \cap F$, respectively. We determine the regular boundary point $b_1 \in$ bd F between the points u and v (Fig. 216). If now the line m is replaced by the supporting line m_1 through b_1 and m' by the supoporting line $m_1' \parallel m_1$, then the point $\{p_1\} = l' \cap m_1'$ cannot be from F and therefore, as above, a circumscribed triangle exists.

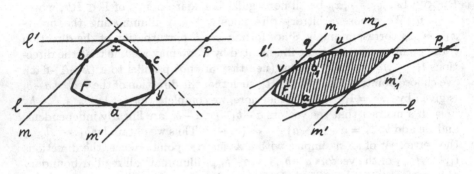

Fig. 215   Fig. 216

Analogously, a circumscribed triangle exists if the point $\{r\} = l \cap m'$ is not from F.

Thus the case $p, q, r \in F$ remains to be investigated. If the point $\{s\} = l \cap m$ is not from F, then the circumscribed triangle exists (Fig. 217). But if all the points p, q, r, s are from F, then F is a parallelogram. ■

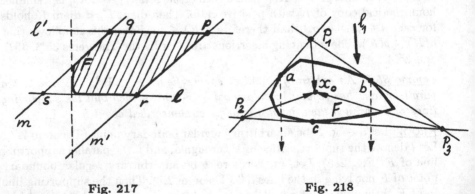

Fig. 217   Fig. 218

Theorem 35.7: For each compact, convex figure $F \subset \mathbf{R}^2$, which is not a parallelogram, we have $c(F) = 3$. If F is a parallelogram, then $c(F) = 4$. □

PROOF: Let F be not a parallelogram and a, b, c be regular boundary points of F such that the supporting lines of F through them form a triangle. circumscribed about F, see Fig. 218. Let p_1, p_2, p_3 be the vertices of the triangle and $x_o \in \text{int } F$. We denote by l_1, l_2, l_3 the directions of the vectors $x_o - p_1, x_o - p_2, x_o - p_3$, respectively. The direction l_1 illuminates the whole arc between a and b, including the endpoints a, b (since these two points are regular, the corresponding supporting cones are half-planes, and therefore the half-lines with starting points a, b and direction l_1 contain interior points of F). The analogous assertion holds for l_2, regarding the arc between a and c, and for l_3, regarding the arc between c and b. Hence $c(F) \leq 3$, and Theorem 35.1 yields $c(F) = 3$.

Let F be a parallelogram. Since no direction l illuminates two vertices of F at the same time, we have $c(F) \geq 4$. Since four directions are sufficient (Fig. 219), we have $c(F) = 4$. ∎

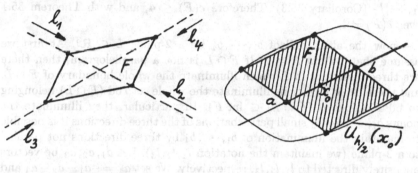

Fig. 219   Fig. 220

By means of Theorem 34.3, the theorem above (and all other results of this section) can be reformulated in terms of the functional $b(K)$ instead of $c(K)$: *for any compact, convex figure $F \subset \mathbf{R}^2$, which is not a parallelogram, the equality $b(F) = 3$ holds; if F is a parallelogram, then $b(F) = 4$ is satisfied.* I. Gohberg and A. Markus [G-M] and H. Hadwiger [Ha 6] formulated the corresponding hypothesis for the space $\mathbf{R}^n$, where n is arbitrary: *for any compact, convex body $F \subset \mathbf{R}^n$ the inequality $b(F) \leq 2^n$ holds, and equality is only attained if F is an n-parallelotope.* Replacing $b(F)$ by $c(F)$, one might formulate the analogous hypothesis in terms of illumination. For $n \geq 3$, the validity of this hypothesis is not proved up to now.

From Theorem 35.7 it is easy to get the proof of Theorem 33.5 in the previous chapter. Namely, let $\sum \subset \Re^2$ be not a parallelogram. If the bounded figure F (whose closedness and convexity can be assumed, too) is also not a parallelogram, then by Theorem 35.7 the inequality $b(F) \leq 3$ holds, and therefore $a_\Sigma(F) \leq 3$ (Corollary 35.5). If now $F \subset \Re^2$ is a parallelogram of diameter h with center x_o, then $F \subset \text{cl } U_{h/2}(x_o)$, where at least one of the pairs of opposite sides of F is not contained in the boundary of $U_{h/2}(x_o)$. If

a, b are the midpoints of these sides, then the line through a and b dissects the figure F into two parts of smaller diameters (Fig. 220), i.e., $a_\Sigma(F) = 2$.

The following theorem of A. Charazishvili [Ch] (see also [Be 3]) shows that Theorem 35.4 can be sharpened for $n = 3$.

Theorem 35.8: For any compact, convex body $F \subset \mathbf{R}^3$ with no more than four nonregular boundary points the equality $c(F) = 4$ holds.   □

PROOF: By Theorem 35.4 it suffices to show that the assertion holds if F has exactly four nonregular boundary points, say b_1, b_2, b_3, b_4.

1) If these points do not lie in a plane, then they form the vertex set of a tetrahedron $T \subset \mathbf{R}^3$. Let $a \in \text{int } T$. Then the directions $l_1, \cdots, l_4$ of the vectors $a - b_1, \cdots, a - b_4$ illuminate the points $b_1, \cdots, b_4$, and since the remaining boundary points of F are regular, they are also illuminated by $l_1, \cdots, l_4$ (Corollary 35.3). Therefore $c(F) \leq 4$, and with Theorem 35.1 even $c(F) = 4$.

2) Let now the affine hull of $b_1, \cdots, b_4$ be a 2-plane $L \subset \mathbf{R}^3$. At first we assume that $L \cap \text{int } F \neq \emptyset$. If $F \cap L$ is not a parallelogram, then there are three directions in L which illuminate the whole boundary of $F \cap L$, and these three directions illuminate the whole set $\text{rbd } (F \cap L)$ belonging to bd F (since ri $(F \cap L) \subset \text{int } F$). In particular, they illuminate the points $b_1, \cdots, b_4$. By small perturbations of the three directions it is possible to maintain the illumination of $b_1, \cdots, b_4$ by three directions not parallel to a 2-plane (we maintain the notation l_1, l_2, l_3). Let a_1, a_2, a_3 be vectors oppositely directed to l_1, l_2, l_3, respectively. We set $a_4 = -a_1 - a_2 - a_3$ and denote by l_4 the direction of $-a_4$. Then the four half-lines $[a_1, o), \cdots, [a_4, o)$ define the directions which coincide with $l_1, \cdots, l_4$. By Corollary 35.3, these directions illuminate all the regular boundary points of F, i.e., the whole set bd $F \setminus \{b_1, \cdots, b_4\}$. Since $\{b_1, \cdots, b_4\}$ is illuminated by l_1, l_2, l_3, we have $c(F) = 4$.

3) Let now (as before) $L \cap \text{int } F \neq \emptyset$, but $F \cap L$ be a parallelogram. If $a_1, \cdots, a_4$ are the vectors from the vertices of this parallelogram to its center, and p is a vector perpendicular to the plane L, then the directions $l_1, \cdots, l_4$ of the vectors $a_1 + \varepsilon p, a_2 + \varepsilon p, a_3 - \varepsilon p, a_4 - \varepsilon p\,(\varepsilon > 0)$ give a direction system from the vertices of a tetrahedron to one of its interior points (since the sum of the four vectors above is o). Hence all the regular boundary points of F are illuminated by $l_1, \cdots, l_4$. Since the center of the parallelogam $F \cap L$ belongs to int F, these directions illuminate the points $b_1, \cdots, b_4$ (for a sufficiently small $\varepsilon > 0$), too.

4) There remains the case $L \cap \text{int } F = \emptyset$, i.e., L is a suppoorting plane of F. Now we can choose a direction l parallel to L and a pair of points from $\{b_1, \cdots, b_4\}$, say b_1 and b_2, such that a translation of $[b_1, b_2]$ in direction l yields a segment $[c_1, c_2]$ contained in $F \cap L \setminus \{b_1, b_2, b_3, b_4\}$ (Fig. 221). Since c_1, c_2 are regular boundary points ($b_1, \cdots, b_4$ are the only nonregular

ones), L is the only supporting plane of F through c_1 (the same holds for c_2). Hence the half-space $\Pi \supset F$ with bounding plane L is the supporting cone of F at c_1 (or c_2). Thus a sufficiently small translation of $[c_1, c_2]$ in Π gives a segment $[e_1, e_2] \subset \text{int } F$. Since the segments $[b_1, b_2]$ and $[e_1, e_2]$ are translates of each other, we have $e_1 - b_1 = e_2 - b_2$, and therefore the direction l_1 of the vector $e_1 - b_1$ illuminates both the points b_1, b_2. We now choose directions l_2, l_3 which illuminate b_3 and b_4. Thus the chosen directions l_1, l_2, l_3 (which illuminate the quadruple $b_1, \cdots, b_4$) are not parallel to a plane, and we can complete the proof as in the case 2 above. ∎

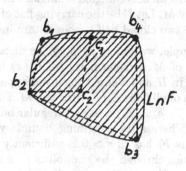

Fig. 221

The following examples will demonstrate that in the Theorems 35.4 and 35.8 the number of nonregular boundary points cannot be increased.

EXAMPLE 35.9: Let b_1, b_2, b_3 be the vertices of an equilateral triangle T in the plane $L \subset \mathbf{R}^3$, whose side-length is 1. Let, furthermore, $[c_1, c_2]$ be a segment of length 1 which is perpendicular to L and whose midpoint is the centroid of T. Then $\| c_i - b_j \| = \sqrt{\frac{7}{12}}$, which shows that each three of the points b_1, b_2, b_3, c_1, c_2 form an acute-angled triangle. Let M be the convex hull of these five points. It is easy to see that each segment with endpoints from $\{b_1, b_2, b_3, c_1, c_2\}$ has the following properties: two planes perpendicular to that segment and containing its endpoints are supporting planes of the polyhedron M and the corresponding intersection with M is a point, in each case.

Let now x be an arbitrary point from M. By $\rho(x)$ we denote the distance of x to the nearest vertex of M. Furthermore, we choose a positive number ε and denote by $E_\varepsilon(x)$ the closed ball centered at x with the radius $\varepsilon \rho(x)$. Let F_ε be the union of all balls $E_\varepsilon(x)$ with $x \in M$. Obviously, for a sufficiently small $\varepsilon > 0$, the points b_1, b_2, b_3, c_1, c_2 are nonregular boundary points of the convex body conv F_ε, whereas all other boundary points of this body are regular. Furthermore, for a sufficiently small $\varepsilon > 0$ the body conv F_ε has the same property as the polyhedron M: if two arbitrary points from $\{b_1, b_2, b_3, c_1, c_2\}$

are taken as endpoints of a segment, then the perpendicular planes through its endpoints support the body conv F_ε. From this it follows that no two of the points b_1, b_2, b_3, c_1, c_2 from bd conv F_ε can be illuminated by one direction. Hence the body conv F_ε (which has only five nonregular boundary points) cannot be illuminated by less than five directions, i.e., $c(\text{conv } F_\varepsilon) > 4$. □

EXAMPLE 35.10: Let $b_1, \cdots, b_{n-1}$ be the vertices of a regular $(n-2)$-simplex $T \subset \mathbf{R}^n$, $n > 3$, whose edge-length is 1. Let, furthermore, $[c_1, c_2]$ be a segment of length 1 which is orthogonal to the affine hull of $\{b_1, \cdots, b_{n-1}\}$ and whose midpoint is the centroid of T. It is easy to show that each triple from $\{b_1, \cdots, b_{n-1}, c_1, c_2\}$ forms an acute-angled triangle. The convex hull of these $n+1$ points is denoted by M. Let Γ be the carrying flat of the $(n-1)$-polytope M and Π be one of the two closed half-spaces determined by Γ.

As in the previous example, we denote by $\rho(x)$ the distance between $x \in M$ and the nearest vertex of M. Furthermore, let $E_\varepsilon(x)$ be the closed ball of radius $\varepsilon \rho(x)$ which lies in Π and whose tangential hyperplane at x is Γ. Finally, let F be the union of all balls $E_\varepsilon(x)$ with $x \in M$. For a sufficiently small $\varepsilon > 0$, all the points $b_1, \cdots, b_{n-1}, c_1, c_2$ are nonregular boundary points of the convex body conv F_ε, whereas its remaining boundary points are regular. Moreover, the body conv F_ε has (if $\varepsilon > 0$ is sufficiently small) the following property: the hyperplanes through the endpoints of a segment spanned by two arbitrary points from $\{b_1, \cdots, b_{n-1}, c_1, c_2\}$, which are also perpendicular to the segment, are supporting hyperplanes of conv F_ε. This implies that no two of the points $b_1, \cdots, b_{n-1}, c_1, c_2 \in$ bd conv F_ε can be illuminated from the same direction, and therefore $n + 1$ directions are necessary for illuminating all these points.

Let, finally, Γ' be the supporting hyperplane of the body conv F_ε which is parallel to Γ and $p \in \Gamma' \cap$ conv F_ε. Since conv F_ε lies between Γ and Γ', a direction which illuminates one of the points $b_1, \cdots, b_{n-1}, c_1, c_2 \in \Gamma$ cannot illuminate p. Thus, except for the $n+1$ directions illuminating $\{b_1, \cdots, b_{n-1}, c_1, c_2\}$, one more is needed for p, showing $c(\text{conv } F_\varepsilon) > n + 1$. □

At first glance, the following assertion of B. Weissbach [We 1] seems to contradict the examples above (cf. also [De 2]). For its formulation, we introduce the *conic hull* of a boundary point x of a compact, convex body $K \subset \mathbf{R}^n$, denoted by con (K, x). If C is the union of all rays emanating from x and meeting interior points of K, then con $(K, x) = C \setminus \{x\}$. The theorem of Weissbach says that, for getting $c(K) = n + 1$, the body K can have arbitrarily many nonregular boundary points, but that (if their number is larger than n) they have to be "good-natured" in a sense defined below.

Theorem 35.11. For a compact, convex body $K \subset \mathbf{R}^n$ and each boundary point x of K, let con (K, x) contain an open rotational cone with the apex x such that $\cos \alpha < \frac{1}{n}$, where α is the angle between the axis and a generator of the cone. Then $c(K) = n + 1$. □

PROOF: Let $T = \text{conv}\,(a_1, \cdots, a_{n+1}) \subset \mathbf{R}^n$ be a regular n-simplex whose circumradius is equal to 1, and H be an arbitrary hyperplane (of $\mathbf{R}^n$) through the centroid of T. First we show that, for each position of H, there exists a point $a \in \{a_1, \cdots, a_{n+1}\}$ such that the Euclidean distance of a and H is not smaller than $\frac{1}{n}$. To see this, we embed T into $\mathbf{R}^{n+1}$ in the usual way, i.e., its vertices are situated on the $n+1$ positive semiaxes of an orthonormal coordinate system $\xi_1, \xi_2, \cdots, \xi_{n+1}$ with equal distances to the origin. Then the simplex T lies in the hyperplane H_1, given by

$$\xi_1 + \xi_2 + \cdots + \xi_{n+1} = c,$$

where $c = \sqrt{\frac{n+1}{n}}$ (to obtain a ball of radius 1, which is circumscribed to T), and the centroid of T is given by $s = \frac{c}{n+1}(1, \cdots, 1)$. Now we can write $H = H_1 \cap H_2$, where H_2 is a suitable n-subspace of $\mathbf{R}^{n+1}$, described by

$$\sum_{i=1}^{n+1} \alpha_i \xi_i = o, \qquad \sum_{i=1}^{n+1} \alpha_i^2 = 1.$$

Since s has to be contained in H_2, we have in addition

$$\sum_{i=1}^{n+1} \alpha_i = 0.$$

Since H_1 and H_2 are orthogonal to each other, the orthogonal projections of $a_i \in H_1$ into $H = H_1 \cap H_2$ and into H_2 coincide. Thus, the oriented distance $d_i(u_i, H)$ is given by $d_i = c\alpha_i$, i.e.,

$$\sum_{i=1}^{n+1} d_i = 0, \qquad \sum_{i=1}^{n+1} d_i^2 = \frac{n+1}{n}.$$

Hence the set $\{d_1, \cdots, d_{n+1}\}$ has to contain at least one positive and at least one negative number, i.e., we can write $d_i \geq 0$, $i = 1, \cdots, k$, and $d_i < 0$, $i = k+1, \cdots, n+1$, with $1 \leq k \leq n$. If we assume that $d_i < \frac{1}{n}$ for $i = 1, \cdots, k$, then

$$0 < -\sum_{i=k+1}^{n+1} d_i = \sum_{i=1}^{k} d_i < \frac{k}{n},$$

and hence

$$\frac{n+1}{n} = \sum_{i=1}^{k} d_i^2 + \sum_{i=k+1}^{n+1} d_i^2 \leq \sum_{i=1}^{k} d_i^2 + \left(-\sum_{i=k+1}^{n+1} d_i\right)^2$$

$$< \frac{k}{n^2} + \frac{k^2}{n^2} \leq \frac{n+n^2}{n^2}$$

would hold, a contradiction.

Thus each sufficiently large rotational cone with apex s contains some vertex a_i of T. This means that for each boundary point x of K the set con (K, x) contains a ray which emanates from x, meets the interior of K, and is parallel to $[o, a_i]$ for some $i \in \{1, \cdots, n + 1\}$. Hence the sufficient conditions for illuminating bd K are satisfied.   ∎

The following assertions (on the illumination of bounded n-polytopes) might help to obtain solutions of the illumination problem for wide classes of convex bodies such as for the family of all convex n-polytopes.

Theorem 35.12: _Let $K \subset \mathbf{R}^n$ be a compact, convex n-polytope and F be one of its facets. If the (closed) facet F cannot be illuminated by one direction, then the following holds: there exist facets $G_o, \cdots, G_m$ of K (one of them possibly coinciding with F) such that $m \leq n$, $G_i \cap F \neq \emptyset$, $i = 0, \cdots, m$, and the outer normals of $\{G_o, \cdots, G_m\}$ are minimally dependent._   □

PROOF: For each facet G, satisfying $G \cap F \neq \emptyset$, we have two closed half-spaces with respect to aff G. Let P_G denote that one which contains K, and $\Pi_G = \mathbf{R}^n \setminus P_G$. Thus Π_G is an open half-space bounded by aff G and satisfying $K \cap \Pi_G = \emptyset$. Obviously, every point $x_i \in$ ri G is illuminated by an arbitrary point source $y \in \Pi_G$. If the intersection of all half-spaces Π_G (with G running over all facets satisfying $G \cap F \neq \emptyset$) would be non-empty, then a point source in this intersection would illuminate all relative interior points of all these facets G, and therefore F (together with its relative boundary) could be illuminated, cf. the proof of $c(K) = c'(K)$ in Theorem 34.1. This contradicts the assumption.

Therefore the intersection of all such half-spaces Π_G has to be empty. Since K is a polytope, their number is finite, and since they are convex, there are $n + 1$ (or less) of them with empty intersection (this follows from Helly's theorem). In other words, the polytope K has facets $G_o, G_1, \cdots, G_k$ $(k \leq n)$ such that $\Pi_{G_o} \cap \cdots \cap \Pi_{G_k} = \emptyset$.

Denote by $\Pi'_{G_o}, \cdots, \Pi'_{G_k}$ the open half-spaces which are translates of Π_{G_o}, $\cdots, \Pi_{G_k}$ such that the origin is contained in each of their bounding hyperplanes. Then also $\Pi'_{G_o} \cap \cdots \cap \Pi'_{G_k} = \emptyset$ holds. Hence the convex cones cl $\Pi_{G_o}, \cdots,$ cl Π_{G_k} are separable, i.e., there exist vectors $a_o \in$ (cl $\Pi_{G_o})^*, \cdots,$ $a_k \in$ (cl $\Pi_{G_k})^*$ (not all equal to o) such that $a_o + a_1 + \cdots + a_k = o$, cf. Theorem 8.4. The vector a_i has the form $-\lambda_i b_i$, where $\lambda_i \geq 0$ and b_i denotes the inner unit normal of cl Π_{G_i}, i.e., b_i is the unit outer normal of the facet G_i $(i = 0, 1, \cdots, k)$. Thus $\lambda_o b_o + \lambda_1 b_1 + \cdots + \lambda_k b_k = o$, where the numbers λ_i are nonnegative and not all equal to 0. Omitting the coefficients equal to 0, we obtain a positive dependence of the remaining system of outer unit normal vectors, and from those we can take (by virtue of Theorem 16.5) the smallest number to get a minimally dependent subsystem of $\{b_1, \cdots, b_m\}$.   ∎

REMARK 35.13: If in the above theorem $m = n$ holds, then $G_o = F$ can be additionally assumed. Indeed, suppose that F is not belonging to the set

$G_o, G_1, \cdots, G_n$, and let $b_o, b_1, \cdots, b_n$ be the outer unit normals of these facets, whereas b denotes that normal of F. Let, furthermore,

$$\lambda_o b_o + \lambda_1 b_1 + \cdots + \lambda_n b_n = o$$

present the positive dependence of the system $\{b_o, \cdots, b_n\}$, and

$$b = \mu_1 b_1 + \cdots + \mu_n b_n$$

be the linear combination of $\{b_1, \cdots, b_n\}$ yielding b. If all the coefficients $\mu_1, \cdots, \mu_n$ are negative, then $b + (-\mu_1)b_1 + \cdots + (-\mu_n)b_n = o$ is a positive dependence (i.e., one can replace G_o by F and obtain the required system $F, G_1, \cdots, G_n$). But if not all the coefficients are negative, then we can choose a number $\alpha > 0$ such that in $b = \mu_1 b_1 + \cdots + \mu_n b_n - \alpha(\lambda_o b_o + \cdots + \lambda_n b_n)$ one of the coefficients $\mu_i - \alpha \lambda_i$ is 0 and the remaining ones are not positive. Then we obtain a positive dependence between b and some $k \leq n$ of the vectors $b_o, \cdots, b_n$. It remains to apply Theorem 16.5.   □

Theorem 35.14: Let $K \subset \mathbf{R}^n$ be a compact, convex n-polytope and F be one of its facets. The facet F cannot be illuminated by one direction if and only if the following holds: there exist supporting hyperplanes $\Gamma_1, \cdots, \Gamma_m$ ($m \leq n$) of K with $F \not\subset \Gamma_i$ and $\Gamma_i \cap F \neq \emptyset$, $i = 1, \cdots, m$, such that the outer normals of the half-spaces $\Pi_1, \cdots, \Pi_m$, bounded by them and containing K in each case, are minimally dependent.   □

PROOF: Assume that F cannot be illuminated by one direction and choose facets $G_o, G_1, \cdots, G_m$ in the sense of Theorem 35.12 (where their existence was also verified). If F is not from the set $\{G_o, \cdots, G_m\}$ and $m < n$, then the affine hulls of the facets $G_o, \cdots, G_m$ present a hyperplane system satisfying the condition of the theorem. On the other hand, for $m = n$ we can assume that $G_o = F$ (cf. Remark 35.13), showing the only case which remains to be verified: $G_o = F$, $m \leq n$. Again, we denote by $b_o, \cdots, b_m$ the unit outer normals of the facets $G_o, \cdots, G_m$ and we obtain the positive dependence $\lambda_o b_o + \lambda_1 b_1 + \cdots + \lambda_m b_m = o$. Now we choose a point $x_o \in G_o \cap G_1 = F \cap G_1$ and set

$$b_1' = \lambda_o b_o + \lambda_1 b_1, \quad \Pi_1 = \{x : \langle b_1', x - x_o \rangle \leq 0\},$$

$\Gamma_1 = \operatorname{bd} \Pi_1$, $\Gamma_i = \operatorname{aff} G_i$, $i = 2, \cdots, m$. Since the number of the facets $G_o, \cdots, G_m$ is larger than 2 (i.e., $m > 1$), we have $\lambda_o b_o + \lambda_1 b_1 \neq o$. Since x_o is from the bounding hyperplane Γ_1 of Π_1, we have $\Gamma_1 \cap F = \emptyset$. For $i \geq 2$, the set $\Gamma_i \cap F \supset G_i \cap F$ is not empty, too. Furthermore, the vectors b_1, b_1 are linearly independent (since $b_o, \cdots, b_m$ is a minimally dependent system with $m > 1$), and therefore b_1', b_o are not proportional, i.e., $\Gamma_1 \neq \operatorname{aff} G_o = \operatorname{aff} F$. This means that F is not contained in Γ_1. It is clear that also the remaining hyperplanes $\Gamma_2, \cdots, \Gamma_m$ do not contain F (since $G_i \neq F$ for $i = 2, \cdots, m$). Finally, the relation $b_1' + \lambda_2 b_2 + \cdots + \lambda_m b_m = o$ shows that the vectors $b_1', b_2, \cdots, b_m$

(i.e., the outer normals of the closed half-spaces correspondingly bounded by $\Gamma_1, \cdots, \Gamma_m$ and containing K in each case) are minimally dependent. Hence the hyperplanes $\Gamma_1, \cdots, \Gamma_m$ satisfy the conditions of the theorem.

Conversely, let there exist hyperplanes $\Gamma_1, \cdots, \Gamma_m$ with the indicated properties.

We show that F cannot be illuminated by one direction. Let $b_1, \cdots, b_m$ be the outer normals of the closed half-spaces $\Pi_1, \cdots, \Pi_m$ bounded by the hyperplanes $\Gamma_1, \cdots, \Gamma_m$ and containing K, in each case. Let $\lambda_1 b_1 + \cdots + \lambda_m b_m = o$ be the minimal dependence of this vector system and l be an arbitrary direction presented by a vector p. Since $\lambda_1 \langle p, b_1 \rangle + \cdots + \lambda_m \langle p, b_m \rangle = 0$, we can find an index $i \in \{1, \cdots, m\}$ for which $\langle p, b_i \rangle \geq 0$ holds. We choose an arbitrary point $x_i \in \Gamma_i \cap F$. The inequality $\langle p, b_i \rangle \geq 0$ shows that the ray l_i, emanating from x_i in the direction of p, does not intersect int Π_i and, moreover, int K. Therefore the point $x_i \in F$ is not illuminated by l, i.e., no direction is illuminating the facet F. ∎

REMARK 35.15: The theorem above shows the following: if the facet F cannot be illuminated by one direction, then the hyperplanes $\Gamma_1, \cdots, \Gamma_m$ (whose existence was clarified by Theorem 35.14) can be chosen in such a way that $\Gamma_i = \text{aff } G_i$, where G_i denotes a facet with $G_i \cap F \neq \emptyset$, $i = 2, \cdots, m$. □

Exercises:

1. Let $M \subset \mathbf{R}^n$ be a compact, convex body and e be a vector. We say that a point $a \in \text{bd } M$ is X-rayed in the direction e, if for a sufficiently small $\lambda > 0$ the point $a - \lambda e$ belongs to int M (Fig. 222). In other words, the ray of direction e coming to $a \in \text{bd } M$ has common points with int M (Fig. 223). Finally, the body M is X-rayed by the directions $e_1, \cdots, e_s$ if every boundary point of M is X-rayed by at least one of the directions $e_1, \cdots, e_s$. Prove that the minimal number $c_X(M)$ of directions X-raying bd M is equal to $c(M)$.

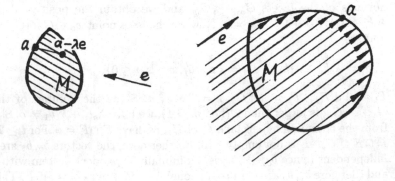

Fig. 222   Fig. 223

2. Let $M \subset \mathbf{R}^n$ be a compact, convex body and $q \notin M$ be a point. We say that a boundary point $a \in \operatorname{bd} M$ is X-rayed from $q \in \mathbf{R}^n$ if the ray emanating from q and passing through $a \in \operatorname{bd} M$ has two common points with $\operatorname{bd} M$ contained in $[q, a]$ (Fig. 224), i.e., $a - \lambda(a - q) \in \operatorname{int} M$ for $\lambda > 0$ small enough. Prove that the minimal number $c'_X(M)$ of points $q_1, \cdots, q_s$, from which the whole boundary of M is X-rayed, satisfies the inequalities $2 \leq c'_X(M) \leq n + 1$. The number $c'_X(M)$ will be investigated (from another point of view) in §37. This problem was introduced in [SP 5].

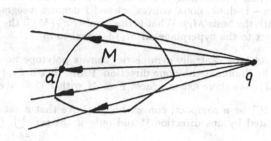

Fig. 224

3. Let $M \subset \mathbf{R}^n$ be a compact, convex body, $q \notin M$ be a point, and $\lambda > 1$ be a number. We say that the body M', obtained from M by the homothety with center q and ratio λ, is a "larger copy" of M (Fig. 225). Denote by $b'_X(M)$ the minimal number of larger copies of M which cover M. Prove that $b'_X(M) = c'_X(M)$.

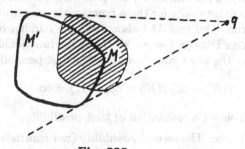

Fig. 225

4. Prove that $c'_X(M) \geq b'_X(M)$ for every closed, convex body M.

5. Prove that for every smooth, compact, convex body $M \subset \mathbf{R}^n$ the equality $c'_X(M) = 2$ holds.

6. What is the number $c'_X(M)$ for the n-dimensional simplex $M \subset \mathbf{R}^n$?

7. Consider possible values of $c'_X(M)$ for unbounded, closed, convex bodies $M \subset \mathbf{R}^n$.

8. Prove that for a two-dimensional, compact, convex figure $M \subset \mathbf{R}^2$ the equality $c'_X(M) = 3$ holds if and only if M is a triangle. (For the n-dimensional generalization we refer to §37.)

9. What is the number $c'_X(M)$ if $M \subset \mathbf{R}^n$ is a cross-polytope?

10. Let a compact, convex body $M \subset \mathbf{R}^n$ have the form $M = M_1 + I$, where M_1 is an $(n-1)$-dimensional convex set and I denotes a segment (thus M is a cylinder with the basis M_1). What is the number $c'_X(M)$ if the number $c_X(M_1)$ (with respect to the hyperplane aff M_1) is equal to m?

11. Let $M \subset \mathbf{R}^3$ be a centrally symmetric, convex polytope no (closed) facet of which can be illuminated by one direction. Prove that if F_1, F_2 are symmetric facets of M, then there exists a facet F of M with $F \cap F_1 \neq \emptyset$, $F \cap F_2 \neq \emptyset$.

12. Let $M \subset \mathbf{R}^n$ be a compact, convex body. Prove that a set $X \subset \operatorname{bd} M$ can be illuminated by one direction if and only if the set $\bigcup_{x \in X} (\operatorname{supcone}_x M)^*$ is contained in an open hemisphere of the unit sphere $S^{n-1} \subset \mathbf{R}^n$.

§36 The infinite values of the functionals b, b', c, c'

As we have seen, for compact, convex bodies $K \subset \mathbf{R}^n$ the functionals $b(K)$, $b'(K)$, $c(K)$, $c'(K)$ can have only finite values (according to Theorem 34.3). The Examples 34.4, 34.5, 34.6, 34.7 show that for unbounded convex bodies these functionals differ, where even infinite values are possible (Example 34.2). This gives a motivation for the following question: What combinations of finite and infinite values for these functionals are possible? The inequalities (1) and (2) from Theorem 34.1 show that the number of such combinations cannot be large. For $c(K) = \infty$, all the other functionals $c'(K), b(K), b'(K)$ have to satisfy the same condition, i.e., the first possibility is given by

$$b(K) = \infty,\ b'(K) = \infty,\ c(K) = \infty,\ c'(K) = \infty. \tag{1}$$

Example 34.2 shows a realization of that possibility.

Let now $c(K) < \infty$. The second possibility (not contradicting the inequalities from Theorem 34.1) consists of

$$c(K) < \infty,\ b(K) = \infty,\ b'(K) = \infty, c'(K) = \infty. \tag{2}$$

If, besides $c(K)$, a second functional has only finite values, then (by Theorem 34.1) this can only be the functional $b'(K)$ or the functional $c'(K)$. This yields two additional possibilities:

$$c(K) < \infty,\ b'(K) < \infty,\ b(K) = \infty,\ c'(K) = \infty, \tag{3}$$

$$c(K) < \infty,\; c'(K) < \infty,\; b(K) = \infty,\; b'(K) = \infty. \tag{4}$$

If even three functionals have finite values, then (again by Theorem 34.1) the only possibility is

$$c(K) < \infty,\; b'(K) < \infty,\; c'(K) < \infty,\; b(K) = \infty. \tag{5}$$

Finally, we have the remaining case

$$c(K) < \infty,\; b'(K) < \infty,\; c'(K) < \infty, b(K) < \infty, \tag{6}$$

which is realizable by bounded and also by unbounded bodies (cf. Examples 34.4, 34.5, 34.6, and 34.7). The following examples will demonstrate the realizability of the cases (3) and (2).

EXAMPLE 36.1: Let $K \subset \mathbf{R}^2$ be a convex figure which is bounded by a parabola, i.e., in a coordinate system (x_1, x_2) the set K is defined by the inequality $x_2 \geq x_1^2$. Obviously, the boundary of K is illuminated by one direction which is given by the vector $(0,1)$, i.e., we have $c(K) = 1$. Analogously, we see that $b'(K) = 1$ (to see this, we translate K by the vector $(0,-1)$, leading to a figure K_1 with $K \subset \operatorname{int} K_1$).

On the other hand, it is impossible to cover K by a finite number of smaller homothetical copies or to illuminate its boundary by finitely many light sources. Namely, let K' be a smaller homothetical copy of K and $y_o \notin K$ the corresponding homothety center (Fig. 226). Furthermore, let a and b be the intersection points of K and its tangents through y_o. Thus the parabola consists of three arcs, the two unbounded of which we denote by $\triangle_1, \triangle_2$. Obviously, K' does not contain any point from $\triangle_1$ or $\triangle_2$. This means that for $y_o \notin K$ the smaller homothetical copy K' contains only a *finite arc* of

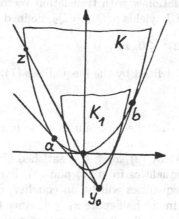

Fig. 226

the parabola $P = $ bd K. For $y_o \in K$, a smaller homothetical copy K' of K cannot contain more than one point of the parabola. This means that it is necessary to use infinitely many smaller homothetical copies, in order to cover P (and, moreover, to cover K). Consequently $b(K) = \infty$. By identifying y_o with the light source, it is easy to see that the same arguments yield $c'(K) = \infty$. Hence the figure K under consideration satisfies (3).

EXAMPLE 36.2: Let $\Gamma \subset \mathbf{R}^3$ be a plane and P one of the closed half-spaces bounded by Γ. Furthermore, let Q be a rotational cone situated as in the Example 34.2 (Fig. 203). Thus the apex of Q lies in int P, and bd $Q \cap \Gamma = L$ is a parabola. Now let Q_1 denote a cone symmetric to Q with respect to Γ, and Q_2 be a translate of Q_1, where the direction of this translation is parallel to the axis of L, with $Q_2 \cap \Gamma \subset$ ri $(Q \cap \Gamma)$. Then for the body $K = P \cap Q \cap Q_2$ we have (2), i.e., $c(K) < \infty$. The functionals $b(K), b'(K), c'(K)$ have infinite values.

In fact, the set $\Lambda = $ bd $Q \cap$ bd Q_2 is completely contained in int P, and thus two opposite directions l_1, l_2 (parallel to the axis of Q_2) illuminate the whole boundary of K, except for the points from Λ. And these remaining points are illuminated from a third direction l_3 which is parallel to the axis of L, leading to $c(K) \leq 3$, i.e., $c(K) < \infty$.

The aforesaid can easily be verified by calculation. In a coordinate system (x_1, x_2, x_3), the cone Q is given by the inequalities

$$x_1^2 + x_2^2 - x_3^2 \leq 0, \; x_3 \geq 0 \tag{7}$$

and the half-space P by $x_1 - x_3 + 1 \geq 0$. Then the reflection of any point (x_1, x_2, x_3) at $\Gamma = $ bd P yields the point $(x_3 - 1, x_2, x_1 + 1)$, and therefore Q_1 has the description

$$-x_1^2 + x_2^2 + x_3^2 - 2x_1 - 2x_3 \leq 0, \; x_1 + 1 \geq 0.$$

Furthermore, a translation π with translation vector $a = (1, 0, 1)$, which is parallel to the axis of L, yields $\pi(Q_1) = Q_2$, defined by the inequalities

$$-x_1^2 + x_2^2 + (x_3 - 2)^2 \leq 0, \; x_1 \geq 0. \tag{8}$$

Hence the body K is defined by the inequalities (7), (8), and

$$x_1 - x_3 + 1 \geq 0. \tag{9}$$

It should be noticed that the inequalities $x_3 \geq 0, x_1 \geq 0$ from (7) and (8) can be omitted, since the sum of the first inequalities in (7) and (8) yields $x_3 \geq 1$, and therefore, by (9), $x_1 \geq 0$ is satisfied. Thus K is defined by (9) and the first two inequalities from (7) and (8). For any point of bd K, at least one of these inequalities will be an equality. Furthermore, we remark that the body K lies in the half-space $x_1 \geq \frac{1}{2}$, since the cone Q_2 is contained in the half-space $x_1 + x_3 - 2 \geq 0$, and together with (9) this gives $2x_1 \geq 1$.

Admit that in a point $z = (x_1, x_2, x_3) \in$ bd K the first inequality of (7) turns into an equality, whereas the first inequality from (8) and the inequality (9) remain true (in the strict sense). Then the direction l_1, defined by the vector $p = (-1, 0, 0)$, illuminates the point z. Indeed, we choose a point $z' = z + \varepsilon p = (x_1 - \varepsilon, x_2, x_3)$. Since z satisfies (9) and the first inequality from (8) in the strict sense, the same holds for z' if ε is sufficiently small. Moreover,

$$(x_1 - \varepsilon)^2 + x_2^2 - x_3^2 = x_1^2 + x_2^2 - x_3^2 - 2x_1\varepsilon + \varepsilon^2 = -2x_1\varepsilon + \varepsilon^2,$$

and therefore this expression is negative if $\varepsilon > 0$ is sufficiently small (since $x_1 \geq \frac{1}{2}$). Hence, if $\varepsilon > 0$ is small enough, for $z' = z + \varepsilon p$ the strict inequalities (7), (8), (9) are satisfied, i.e, $z' \in$ int K. Thus z is illuminated by l_1.

Furthermore, if for $z = (x_1, x_2, x_3) \in$ bd K the first inequality from (7) is strict, then z is illuminated by the direction l_2, defined by the vector $-p = (-1, 0, 0)$. In fact, we choose a point $z' = z - \varepsilon p = (x_1 + \varepsilon, x_2, x_3)$. Since at z the first inequality of (7) is strict, this inequality is strict also for z' if $\varepsilon > 0$ is small enough. In addition, we have

$$-(x_1 + \varepsilon)^2 + x_2^2 + (x_3 - 2)^2$$
$$= -x_1^2 + x_2^2 + (x_3 - 2)^2 - 2x_1\varepsilon - \varepsilon^2 \leq -2x_1\varepsilon - \varepsilon^2 < 0;$$
$$(x_1 + \varepsilon) - x_3 + 1 = (x_1 - x_3 + 1) + \varepsilon \geq \varepsilon > 0.$$

Hence for a sufficiently small $\varepsilon < 0$ the strict inequalities from (7), (8), (9) are satisfied at $z' = z - \varepsilon p$, i.e., z is illuminated by l_2.

Now we remark that the body K does not contain points at which in (9) and in the first inequality from (7) equalities hold. Indeed, the points of such a type (i.e., points from the parabola L, cf. Example 34.2) can be described by $x_1 = \frac{\xi^2-1}{2}, x_2 = \xi, x_3 = \frac{\xi^2+1}{2}$, and the explicit description shows that at these points the first inequality from (8) is not satisfied.

Thus there remain the points $z \in$ bd K at which the first inequalities from (7) and (8) become equalities. They are the points from Λ. It is easy to show that these points are illuminated by the direction l_3 defined by $a = (1, 0, 1)$. In fact, let $z = (x_1, x_2, x_3) \in \Lambda$, i.e., at z the first inequalities from (7) and (8) are equalities and (9) is a strict inequality. We choose a point $z' = z + \varepsilon a = (x_1 + \varepsilon, x_2, x_3 + \varepsilon)$. Since at z the inequality (9) is strict, the same holds for z'. Furthermore,

$$(x_1 + \varepsilon)^2 + x_2^2 - (x_3 + \varepsilon)^2 = x_1^2 + x_2^2 - x_3^2 + 2\varepsilon(x_1 - x_3) = 2\varepsilon(x_1 - x_3);$$

$$-(x_1 + \varepsilon)^2 + x_2^2 + (x_3 + \varepsilon - 2)^2 = -x_1^2 + x_2^2$$
$$+ (x_3 - 2)^2 + 2\varepsilon(-x_1 + x_3 - 2)$$
$$= 2\varepsilon(-x_1 + x_3 - 2).$$

For $\varepsilon > 0$, these two expressions are negative, since the body K lies in the strip given by $0 \leq x_3 - x_1 \leq 1$, where $x_1 \neq x_3$ is satisfied for the points of Λ. Hence for $\varepsilon > 0$, at the point $z' = z + \varepsilon a$ the inequalities (7), (8), (9) are strict, i.e., $z' \in$ int K. This means that z is illuminated by the direction l_3.

Summarizing the above statements we see that the whole boundary of K is illuminated by the system l_1, l_2, l_3, and this implies $c(K) < \infty$.

Now we will show that $b'(K) = \infty$. Let K' be a translate of K, where the translation vector is denoted $q = (\alpha, \beta, \gamma)$. We are interested in the points of $F = K \cap \Gamma$ which are contained in int K'. Since K is contained in the half-space $x_1 - x_3 + 1 \geq 0$, the set int K' belongs to the half-space $(x_1 - \alpha) - (x_3 - \gamma) > 0$. For $\gamma - \alpha \leq 0$, this open half-space does not contain any point of the plane $\Gamma \supset F$, i.e., (int $K') \cap F = \emptyset$. Analogously, int K' is contained in the open half-space $(x_3 - \gamma) - (x_1 - \alpha) > 0$ (since K belongs to the half-space $x_3 - x_1 \geq 0$). For $\gamma - \alpha \geq 1$, this open half-space does not contain any point from $\Gamma \supset F$, i.e., (int $K') \cap F = \emptyset$. Let now $0 < \gamma - \alpha < 1$. By $K \subset Q$ (cf. (7)), the body K' is contained in the cone Q' defined by the inequalities

$$(x_1 - \alpha)^2 + (x_2 - \beta)^2 - (x_3 - \gamma)^2 \leq 0; \; x_3 - \gamma \geq 0.$$

Denote by x_o the point $\left(\frac{\alpha}{\gamma-\alpha}, \frac{\beta}{\gamma-\alpha}, \frac{\gamma}{\gamma-\alpha}\right)$. Since the apex of Q is $(0,0,0)$ and that of Q' is (α, β, γ), the cone Q' is obtained from Q by a homothety g with the center x_o and ratio $k = 1 - (\gamma - \alpha)$, where $0 < k < 1$. Furthermore, by $x_o \in \Gamma$, the plane Γ is fixed regarding g. Hence,

$$(\text{int } K') \cap \Gamma \subset K' \cap \Gamma \subset Q' \cap \Gamma = g(Q \cap \Gamma).$$

But the set $Q \cap \Gamma$ is obtained from $F = K \cap \Gamma$ by the translation π^{-1}. Thus (int $K') \cap \Gamma$ is contained in the homothetical copy $g'(F)$ of F, where $g' = g \circ \pi^{-1}$ (with $k < 1$ as its ratio). Since F is a plane figure bounded by a parabola, we have $b(F) = \infty$ (Example 36.1). Hence the interiors of finitely many translates of K (such as K') cannot cover the set $F \subset K$ and hence cannot cover K. This shows that $b'(K) = \infty$.

By Theorem 34.1 we obtain $b(K) = \infty$. The relation $c'(K) = \infty$ can be obtained in the same way as in Example 34.2; we omit the proof, since $c'(K) = \infty$ is also a consequence of the Theorem 36.3 which will be proved below. Therefore the body $K \subset \mathbf{R}^3$ satisfies the possibility (2). $\square$

The examples above show realizations of the possibilities (1), (2), (3), and (6). The following theorem shows that the remaining possibilities (4) and (5) are not realizable.

Theorem 36.3: The equality $b(K) = \infty$ holds if and only if $c'(K) = \infty$. $\square$

This follows immediately from the next theorem. To formulate it, we give the following definition, cf. [SP 3]. Let K be an unbounded convex body and Q be its inscribed cone with an apex $x_o \in K$. We say that K is *nearly conic* if there is a number $r > 0$ such that $K \subset U_r(Q)$. Now a more precise formulation of Theorem 36.3 can be given.

Theorem 36.4: For an unbounded, closed, convex body $K \subset \mathbf{R}^n$ the following three conditions are equivalent:

(a) $b(K) < \infty$,

(b) $c'(K) < \infty$,

(c) *the body K is nearly conic.*   □

PROOF: It is sufficient to show the implications (a) $\Rightarrow$ (b), (b) $\Rightarrow$ (c), and (c) $\Rightarrow$ (a). The first one is an immediate consequence of Theorem 34.1.

(b) $\Rightarrow$ (c): Assume that $c'(K) < \infty$. Let $y_1, \cdots, y_m \in \mathbf{R}^n \setminus K$ be light sources illuminating the whole boundary of K. Let, furthermore, Q denote the inscribed cone of K with an apex $x_o \in \operatorname{int} K$ and $Q_1, \cdots, Q_m$ be its translates with apices $y_1, \cdots, y_m$, respectively. Furthermore, introduce $F = \operatorname{conv}(Q_1 \cup \cdots \cup Q_m)$. We will show that $K \subset F$, which implies that F is nearly conic. For the proof, at first we assume that the cone K is pointed (i.e., it does not contain a complete line). Admit that $K \not\subset F$, i.e., there is a point $z \in K$ with $z \notin F$. Since F is closed, a strict separation of $z \notin F$ and F has to be realizable. This means that there is a closed half-space $P \supset F$ with $z \notin P$. By a small motion of the hyperplane $\Gamma = \operatorname{bd} P$ it is realizable that the hyperplane $\Gamma' \parallel \Gamma$ through x_o intersects the cone Q only at the point x_o.

The convex set $K \setminus P$ is bounded (otherwise, the half-space $\mathbf{R}^n \setminus P$ would contain a ray situated in $K \setminus P$, what is impossible since Q does not contain such a ray). Hence there is a supporting hyperplane $\Gamma_1 \parallel \Gamma$ of K having the sets K and F in the same closed half-space with respect to itself (Fig. 227). Let $x_1 \in \Gamma_1 \cap K$. Since K and the light sources $y_1, \cdots, y_m \in F$ lie in the same closed half-space with respect to Γ_1, the point x_1 cannot be illuminated by one of the points $y_1, \cdots, y_m$, a contradiction. Thus we have $K \subset F$, i.e., K is nearly conic.

We now suppose that the cone Q is not pointed. Let L be the minimal face of Q (i.e., L is the union of all apices of Q). Then, together with each $x \in K$, the body K contains the whole flat which is parallel to L and contains x. Denote by $\mathbf{R}'$ the orthocomplement of L in $\mathbf{R}^n$ and by π the projection of $\mathbf{R}^n$ onto $\mathbf{R}'$ parallel to L. If the points $y_1, \cdots, y_m \in \mathbf{R}^n \setminus K$ illuminate the whole boundary of K, then the points $\pi(y_1), \cdots, \pi(y_m) \in \mathbf{R}' \setminus \pi(K)$ will do the same with $\pi(K)$. Since the body $\pi(K) \subset \mathbf{R}'$ has a pointed inscribed cone, the aforesaid implies that $K' = \pi(K)$ is nearly conic. But then also $K = \pi^{-1}(K')$ is nearly conic, and (b) $\Rightarrow$ (c) is proved.

(c) $\Rightarrow$ (a): Let K be a nearly conic body, i.e., $K \subset U_r(Q)$ with respect to a cone Q, which is inscribed to K and has a point $x_o \in \operatorname{int} K$ as its apex. Let, furthermore, $\rho > 0$ be a number such that $U_{2\rho}(x_o) \subset K$. For $a > 0$, denote by Σ_a the union of all cones obtained from Q by translation and having points from the ball $U_a(x_o)$ as their apices. Thus,

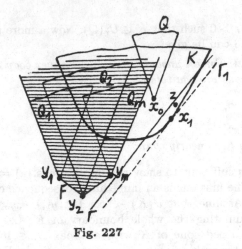

Fig. 227

$$\textstyle\sum_a = \bigcup_{\|x\|<a} (x + Q).$$

In particular, one might consider the bodies $\sum_r, \sum_{2\rho}$ and $\sum_\rho$. From $U_{2\rho}(x_o)$ $\subset K$ it follows that $\sum_{2\rho} \subset K$. Furthermore, we have

$$\textstyle\sum_r = \bigcup_{\|x\|<r} (x + Q) = \bigcup_{y \in Q} U_r(y) = U_r(Q) \supset K.$$

This yields the inclusion $\sum_{2\rho} \subset K \subset \sum_r$. Finally we remark that $\sum_\rho$ is obtainable from $\sum_{2\rho}$ by a homothety with $\frac{1}{2}$ as its ratio and x_o as its center. Let now vectors $z_1, \cdots, z_q \in \mathbf{R}^n$ be given such that the balls $z_1 + U_\rho(x_o), \cdots,$ $z_q + U_\rho(x_o)$ cover the ball $U_r(x_o)$. Then the sets $z_1 + \sum_\rho, \cdots, z_q + \sum_\rho$ cover the whole set $\sum_r$, i.e.,

$$\textstyle(z_1 + \sum_\rho) \cup \cdots \cup (z_q + \sum_\rho) \supset \sum_r \supset K.$$

Since $z_i + \sum_\rho$ is a translate of $\sum_\rho$, the set $z_i + \sum_\rho$ is obtainable from $\sum_{2\rho}$ under a homothety g_i with ratio $\frac{1}{2}$. Hence we have

$$\textstyle g_i(K) \supset g_i(\sum_{2\rho}) \supset z_i + \sum_\rho,$$

and therefore $g_1(K) \cup \cdots \cup g_q(K) \supset K$. This shows that $b(K) \le q < \infty$. ∎

In particular, the proved theorem implies that for any nearly conic body K the value of the functional $b(K)$ is finite. The following theorem refers to a way of *calculation* of $b(K)$.

Theorem 36.5: *Let $K \subset \mathbf{R}^n$ be a nearly conic, convex body, Q its inscribed cone with an apex $x_o \in \text{int } K$, and $\mathbf{R}'$ the orthocomplement of the flat aff Q. Furthermore, let π denote the projection of $\mathbf{R}^n$ onto $\mathbf{R}'$ parallel to aff Q,* and $F = \text{cl } \pi(K)$. *Then $F \subset \mathbf{R}'$ is a compact, convex body and the equality $b(K) = b(F)$ holds.* $\square$

PROOF: Since $\pi(Q)$ is a point in the space $\mathbf{R}'$, the set $\pi(U_r(Q)) = \pi(U_r(x_o))$ is bounded. The inclusion $K \subset U_r(Q)$ implies $\pi(K) \subset \pi(U_r(x_o))$, and therefore $\pi(K)$ and even $F = \text{cl } \pi(K)$ are bounded sets. Since K is a body, the convex set $F = \text{cl } \pi(K)$ is also a body in $\mathbf{R}'$.

Let $b(K) = m$ (we notice that $b(K) < \infty$, cf. Theorem 36.4) and $K_1, \cdots, K_m$ be smaller homothetical copies of K such that $K \subset K_1 \cup \cdots \cup K_m$. Let k_i be the ratio of the homothety transforming K into K_i (so $0 < k_i < 1$). Then a homothety g_i with k_i as its ratio transforms $\pi(K) \subset \mathbf{R}'$ into $\pi(K_i) \subset \mathbf{R}'$, $i = 1, \cdots, m$. Furthermore, let a_i be an arbitrary interior point of $\pi(K_i)$ and μ_i be a number satisfying $1 < \mu_i < \frac{1}{k_i}$. By f_i denote the homothety with center a_i and ratio μ_i. Since $\mu_i > 1$, the body $f_i(\pi(K_i))$ contains $\text{cl } \pi(K_i)$ (even, if $\pi(K_i)$ is not closed). Hence the homothety $h_i = f_i \circ g_i$ satisfies the condition

$$h_i(\text{cl } \pi(K)) = f_i(g_i(\text{cl } \pi(K))) = f_i(\text{cl } \pi(K_i)) \supset f_i(\pi(K_i)) \supset \text{cl } \pi(K_i),$$

and therefore

$$h_1(\text{cl } \pi(K)) \cup \cdots \cup h_m(\text{cl } \pi(K)) \supset \text{cl } \pi(K_1) \cup \cdots \cup \text{cl } \pi(K_m) =$$

$$= \text{cl } \pi(K_1 \cup \cdots \cup K_m) \supset \text{cl } \pi(K).$$

Since the ratio $\mu_i k_i$ of h_i satisfies $0 < \mu_i k_i < 1 \, (i = 1, \cdots, m)$, we have $b(\text{cl } \pi(K)) \leq m$. Thus $b(F) \leq b(K)$.

Now we will prove the converse inequality. Let $b(F) = s$ and $F \subset F_1 \cup \cdots \cup F_s$, where F_i is the image of F under a homothety ρ_i with center y_i and ratio $\lambda_i (0 < \lambda_i < 1, i = 1, \cdots, s)$. Without loss of generality we may assume that aff Q and $\mathbf{R}'$ are subspaces (i.e., their intersection point is the origin of $\mathbf{R}^n$). Denote by δ the projection of $\mathbf{R}^n$ onto aff Q parallel to $\mathbf{R}'$. Since for arbitrary points $x \in \mathbf{R}^n$ the relation $x = \delta(x) + \pi(x)$ holds, we have

$$K \subset \delta(K) + \pi(K) \subset \delta(K) + F \subset \delta(K) + (F_1 \cup \cdots \cup F_s)$$

$$= (\delta(K) + F_1) \cup \cdots \cup (\delta(K) + F_s).$$

Thus, for showing $b(K) \leq s$ it suffices to prove that, for each $i \in \{1, \cdots, s\}$, the set $\delta(K) + F_i$ is contained in a smaller homothetical copy of K.

It is easy to see that there exists a translation t such that $\delta(K)$ is contained in the cone $t(Q)$. Namely, the cone Q is a body with respect to aff Q and the set $\delta(K) \subset \text{aff } Q$ is nearly conic (with the inscribed cone Q). If now $y_o \in Q$ is a point such that its ε-neighbourhood (with respect to aff Q) is contained in Q, then the cone $t(Q) = Q + (x_o - y_o)$ contains the ε-neighbourhood of Q (here the translation vector of t is given by $x_o - y_o$). Hence $t(Q) \supset \delta(K)$.

From this inclusion we obtain $t(Q) + F_i \supset \delta(K) + F_i$, and therefore it remains to verify that $t(Q) + F_i$ (or, equivalently, $Q + F_i$) is contained in a smaller homothetical copy of K.

Let a be an arbitrary interior point of the body F and Ψ be a homothety with center a and ratio λ, where $\lambda_i < \lambda < 1$. Then $\Psi(F) \subset \mathrm{ri}\, F = \mathrm{ri}\, \pi(K)$. Obviously, the homothety $\rho_i \circ \Psi^{-1}$, whose ratio $\gamma_i = \frac{\lambda_i}{\lambda}$ satisfies $0 < \gamma_i < 1$, transforms the body $\Psi(F) \subset \mathbf{R}'$ into F_i. We will show that K contains a translate of $\Psi(F)$.

Let $\xi \in \Psi(F)$, yielding $\xi \in \mathrm{ri}\, \pi(K)$. Hence there exists a point $x(\xi) \in \mathrm{int}\, K$ such that $\pi(x(\xi)) = \xi$. In other words, we have $x(\xi) = \delta(x(\xi)) + \xi$. By $x(\xi) \in \mathrm{int}\, K$ there exists a neighbourhood $V(\xi)$ of the point ξ in $\mathbf{R}'$ such that $\delta(x(\xi)) + V(\xi) \subset K$. We can choose such neighbourhoods $V(\xi)$ and points $x(\xi)$ for each $\xi \in \Psi(F)$. Since $\Psi(F)$ is compact, there exist finitely many neighbourhoods of the type $V(\xi)$ which cover $\Psi(F)$. We assume that such a finite system is given by the corresponding neighbourhoods of the points $\xi_1, \cdots, \xi_i \in \Psi(F)$, i.e., that $V(\xi_1) \cup \cdots \cup V(\xi_k) \supset \Psi(F)$, where $\delta(x(\xi_i)) + V(\xi_i) \subset K$ for $i = 1, \cdots, k$. We choose a point $y_o \in \mathrm{ri}\, Q$, which does not coincide with the apex x_o of the cone Q. Let $\varepsilon > 0$ be a number such that the ε-neighbourhood of y_o in aff Q is contained in Q, but does not contain x_o. Furthermore, we choose a number t which is larger than each of the numbers $\frac{1}{\varepsilon} \parallel \delta(x(\xi_i)) - x_o \parallel, i = 1, \cdots, k$. Then, for each $i \in \{1, \cdots, k\}$, the point $x' = x_o + t(y_o - x_o)$ belongs to the cone $Q(x(\xi_i))$ with the apex $\delta(x(\xi_i))$ which is inscribed to the body K. In fact, the distance of the points y_o and $y_o + \frac{x_o - \delta(x(\xi_i))}{t}$ is given by $\frac{1}{t} \parallel x_o - \delta(x(\xi_i)) \parallel < \varepsilon$, and therefore the latter point is contained in Q. Hence the point $x_o + t(y_o + \frac{x_o - \delta(x(\xi_i))}{t} - x_o) = x_o + t(y_o - x_o) + x - \delta(x(\xi_i))$ belongs to Q. This means that the point $x' = x_o + t(y_o - x)$ (obtained by translation along the vector $\delta(x(\xi_i)) - x_o$) is contained in the cone $Q(x(\xi_i))$, $i = 1, \cdots, k$. Consequently for an arbitrary point $\xi \in V(\xi_i)$ we have $x' + \xi \in Q(x(\xi_i)) + \xi$, i.e., $x' + \xi$ lies in the cone $Q(x(\xi_i)) + \xi$, which is a translate of Q and has the apex $\delta(x(\xi_i)) + \xi$. Since $\delta(x(\xi_i)) + \xi \in \delta(x(\xi_i)) + V(\xi_i) \subset K$, the cone $Q(x(\xi_i)) + \xi$ is contained in K and, in particular, the inclusion $x' + \xi \in K$ holds. Thus $x' + V(\xi_i) \subset K$ holds for arbitrary $i \in \{1, \cdots, k\}$, giving

$$K \supset (x' + V(\xi_1)) \cup \cdots \cup (x' + (V(\xi_k)$$
$$= x' + V(\xi_1)) \cup \cdots \cup V(\xi_k)) \supset x' + \Psi(F).$$

This shows that also the translate $x' + \Psi(F)$ of $\Psi(F)$ is contained in K.

Finally, let $\sum$ denote the union of all cones which are translates of Q and whose apices are from the set $x' + \Psi(F)$, i.e., $\sum = x' + \Psi(F) + (-x_o + Q)$. Since $x' + \Psi(F) \subset K$, the union of all these cones is contained in K.

Since the cone $\Psi(\rho_i^{-1}(Q))$ is a homothetical copy of Q (by means of a positive ratio), it can be obtained from Q also by a translation, i.e., the equalities $\Psi(\rho_i^{-1}(Q)) = Q + b_i$ and $Q = \rho_i(\Psi^{-1}(Q + b_i))$ hold. Therefore

$$
\begin{aligned}
F_i + Q &= \rho_i(F) + Q = \rho_i(\Psi^{-1}(\Psi(F) + Q + b_i)) \\
&= \rho_i(\Psi^{-1}(\textstyle\sum - x' + x_o + b_i)) \subset \rho_i(\Psi^{-1}(K - x' + x_o + b_i)).
\end{aligned}
$$

Thus the set $F_i + Q$ is contained in $\rho_i(\Psi^{-1}(K - x' + x_o + b_i))$, which is a homothetical copy of K under the ratio $\gamma_i < 1$. ∎

Corollary 36.6: If for the inscribed cone Q of a nearly conic body $K \subset \mathbf{R}^n$ the equality $\dim Q = n$ holds, then $b(K) = b'(K) = c(K) = c'(K) = 1$. □

In this case aff $Q = \mathbf{R}^n$ (i.e., the body F from the previous theorem is a point. Hence $b(K) = b(F) = 1$. By Theorem 34.1, this implies $c(K) = c'(K) = b'(K) = 1$.

Corollary 36.7: If for the inscribed cone Q of a nearly conic body $K \subset \mathbf{R}^n$ the equality $\dim Q = n - 1$ holds, then $b(K) = c'(K) = 2$. □

In fact, in this case F is a segment and therefore $b(K) = b(F) = 2$. This implies $c'(K) \leq 2$ (Theorem 34.1), and it is easy to see that $c'(K) \neq 1$ (see the implication (b) $\Rightarrow$ (c) in Theorem 36.4, in which the constructed body $F = \operatorname{conv}(Q_1 \cup \cdots \cup Q_m)$ for $c'(K) = 1$ is reduced to a single cone Q_1, i.e., F is a translate of Q, contradicting the inclusion $K \subset F$).

Corollary 36.8: If for the inscribed cone Q of a nearly conic body $K \subset \mathbf{R}^n$ the equality $\dim Q = n - 2$ holds, then $b(K) \leq 4$. □

Indeed, in this case F is a two-dimensional, convex, bounded figure. By Theorem 35.7 we have $b(K) = b(F) = 4$ if F is a parallelogram, and otherwise $b(K) = b(F) = 3$ holds. ∎

The previous three corollaries allow the complete description of the values of $b(K)$ for unbounded convex bodies in $\mathbf{R}^2$ and $\mathbf{R}^3$. Namely, let $K \subset \mathbf{R}^2$ (with $K \neq \mathbf{R}^2$) be an unbounded convex figure and Q its inscribed cone. If K is not nearly conic, then $b(K) = \infty$ (Example 36.1). If K is nearly conic and $\dim Q = 1$, then $b(K) = 2$. Furthermore, let $K \subset \mathbf{R}^3$ (with $K \neq \mathbf{R}^3$) be an unbounded, convex body and Q be its inscribed cone. If K is not nearly conic, then $b(K) = \infty$. If K is nearly conic, then $b(K) = 1$ for $\dim Q = 3$, $b(K) = 2$ for $\dim Q = 2$, and $b(K) = 3$ or 4 for $\dim Q = 1$ (depending on the fact whether F is a parallelogram, cf. Theorem 36.5).

For example, in $\mathbf{R}^3$ this can be also formulated in the following manner: The quantity $\max b(K)$ is equal to ∞ if K runs through the set of all convex bodies in $\mathbf{R}^3$, and it is equal to 4 if K runs through the set of all nearly conic bodies.

Theorem 36.9: Let $K \subset \mathbf{R}^n$ be an unbounded, nearly conic, convex body and π the projecton introduced in Theorem 36.5. If the set $\pi(K)$ is closed (i.e., $\pi(K) = F$, cf. Theorem 36.5), then $b(K) = b'(K) = c(K) = c'(K) = b(\pi(K))$. □

PROOF: Theorems 34.1 and 36.5 imply that $c(K) \leq b(K) < \infty$. We set $c(K) = m$ and choose m directions $l_1, \cdots, l_m$ illuminating the whole boundary of K. Denote by $l_1', \cdots, l_m'$ the images of $l_1, \cdots, l_m$ under the projection π onto the space $\mathbf{R}'$ (if a direction l_i is parallel to the projecting flat aff Q, then its image l_i' degenerates, and thus the number of the images of $l_1, \cdots, l_m$ in $\mathbf{R}'$ can be smaller than m).

Let now x be an arbitrary boundary point of the body $F \subset \pi(K) \subset \mathbf{R}'$. By $x \in \pi(K)$, there exists a point $y \in K$ with $\pi(y) = x$. It is obvious that $y \notin \operatorname{int} K$, i.e., $y \in \operatorname{bd} K$. Hence there exists an index $i \in \{1, \cdots, m\}$ such that the direction l_i illuminates the point $y \in \operatorname{bd} K$. Consequently there is a point $z \in \operatorname{int} K$ such that l_i presents the direction of the ray $[y, z)$. By $u = \pi(z) \in \operatorname{int} F$ we have $u \neq x$, and therefore the direction l_i is not parallel to the projecting flat aff Q. The direction l_i' is determinded by the ray $[x, u)$, and therefore the point $x \in \operatorname{bd} F$ is illuminated by l_i' in $\mathbf{R}'$ (since $u \in \operatorname{int} F$). This shows that the whole boundary of F is illuminated by m directions $l_1', \cdots, l_m'$ (or even less), i.e., $b(F) \leq b(K)$. In other words, we have $b(K) \leq c(K)$, leading to $b(K) = c(K) = b'(K) = c'(K)$ (cf. Theorem 34.1). ∎

EXAMPLE 36.10: Let K be the body introduced in Example 34.5. The inscribed cone of K is the positive x_3-axis, and the plane $x_3 = 0$ can be considered as $\mathbf{R}'$, cf. the formulation of Theorem 36.5. Then π is the projection of $\mathbf{R}^3$ onto $x_3 = 0$, parallel to the x_3-axis. The set $F = \operatorname{cl} \pi(K)$ is a square, described in $\mathbf{R}'$ by the inequalities $0 \leq x_1 \leq 1$, $0 \leq x_2 \leq 1$. The set $\pi(K)$ is obtained from the square F by omitting the vertex $(1, 1, 0) \in \mathbf{R}^3$. Obviously, the relations $c(K) = c'(K) = 3$ and $b(K) = b'(K) = 4$ hold (cf. Example 34.7). According to Theorem 36.5, this is justified since $\pi(K)$ is not closed. □

EXAMPLE 36.11: Theorem 36.5 gives sufficient conditions for $b(K) = c(K)$ (and hence for $b(K) = b'(K) = c'(K) = c(K)$). In the previous example $b(K) \neq c(K)$. But the conditions given in Theorem 36.5 are not necessary for $b(K) = c(K)$, even if K is nearly conic. We will give an example showing this.

Let $K \subset \mathbf{R}^3$ be a convex body which, in a coordinate system (x_1, x_2, x_3), is given by the infinite system of inequalities:

$$-1 \leq x_1 \leq 1, \; -1 \leq x_2 \leq 1;$$

$$(4k - 1)x_1 + x_2 - \frac{1}{2k}x_3 \leq 4k - 2, \; k = 1, 2, \cdots;$$

$$x_1 + (4k + 1)x_2 - \frac{1}{2k + 1}x_3 \leq 4k, \; k = 1, 2, \cdots.$$

It is easy to check that for the point y_k with the coordinates $x_1 = x_2 = 1 - \frac{1}{4k+1}$, $x_3 = 2k(1 + \frac{1}{4k+1})$, the inequalities

$$(4k - 1)x_1 + x_2 - \frac{1}{2k}x_3 \leq 4k - 2; \; x_1 + (4k + 1)x_2 - \frac{1}{2k + 1}x_3 \leq 4k \quad (10)$$

become equalities, whereas all other inequalities defining K are strict at this point. Hence $y_k \in$ bd K.

Now it is easy to show that $c(K) = 4$. Let l_1, l_2, l_3 be arbitrary directions determined by vectors $(\alpha_1, \beta_1, \gamma_1)$, $(\alpha_2, \beta_2, \gamma_2)$, $(\alpha_3, \beta_3, \gamma_3)$. Assume that bd K is completely illuminated by the system l_1, l_2, l_3. Thus also $(-1, -1, -1) \in$ bd K has to be illuminated by this system. Let l_1 be a direction illuminating that point. Since K is contained in the half-spaces $x_1 \geq -1$, $x_2 \geq -1$, the point $(-1, -1, -1)$ can only be illuminated by l_1 if $\alpha_1 > 0$, $\beta_1 > 0$. Furthermore, the point $(1, -1, 0) \in$ bd K has to be illuminated by l_2 or l_3 (the direction l_1 cannot do this, since $\alpha_1 > 0$). If the point $(1, -1, 0)$ is assumed to be illuminated by l_2, then $\alpha_2 < 0$, $\beta_2 > 0$ must hold, and analogously $(-1, 1, 0) \in$ bd K is illuminated by l_3 if $\alpha_3 > 0$, $\beta_3 < 0$.

Finally we show that, for k sufficiently large, the point $y_k \in$ bd K cannot be illuminated by the system l_1, l_2, l_3. Since the body K has points in both the half-spaces (10) and (11) and $y_k \in$ bd K lies in the intersection of the corresponding bounding planes, this point can only be illuminated by a direction l of the following type: The direction vector (α, β, γ) of l has to satisfy

$$(4k - 1)\alpha + \beta - \frac{1}{2k}\gamma < 0; \quad \alpha + (4k + 1)\beta - \frac{1}{2k + 1}\gamma < 0,$$

i.e., (α, β, γ) yields a negative scalar product with each of the outward normal vectors of the half-spaces.

For $\alpha > 0$, the first of these inequalities is not satisfied if k is large enough, and for $\beta > 0$, the second one is not satisfied. Since $\alpha_1 > 0$, $\beta_2 > 0$, $\alpha_3 > 0$, the point y_k cannot be illuminated by any of the directions l_1, l_2, l_3 if k is large enough.

Thus we have verified that $c(K) \geq 4$. By Theorem 34.1 and Corollary 36.8, the relations $c(K) \leq b(K) \leq 4$ hold, and consequently $b(K) = b'(K) = c'(K) = c(K) = 4$. But the image $\pi(K)$ of K (for the projection onto the plane $x_3 = 0$ and parallel to the x_3-axis) is not a closed set: the image is a square $-1 \leq x_1 \leq 1, -1 \leq x_2 \leq 1$ from which all boundary points with $x_1 + x_2 > 0$ are omitted. By a slight modification of Example 34.5, it is easy to construct an analogous body K' (with $c(K') = 4$) having the same image $\pi(K')$ in the plane $x_3 = 0$ as the body K described in Example 34.5. In other words, we have $\pi(K) = \pi(K')$, but $c(K) \neq c(K')$. This shows that one cannot calculate $c(K)$ by means of properties of $\pi(K)$ if $\pi(K)$ is not closed (cf. Theorem 36.5).

EXAMPLE 36.12: The nearly conic body K, considered in the two previous examples, has the following property: its asymptotic scope $\tilde{K}$ has a closed image $\pi(\tilde{K})$ under the corresponding projection onto $\mathbf{R}'$. By Theorem 36.9, this implies $b(\tilde{K}) = b'(\tilde{K}) = c'(\tilde{K}) = c(\tilde{K})$. It is easy to show that this holds for all convex bodies which are nearly conic and have a *one-dimensional* inscribed cone. But if the dimension of the inscribed cone is greater than one, then the image $\pi(\tilde{K})$ can be non-closed, i.e., it is possible that the equalities $b(\tilde{K}) = b'(\tilde{K}) = c'(\tilde{K} = c(\tilde{K})$ are not satisfied. This is demonstrated by the following example.

In $\mathbf{R}^3$ with a coordinate system (x_1, x_2, x_3), let a convex body K be given by

$$(x_3 + x_2)^2(1 - x_1^2) \geq 1; \ (x_3 - x_2)^2(1 - x_1^2) \geq 1; \ x_3 \geq |x_2| + 1.$$

Thus K contains a two-dimensional cone (angle) Q defined by $x_1 = 0$, $x_3 \geq |x_2| + 1$. On the other hand, K is contained in the set defined by

$$x_3 \geq |x_2| + 1; \ |x_1| \leq 1,$$

and hence it is contained in the 1-neighbourhood of the cone Q. Therefore K is nearly conic and Q is its inscribed cone.

Taking all points of K as carrier points of lines parallel to the vector $(0, 1, 1)$, we get as union of all these lines the body M_1, described by

$$(x_3 - x_2)^2(1 - x_1^2) \geq 1; \ x_3 \geq x_2 + 1.$$

The arguments from the proof of Theorem 10.12 show that $\tilde{K} \subset M_1$. Taking now all the points of K as carrier points of lines parallel to $(0, -1, 1)$, in an analogous manner we get the body M_2 satisfying

$$(x_3 + x_2)^2(1 - x_1^2) \geq 1; \ x_3 \geq -x_2 + 1,$$

where $\tilde{K} \subset M_2$. Hence $\tilde{K} \subset M_1 \cap M_2$. But it is easy to see that even $M_1 \cap M_2 = K$ holds; i.e., the nearly conic body K *coincides* with its asymptotic scope. Since aff Q is given by the plane $x_1 = 0$, we may choose the x_1-axis as $\mathbf{R}'$ (cf. Theorem 36.5) such that π coincides with the projection along the plane $x_1 = 0$ onto the x_1-axis. Thus $\pi(K)$ is the nonclosed interval $(-1, 1)$ of the x_1-axis, and Corollary 36.7 implies that $b(K) = c'(K) = 2$. In this case we have $c(K) = 1$, since the direction l, corresponding to the vector $(0, 0, 1)$, illuminates the whole boundary of K. Analogously we get $b'(K) = 1$. Consequently the body $K = \tilde{K}$ does not satisfy $b(K) = b'(K) = c'(K) = c(K)$. $\square$

Exercises:

1. Which possibilities from (1) – (3), (6) are realizable for unbounded, closed, convex bodies $M \subset \mathbf{R}^2$?

2. Which possibilities from (1) – (3), (6) are realizable for unbounded, closed, convex bodies $M \subset \mathbf{R}^3$ with one-dimensinal (two-dimensional, three-dimensional) inscribed cones?

3. Let $M \subset \mathbf{R}^n$ be an n-dimensional convex set (not necessarily closed) and $\pi : \mathbf{R}^n \to L$ be the orthogonal projection onto a subspace L. Prove that $b'(M) \geq b'(\pi(M))$. Give an example for strict inequality.

4. Let $M \subset \mathbf{R}^n$ be an n-dimensional convex set. Prove that $b'(M) \leq b'(\mathrm{cl}\ M)$. Give an example for strict inequality.

5. Let $M \subset \mathbf{R}^n$ be an unbounded convex body and $L \subset \mathbf{R}^n$ be the subspace of maximal dimension for which the image $\pi(M)$ under the orthogonal projection $\pi : \mathbf{R}^n \to L$ is a bounded convex set (in general non-closed). Show by examples that all the possibilities $b'(M) < b'(\mathrm{cl}\ \pi(M))$, $b'(M) = b'(\mathrm{cl}\ \pi(M))$, $b'(M) > b'(\mathrm{cl}\ \pi(M))$ are realizble.

6. Let M and π be given as in the previous exercise. Prove that $b(M) \geq b(\mathrm{cl}\ \pi(M))$. Find an example for strict inequality.

7. Investigate the questions as in Exercises 5 and 6 for the functionals $c(M)$ and $c'(M)$.

8. Let $K \subset \mathbf{R}^n$ be an unbounded, convex body which contains no line. Prove that K is nearly conic if and only if the inscribed cones of K and of the asymptotic scope $\tilde{K}$ either are one-dimensional or (up to a translation) coincide.

9. Let M and π be as in Exercise 5. Prove that md $M \leq$ md cl $\pi(M)$.

10. Let M be an unbounded, convex body. Prove that $c(M) \geq$ md $M + 1$.

11. Let $M \subset \mathbf{R}^n$ be an nearly conic, convex body. Prove that the subspace L (given as in Exercise 5) is the orthogonal complement of the affine hull of a cone inscribed to M.

§37 Inner illumination of convex bodies

Let $K \subset \mathbf{R}^n$ be a closed (possibly unbounded) convex body with bd $K \neq \emptyset$, i.e., $K \neq \mathbf{R}^n$. We will assume that each point $b \in$ bd K can be considered as a "light source", and we look for those boundary points of K which are illuminated by b. More precisely, a point $x \in$ bd K is said to be *inner illuminated* by the light source $b \in$ bd K if $x \neq b$ and $]x, b[\subset$ int K, see Fig. 228. Furthermore, a set $N \subset$ bd K is inner illuminated by a set $M \subset$ bd K, if each point $x \in N$ is inner illuminated by at least one point $b \in M$. The *problem of inner illumination* is the following: what is the smallest positive integer m such that there exists a family of light sources $b_1, \cdots, b_m \in$ bd K from which the whole boundary of K is inner illuminated? This smallest integer we denote by $p(K)$.

We now formulate a covering problem which is strongly connected with the question above: what is the smallest positive integer m such that there exist larger homothetical copies $K_1, \cdots, K_m \subset \mathbf{R}^n$ of K whose union forms a covering of K, i.e., $K \subset K_1 \cup \cdots \cup K_m$. We underline that the corresponding homothety centers have to lie in $\mathbf{R}^n \setminus K$ (and the ratios have to be larger than one), see Fig. 229. We denote this smallest integer by $q(K)$.

Another covering problem is obtained by the following modification: what is the smallest positive integer m such that there exist larger homothetical

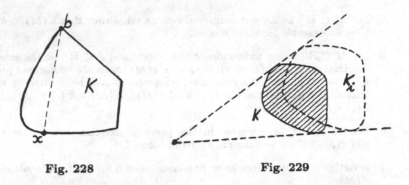

Fig. 228   Fig. 229

copies $K_1, \cdots, K_m$ of K with $K \subset \text{int } K_1 \cup \cdots \cup \text{int } K_m$, where the homothety centers are taken from bd K. This smallest integer is denoted by $q'(K)$.

Finally, we formulate the *problem of X-raying* a body K. Again, the light sources are taken from $\mathbf{R}^n \setminus K$. We say that a point $x \in$ bd K is *X-rayed* by a light source $y \in \mathbf{R}^n \setminus K$ if the intersection of $]x, y[$ and int K is nonempty, see Fig. 230. Furthermore, the set $N \subset$ bd K is X-rayed by the set $M \subset \mathbf{R}^n \setminus K$ if each point from N is X-rayed by at least one of the points from M. What is the smallest positive integer m such that there is a family $\{y_1, \cdots, y_m\} \subset \mathbf{R}^n \setminus K$ of light sources X-raying the whole boundary of K? This smallest integer is denoted by $p'(K)$.

Problems and results about inner illumination and related topics can be found in [Grü 5], [Ha 8], [B-SP 1], [SP 1,4,5], [Man 2], [Ros], and [SV 5]. They are mainly described in §37 and §38. In addition the last three papers contain the following statements: A convex n-polytope P is inner illuminated by its vertices if for each of its vertices there is another vertex such that the joining segment meets the interior of P. H. Hadwiger [Ha 8] asked whether such polytopes must have at least $2n$ vertices. P. Mani [Man 2] showed that this is true for $n \leq 7$, and for $n > 7$ there exist convex n-polytopes with this property having about $n + 2\sqrt{n}$ vertices, see also [Ros].

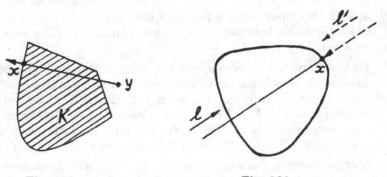

Fig. 230   Fig. 231

For a compact, convex body $M \subset \mathbf{R}^n$, an inner illuminating subsystem of bd M is said to be *primitive* if it has no proper subsystem realizing inner illumination of bd M. V. Soltan [SV 5] showed that any primitive, inner illuminating system of a compact, convex body $M \subset \mathbf{R}^3$ has at most eight points from bd M; this upper bound is only attained by convex 3-polytopes combinatorially equivalent to the 3-cube (with the vertices as light sources). For $n \geq 4$, the problem is still unsolved.

It should be noticed that we do not consider the problem of X-raying bd K by means of parallel bundles. This is obvious by the following reflection: if the point $x \in$ bd K is X-rayed by a direction l (Fig. 231), then x is also illuminated by the opposite direction l' and vice versa, which was investigated in previous sections.

Theorem 37.1: Each of the quantities $p(K), p'(K), q(K), q'(K)$ is defined if and only if the body $K \subset \mathbf{R}^n$ (with $K \neq \mathbf{R}^n$) is not a cone. □

PROOF: Let $K \subset \mathbf{R}^n$ be a cone with apex x_o. If $b \neq x_o$ is an arbitrary boundary point of K, then the segment $[b, x_o]$ is contained in bd K and the point x_o cannot be inner illuminated by $b \in$ bd K. By definition, the point x_o is also not illuminated by $b = x_o$. Thus also an arbitrary subset M of bd K cannot illuminate x_o in such a manner. Hence for this case $p(K)$ is not defined.

Furthermore, let $y \in \mathbf{R}^n \setminus K$. Then int K is not intersected by $]x_o, y[$ (since otherwise $y \in$ int K would hold). In other words, the point x_o cannot be X-rayed by any $y \in \mathbf{R}^n \setminus K$, i.e., $p'(K)$ is not defined.

Let now $c \in \mathbf{R}^n \setminus K$ and $k > 1$ be a positive integer. We denote by f a homothety with center c and ratio k. Then we have $f^{-1}(x_o) \in]c, x_o[$, since $k > 1$. But the open segment $]c, x_o[$ does not intersect the body K (since $c \notin K$). Consequently the point x_o cannot be contained in a body which is obtained from K by such a homothety whose ratio is larger than 1. Thus also $q(K)$ is not defined for this case.

Finally, let $b \in$ bd K, $k > 1$ be a natural number, and g be a homothety with center b and ratio k. Then $g(K)$ is a cone with apex $g(x_o)$. Since b is also a boundary point of $g(K)$, the segment $[g(x_o), b]$ belongs to the boundary of the cone. This menas that also $x_o \in [g(x_o), b]$ is a boundary point of $g(K)$, and therefore $x_o \notin$ int $g(K)$. In other words, there is no larger homothetical copy of K (where the homothety center is taken from bd K) with x_o in its interior. Hence also $q'(K)$ is not defined in this situation.

To show the converse statements, let the body K be not a cone. We choose an arbitrary point $x \in$ bd K and consider the inscribed cone Q with the apex x (if the body K is bounded, then $Q = \{x\}$). Clearly, we have $Q \neq K$, and hence there exists some $z \in$ int K with $z \notin Q$. Then the ray $[x, z)$ cannot be contained in Q and also not in K. Since $z \in [x, z)$ belongs to int K, we have $[x, z) \cap K = [x, b]$ with $b \in$ bd K and $b \neq x$. From the inclusions $z \in]x, b[, z \in$ int K it follows that x is inner illuminated by $b \in$ bd K. Hence each point $x \in$ bd K is inner illuminated by some $b \in$ bd K, i.e., the whole boundary of K is inner illuminated by the set $M =$ bd K. Thus $p(K)$ is defined; it is finite if even a finite set $b_1, \cdots, b_m \in$ bd K of sources is sufficient (otherwise, we have $p(K) = \infty$).

Furthermore, let $x \in$ bd K, $b \in$ bd K be points such that int $K \cap [x, b] \neq \emptyset$. We choose the points c, z on the line (b, x) such that $z \in]b, x[$ and $b \in]c, x[$ (Fig. 232). Since the open segment $]c, x[$ contains $z \in$ int K, the point $x \in$ bd K is X-rayed by the source $c \in \mathbf{R}^n \setminus K$. Thus the whole boundary of K can be X-rayed by the set $\mathbf{R}^n \setminus K$, and therefore $p'(K)$ is defined (again, the value of the functional can be finite or infinite).

Finally, let x be an arbitrary point from K and $b \in$ bd K a suitable boundary point of K such that the open segment $]b, x[$ intersects the interior of K (if $x \in$ bd K, then the point b is constructed as above, and for $x \in$ int K the point $b \in$ bd K can be arbitrarily chosen). Let z be an arbitrary point from $]b, x[$ and a point c will be chosen such that $b \in]c, x[$, see Fig. 232. Let f be the homothety with center c such that $f(b) = x$, and g be the homothety with center b such that $g(z) = x$. The ratios of f and g are greater than one. Finally, we put $K_1 = f(K)$ and $K' = g(K)$. Since $b \in K$, we have $x = f(b) \in f(K) = K_1$. Hence for each point $x \in K$ there exists a larger homothetical copy K_1 of K (where the center of the homothety is taken from $\mathbf{R}^n \setminus K$) with $x \in K_1$, and therefore $q(K)$ is defined. Since also $x = g(z) \in$ int $g(K) =$ int K', analogous arguments show that $q'(K)$ is defined as well. ∎

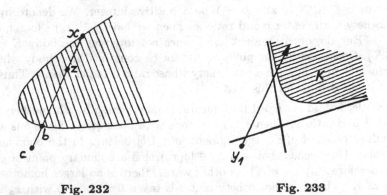

Fig. 232   Fig. 233

Theorem 37.2: For an arbitrary convex body $K \subset \mathbf{R}^n$, which is not a cone, the relations

$$p(K) = q'(K) \leq p'(K) \leq q(K) \tag{1}$$

hold. □

PROOF: Let $K_1, \cdots, K_m$ be homothetical copies of K, where for every $i = 1, \cdots, m$ the center b_i of the corresponding homothety satisfies $b_i \in \text{bd } K$, the ratio satisfies $k_i > 1$, and $K \subset \bigcup \text{int } K_i$. Let x be an arbitrary boundary point of K. Then there is some $i \in \{1, \cdots, m\}$ such that $x \in \text{int } K_i$. Denote by z the point whose image regarding the homothety f_i (with the center b_i and ratio k_i) is given by x, i.e.,

$$z = f_i^{-1}(x) = b_i + \frac{x - b_i}{k_i}.$$

Since $x \in \text{int } K_i$, we have $z = f_i^{-1}(x) \in \text{int } f_i^{-1}(K_i) = \text{int } K$. Moreover, $z \in]b_i, x[$ by $k_i > 1$. From $b_i \in \text{bd } K$, $x \in \text{bd } K$, $z \in \text{int } K$ it follows that $]b_i, x[\subset \text{int } K$, and therefore x is inner illuminated by b_i. Hence the whole boundary of K is inner illuminated by the system $\{b_1, \cdots, b_m\}$. Thus the inequality $p(K) \le q'(K)$ holds (for $q'(K) < \infty$).

Conversely, let $b_1, \cdots, b_m \in \text{bd } K$ be light sources illuminating bd K from inside. Let f_i be the homothety with center b_i and ratio $k = 2$. We will prove that $K \subset \text{int } f_1(K) \cup \cdots \cup \text{int } f_m(K)$. In fact, let $x \in K$. For $x \in \text{int } K$, the point $z = \frac{1}{2}(x + b_i)$ lies in int K for aribtrary $i \in \{1, \cdots, m\}$ (since $x \in \text{int } K$, $b_i \in \text{bd } K$); hence $x = f_i(z) \in \text{int } f_i(K)$. For $x \in \text{bd } K$, there is some $i \in \{1, \cdots, m\}$ for which the point x is inner illuminated by b_i. Therefore, in this case, $z = \frac{1}{2}(x + b_i) \in \text{int } K$, i.e., $x \in f_i(z) \in \text{int } f_i(K)$. This yields the inclusion $K \subset \text{int } K_1 \cup \cdots \cup \text{int } K_m$ for $K_i = f_i(K)$, $i = 1, \cdots, m$. We obtain $q'(K) \le p(K)$ if $p(K) < \infty$. The derived inequalities show that $p(K) = q'(K)$.

Let now $y_1, \cdots, y_m \in \mathbf{R}^n \setminus K$ be exterior light sources from which the whole boundary of K is X-rayed. Let a_i be an arbitrary boundary point of K which is X-rayed by y_i (if no point is X-rayed, as in Fig. 233, then y_i is superfluous and can be omitted). According to the definition, this means that int $K \cap]a_i, y_i[\ne \emptyset$. For $z_i \in \text{int } K \cap]a_i, y_i[$, the relations $y_i \notin K$ and $z_i \in \text{int } K$ show that the intersection $[y_i, z_i] \cap \text{bd } K$ is a point $b_i \ne a_i$ (Fig. 234). Constructing this configuration for each $i \in \{1, \cdots, m\}$, we obtain points $b_1, \cdots, b_m$. We will show that the sources $b_1, \cdots, b_m \in \text{bd } K$ illuminate the whole boundary of K from inside.

Indeed, let $x \in \text{bd } K$. Then x is X-rayed by one of the sources $y_1, \cdots, y_m$, i.e., there exists an $i \in \{1, \cdots, m\}$ such that int $K \cap [x, y_i] \ne \emptyset$. Let $u \in \text{int } K \cap [x, y_i]$. If $x = a_i$, then x is inner illuminated by b_i, since $z_i \in [a_i, b_i] \cap \text{int } K$ and hence $]a_i, b_i[\subset \text{int } K$. For $x \ne a_i$, the points a_i, b_i, y_i, x, u lie in a two-dimensional plane (they are situated on two rays emanating from y_i), where the points a_i, x, y_i are vertices of a triangle and b_i, u lie on the sides of this triangle, Fig. 235. Let v denote the intersection point of the segments $[b_i, x]$ and $[a_i, u]$. By $u \in \text{int } K$, $a_i \in \text{bd } K$ we have $]a_i, u[\subset \text{int } K$ and therefore $v \in \text{int } K$. By $v \in [b_i, x]$, this implies the inclusion $]b_i, x[\subset \text{int } K$, i.e., the point x is inner illuminated by $b_i \in \text{bd } K$. Consequently the whole boundary of K is inner illuminated by the sources $b_1, \cdots, b_m \in \text{bd } K$, leading to $p(K) \le p'(K)$.

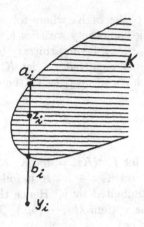

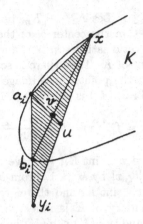

Fig. 234   Fig. 235

Now we will verify the inequality $p'(K) \leq q(K)$. Let $K_1, \cdots, K_m$ be a family of bodies covering K with $K_i = h_i(K)$, where h_i is a homothety with center $c_i \in \mathbf{R}^n \setminus K$ and ratio $k_i > 1$, $i = 1, \cdots, m$. Let $\triangle_i$ denote the intersection $K_i \cap \operatorname{bd} K$; then $\triangle_i \subset K_i$ and $\bigcup_i \triangle_i = \operatorname{bd} K$, $i = 1, \cdots, m$.

For w an arbitrary interior point of K, the relation $c_i \notin K$ implies that bd K is intersected by $[c_i, w]$ at a point, say d_i (Fig. 236). Let y_i be an aribtrary point from $]c_i, d_i[$. Then the points $h_i(y_i)$ and $h_i(w)$ lie on the ray $[c_i, d_i)$, where $h_i(y_i) \in]y_i, h_i(w)[$, since $k_i > 1$. Hence,

$$\lambda_i = \frac{\| y_i - h_i(w) \|}{\| h_i(y_i) - h_i(w) \|} > 1.$$

Now let g_i denote a homothety with center $h_i(w)$ and ratio λ_i. Then we have $g_i(h_i(y_i)) = y_i$, showing that $f_i = g_i \circ h_i$ is a homothety with center $y_i \in \mathbf{R}^n \setminus K$ and ratio $k_i \lambda_i > 1$.

Since $w \in \operatorname{int} K$, we have $h_i(w) \in \operatorname{int} h_i(K) = \operatorname{int} K_i$; and $K_i \subset \operatorname{int} g_i(K_i)$, since $\lambda_i > 1$. In other words, $K_i \subset \operatorname{int} g_i(h_i(K)) = \operatorname{int} f_i(K)$. In particular, this yields $\triangle_i \subset \operatorname{int} f_i(K)$.

Finally, we show that the whole boundary of K is X-rayed by the sources $y_1, \cdots, y_m \in \mathbf{R}^n \setminus K$. In fact, let $x \in \operatorname{bd} K$. Then $x \in \triangle_i$ for some $i \in \{1, \cdots, m\}$. Since $\triangle_i \subset \operatorname{int} f_i(K)$, the inclusion $x \in \operatorname{int} f_i(K)$ holds and therefore $z = f_i^{-1}(x) \in \operatorname{int} K$. Because y_i is the center of the homothety f_i with ratio greater than 1, we have $z \in]y_i, f_i(z)[$ and $z \in]y_i, x[$. Hence the open segment $]y_i, x[$ contains an interior point z of K, and the point x is X-rayed by y_i. Consequently the whole boundary of K is X-rayed by the set $\{y_1, \cdots, y_m\} \subset \mathbf{R}^n \setminus K$, showing $p'(K) \leq q(K)$. ∎

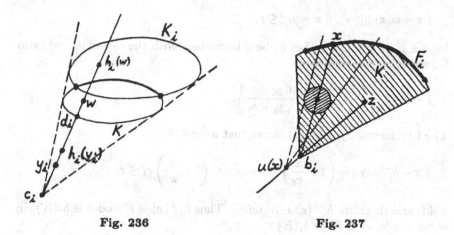

Fig. 236   **Fig. 237**

Theorem 37.3: *If the body $K \subset \mathbf{R}^n$ is compact, then the equalities*

$$p(K) = p'(K) = q(K) = q'(K)$$

hold.   □

PROOF: By the previous theorem, it suffices to show that $q(K) \leq p(K)$. Let $b_1, \cdots, b_m \in$ bd K be sources illuminating the whole boundary of K from inside. Denote by $\triangle_1, \cdots, \triangle_m$ the regions correspondingly illuminated by $b_1, \cdots, b_m$ (i.e., $x \in \triangle_i$ if $x \in$ bd K is inner illuminated by b_i, $i = 1, \cdots, m$). Then $\triangle_1 \cup \cdots \cup \triangle_m =$ bd K, where $\triangle_i$ is an *open* set in bd K for each $i \in \{1, \cdots, m\}$. Hence there exist closed sets $F_1, \cdots, F_m$ with $F_1 \cup \cdots \cup F_m =$ bd K and $F_i \subset \triangle_i$, $i = 1, \cdots, m$ (cf. the proof of Theorem 34.3).

We fix a point $z \in$ int K and consider the ray $[z, b_i)$. For every point $x \in F_i$ one can find a neighbourhood $V(x)$ of x in F_i and a point $u(x) \in [z, b_i)$ such that $b_i \in]z, u(x)[$ and, for $x' \in V(x)$ and $u' \in]b_i, u(x)[$, the segment $[x', u]$ intersects the interior of K (Fig. 237). Since F_i is compact, one can find a finite number of neighbourhoods like $V(x)$ which cover the set F_i. Assume that this finite family is given by neighbourhoods of the points $x_1, \cdots, x_k \in F_i$. Among the set of all open segments $]b_i, u(x_i)[$, $i = 1, \cdots, k$, we now consider the *smallest* one and choose a point y_i from it. Then $y_i \notin K$, and therefore the segment $[x', y_i]$ (with $x' \in V(x_j), ; j = 1, \cdots, k$) intersects the interior of K. In other words, since $V(x_1) \cup \cdots \cup V(x_k) \supset F_i$, we have $[x, y_i] \cap$ int $K \neq \emptyset$ for any point $x \in F_i$.

Now we will construct such a point y_i for every $i = 1, \cdots, m$. For $x \in F_i$ and $y_i \notin K$, the intersection int $K \cap [x, y_i]$ is nonempty. Therefore the open segment $]x, y_i[$ intersects bd K at a point which we denote by $w(x)$. The functions $\| x - w(x) \|$ and $\| x - y_i \|$ on the variable vector $x \in F_i$ are continuous and positive, and hence there are positive constants ε_i and r_i such that

$$\| x - w(x) \| \geq \varepsilon_i, \| x - y_i \| \leq r_i$$

for $x \in F_i$, $i = 1, \cdots, m$. Let h_i be a homothety with the center y_i and ratio k_i, satisfying the conditions

$$1 < k_i \leq \frac{r_i}{r_i - \varepsilon_i}; \quad k_i \leq \frac{\| y_i - z \|}{\| y_i - b_i \|}. \tag{2}$$

The first inequality from (2) shows that for $x \in F_i$

$$\| x - h_i^{-1}(x) \| = \left(1 - \frac{1}{k_i}\right) \| x - y_i \| \leq \left(1 - \frac{1}{k_i}\right) r_i \leq \varepsilon_i$$

holds and therefore $h_i^{-1}(x) \in [x, w(x)]$. Thus $h_i^{-1}(x) \in K$ and $x \in h_i(K)$. In other words, $F_i \subset K_i = h_i(K)$.

The second inequality from (2) shows that

$$\| z - h_i^{-1}(z) \| = \left(1 - \frac{1}{k_i}\right) \| z - y_i \| \leq \| y_i - z \| - \| y_i - b_i \| = \| z - b_i \|$$

and hence $h_i^{-1}(z) \in [z, b_i] \subset K$, leading to $z \in h_i(K) = K_i$.

Since K_i is convex, we have $K_i \supset \mathrm{conv}\,(F_i \cup \{z\})$ for each $i \in \{1, \cdots, m\}$, i.e., $K_1 \cup \cdots \cup K_m \supset \mathrm{conv}\,(F_1 \cup \{z\}) \cup \cdots \cup \mathrm{conv}\,(F_m \cup \{z\}) = K$. Thus the body K is covered by its larger homothetical copies $K_1, \cdots, K_m$, and $q(K) \leq m$ is shown. ∎

The following example demonstrates that, for unbounded bodies K, the values of the functionals $p(K)$ and $p'(K)$ as well as $p(K)$ and $q(K)$ can be different. The question whether the functionals $p'(K)$ and $q(K)$ coincide is open.

EXAMPLE 37.4: Again we consider the body K from Example 34.2. Since $b_1 = (0, 0, 1) \in \mathrm{bd}\,K$ lies in the interior of the cone Q, all points $x \in \mathrm{bd}\,K$ are inner illuminated by b_1 if they do not lie in the plane $\Gamma = \mathrm{bd}\,P$. By $b_2 = (1, 0, 1) \in \mathrm{bd}\,K$, the remaining boundary points of K are inner illuminated (i.e., the whole relative boundary of $\Gamma \cap Q$ is inner illuminated by b_2). Since the equality $p(K) = 1$ cannot hold for a body K, this yields $p(K) = 2$.

We now turn to the functional $p'(K)$. It is clear that $(0, 0, 0) \in \mathrm{bd}\,K$, i.e., the apex of the cone Q can only be X-rayed by some $y_1 \in \mathbf{R}^3 \setminus K$ if $y_1 \in \mathrm{int}\,Q$ holds. Besides $(0, 0, 0)$, all the points $x \in \mathrm{bd}\,K$ with $x \notin \Gamma$ are X-rayed by y_1, but no point from $\Gamma \cap Q$. Therefore it remains to look for further exterior sources such that $\Gamma \cap Q$ is X-rayed. It is obvious that one source is not enough. But two sources are sufficient; the choice of the points $(-\frac{1}{2}, 0, 0)$ and $(100, 0, 98)$ is showing this. Therefore $p'(K) = 3$, i.e., the values of the functionals $p(K)$ and $p'(K)$ differ. By $q(K) \geq p'(K)$, i.e., $q(K) \geq 3$, the same is true for the functionals $p(K)$ and $q(K)$. □

Exercises:

1. Prove that for the body $K \subset \mathbf{R}^2$ considered in Example 34.4 the equalities $p(K) = p'(K) = q(K) = 2$ hold.

2. Prove that for the body $K \subset \mathbf{R}^3$ considered in Example 34.5 the equalities $p(K) = p'(K) = q(K) = 2$ hold.

3. Prove that for the body $K \subset \mathbf{R}^2$ considered in Example 36.1 the equalities $p(K) = p'(K) = q(K) = 2$ hold.

4. What are the values of $p(K), p'(K), q(K)$ for the body $K \subset \mathbf{R}^3$ considered in Example 36.2?

5. What are the values of $p(K), p'(K), q(K)$ for the body $K \subset \mathbf{R}^3$ considered in Example 36.11?

6. What are the values of $p(K), p'(K), q(K)$ for the body $K \subset \mathbf{R}^3$ considered in Example 36.12?

7. Prove that for the body $K \subset \mathbf{R}^3$ considered in Example 34.2 the equality $q(K) = 3$ holds.

8. Let $\mathbf{R}^n = L_1 \oplus L_2$ and, furthermore, M_1 be a convex body in the subspace L_1 with the origin in its relative interior; by M_2 denote the unit ball of the subspace L_2. Prove that $p(M_2 \vee M_2) = p(M_1)$.

9. In the notation of the previous exercise, find the number $p(M_1 \oplus M_2)$.

10. Let $\mathbf{R}^n = L_1 \oplus L_2 \oplus L_3$, where $L_3 = \mathrm{lin}\,(e)$ is one-dimensional. Denote by L_2' the flat $L_2 + e$ and by M_1, M_2' the unit balls in the flats L_1 and L_2', respectively. Prove that $p(\mathrm{conv}\,(M_1 \cup M_2')) = 4$. Consider the particular case when dim $L_1 = \dim L_2 = 1$.

11. In Example 37.4 a body K with $p(K) = 2, p'(K) = 3$ is described. Can you construct an example for which the difference $p'(K) - p(K)$ is equal to a given positive integer r?

§38 Estimates for the value of the functional $p(K)$

Theorem 38.1: Let $K \subset \mathbf{R}^n$ be a closed, convex body which is not a cone. If $\mathbb{K}$ denotes the family of all maximal faces of K, then $p(K) \leq \mathrm{him}\,\mathbb{K} + 1$.   □

PROOF: We set him $K = h$. Let $F_1, \cdots, F_{h+1}$ be a family of maximal faces of K such that each h of them have nonempty intersection, but $F_1 \cap \cdots \cap F_{h+1} = \emptyset$. For each $i \in \{1, \cdots, h+1\}$, choose a point $b_i \in \mathrm{ri}\,F_i$. We will prove that the whole boundary of K is inner illuminated by the sources $b_1, \cdots, b_{h+1}$. Let $x \in \mathrm{bd}\,K$. Since $F_1 \cap \cdots \cap F_{h+1} = \emptyset$, there exists an index $i \in \{1, \cdots, h+1\}$ such that $x \notin F_i$. Assume that int $K \cap [b_i, x] = \emptyset$, i.e., $[b_i, x] \subset \mathrm{bd}\,K$. Theorem 2.1 implies that for any point $y \in]b_i, x[$ the face F_y contains the face of b_i, i.e., $F_y \supset F_i$. By the maximality of F_i, even $F_y = F_i$ holds, leading to $x \in F_i$, a contradiction. Consequently $[b_i, x] \cap \text{int } K \neq \emptyset$, i.e., x is inner illuminated by b_i. We see that bd K is inner illuminated by the system $b_1, \cdots, b_{h+1}$, and this shows that $p(K) \leq \text{him } \mathbb{K} + 1$.   ∎

EXAMPLE 38.2: Let $T \subset \mathbf{R}^n$ denote an arbitrary n-simplex and $S_1, \cdots, S_{n+1}$ be its facets. Furthermore, we write K for the $(n+1)$-dimensional direct sum of T and a segment $I = [a_1, a_2]$. Then the sets $S_1 \oplus I, \cdots, S_{n+1} \oplus I$ are maximal faces of the polytope $K = S \oplus I$, and each n of them have nonempty intersection, whereas $S_1 \oplus I \cap \cdots \cap S_{n+1} \oplus I = \emptyset$. Consequently the family $\mathbb{K}$ of all maximal faces of $K = S \oplus I$ has to satisfy him $\mathbb{K} \geq n$. (It is easily shown that him $K = n$.) Thus we have $p(K) = 2$. This is satisfied by the choice of two light sources $b_1 \in \text{ri } (T \oplus a_1)$, $b_2 \in \text{ri } (T \oplus a_2)$. The considered example shows that, in general, Theorem 38.1 gives an estimate and not an equality.   □

Theorem 38.1 and the previous example yield the motivation for the introduction of a combinatorial invariant. Let $\mathbb{F}$ be a family of sets. By bim $\mathbb{F}$ we denote the smallest of the positive integers $k \geq 1$ such that $\mathbb{F}$ has a subfamily $\{F_1, \cdots, F_{k+1}\}$ with $F_1 \cap \cdots \cap F_{k+1} = \emptyset$ (hence each k of the sets $F_1, \cdots, F_{k+1}$ have nonempty intersection). Therefore bim $\mathbb{F} \leq$ him $\mathbb{F}$. In particular, we can consider the number bim $\mathbb{K}$ for $\mathbb{K}$ denoting the family of all maximal faces of a convex body $K \subset \mathbf{R}^n$; then bim $\mathbb{K}$ can be denoted by bim K. The proof of Theorem 38.1 gives the inequality

$$p(K) \leq \text{bim } K + 1, \tag{1}$$

which is even stronger than that given in the theorem. (Actually, in the proof an *arbitrary* family $\{F_1, \cdots, F_{h+1}\}$ with $F_1 \cap \cdots \cap F_{h+1} = \emptyset$ is used, showing (1).) Moreover, we have *equality* in (1), i.e., the following theorem holds.

Theorem 38.3: Let $K \subset R^n$ be a closed, convex body which is not a cone. Then

$$p(K) = \text{bim } K + 1. \quad □$$

PROOF: By (1) it is enough to verify the inequality $p(K) \geq \text{bim } K + 1$. We set $p(K) = p$ and denote by $c_1, \cdots, c_p \in \text{bd } K$ a family of light sources by which the whole boundary of K is inner illuminated. Furthermore, let G_i be the face of c_i and H_i be a maximal face containing G_i, $i = 1, \cdots, p$. Then $c_i \in H_i$, and any point $x \in H_i$ cannot be inner illuminated by c_i. Hence $H_1 \cap \cdots \cap H_p = \emptyset$ (since the whole boundary of K is inner illuminated by the system $c_1, \cdots, c_p$), and therefore bim $K \leq p - 1$.   ∎

Theorem 38.4: For an arbitrary convex body $K \subset \mathbf{R}^n$, which is not a cone, the inequality

$$p(K) \leq \min_F \dim \text{aff } F + 2$$

holds, where the minimum is taken over all maximal faces F of K.

PROOF: Let F_o be a maximal face of K such that dim aff $F_o = \min\limits_F$ dim aff F. Let, furthermore, $\mathbb{M}$ denote the family of all sets $F_o \cap F$, where again F runs over the set of all maximal faces of K. Thus the set family $\mathbb{M}$ is contained in aff F_o, and therefore

him $\mathbb{M} \leq$ dim aff F_o.

Moreover, the intersection of all sets from $\mathbb{M}$ is empty (since K is not a cone), and therefore him $\mathbb{M} > 0$. We set $h = $ him $\mathbb{M}$. Thus there exist sets $F_o \cap F_1, \cdots, F_o \cap F_{h+1}$ in $\mathbb{M}$ such that the intersection of each h of them is nonempty, but $F_o \cap F_1 \cap \cdots \cap F_{h+1} = \emptyset$. This means (since the faces $F_1, \cdots, F_{h+1}$ are maximal) that there exist $h + 2$ maximal faces whose intersection is empty. According to the proof of Theorem 38.1, this implies $p(K) \leq h + 2$ and hence

$$p(K) \leq \text{him } \mathbb{M} + 2 \leq \text{dim aff } F_o + 2 = \min_F \text{dim aff } F + 2. \quad \blacksquare$$

Corollary 38.5: If the body $K \subset \mathbf{R}^n$, which again is not a cone, has at least one regular, exposed point $x_o \in$ bd K, then $p(K) = 2$. $\square$

In fact, since x_o is its own face and this face is maximal, the previous theorem shows $p(K) \leq 2$. But the strict inequality $p(K) < 2$ is impossible.

Corollary 38.6: If all boundary points of the body $K \subset \mathbf{R}^n$, which is not a cone, are regular, then $p(K) = 2$. $\square$

Theorem 38.7: For any convex body $K \subset \mathbf{R}^n$, which is not a cone, the inequality $p(K) \leq n + 1$ holds. Equality occurs only if K is an n-dimensional simplex. $\square$

PROOF: Since min dim aff $F \leq n - 1$, the inequality $p(K) \leq n + 1$ is an immediate consequence of the previous theorem. Let now $p(K)$ be equal to $n + 1$. Then, by Theorem 38.3, the equality bim $K = n$ holds, i.e., any n maximal faces of K have a nonempty intersection, but there exist maximal faces $G_1, \cdots, G_{n+1}$ whose intersection is empty. For each $i \in \{1, \cdots, n+1\}$, we choose a point p_i such that $p_i \in G_j$ if $i \neq j$. In other words, p_i belongs to the intersection of all sets $G_1, \cdots, G_{n+1}$, except for G_i. If all the points $p_1, \cdots, p_{n+1}$ would lie in an $(n-1)$-dimensional flat L, then (by the classical Helly theorem) the convex sets $L \cap G_1, \cdots, L \cap G_{n+1}$ would have a nonempty intersection, contradicting the relation $G_1 \cap \cdots \cap G_{n+1} = \emptyset$. Therefore the points $p_1, \cdots, p_{n+1}$ are the vertices of an n-simplex T. The facet of T, which is spanned by all points $p_1, \cdots, p_{n+1}$ except for p_i, is contained in G_i and therefore in bd K. By bd $T \subset$ bd K, the bodies T and K coincide. $\blacksquare$

Theorem 38.8: Let the space $\mathbf{R}^n$ be a direct sum of its subspaces $\mathbf{R}^k$ and $\mathbf{R}^{n-k}$. Let, furthermore, $M \subset \mathbf{R}^k$, $N \subset \mathbf{R}^{n-k}$ be convex bodies in these subspaces which are not cones and satisfy $o \in$ ri M, $o \in$ ri N. Then for the n-dimensional convex set conv $(M \cup N)$ the relation

$$p(\text{conv } (M \cup N)) = \max(p(M), p(N))$$

holds. □

PROOF: Assume bim $M = r$, bim $N = s$, bim conv $(M \cup N)) = m$ and, for example, $r \geq s$. Let $F_1, \cdots, F_{r+1}$ be maximal faces of M with $F_1 \cap \cdots \cap F_{r+1} = \emptyset$ and $G_1, \cdots, G_{s+1}$ be maximal faces of N with $G_1 \cap \cdots \cap G_{s+1} = \emptyset$. For $r > s$, we set $G_{s+1} = \cdots = G_{r+1}$. Furthermore, we set

$$H_1 = \text{conv } (F_1 \cup G_1), \cdots,$$
$$H_r = \text{conv } (F_r \cup G_r), \; H_{r+1} = \text{conv } (F_{r+1} \cup G_{r+1}).$$

Then the sets $H_1, \cdots, H_{r+1}$ are maximal faces of conv $(M \cup N)$, satisfying $H_1 \cap \cdots \cap H_{r+1} = \emptyset$. This implies that $m \leq r$.

Conversely, let $K_1, \cdots, K_{m+1}$ be maximal faces of conv $(M \cup N)$, whose intersection is empty. Each of these faces is representable by $K_i = \text{conv } (F_i' \cup G_i')$, where F_i', G_i' are maximal faces of M and N, respectively. Hence $F_1' \cap \cdots \cap F_{m+1}' = \emptyset$, i.e., $r \leq m$.

This shows that $r = m$, i.e., $m = \max(r, s)$. By Theorem 38.3 we have

$$\begin{aligned} p(\text{conv } (M \cup N)) &= m + 1 = \max(r, s) + 1 = \max(r + 1, s + 1) \\ &= \max(p(M), p(N)). \; \blacksquare \end{aligned}$$

Corollary 38.9: Let $K \subset \mathbf{R}^{n-1}$ be an arbitrary convex body which is not a cone and $M \subset \mathbf{R}^n$ be a suspension over K, i.e., $M = \text{conv } (K \cup I)$, where I is a segment with ri $I \cap$ ri $K \neq \emptyset$, dim $M = n$. Then $p(K) = p(M)$. □

Indeed, it suffices to apply Theorem 30.9 for $k = n - 1$.

Theorem 38.10: Let M, N be arbitrary convex sets in $\mathbf{R}^n$ such that the flats aff M and aff N are in general position, with dim aff $M = k > 0$, dim aff $N = l > 0$, and $k + l = n - 1$. Then for the n-dimensional convex body conv $(M \cup N)$ the equality

$$p(\text{conv } (M \cup N)) = p(M) + q(N)$$

holds. If M is a point, then (by agreement) this equality is also true for $p(M)) = 1$. □

PROOF: First we assume that dim $M > 0$ and dim $N > 0$. Then the maximal faces of the body conv $(M \cup N)$ are representable by conv $(M \cup G)$ and conv $(F \cup N)$, where F and G are maximal faces of M and N, respectively. We set bim $M = r$, bim $N = s$, and bim $(\text{conv } (M \cup N)) = m$. Furthermore, let $F_1, \cdots, F_{r+1}$ with $F_1 \cap \cdots \cap F_{r+1} = \emptyset$ be maximal faces of M and $G_1, \cdots, G_{s+1}$ with $G_1 \cap \cdots \cap G_{s+1} = \emptyset$ be maximal faces of N. Then the sets

$$\text{conv } (F_1 \cup N), \cdots, \text{conv } (F_{r+1} \cup N),$$
$$\text{conv } (M \cup G_1), \cdots, \text{conv } (M \cup G_{s+1})$$

are maximal faces of conv $(M \cup N)$ and their intersection is empty. Since their total number is given by $r+s+2$, we get $m = $ bim (conv $(M \cup N)) \leq r+s+1$.

Conversely, let $H_1, \cdots, H_{m+1}$ be maximal faces of conv $(M \cup N)$, satisfying $H_1 \cap \cdots \cap H_{m+1} = \emptyset$. Then (changing the numeration of the faces, if necessary) we may write

$$H_1 = \text{conv } (F_1' \cup N), \cdots; \ H_q = \text{conv } (F_q' \cup N);$$

$$H_{q+1} = \text{conv } (M \cup G_1'), \cdots; \ H_{m+1} = \text{conv } (M \cup G_{m+1-q}'),$$

where F_i', G_j' are maximal faces of M and N, respectively. Then $F_1' \cap \cdots \cap F_q' = \emptyset$ and $G_1' \cap \cdots \cap G_{m+1-q}' = \emptyset$. Therefore $r \leq q-1$, $s \leq m-q$, i.e., $r+s \leq m-1$. Thus $m = r + s + 1$. By Theorem 38.3 we obtain

$$
\begin{aligned}
p(\text{conv } (M \cup N)) &= m + 1 = r + s + 2 = (r+1) + (s+1) \\
&= p(M) + p(N).
\end{aligned}
$$

For the case dim $M = 0$, analogous arguments are sufficient, since maximal faces of conv $(M \cup N)$ are given by N and by sets of the form conv $(M \cup G)$, where G is a maximal face of the set N. ∎

Theorem 38.11: Let M, N be convex sets and $M \oplus N$ denote their direct vector sum. Then

$$p(M \oplus N) = \min(p(M), p(N)). \ \square$$

PROOF: Let bim $M = r$, bim $N = s$, and bim $(M \oplus N) = m$, where (for example) $r \leq s$. We choose maximal faces $F_1, \cdots, F_{r+1}$ of M, whose intersection is empty. Then the sets $F_1 \oplus N, \cdots, F_{r+1} \oplus N$ are maximal faces of $M \oplus N$, having also empty intersection. This implies that $m = $ bim $(M \oplus N) \leq r = \min(r,s)$.

Conversely, let $H_1, \cdots, H_{m+1}$ be maximal faces of $M \oplus N$. Each of them is representable either by $F \oplus N$ or by $M \oplus G$, where F, G are maximal faces of M and N, respectively. We may suppose that

$$H_1 = F_1' \oplus N, \cdots; \ H_q = F_q' \oplus N;$$

$$H_{q+1} = M \oplus G_1', \cdots; \ H_{m+1} = M \oplus G_{m+1-q}',$$

where F_i', G_j' are maximal faces of M and N, respectively. By $H_1 \cap \cdots \cap H_{m+1} = \emptyset$, at least one of the intersections $F_1' \cap \cdots \cap F_q'$, $G_1' \cap \cdots \cap G_{m+1-q}'$ is empty. Consequently at least one of the inequalities $r \leq q-1$, $s \leq m-q$ holds. But we have $0 \leq q \leq m+1$ and therefore $q-1 \leq m$, $m-q \leq m$. Hence at least one of the inequalities $r \leq m$, $s \leq m$ holds, i.e., min $(r,s) \leq m$.

Summarizing the statements, we obtain $m = \min(r,s)$, yielding by Theorem 38.3 the desired equality. ∎

314 VI. Homothetic covering and illumination

Corollary 38.12: _For an arbitrary $k \in \{2, \cdots, n+1\}$, there exists a convex body K in $\mathbf{R}^n$ satisfying_

$$p(K) = k. \quad \square$$

It suffices to consider $(n-k+1)$-fold suspensions over a $(k-1)$-dimensional simplex T. By Corollary 38.9, $p(K) = p(T) = k$. ∎

Finally, we formulate some theorems on the inner illumination of d-convex and H-convex bodies.

Theorem 38.13: _Let $K \subset \Re^n$ be an arbitrary closed, d-convex body which is not a cone. Then the inequality_

$$p(K) \leq \operatorname{md} \Re^n + 1$$

holds. $\square$

PROOF: By Theorem 24.1, the body K is $H(\Re^n)$-convex and, by virtue of Theorem 21.10 and the closedness of $H(\mathbf{R}^n)$, the maximal faces of K are $H(\Re^n)$-convex sets, too (cf. §24). Consequently, denoting by $\mathbb{K}$ the family of all maximal faces of K, we have

$$\operatorname{him} \mathbb{K} \leq \operatorname{him} \mathbf{V}_H = \operatorname{md} H(\Re^n),$$

where $\mathbf{V}_H$ is written for the family of all $H(\Re^n)$-convex sets in $\Re^n$. Moreover, since K is not a cone (and the intersection of all maximal faces is empty), $\operatorname{him} \mathbb{K} > 0$ holds. We get $\operatorname{bim} \mathbb{K} \leq \operatorname{him} \mathbb{K} \leq \operatorname{md} H(\Re^n)$, and therefore $p(K) \leq \operatorname{md} H(\Re^n) + 1$ (Theorem 38.3). ∎

Analogously, we can prove the following

Theorem 38.14: _Let $H \subset S^{n-1}$ be a closed set symmetric with respect to the origin. For any H-convex body $K \subset \mathbf{R}^n$, which is not a cone, the inequality_

$$p(K) \leq \operatorname{md} H + 1$$

holds. $\square$

Theorem 38.15: _Let $K \subset \Re^n$ be any closed, d-convex body, which is not a cone. Then the inequality_

$$p(K) \leq \min_F \operatorname{md} (\operatorname{aff} F) + 2$$

holds, where the minimum is taken over all maximal faces of K. $\square$

PROOF: Let F_o be a maximal face of K such that $\operatorname{md} (\operatorname{aff} F_o) = \min_F \operatorname{md} (\operatorname{aff} F)$. Denote by $\mathbb{M}$ the family of all sets $F_o \cap F$, where F runs over the family of all maximal faces of K. The sets from $\mathbb{M}$ are closed and $H(\Re^n)$-convex and therefore $H(\operatorname{aff} F_o)$-convex. Consequently we have $\operatorname{him} \mathbb{M} \leq \operatorname{md} (\operatorname{aff} F_o)$. The further steps of the proof are analogous to those of the proof of Theorem 38.4. ∎

Finally, we will prove a theorem (see [Bo-So]) on the illumination of two-sided cones. Let $C \subset \mathbf{R}^n$ be a convex cone with apex o. Then the union $C \cup (-C)$ is said to be a *two-sided cone* in $\mathbf{R}^n$. If C is a pointed cone, then $C \cup (-C)$ is said to be a *pointed two-sided cone*.

Theorem 38.16: Let $C \cup (-C) \subset \mathbf{R}^n$ be a pointed two-sided cone with apex o. Then $c'(C \cup (-C) \setminus \{o\}) \leq 2n$ and $p'(C \cup (-C) \setminus \{o\}) \leq 2n$. More detailed, there exist $2n$ points $s_1, \cdots, s_{2n} \in \mathbf{R}^n \setminus (C \cup (-C))$ which illuminate the body $C \cup (-C) \setminus \{o\}$ and, in addition, the body $C \cup (-C) \setminus \{o\}$ is X-rayed by the same points $s_1, \cdots, s_{2n}$. □

PROOF: Let Γ be a hyperplane (not through the origin) whose intersection with C is a compact, convex body M with respect to Γ. According to Theorems 37.3 and 38.7, for the body $M \subset \Gamma$ we have $p'(M) = p(M) \leq n$. This means that there exist points $s_1, \cdots, s_n \in \Gamma \setminus M$ from which the boundary of M (with respect to Γ) is X-rayed. Consequently the whole set $C \setminus \{o\}$ is X-rayed by these points $s_1, \cdots, s_n$ (cf. Exercise 7). We put $s_{n+i} = -s_i$, $i = 1, \cdots, n$. Then the set bd $(-C) \setminus \{o\}$ is X-rayed by the points $s_1, \cdots, s_{2n}$, i.e., the points $s_1, \cdots, s_{2n}$ form a system by which the body $C \cup (-C) \setminus \{o\}$ is X-rayed. At the same time, these $2n$ points illuminate the whole body $C \cup (-C) \setminus \{o\}$ (see Exercise 8). ∎

Corollary 38.17: In the notation of Theorem 38.16, the equality $p(C \cup (-C) \setminus \{o\}) = p'(C \cup (-C) \setminus \{o\}) = 2n$ holds if and only if C is an n-faced polyhedral cone, i.e., it is spanned by a simplex M not containing the origin. □

In fact, the equality $p'(M) = p(M) = n$ holds if and only if M is a simplex, cf. Theorem 38.7. ∎

Exercises:

1. Let $M \subset \mathbf{R}^n$ be a convex body distinct from $\mathbf{R}^n$ and $\mathfrak{M}$ be the family of all convex sets contained in bd M. Prove that

$$\max_{F} \dim \text{aff } F \leq \text{him } \mathfrak{M} \leq \max_{F} \dim \text{aff } F + 1,$$

where the maximum is taken over all proper faces F of M.

2. Prove that the upper estimate in the previous exercise is attained if and only if there exists a simplex T such that $\dim T = \max_{F} \dim \text{aff } F + 1$ and rbd $T \subset \text{bd } M$.

3. Let $M \subset \mathbf{R}^n$ be a convex body distinct from $\mathbf{R}^n$ and from a simplex. In the notation of Exercise 1, prove that him $\mathfrak{M} < n$.

4. Let $M \subset \mathbf{R}^n$ be a convex polytope distinct from a simplex. In the notation of Exercise 1, prove that him $\mathfrak{M} = n - 1$.

5. In the Euclidean 3-space with the orthonormal basis e_1, e_2, e_3 we denote by f_α the homothety with the center o and ratio $\cos\alpha$, and by g_α the translation with the translation vector $(\sin\alpha)\,e_3$. Let $N \subset \operatorname{lin}(e_1, e_2)$ and $Q(N) = \bigcup g_\alpha(f_\alpha(N))$, where the union is taken over all $\alpha \in [-\frac{\pi}{2}, \frac{\pi}{2}]$. Prove that if N is an equilateral triangle with center o, then the convex body $M = Q(N)$ (Fig. 238) satisfies the conditions $\max_{F} \dim \operatorname{aff} F = 1$ and him $\mathfrak{M} = 2$ (cf. the notation in Exercise 1). If even $N \subset \operatorname{lin}(e_1, e_2)$ is a polygon distinct from a triangle and satisfying $o \in \operatorname{ri} N$, then him $\mathfrak{M} = 1$.

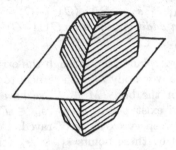

Fig. 238

6. Prove that in Example 38.2 the equality him $\mathbb{K} = n$ holds.

7. Prove that if a boundary point x of a convex, n-dimensional cone $C \subset \mathbb{R}^n$ (distinct from the apex o of the cone) is illuminated by a point $s \in \mathbb{R}^n \setminus C$, then all the points of the ray $[o, x) \subset C$ are illuminated by the point s.

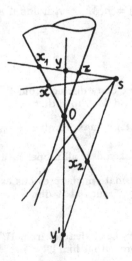

Fig. 239

8. Let $C \subset \mathbf{R}^n$ be a convex, n-dimensional cone with apex o. Prove that a point $x \in \operatorname{bd} C \setminus \{o\}$ is illuminated by a point $s \in \mathbf{R}^n \setminus (C \cup (-C))$ if and only if the point $-x \in (-C)$ is X-rayed by the same point s (Fig. 239).

9. Let $C \subset \mathbf{R}^n$ be a convex, n-dimensional cone with r-dimensional minimal face F. Prove that $p(C \cup (-C) \setminus F) = p'(C \cup (-C) \setminus F) \le 2(n - r)$.

10. For $n \ge 2$, let $C \subset \mathbf{R}^n$ be a convex, n-dimensional cone with apex o. Prove that $p(C \cup (-C) \setminus \{o\}) = p'(C \cup (-C) \setminus \{o\})$.

11. Let $C \subset \mathbf{R}^n$ be a pointed, convex, n-dimensional cone with the apex o and Γ be a hyperplane not containing o and intersecting C in the compact set $M = C \cap \Gamma$. Prove that if $\operatorname{rbd} M$ contains at least two points which are simultaneously regular and extreme (with respect to the body $M \subset \Gamma$), then $p'(C \cup (-C) \setminus \{o\}) = 3$.

12. Prove that if C is a convex, n-dimensional cone with the apex o and distinct from $\mathbf{R}^n$, then $q(C \cup (-C) \setminus \{o\}) = \infty$.

VII. Combinatorial geometry of belt bodies

In this chapter we consider a new class of convex bodies which was introduced in [Bt 14] and [B-B 2]. This is the class of *belt bodies* which is much larger than the class of zonoids. (For zonoids and their fascinating properties, the reader is referred to the surveys [S-W], [G-W], [Bk 1], and [Mar 4].) Moreover, the class of belt bodies is dense in the family of all compact, convex bodies. Nevertheless, solutions of combinatorial problems for zonoids [Ba 1, Ba 2, Mar 2, B-SP 5, B-SP 6] can be extended to belt bodies. The aim of this chapter is the explanation of combinatorial properties of belt bodies, cf. also [B-M 1].

§39 The integral respresentation of zonoids

At first we consider some properties of *zonoids*. This will lead us to the definition of belt bodies.

A vector sum (not necessarily direct) of a finite number of segments is said to be a *zonotope*. For example, every centrally symmetric polygon in the plane is a two-dimensional zonotope (or, for this dimension, a zonogon). In Fig. 240, a three-dimensional zonotope (rhombic dodecahedron) is shown, which is the vector sum of four segments each three of which have linearly independent direction vectors. A compact, convex set $M \subset \mathbf{R}^n$ is said to be a *zonoid* if it is the limit of a convergent sequence of zonotopes (in the Hausdorff metric).

In [Mar 2], the *homothetic covering problem* is solved for zonotopes, and in [Ba 1] the *Szökefalvi-Nagy problem* is solved for them. In both cases, the solution was later extended to the class of zonoids, cf. [Ba 2, B-SP 5, B-SP 6]. And in both cases, a direct limit passing from zonotopes to zonoids is unrealizable. In this connection, a suitable indirect method is used, applying approximate continuity and *tangential zonotopes* introduced in [Ba 2, B-SP 6]. As a starting point, we use the following theorem of Zalgaller and Reshetnyak [Z-R].

Theorem 39.1: Let $\varphi(s), 0 \leq s \leq l$, be a vector-parametric representation of a rectifiable curve $K \subset \mathbf{R}^n$ such that s is the arc length along K, i.e., the arc

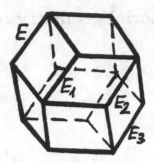

Fig. 240

of K with endpoints $\varphi(0)$ and $\varphi(s)$ has length s. Consider the integral (in the sense of Lebesgue)

$$z(\mu) = \int_0^1 \mu(s)\varphi'(s)ds, \qquad (1)$$

where l is the length of K. If $\mu(s)$ runs through the family of all measurable functions such that $|\mu(s)| \leq 1$, $s \in [0,l]$, then the set of all points $z(\mu)$ is a zonoid in $\mathbf{R}^n$. The curve K is said to be a generating curve of the zonoid. Conversely, each zonoid can be represented in such a form (up to a parallel translation).

We notice that a generating curve K is not uniquely defined by the zonoid. For example, if we break K into pieces and rearrange them, then we obtain a new curve which generates the same zonoid (by virtue of the additive property of integrals).

The idea of this integral representation is rather clear. Indeed, let $M = I_1 + \cdots + I_s$ be a zonotope in $\mathbf{R}^n$, i.e., the vector sum of the segments $I_1, \cdots, I_s$. We may assume (using parallel translations, if necessary) that each segment I_i has its midpoint at the origin. Let, furthermore, $B_s \subset \mathbf{R}^s$ be the cube determined in a cartesian coordinate system $y_1, \cdots, y_s$ by the inequalities $-1 \leq y_i \leq 1, i = 1, \cdots, s$. The cube B_s is the vector sum of s segments of the length 2, each of which is situated in a coordinate axis and has its midpoint at the origin. It is clear that $M = F(B_s)$, where $F : \mathbf{R}^s \rightarrow \mathbf{R}^n$ is a suitable linear operator (since the vector sum of sets passes to the vector sum of their images). The cube B_s is the unit ball of the Minkowski metric, such that the norm of a vector $y = (y_1, \cdots, y_s)$ is equal to $\| y \| = \max(|y_1|, \cdots, |y_s|)$. Denote this Minkowski space by $L_\infty^{(s)}$. Thus every zonotope $M \subset \mathbf{R}^n$ is the image of the unit ball $B_s \subset L_\infty^{(s)}$ for s large enough and a suitable linear operator.

Furthermore, each zonoid $Z \subset \mathbf{R}^n$ is the limit of a convergent sequence of zonotopes. It is natural to expect that Z is the image of the unit ball B_∞ of the infinite-dimensional space $L_\infty[0,1]$ for a suitable continuous linear operator $F : L_\infty[0,1] \to \mathbf{R}^n$, where $L_\infty[0,1]$ is the space of bounded measurable functions $\nu(t)$ on $[0,1]$ with the norm

$$\| \nu(t) \| = \operatorname*{vrai\ max}_{[0,1]} |\nu(t)|.$$

This is, indeed, true (see, for example, [Bk 1]).

Introduce now a coordinate system $x_1, \cdots, x_n$ in $\mathbf{R}^n$. Then for every element $\nu(t) \in B_\infty$ we have $F(\nu(t)) = (f_1(\nu(t)), \cdots, f_n(\nu(t)))$, where $f_i : L_\infty[0,1] \to \mathbf{R}$ is a continuous linear functional, $i = 1, \cdots, n$. But every continuous linear functional on $L_\infty[0,1]$ has the form $f(\nu(t)) = \int_o^1 \nu(t)\psi'(t)dt$, where $\psi(t)$ is an absolutely continuous function. Thus,

$$F(\nu(t)) = \left(\int_0^1 \nu(t)\psi_1'(t)dt, \cdots, \int_0^1 \nu(t)\psi_n'(t)dt \right) = \int_0^1 \nu(t)\psi'(t)dt,$$

where $\psi(t) = (\psi_1(t), \cdots, \psi_n(t))$ is an absolutely continuous vector-function on $[0,1]$ with values in $\mathbf{R}^n$. Consequently, $\psi(t), 0 \le t \le 1$, is a vector parametric representation of a rectifiable curve $K \subset \mathbf{R}^n$. Now, to obtain the integral representation (1), it is sufficient to replace the parameter t along K by the arc length s. We limit ourselves by these remarks and refer the reader to the paper [Z-R].

EXAMPLE 39.2: Consider the case when K is a broken (polygonal) line, i.e., $K = C_1 \cup \cdots \cup C_s$, where $C_1, \cdots, C_s$ are segments such that, when we move along K, the initial point of C_{i+1} coincides with the terminal endpoint of C_i, $i = 1, \cdots, s-1$ (Fig. 241). Then the zonoid indicated in Theorem 39.1 is a zonotope. Indeed, let l_i be the length of the segment C_i and e_i be the unit vector directed along C_i, $i = 1, \cdots, s$. We put $a_o = 0$ and $a_i = l_1 + \cdots + l_i$ for $i = 1, \cdots, s$. Then

$$z(\mu) = \int_0^l \mu(s)\varphi'(s)ds = \sum_{i=1}^s \int_{a_{i-1}}^{a_i} \mu(s)e_i ds = \sum_{i=1}^s \left(\int_{a_{i-1}}^{a_i} \mu(s)ds \right) e_i.$$

Since $| \int_{a_{i-1}}^{a_i} \mu(s)ds | \le l_i$, the point $z(\mu)$ has the form

$$z(\mu) = \lambda_1 e_1 + \cdots + \lambda_s e_s,$$

where $|\lambda_i| \le l_i$ for each $i = 1, \cdots, s$. This means that the zonoid Z coincides with the zonotope $I_1 + \cdots + I_s$, where the segment I_i has the length $2l_i$ and is parallel to e_i, $i = 1, \cdots, s$.   □

EXAMPLE 39.3: Since the parameter s in Theorem 39.1 denotes the arc length along K, the length of the arc of K between the endpoints $\varphi(s_1)$ and $\varphi(s_2)$ is equal to $|s_1 - s_2|$. On the other hand, $\|\varphi(s_1) - \varphi(s_2)\|$ is the length of the chord with the same endpoints (Fig. 242). Consequently,

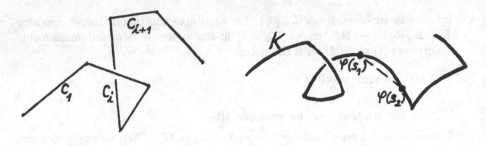

Fig. 241   Fig. 242

$$\|\varphi(s_1) - \varphi(s_2)\| \leq |s_1 - s_2|$$

for every $s_1, s_2 \in [0, l]$. This means that $\varphi(s)$ is a Lipschitz function with the Lipschitz constant 1, and therefore $\|\varphi'(s)\| \leq 1$ if the derivative $\varphi'(s)$ exists. Moreover, since s is the arc length, we have $\|\varphi'(s)\| = 1$ almost everywhere on $[0, 1]$. But the existence of the derivative $\varphi'(s)$ does not imply its continuity. Moreover, in general the derivative $\varphi'(s)$ is *nowhere* continuous. We will give an example.

Divide the segment $[0, 1]$ into 2^k equal parts $I_1^{(k)}, \cdots, I_{2^k}^{(k)}$. We will inductively define perfect, nowhere dense, pairwise disjoint sets $M_i^{(k)} \subset I_i^{(k)}, N_i^{(k)} \subset I_i^{(k)}$ (for $k = 1, 2, \cdots$, and $i = 1, \cdots 2^k$) each of which has a positive measure. It is easy to construct first four sets $M_1^{(1)}, M_2^{(1)}, N_1^{(1)}, N_2^{(1)}$. Admit that for an integer k the sets $M_i^{(j)}, N_i^{(j)}$ with $j < k$ are already defined. The union

$$Q_{k-1} = \bigcup_{j<k} \bigcup_i \left(M_i^{(j)} \cup N_i^{(j)} \right)$$

is a perfect and nowhere dense set. Consequently it is possible to choose for every $i = 1, \cdots, 2^k$ a segment $A_i^{(k)} \subset I_i^{(k)}$ that has no common points with Q_{k-1}. Then for $i = 1, \cdots, 2^k$ we can choose perfect, nowhere dense, disjoint sets $M_i^{(k)} \subset A_i^{(k)}, N_i^{(k)} \subset A_i^{(k)}$ each of which has a positive measure. The sets $M_i^{(k)}, N_i^{(k)}, i = 1, \cdots, 2^k$, are pairwise disjoint, and they have no common points with the sets constructed before.

After this inductive construction we put

$$M = \bigcup_{k=1}^\infty \bigcup_i M_i^{(k)}, N = [0, 1] \setminus M.$$

Then for every segment $I \subset [0, 1]$, both sets $I \cap M, I \cap N$ have positive measures. Indeed, if k, i are indices such that $I_i^{(k)} \subset I$, then $M_i^{(k)} \subset I_i^{(k)} \subset I$, $N_i^{(k)} \subset I_i^{(k)} \subset I$ and hence $I \cap M \supset M_i^{(k)}$, $I \cap N \supset N_i^{(k)}$, i.e., mes $(I \cap M) > 0$, mes $(I \cap N) > 0$.

We now define a vector-function $\psi(s)$ on $[0,1]$ with values in $\mathbf{R}^2$ in the following way:

$$\psi(s) = e_1 = (1,0) \text{ for } s \in M, \quad \psi(s) = e_2 = (0,1) \text{ for } s \in N.$$

The vector-function is bounded and measurable. The curve K with the parametric representation

$$\varphi(s) = \int_0^s \psi(s)ds, \ 0 \le s \le 1,$$

is rectifiable and s is the arc length along K, since $\|\varphi'(s)\| = 1$ holds almost everywhere. The corresponding zonoid (as in Theorem 39.1) is the rectangle with the vertices $(\pm a, \pm b)$, where $a = \text{mes } M$, $b = \text{mes } N$. Indeed, we have $\varphi'(s) = e_1$ almost everywhere on M and $\varphi'(s) = e_2$ almost everywhere on N. At the same time, the derivative $\varphi'(s)$ is nowhere continuous on $[0,1]$, since in every segment $I \subset [0,1]$ there are points at which $\varphi'(s) = e_1$ and at which $\varphi'(s) = e_2$. $\square$

The previous example shows that if we wish to use properties of continuous functions, then we have to replace the usual definition of continuity by another one with weaker requirements. The *approximate continuity* applied in the theory of Denjoy's integral (cf., for example, [Sak]) is just a notion convenient for our aims. We recall the corresponding definitions and facts from Real Analysis.

Let $E \subset \mathbf{R}$ be a measurable set. A point x_o is said to be a *density point* of E if

$$\lim_{\text{mes } \Delta \to 0} \frac{\text{mes } (E \cap \Delta)}{\text{mes } \Delta} = 1,$$

where Δ runs over the family of all segments containing x_o.

The following proposition is well-known (see, for example, the monographs [Sak, Nat]): *Almost all points of any measurable set $E \subset \mathbf{R}$ are density points of E.*

Let now $f(x)$ be a real function defined on a set $D \subset \mathbf{R}$. A number y_o is said to be the *approximate limit* of the function $f(x)$ as $x \to x_o$ if there exists a measurable set $E \subset D$ such that x_o is a density point of E and

$$\lim_{x \to x_o, x \in E} f(x) = y_o.$$

This limit is denoted by $\text{limap}_{x \to x_o} f(x)$.

The approximate limit (if it exists) is defined correctly, i.e., it does not depend on the choice of the set E. More exactly, if $E_1 \subset D, E_2 \subset D$ are measurable sets such that x_o is a density point for each of the sets E_1, E_2 and the limits

$$\lim_{x \to x_o, x \in E_1} f(x), \quad \lim_{x \to x_o, x \in E_2} f(x)$$

exist, then the limits coincide.

We notice that if the usual limit $\lim_{x \to x_o} f(x)$ exists, then also the approximate limit exists, and $\text{limap}_{x \to x_o} f(x) = \lim_{x \to x_o} f(x)$.

A function $f(x)$ is *approximately continuous* at a point x_o if $f(x_o) = \text{limap}_{x \to x_o} f(x)$.

The following proposition is known in Real Analysis as Stepanoff-Denjoy's Theorem (see, for example, the monographs [Sak, Nat]): *If a real function $f(x)$ is measurable (in the sense of Lebesgue) on a set $D \subset \mathbf{R}$, then it is approximately continuous almost everywhere on D.*

All the facts mentioned above are true for vector-functions as well, since each coordinate of a vector-function is a real function. Let, in particular, $Z \subset \mathbf{R}^n$ be a zonoid, K be its generating curve, and $\varphi(s), 0 \le s \le l$, be the vector-parametric representation of K such that s is the arc length along K. Then $\varphi(s)$ is a Lipschitz function (consequently an absolutely continuous function). Its derivative $\varphi'(s)$ exists almost everywhere, being a bounded, measurable function. According to Stepanoff-Denjoy's Theorem, the derivative $\varphi'(s)$ is approximately continuous almost everywhere on $[0, l]$. This is a basis of our further consideration.

Let $Z \subset \mathbf{R}^n$ be a zonoid and $\varphi(s), 0 \le s \le l$, be the vector-parametric representation of its generating curve K (s is the arc length along K). Let, furthermore, $s_1, \cdots, s_p$ be points of the segment $[0, l]$ such that at each of them the derivative $\varphi'(s)$ exists and is approximately continuous. Then the vector sum of p segments parallel to the vectors $\varphi'(s_1), \cdots, \varphi'(s_p)$ is called a *tangential zonotope* of Z (Fig. 243).

We notice that in this definition the attention is paid only to *directions* of segments (not to their lengths).

The following Theorem is proved in [Ba 2]. In its proof given below we follow [B-SP 6].

Theorem 39.4: Each zonoid $Z \subset \mathbf{R}^n$ can be represented as the limit of a convergent sequence of its tangential zonotopes.

PROOF: Let $\varphi(s), 0 \le s \le l$, be the vector-parametric representation of a generating curve K of Z (s is the arc length along K). We choose a positive number ε and construct an ε-net $\{e_1, \cdots, e_p\}$ on the unit sphere $S^{n-1} \subset \mathbf{R}^n$, i.e., for every point $x \in S^{n-1}$ there is an index $i \in \{1, \cdots, p\}$ such that $\|x - e_i\| < \varepsilon$. Let E_i $(i = 1, \cdots, p)$ be the set of the points $s \in [0, l]$, at which $\varphi'(s)$ exists, is approximately continuous, and satisfies the condition $\|\varphi'(s) - e_i\| < \varepsilon$ and $\|\varphi'(s)\| = 1$ (almost everywhere). Almost all points of the segment $[0, l]$ belong to $E_1 \cup \cdots \cup E_p$. We now put

$$F_1 = E_1; F_i = E_i \setminus (E_1 \cup \cdots \cup E_{i-1}) \text{ for } i = 2, \cdots, p.$$

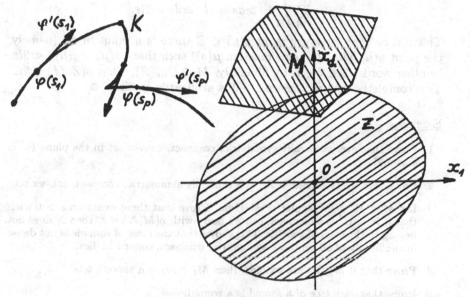

Fig. 243

Then each two of the sets $F_1, \cdots, F_p$ are disjoint, and almost all points of $[0, l]$ belong to $F_1 \cup \cdots \cup F_p$. We will assume that the sets $F_1, \cdots, F_p$ are nonempty (empty sets may be omitted).

For every $i = 1, \cdots, p$ we fix a point $s_i \in F_i$ and put

$$z(\mu) = \int_0^l \mu(s)\varphi'(s)ds = \sum_{i=1}^p \int_{F_i} \mu(s)\varphi'(s)ds,$$

$$y(\mu) = \sum_{i=1}^p \int_{F_i} \mu(s)\varphi'(s_i)ds = \sum_{i=1}^p \left(\int_{F_i} \mu(s)ds \right) \varphi'(s_i),$$

where $|\mu(s)| \leq 1$. Then the set of the points $z(\mu)$ is the given zonoid Z and the set of the points $y(\mu)$ is a tangential zonotope $M = I_1 + \cdots + I_p$, where I_i is a segment which is parallel to $\varphi'(s_i)$ and has the length 2 mes F_i. We have:

$$\|z(\mu) - y(\mu)\| = \|\sum_{i=1}^p \left(\int_{F_i} \mu(s)\varphi'(s)ds - \int_{F_i} \mu(s)\varphi'(s_i)ds \right) \|$$

$$\leq \sum_{i=1}^p \int_{F_i} |\mu(s)| \cdot \|\varphi'(s) - \varphi'(s_i)\|ds$$

$$\leq \sum_{i=1}^p \int_{F_i} |\mu(s)| \cdot (\|\varphi'(s) - e_i\| + \|e_i - \varphi'(s_i)\|)ds$$

$$< \sum_{i=1}^{p} \int_{F_i} 1 \cdot 2\varepsilon ds = \int_0^l 2\varepsilon ds = 2l\varepsilon.$$

This means that for every point $z(\mu) \in Z$ there is a point in M (namely, the point $y(\mu)$ with the same function $\mu(s)$) such that $||z(\mu) - y(\mu)|| < 2l\varepsilon$. In other words, $Z \subset U_{2l\varepsilon}(M)$. Similarly, $M \subset U_{2l\varepsilon}(Z)$. Thus $\varrho(Z, M) < 2l\varepsilon$. This completes the proof, since $\varepsilon > 0$ is arbitrarily small. ∎

Exercises:

1. Prove that every centrally symmetric, compact, convex set in the plane $\mathbf{R}^2$ is a zonoid.

2. Prove that every zonoid in $\mathbf{R}^n$ is a centrally symmetric, compact, convex set.

3. Let $M \subset \mathbf{R}^3$ be a regular octahedron. Prove that there exists an $\varepsilon > 0$ such that if $N \subset \mathbf{R}^3$ is a compact, convex body with $\varrho(M, N) < \varepsilon$, then N does not belong to the class of zonoids. This shows that the class of zonoids is not dense in the family of all centrally symmetric, compact, convex bodies.

4. Prove that if M_1, M_2 are zonoids, then $M_1 + M_2$ is a zonoid, too.

5. Prove that each face of a zonoid is a zonoid.

6. Prove that every affine image of a zonoid (in particular, every of its parallel projections) is a zonoid.

7. Show by an example that the following assertion is false: If a zonoid M is represented in the form $M = M_1 + \cdots + M_s$, then each convex set M_i, $i = 1, \cdots, s$, is a zonoid. Can you find an example for zonotopes in $\mathbf{R}^2$? Is the aforesaid assertion true for a *direct* vector sum $M = M_1 \oplus \cdots \oplus M_s$?

8. Prove that each zonotope is a linear image of a cube (which has sufficiently large dimension).

9. Prove that if each two-dimensional face of a convex polytope $M \subset \mathbf{R}^n$ is a centrally symmetric polygon, then M is a zonotope.

10. Prove that if $M \subset \mathbf{R}^n$ is a zonotope distinct from a parallelotope, then $M = M_1 + M_2$, where M_1, M_2 are zonotopes such that $0 \leq \dim M_2 \leq n$, and M_1 is an n-dimensional zonotope representable as the vector sum of $n + 1$ pairwise nonparallel segments.

11. A *simple arc* is the $1 - 1$-image of the segment $[0, 1] \subset \mathbf{R}$ into $\mathbf{R}^n$. Prove that the length $l(A)$ of the arc A (in Jordan's sense) is a lower semicontinuous function on the family of all rectifiable simple arcs in $\mathbf{R}^n$. Show by an example that the corresponding upper semicontinuity property does not hold.

12. Prove that if segments $I_1, \cdots, I_s$ are not parallel to a hyperplane of the space $\mathbf{R}^n$, then the zonotope $M = I_1 + \cdots + I_s$ is a body in $\mathbf{R}^n$.

13. Prove that if there are points $s_1, \cdots, s_n \in [0, l]$ such that the derivative $\varphi'(s)$ exists and is approximately continuous at each of the points, and $\varphi'(s_1), \cdots, \varphi'(s_n)$ are not parallel to a hyperplane in $\mathbf{R}^n$, then the zonoid Z with the generating curve $\varphi(s)$ (cf. Theorem 39.1) is a body in $\mathbf{R}^n$.

14. Show by an example that the following assertion is false: If $M \subset \mathbf{R}^n$ is a zonotope and L is a hyperplane through its center, then $M \cap L$ is an $(n-1)$-dimensional zonotope in L. For which dimensions n is the statement true?

§40 Belt vectors of a compact, convex body

Together with each of its edges E, every zonotope $M \subset \mathbf{R}^n$ contains a *belt* of edges, each of which is parallel and congruent to E (Fig. 240). More detailed, denote by $\triangle$ a hyperplane orthogonal to the edge E and by $\pi : \mathbf{R}^n \to \triangle$ the orthogonal projection onto $\triangle$. Then, for every vertex b of the polytope $\pi(M)$, the set $\pi^{-1}(b) \cap M$ is an edge of M that is parallel and congruent to E. All the edges $\pi^{-1}(b) \cap M$ form the *belt* of edges in M. Moreover, for every point $b \in \text{rbd}\, \pi(M)$ (not only for a vertex of $\pi(M)$) the intersection $\pi^{-1}(b) \cap M$ is a segment, parallel to E (maybe, not congruent to E), and the family of all the segments $\pi^{-1}(b) \cap M$, $b \in \text{rbd}\, \pi(M)$, is the belt defined by E. In other words, every supporting line of M parallel to E meets the body M in a segment which is parallel to E.

EXAMPLE 40.1: Let $M = B \cap \Pi$, where $B \subset \mathbf{R}^n$ is a closed ball centered at o and Π is a half-space with its boundary hyperplane passing through the origin. Denote by e the unit outward normal of the half-space Π. Then every supporting line L of M parallel to e meets M at a unique point a, and one of the rays defined on L by the point a is a *tangential ray* of M at a. We say that e is a *belt vector* of the body M. Furthermore, if e' is a nonzero vector orthogonal to e, then each supporting line of M parallel to e' meets M either by a segment parallel to e' or at a point a such that each of the rays defined on L by the point a is a tangential ray of M at a. Thus e' is also a belt vector of M. If, however, e'' is a nonzero vector that is neither parallel nor orthogonal to e, then the analogous assertion is false, i.e., e'' is not a belt vector of M. Namely, there is a supporting line L'' of M parallel to e'' such that none of the rays defined on L'' by the common point of L'' and M is a tangential ray of M. In fact, if $\alpha > 0$ and β are real numbers such that $x_o = \alpha e + \beta e'' \in \text{bd}\, B \cap \text{bd}\, \Pi$, then none of the rays, emanating from x_o and having the directions $\pm e''$, is a tangential ray of M (Fig. 244).

We now give the exact definition of belt vectors. Let $M \subset \mathbf{R}^n$ be a compact, convex body and e be a nonzero vector. We say that e is a *belt vector* of M if every supporting line L of M, parallel to e, meets M either by a segment, or at a unique point a such that at least one of the rays, defined on L by a, is a tangential ray of M at a.

To clarify this, we remark that a ray r emanating from a point $a \in \text{bd}\, M$ is a tangential ray of M if and only if $r \cap \text{int}\, M = \emptyset$ and $r \subset \text{sup cone}_a M$. In other words, r is a tangential ray if and only if $r \cap \text{int}\, M = \emptyset$ and, moreover, for every $\varepsilon > 0$ there is a point $x \in M$ distinct from a such that the angle between the ray r and the vector $x - a$ is less than ε.

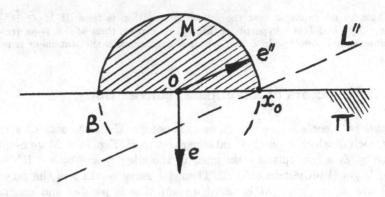

Fig. 244

Consider an interpretation similar to that which was given for a zonotope. Let $e \in \mathbf{R}^n$ be a nonzero vector and $\pi : \mathbf{R}^n \to \triangle$ be the orthogonal projection parallel to e onto a hyperplane $\triangle$. The vector e is a belt vector of a compact, convex body $M \subset \mathbf{R}^n$ if and only if for every $b \in \operatorname{rbd} \pi(M)$ the line $L = \pi^{-1}(b)$ meets M either by a segment or at a unique point a such that at least one of the rays, defined on $\pi^{-1}(b)$ by this point a, is a tangential ray of M at the point a; i.e., $L \cap \sup \operatorname{cone}_a M \neq \{a\}$. We remark that the inclusion $b \in \operatorname{rbd} \pi(M)$ replaces the condition "L is a supporting line of M parallel to e". The set $\pi^{-1}(\operatorname{rbd} \pi(M)) \cap M$ is the *belt* defined by the belt vector e. Finally, we denote by $B(M)$ the set of all belt vectors of the body M.

EXAMPLE 40.2: In the plane $x_2 = 0$ we consider the curve $x_1 = \cos \frac{\pi x_3}{2}$, $-1 \leq x_3 \leq 1$, and denote by $M \subset \mathbf{R}^3$ the convex body which is bounded by the surface obtained by rotation of the curve around the x_3-axis, see Fig. 245. This rotational body has two spinodes $(0, 0, 1)$ and $(0, 0, -1)$, and all other boundary points of M are *regular*. Any unit vector e, forming with one of the vectors $(0, 0, \pm 1)$ an angle not larger than $\operatorname{arctg} \frac{\pi}{2}$, is a belt vector of M. □

Theorem 40.3: For any compact, convex body $M \subset \mathbf{R}^n$ the set of all its unit belt vectors is a closed set.

PROOF: We will prove that the set of all unit vectors, which are not belt vectors of M, is an open subset of the unit sphere $S^{n-1} \subset \mathbf{R}^n$.

Let e_o be a unit vector which is not a belt vector of M, i.e., $e_o \notin B(M)$. Then there exists a supporting line L_o of the body M which is parallel to e_o and has the following property: for a point $x_o \in M \cap L_o$, neither the ray r_o (starting from x_o in the direction e_o) nor the ray r_o' (starting from x_o in the direction $-e_o$) is a tangential ray of M at x_o. We denote by $b_o \in r_o$, $b_o' \in r_o'$ the points at distance 1 from x_o. They do not belong to the supporting cone $K_o = \sup \operatorname{conv}_{x_o} M$, and hence there exist supporting hyperplanes $\triangle_o, \triangle_o'$ of K_o which separate these points from K_o. In other words, if Π_o, Π_o' are

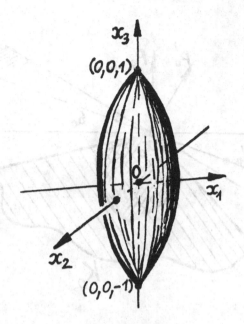

Fig. 245

the closed half-spaces with bounding hyperplanes $\triangle_o, \triangle_o'$ and containing the cone K_o, then $b_o \notin \Pi_o, b_o' \notin \Pi_o'$. The outer normal unit vectors of Π_o, Π_o' will be denoted by p_o and p_o', respectively.

Any supporting hyperplane of the cone K_o passes through its apex x_o. In particular, we have $x_o \in \triangle_o, x_o \in \triangle_o'$. By $b_o \notin \Pi_o$, the scalar product $\langle p_o, b_o - x_o \rangle$ is positive (Fig. 246) and analogously $\langle p_o', b_o' - x_o \rangle > 0$. Since $b_o - x_o = e_o$ and $b_o' - x_o = -e_o$, we obtain

$$\langle p_o, e_o \rangle > 0, \quad \langle p_o', e_o \rangle < 0. \tag{1}$$

Furthermore, since $K_o \subset \Pi_o \cap \Pi_o'$, for an arbitrary point $x \in K_o$ the inequalities

$$\langle p_o, x - x_o \rangle \le 0, \quad \langle p_o', x - x_o \rangle \le 0 \tag{2}$$

hold.

Let now ε be a positive number such that the ε-neighbourhood of each of the points b_o, b_o' does not intersect the cone K_o, and for any unit vector e, forming with e_o an angle smaller than ε, the following inequalities are satisfied:

$$\langle p_o, e \rangle > 0, \quad \langle p_o', e \rangle < 0. \tag{3}$$

Now it suffices to show that, if e and e_o form an angle smaller than ε, then $e \notin B(M)$. In fact, by (3) there exist positive coefficients α, α' such that

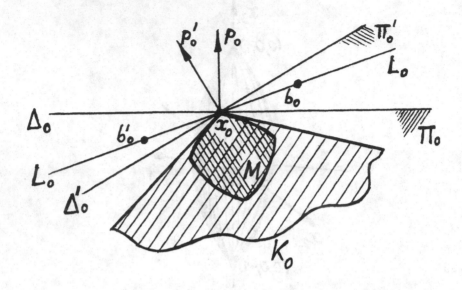

Fig. 246

$p = \alpha p_o + \alpha' p'_o$ is a unit vector satisfying $\langle p, e \rangle = 0$ (we note that $p'_o \neq p_o$ by (2)).

Let $\Pi = \{x : \langle p, x - x_o \rangle \leq 0\}$. By $\alpha, \alpha' > 0$ the relation (2) implies that for any point $x \in K_o$ the inequality $\langle p, x - x_o \rangle \leq 0$ holds, i.e., $K_o \subset \Pi$. Therefore the set $\Gamma = \mathrm{bd}\,\Pi$ is a supporting hyperplane of the cone K_o. Hence Γ is a supporting hyperplane of the body $M \subset K_o$ passing through $x_o \in M$ and parallel to the vector e (since $\langle p, e \rangle = 0$). This means that the line L, which is parallel to e with $x_o \in L$, lies in Γ and is a supporting line of M.

Denote by b, b' the points defined by $b - x_o = e$, $b' - x_o = -e$. Then $\| b - b_o \| < \varepsilon$, since the angle formed by e and e_o is smaller than ε, and hence $b \notin K_o$ (by the choice of ε). Analogously, $b' \notin K_o$. This means that none of the rays, emanating from x_o with the direction vectors $\pm e$, can be a tangential ray of M at x_o, and therefore e is not a belt vector of M. ∎

Theorem 40.4: Let Z be a zonoid, K be its generating curve, and $x = \varphi(s), 0 \leq s \leq l$, be its vector-parametric representation (where s is the arc length along K). If at a point $s_o \in [0, l]$ the derivative $\varphi'(s)$ exists and is approximatively continuous, then $\pm\varphi'(s_o)$ are belt vectors of Z. □

<u>PROOF:</u> Let $\varepsilon > 0$ be given. Since $\varphi'(s)$ is approximately continuous at s_o, we can choose a measurable set $E \subset [0, l]$ with s_o as its density point such that $|\varphi'(s) - \varphi'(s_o)| < \varepsilon$ for $s \in E$. Furthermore, we can choose a segment

$\Delta \subset [0,l]$ such that $\frac{\text{mes}(E \cap \Delta)}{\text{mes}\,\Delta} > 1 - \varepsilon$, i.e., mes $(E - \Delta) > (1 - \varepsilon)h$, where h is the length of the segment Δ.

Let now L be a supporting line of Z which is parallel to the vector $e = \varphi'(s_o)$ and $x_o \in Z \cap L$. Since $x_o \in Z$, there exists a measurable function $\mu_o(s)$, satisfying $|\mu_o(s)| \leq 1$ for all $s \in [0,1]$ such that $z(\mu_o) = x_o$ (cf. (1) in Theorem 39.1). Denote by F^+, F^- the set of the points $s \in E \cap \Delta$ for which $\mu_o(s) \geq 0$ and $\mu_o(s) < 0$, respectively. Then

$$\text{mes } F^+ + \text{mes } F^- = \text{mes } (E \cap \Delta) > (1 - \varepsilon)h$$

and therefore the measure of one of the sets F^+, F^- has to be greater than $\frac{1}{2}(1 - \varepsilon)h$, say mes $F^- > \frac{1}{2}(1 - \varepsilon)h$. We put

$$\mu(s) = \begin{cases} 1 & \text{for} \quad s \in F^-, \\ \mu_o(s) & \text{for} \quad s \notin F^-. \end{cases}$$

Then

$$
\begin{aligned}
z(\mu) - x_o &= \int_0^l (\mu(s) - \mu_o(s))\varphi'(s)ds \\
&= \int_{F^-} (\mu(s) - \mu_o(s))\varphi'(s)ds \\
&= \int_{F^-} (\mu(s) - \mu_o(s))\varphi'(s_o)ds \\
&\quad + \int_{F^-} (\mu(s) - \mu_o(s))(\varphi'(s) - \varphi'(s_o))ds \\
&= ke + \int_{F^-} (\mu(s) - \mu_o(s))(\varphi'(s) - \varphi'(s_o))ds,
\end{aligned}
$$

where

$$k = \int_{F^-} (\mu(s) - \mu_o(s))ds \geq \int_{F^-} (1-0)ds = \text{mes } F^- > \frac{1}{2}(1 - \varepsilon)h. \quad (4)$$

Furthermore,

$$\| (z(\mu) - x_o) - ke \| \leq \int_{F^-} |\mu(s) - \mu_o(s)| \cdot |\varphi'(s) - \varphi'(s_o)|ds \quad (5)$$

$$\leq \int_{F^-} (1 - (-1))\varepsilon ds \quad (6)$$

$$= 2\varepsilon \text{ mes } F^- \leq 2\varepsilon \text{ mes } \Delta = 2\varepsilon h.$$

We note that by (4) and (5) the point $z(\mu)$ does not coincide with x_o if ε is small enough. Besides, (4) and (5) imply that the angle between the vectors ke and $z(\mu) - x_o$ (i.e., between e and $z(\mu) - x_o$) is smaller than

$$\arcsin \frac{2\varepsilon h}{\frac{1}{2}(1 - \varepsilon)h} = \arcsin \frac{4\varepsilon}{1 - \varepsilon},$$

i.e., for $\varepsilon < \frac{1}{5}$ this angle is smaller than 5ε.

Finally, we observe that the ray l_μ, emanating from x_o and passing through $z(\mu)$, is contained in the supporting cone $K_o = \sup \mathrm{cone}_{x_o} Z$ (since $z(\mu) \in Z$). From this we conclude that the ray l, emanating from x_o in the direction e, forms an angle with some ray from K_o which is smaller than 5ε.

Analogously, if mes $F^+ > \frac{1}{2}(1-\varepsilon)$, then the ray l', emanating from x_o in the direction $-e$, forms an angle with some ray from K_o which is smaller than 5ε.

In view of the arbitrariness of $\varepsilon > 0$, it follows that at least of of the rays l, l' is contained in K_o, that is, it is a tangential ray of the zonoid Z at x_o.   ∎

EXAMPLE 40.5: Let $K \subset \mathbf{R}^3$ be a winding curve given in an orthonormal coordinate system by the parametric equation

$$x = \varphi(s) \quad = \quad (\varphi_1(s), \varphi_2(s), \varphi_3(s)) \text{ with} \tag{7}$$
$$\varphi_1(s) \quad = \quad a \cos s, \; \varphi_2(s) = a \sin s, \; \varphi_3(s) = bs,$$

where $0 \le s \le 2\pi$, $0 < a < 1$, and $b = \sqrt{1-a^2}$. It is easy to see that the parameter s is the arc length along the curve K. The tangential vector

$$\varphi'(s) = (\varphi_1'(s), \varphi_2'(s), \varphi_3'(s)) = (-a \sin s, a \cos s, b) \tag{8}$$

forms an angle given by arctg $\frac{b}{a}$ with the x_3-axis. Moreover, every vector forming such an angle with the x_3-axis has the form $\varphi'(s)$ for an arbitrary choice of the parameter s. In view of Theorem 40.4, all these vectors are belt vectors of the zonoid $Z \subset \mathbf{R}^3$ with generating curve K. It is easy to calculate that this zonoid Z is a body of revolution whose boundary is obtained by rotation of the curve

$$x_1 = a \cos \frac{x_3}{4b}, \; -2\pi b \le x_3 \le 2\pi b$$

around the x_3-axis.

Each unit vector forming with the x_3-axis an angle smaller than arctg $\frac{b}{a}$ (and its opposite vector) is a belt vector of Z. This example shows that, in general, not all belt vectors of a zonoid are described by Theorem 40.4.   □

Exercises:

1. Let $M \subset \mathbf{R}^n$ be a compact, convex body and $B(M)$ be the set of all its belt vectors. Prove that $B(M)$ is centrally symmetric, i.e., $-B(M) = B(M)$.

2. Prove that if $M \subset \mathbf{R}^n$ is a simplex, then the set $B(M)$ of all its belt vectors is empty.

3. Prove that a nonzero vector e is a belt vector of a compact, convex body $M \subset \mathbf{R}^n$ if and only if the following condition holds: for every supporting hyperplane $\Gamma \parallel e$ of M and every point $a \in \Gamma \cap M$, at least one of the rays emanating from a in the directions $\pm e$ is contained in $\sup \mathrm{cone}_a M$.

4. Prove that in Example 40.2 a unit vector e is a belt vector of the body M if and only if it forms an angle with one of the vectors $(0, 0, \pm 1)$ which is not greater than arctg $\frac{\pi}{2}$.

5. Construct a compact, convex body $M \subset \mathbf{R}^n$ and a nonzero vector $e \in \mathbf{R}^n$ such that there is only one point $b \in \operatorname{rbd} \pi(M)$ (with $\pi : \mathbf{R}^n \to \triangle$ denoting the orthogonal projection onto a hyperplane having e as its normal vector) for which the line $L = \pi^{-1}(b)$ meets M at a unique point a with $L \cap \sup \operatorname{cone}_a M = \{a\}$.

6. Show by a counterexample that the following assertion is false: If $s_1, \cdots, s_k \in [0, l]$ are points at which the derivative $\varphi'(s)$ exists and is approximately continuous, then for any positive real numbers $\lambda_1, \cdots, \lambda_k$ the vector $e = \lambda_1 \varphi'(s_1) + \cdots + \lambda_k \varphi'(s_k)$ (if nonzero) is a belt vector of the zonoid Z with the generating curve $x = \varphi(s)$, $s \in [0, l]$.

7. Prove that for each pair of compact, convex bodies $M_1, M_2 \subset \mathbf{R}^n$ the inclusion $B(M_1 + M_2) \supset B(M_1) \cup (M_2)$ holds.

8. Show by an example that the inclusion, which is opposite to that one indicated in the previous exercise, in general, is false.

9. Prove that for every closed set $Q \subset S^{n-1}$ there exists a compact, convex body $M \subset \mathbf{R}^n$ with $B(M) = Q \cup (-Q)$.

§41 Definition of belt bodies

In [Bk 2], [Mar 2], [Ba 4] a class of polytopes is considered that is wider than the class of zonotopes. In [Bk 2], [Mar 2] these polytopes are called *planets*, and in [Ba 4] the term *belt polytopes* is used. We will utilize the second of these terms. Namely, a convex polytope $M \subset \mathbf{R}^n$ is said to be a *belt polytope* if every its 2-face has pairwise parallel (not necessarily congruent) opposite sides, see Fig. 247. Equivalently, M is a belt polytope if and only if it has a segment summand parallel to each of its edges (cf. Exercise 7), i.e., for every edge E of M and the projection $\pi : \mathbf{R}^n \to \triangle$ (as at the beginning of the previous section), the set $\pi^{-1}(b) \cap M$, for every vertex b of the $(n-1)$-polytope $\pi(M)$, is an *edge* (not a vertex!) of M. The family of all edges $\pi^{-1}(b) \cap M$, with b running through all vertices of $\pi(M)$, form the *belt* of edges parallel to e. In the case of a zonotope, the edges of a belt are parallel and congruent. For the history of belt polytopes, going back to the crystallographer Fedorov, we also refer to [T]. The set of polars of belt polytopes was investigated in [Bk 1,2] and [Wi]. In particular, within this set the polars of zonotopes can be geometrically characterized, see [Wi]. It is natural to ask for *belt bodies* as a generalization of *zonoids*, like belt polytopes generalize zonotopes. The simplest way to do this would be to define such bodies as limits of convergent sequences of belt polytopes. But this idea is incorrect. In fact, the following assertion is true.

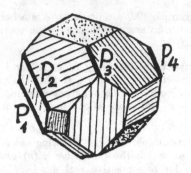

Fig. 247

Theorem 41.1: Every compact, convex body M can be represented as a limit of a convergent sequence of belt polytopes.   □

PROOF: Let ε be an arbitrary positive number and N be a convex polytope such that $\varrho(M, N) < \frac{\varepsilon}{2}$, where ϱ denotes the Hausdorff distance. Let $E_1, \cdots, E_q$ be the edges of N. By $I_j \, (j = 1, \cdots, q)$ denote the segment of length $\frac{\varepsilon}{2q}$ which is parallel to E_j and has its midpoint at the origin. Then

$$P = N + (I_1 + \cdots + I_q)$$

is a *belt polytope*. Since the zonotope $I_1 + \cdots + I_q$ is centered about the origin and its diameter does not exceed $q \cdot \frac{\varepsilon}{2q} = \frac{\varepsilon}{2}$, we have $\varrho(N, P) < \frac{\varepsilon}{2}$. Consequently $\varrho(M, P) < \varepsilon$. ∎

Thus a solution of the problem to introduce belt bodies has to be more delicate than the direct consideration of limits of belt polytopes. This problem is solved in the present secion. First, we consider some preliminary statements.

Theorem 41.2: Let $M \subset \mathbf{R}^n$ be a compact, convex body, $e_1, \cdots, e_{n-1}$ be linearly independent unit belt vectors of M, and p be the unit normal of the hyperplane Γ spanned by these belt vectors. Then the vectors $\pm p$ belong to cl $H(M)$.   □

PROOF: We conduct the proof for the vector p (for $-p$ the reasoning is analogous). Without loss of generality we may assume that Γ is a supporting hyperplane of M and p is the unit outward normal of the half-space Π, satisfying $M \subset \Pi$ and having Γ as its bounding hyperplane. Let $x_o \in M \cap \Gamma$. Since e_i is a belt vector of M, at least one of the two rays emanating from x_o and having the direction vectors $\pm e_i$ is a tangential ray of M at x_o, $i = 1, \cdots, n-1$. We may suppose (replacing, if necessary, some of the vectors $e_1, \cdots, e_{n-1}$ by their opposite ones) that the rays $l_1, \cdots, l_{n-1}$, starting from x_o with the direction vectors $e_1, \cdots, e_{n-1}$, are tangential rays of M at x_o, i.e., each of them is contained in the supporting cone $K_o = \sup \text{cone}_{x_o} M$. The ray l_o, emanating from x_o with the direction vector

$$e = \frac{1}{n}(e_1 + \cdots + e_{n-1}),$$

belongs to K_o as well. Therefore l_o is a tangential ray of M at x_o. Let ε be a positive number. We put

$$x_1 = x_o + e_1, \cdots, x_{n-1} = x_o + e_{n-1}, a = x_o + e.$$

Then a is the barycenter of the $(n-1)$-simplex T with the vertices $x_o, x_1, \cdots, x_{n-1}$, i.e., a belongs to the relative interior of T. We fix a point $s \in \text{int } M$. The unique closed half-space with the boundary through a, which contains the points $x_o, x_1, \cdots, x_{n-1}, s$, coincides with Π. Consequently there exists a positive number δ with the following property: if for certain points $x'_o, x'_1, \cdots, x'_{n-1}$ the inequalities

$$\| x'_o - x_o \| < \delta, \| x'_1 - x_1 \| < \delta, \cdots, \| x'_{n-1} - x_{n-1} \| < \delta, \| a' - a \| < \delta \quad (1)$$

hold and if a closed half-space Π' with its boundary through a' contains all the points $x'_o, x'_1, \cdots, x'_{n-1}, s$, then its unit outward normal q satisfies the condition $\| p - q \| < \varepsilon$.

Denote by g_λ the homothety with the center x_o and ratio λ. Since $x_i \in K_o$, for every $i \in \{1, \cdots, n-1\}$ there exists a point $x'_i \in \text{int } K_o$ and a positive number ρ_i such that $\| x'_i - x_i \| < \delta$ and $g_\lambda(x'_i) \in \text{int } M$ for any positive $\lambda < \rho_i$ (Fig. 248). We choose a positive number $\lambda < 1$ which is smaller than each of the numbers $\rho_1, \cdots \rho_{n-1}$. Then $g_\lambda(x'_i) \in \text{int } M$ for $i = 1, \cdots, n-1$.

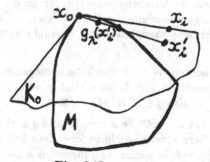

Fig. 248

Denote by T' the $(n-1)$-dimensional simplex with the vertices $x'_o = x_o, x'_1, \cdots, x'_{n-1}$ and by c its barycenter. Since $\| x'_i - x_i \| < \delta$ for $i = 0, 1, \cdots, n-1$, we conclude that $\| a - c \| < \delta$. Furthermore, since $x_o = x'_o = g_\lambda(x'_o) \in M$ and $g_\lambda(x'_i) \in \text{int } M$ for $i = 1, \cdots, n-1$, the point $g_\lambda(c)$ belongs to int M. At the same time, the point $g_\lambda(a) \in \Gamma$ does not belong to int M. Consequently there is a point $b \in [g_\lambda(a), g_\lambda(c)]$ which belongs to bd M. We have

$$\| \, g_\lambda(a) - b \, \| < \| \, g_\lambda(a) - g_\lambda(c) \, \| = \lambda \, \| \, a - c \, \| < \lambda\delta. \tag{2}$$

It follows (since $b \in$ bd M and the set of the regular boundary points of M is dense in bd M) that there exists a *regular* boundary point b_1 of the body M such that $\| \, g_\lambda(a) - b_1 \, \| < \lambda\delta$. We now denote by Π_1 the supporting half-space of the body M at the point b_1.

Finally, we put $\Pi' = g_\lambda^{-1}(\Pi_1)$. It follows from the inclusion $g_\lambda(x_i') \in M$ that $g_\lambda(x_i') \in \Pi_1$ and hence $x_i' \in g_\lambda^{-1}(\Pi_1) = \Pi'$, $i = 0, 1, \cdots, n-1$. Furthermore, since $x_o \in \Pi_1$ and $\lambda < 1$, the inclusion $g_\lambda^{-1}(\Pi_1) \supset \Pi_1$ holds, i.e., $\Pi' \supset \Pi_1$ and therefore $s \in \Pi_1 \subset \Pi'$. The point $a' = g_\lambda^{-1}(b_1)$ satisfies the condition

$$\| \, a - a' \, \| = \frac{1}{\lambda} \, \| \, g_\lambda(a) - g_\lambda(a') \, \| = \frac{1}{\lambda} \, \| \, g_\lambda(a) - b_1 \, \| < \frac{1}{\lambda} \cdot \lambda\delta = \delta$$

(cf. (2)). Thus the points $x_o' = x_o, x_1', \cdots, x_{n-1}', a'$ satisfy the condition (1). Moreover, the points $x_o', x_1', \cdots, x_{n-1}', s$ belong to Π' and $a' \in$ bd Π' (since $b_1 \in$ bd Π_1 and $a' = g_\lambda^{-1}(b_1), \Pi' = g_\lambda^{-1}(\Pi_1)$). This means (according to the choice of the number δ) that the unit outward normal q of the half-space Π' satisfies the condition $\| \, p - q \, \| < \varepsilon$. But at the same time q is the unit outward normal of the half-space $\Pi_1 = g_\lambda(\Pi')$. Moreover, $q \in H(M)$, since Π_1 is the supporting half-space of the body M at the *regular* boundary point b_1.

We have verified that for every $\varepsilon > 0$ there exists a vector $q \in H(M)$ such that $\| \, p - q \, \| < \varepsilon$. Consequently $p \in$ cl $H(M)$. ∎

Theorem 41.3: Let $M \subset \mathbf{R}^n$ be a compact, convex body. Let, furthermore, q be a unit vector with the following property: for an arbitrary real number $\varepsilon > 0$, there exist linearly independent belt vectors $e_1, \cdots, e_{n-1}$ of M such that the hyperplane spanned by them has a unit normal p satisfying $\| \, p - q \, \| < \varepsilon$. Then $q \in$ cl $H(M)$. □

PROOF: This theorem is an immediate consequence of the previous one. In fact, from Theorem 41.2 it follows that $p \in$ cl $H(M)$, and therefore (by arbitrariness of ε) also $q \in$ cl $H(M)$. ∎

Theorem 41.4: Let $Z \subset \mathbf{R}^n$ be a zonoid and $q \in$ cl $H(Z)$. Then for any real number $\varepsilon > 0$ there exist linearly independent belt vectors $e_1, \cdots, e_{n-1}$ of Z such that the hyperplane spanned by them has a unit normal p which satisfies $\| \, p - q \, \| < \varepsilon$. □

PROOF: Let K be a generating curve of Z and $x = \varphi(s), 0 \leq s \leq l$, be its parametric vector equation (with the parameter s denoting the arc length along K). Let, furthermore, q be a unit vector from cl $H(Z)$. Then there exists a vector $q_1 \in H(Z)$ such that $\| \, q - q_1 \, \| < \frac{\varepsilon}{2}$. Let now x_o be a regular boundary point of Z having q_1 as its outward normal, and Γ denotes the supporting hyperplane of Z containing x_o. We choose a number $\delta > 0$ such that for each convex polytope M with $\varrho(Z, M) < \delta$ there is a facet of M, whose unit outward normal p satisfies $\| p - q_1 \| < \frac{\varepsilon}{2}$ (such a number δ exists since x_o is a *regular* boundary point of Z).

In view of Theorem 39.4, there are points $s_1, \cdots, s_q \in [0, l]$, at which the derivative $\varphi'(s)$ exists and is approximately continuous, and numbers $\lambda_1, \cdots, \lambda_q$ such that the vector sum of the segments $I_1, \cdots, I_q$ (which are parallel to $\varphi'(s_1), \cdots, \varphi'(s_q)$ and have the lengths $\lambda_1, \cdots, \lambda_q$, respectively) is a zonotope M satisfying $\varrho(M, Z) < \delta$. In other words, by Theorem 40.4 there exist line segments $I_1, \cdots, I_q$ parallel to some belt vectors of a zonoid Z such that the zonotope $M = I_1 + \cdots + I_q$ satisfies the condition $\varrho(Z, M) < \delta$. Consequently by the choice of δ there exists a facet $\triangle$ of the zonotope M such that for its unit outward normal p the inequality $\| q_1 - p \| < \frac{\varepsilon}{2}$ holds, implying also $\| q - p \| < \varepsilon$. It remains to note that $\triangle$ is an $(n-1)$-zonotope whose edges are parallel to some of the segments $I_1, \cdots, I_q$, i.e., they are parallel to some belt vectors of Z. In other words, there exist linearly independent belt vectors $e_1, \cdots, e_{n-1}$ of Z which are parallel to the hyperplane $\Gamma = \operatorname{aff} \triangle$. ∎

Theorems 41.2 and 41.3 give us the base to introduce the class of compact, convex *belt bodies*.

<u>DEFINITION 41.5:</u> A compact, convex body $M \subset \mathbf{R}^n$ is said to be a *belt body* if for every $q \in \operatorname{cl} H(M)$ and any $\varepsilon > 0$ there exist linearly independent belt vectors $e_1, \cdots, e_{n-1}$ of M such that the hyperplane spanned by them has a unit normal p satisfying $\| p - q \| < \varepsilon$. □

Theorem 41.6: Every zonoid is a belt body. □

This is an immediate consequence of Theorem 41.4 and the definition above.

By the following three theorems, belt bodies are described in terms of their polar bodies.

Theorem 41.7: Let $M \subset \mathbf{R}^n$ be a compact, convex body containing the origin in its interior. Let, furthermore, $e \neq o$ be a vector, $\Gamma' = e^*$ be its polar hyperplane, and Γ be the $(n-1)$-subspace parallel to Γ'. The vector e is a belt vector of M if and only if Γ is an equatorial $(n-1)$-subspace of the polar body M^*, i.e., for each point $x \in \Gamma \cap \operatorname{bd} M^*$ its face (in M^*) is contained in Γ.

<u>PROOF:</u> Denote by Φ the face of the point $x \in \Gamma \cap M^*$ with respect to M^*. Assume that Φ is not contained in Γ, i.e., there are points $y_1, y_2 \in \Phi$ not belonging to Γ such that $x \in]y_1, y_2[$. Since $\langle e, z \rangle = 1$ for all $z \in \Gamma'$, we have $\langle e, z \rangle = 0$ for every $z \in \Gamma$, and vice versa. In particular, $\langle e, x \rangle = 0$. As far as $y_1 \notin \Gamma$, the inequality $\langle e, y_1 \rangle \neq 0$ holds, and hence the numbers $\langle e, y_1 \rangle, \langle e, y_2 \rangle$ have different signs, say $\langle e, y_1 \rangle < 0, \langle e, y_2 \rangle > 0$. Let now $a \in M$ be a point such that $\langle a, x \rangle = 1$. Since $y_1, y_2 \in \Phi \subset M^*$, the equalities $\langle a, y_1 \rangle \leq 1, \langle a, y_2 \rangle \leq 1$ hold and hence $\langle a, y_1 \rangle = \langle a, y_2 \rangle = 1$ (since $\langle a, x \rangle = 1$). Now $\langle a + e, y_2 \rangle = \langle a, y_2 \rangle + \langle e, y_2 \rangle > 1, \langle a - e, y_1 \rangle = \langle a, y_1 \rangle - \langle e, y_1 \rangle > 1$. This means that the ray r_1, emanating from a with the direction vector e, is not contained in the half-space $\{ w : \langle w, y_2 \rangle \leq 1 \}$, and the opposite ray is

not contained in $\{w : \langle w, y_1 \rangle \leq 1\}$. Since both the half-spaces are supporting half-spaces of the body M at a, the vector e is not a belt vector of M.

Conducting this reasoning in the opposite direction, we obtain the converse assertion: if e is not a belt vector of M, then Γ is not an equatorial $(n-1)$-subspace of M^*. ∎

By virtue of the established theorem, we obtain two more propositions which follow immediately from Theorem 41.2 and Definition 41.5 (cf. also Theorem 3.2). In other words, the following two theorems are natural translations of Theorem 41.2 and Definition 41.5 to the "polar language".

Theorem 41.8: Let $M \subset \mathbf{R}^n$ be a compact, convex body which contains the origin in its interior. Let, furthermore, $\Gamma_1, \cdots, \Gamma_{n-1}$ be equatorial $(n-1)$-subspaces of the polar body M^* with one-dimensional intersection. Then both the points $(\Gamma_1 \cap \cdots \cap \Gamma_{n-1}) \cap \operatorname{bd} M^*$ are contained in $\operatorname{cl} \exp M^*$. □

Theorem 41.9: Let $M \subset \mathbf{R}^n$ be a compact, convex body containing the origin in its interior. The body M is a belt body if and only if for every exposed point p of M^* and every positive ε' there are $n-1$ equatorial $(n-1)$-subspaces $\Gamma_1, \cdots, \Gamma_{n-1}$ of M^* with one-dimensional intersection such that a point $q \in (\Gamma_1 \cap \cdots \cap \Gamma_{n-1}) \cap \operatorname{bd} M^*$ satisfies the condition $\| p - q \| < \varepsilon'$. □

<u>Exercises:</u>

1. Prove that if a compact, convex body $M \subset \mathbf{R}^n$ has a smooth boundary (i.e., each boundary point of M is regular), then M is a belt body.

2. Prove that if a polytope $M \subset \mathbf{R}^n$ is a belt body, then M is a belt polytope (in the sense considered at the beginning of the section).

3. Prove that if a belt body $M \subset \mathbf{R}^n$ has only n pairwise non-parallel belt vectors, then M is a parallelotope.

4. Prove that if $M \subset \mathbf{R}^n$ is a belt body, then the set $\operatorname{cl} H(M)$ is centrally symmetric with respect to the origin. Show by an example that, in general, $H(M)$ is not centrally symmetric.

5. Prove that a compact, convex figure $M \subset \mathbf{R}^2$ is a belt body if and only if the set $\operatorname{cl} H(M)$ is symmetric with respect to the orign.

6. Give an example of a compact, convex body $M \subset \mathbf{R}^3$ with centrally symmetric set $\operatorname{cl} H(M)$ which is not a belt body.

7. Prove that each belt polytope $M \subset \mathbf{R}^n$ can be represented by $M = Z + N$, where Z is a zonotope, N is an r-dimensional convex polytope ($r \in \{0, \cdots, n\}$) each edge of which is parallel to a segment summand of Z and N has no belt of parallel edges.

8. We say that an n-dimensional zonoid Z is translatively moveable in a compact, convex body $M \subset \mathbf{R}^n$ if for every boundary point a of M there exists a translate Z_a of Z such that $a \in Z_a \subset M$ and $\sup \operatorname{cone}_a Z_a = \sup \operatorname{cone}_a M$. Prove that if there exists a zonoid Z which is translatively moveable in M, then M is a belt body.

9. Let A be the arc of the circle $x_1^2 + x_2^2 = 16$ contained in the first quadrant $x_1 \geq 0, x_2 \geq 0$ and B be the circle $(x_1 - 1)^2 + (x_2 - 1)^2 = 1$. Prove that $M = \text{conv}\,(A \cup B)$ is a belt body in $\mathbf{R}^2$ which has no zonoid translatively movable in M.

10. Let $M \subset \mathbf{R}^n$ be a belt body. Prove that there are n belt vectors $e_1, \cdots, e_n$ of the body M which form a basis in $\mathbf{R}^n$.

11. Show by a counterexample that the following assertion is false: Let $M \subset \mathbf{R}^n$ be a belt body, a be its regular boundary point, and Γ be the suppoorting hyperplane of M at a; then there exist $n - 1$ linearly independent belt vectors $e_1, \cdots, e_{n-1}$ of M parallel to Γ.

12. Prove that if $M \subset \mathbf{R}^n$ is a belt body and $a \in \text{bd}\,M$, then there exists a belt vector $e \in B(M)$ such that the line $L \parallel e$ through a is a supporting line of M.

13. We say that a compact, convex body $M \subset \mathbf{R}^n$ has the "belt-tangential property" if for every $a \in \text{bd}\,M$ there exists a belt vector $e \in B(M)$ such that the line L through a parallel to e is a supporting line of M (consequently L is a semitangent line, i.e., at least one of the rays, in which a divides L, is a tangential ray of M at a). Exercise 12 says that the belt-tangential property is a *necessary* condition for M being a belt body. Show by a counterexample that this condition is not sufficient, i.e., there exists a body M with the belt-tangential property which is not a belt body.

14. Let $M_1, \cdots, M_s$ be belt bodies. Prove that the vector sum $M_1 + \cdots + M_s$ is a belt body, too.

15. Show by a counterexample that the following assertion is false: If a belt body M is represented in the form $M = M_1 + \cdots + M_s$, then each convex set M_i, $i = 1, \cdots, s$, is a belt body. Is the above assertion true for a *direct vector sum* $M = M_1 \oplus \cdots \oplus M_s$?

16. Is it true that every centrally symmetric belt polytope is a zonotope ?

§42 Solution of the illumination problem for belt bodies

The problem of illumination (by parallel bundles), formulated at the beginning of the previous chapter, was solved for zonotopes in [Mar 2] and the bound given there was verified also for zonoids in [B-SP 5], [B-SP 6]. Thus from the latter paper we take the following

Theorem 42.1: If a zonoid $M \subset \mathbf{R}^n$ is not a parallelotope, then

$$c(M) \leq 3 \cdot 2^{n-2} = \frac{3}{4} \cdot 2^n. \quad \square$$

In this section we establish that for every belt body the same estimate holds. First we show the way to establish Theorem 42.1 for zonotopes.

PROOF: There are nonzero vectors $e_1, \cdots, e_p$ such that, up to a translation, M is the set of the points

$$x_1e_1 + \cdots + x_pe_p \quad (\|x_i\| \le 1 \text{ for all } i \in \{1, \cdots, p\}).$$

We may assume that e_i is not parallel to e_j for $i \ne j$. Furthermore, since M is n-dimensional and is not a parallelotope, the inequality $p > h$ holds, and we may assume that $e_1, \cdots, e_n$ is a basis in $\mathbf{R}^n$. Denote by M_1 the set of the points $x_1e_1 + \cdots + x_ne_n + x_{n+1}e_{n+1}$ and by M_2 the set of the points $x_{n+2}e_{n+2} + \cdots + x_pe_p$ (where $|x_i| \le 1$ for all i). Then $M = M_1 + M_2$, and M_1 is a body. Therefore, $c(M) \le c(M_1)$ (by virtue of Theorem 34.8), i.e., it is sufficient to prove that $c(M_1) \le 3 \cdot 2^{n-2}$. Let

$$\lambda_1 e_1 + \cdots + \lambda_n e_n + \lambda_{n+1}e_{n+1} = o \tag{1}$$

be a nontrivial linear dependence. Then $\lambda_{n+1} \ne 0$, and at least two of the numbers $\lambda_1, \cdots, \lambda_n$ are not equal to 0 (since e_{n+1} is not parallel to e_i for any $i = 1, \cdots, n$). Without loss of generality we may assume (replacing, if necessary, e_i by $-e_i$) that all λ_i are nonnegative and

$$\lambda_{n-1} > 0, \ \lambda_n > 0, \ \lambda_{n+1} > 0.$$

Let $x = x_1e_1 + \cdots + x_ne_n + x_{n+1}e_{n+1}$ be a point of the body M_1 (where $|x_i| \le 1$ for $i = 1, \cdots, n+1$). If $|x_i| < 1$ for at least one index $i \in \{1, \cdots, n+1\}$, then x is not at vertex of M_1 (since both the points $x - \varepsilon e_i, x + \varepsilon e_i$ belong to M_1 for ε small enough, i.e., x is an interior point of the segment $[x - \varepsilon e_i, x + \varepsilon e_i]$ contained in M_1). Thus every vertex q of M_1 has the form

$$q = \sigma_1 e_1 + \cdots + \sigma_n e_n + \sigma_{n+1}e_{n+1}, \text{ where } |\sigma_i| = 1 \text{ for all } i. \tag{2}$$

We remark, however, that not each of the points (2) is a vertex of M_1. For example, the point $q' = e_1 + \cdots + e_n + e_{n+1}$ can be represented (by virtue of (1)) in the form

$$q' = (1 - \varepsilon\lambda_1)e_1 + \cdots + (1 - \varepsilon\lambda_n)e_n + (1 - \varepsilon\lambda_{n+1})e_{n+1}.$$

If $\varepsilon > 0$ is small enough, then all the coefficients in this equality belong to $[0, 1]$, but the last three of them become less than 1, i.e. q' is not vertex.

We fix a combination of signs $\sigma_1 = \pm 1, \cdots, \sigma_{n-2} = \pm 1$. There are at most eight vertices of M_1 with this combination of signs (because, in order to get a vertex, it is sufficient to choose in addition $\sigma_{n-1} = \pm 1, \sigma_n = \pm 1, \sigma_{n+1} = \pm 1$). We will show that the eight points (if they are vertices) are illuminated by the folloowing *three* directions:

$$d_1 = f + (e_n - e_{n-1}), \ d_2 = f + (e_{n+1} - e_n), \ d_3 = f + (e_{n-1} - e_{n+1}),$$

where $f = -(\sigma_1 e_1 + \cdots + \sigma_{n-2}e_{n-2})$. More detailed, if the point (2) is a vertex of M_1, then

a) for $\sigma_{n-1} = 1, \sigma_n = -1$ it is illuminated by d_1;

b) for $\sigma_n = 1, \sigma_{n+1} = -1$ it is illuminated by d_2;

c) for $\sigma_{n+1} = 1, \sigma_{n-1} = -1$ it is illuminated by d_3;

d) for $\sigma_{n-1} = \sigma_n = \sigma_{n+1} = \pm 1$ it is illuminated by any d_i.

We prove the assertion a); the other ones can be proved similarly. Thus let q be the vertex which has the form (2) with $\sigma_{n-1} = 1, \sigma_n = -1, \sigma_{n+1} = \pm 1$. Then

$$
\begin{aligned}
q + \varepsilon d_1 &= \sigma_1 e_1 + \cdots + \sigma_{n-2} e_{n-2} + e_{n-1} - e_n + \sigma_{n+1} e_{n+1} \\
&\quad + \varepsilon(f + e_n - e_{n-1}) \\
&= (1 - \varepsilon)\sigma_1 e_1 + \cdots + (1 - \varepsilon)\sigma_{n-2} e_{n-2} + (1 - \varepsilon) e_{n-1} \\
&\quad - (1 - \varepsilon) e_n + \sigma_{n+1} e_{n+1}.
\end{aligned}
$$

All the coefficients here, except for the last one, are equal to $\pm(1 - \varepsilon)$, i.e., for $0 < \varepsilon < 2$ they are interior points of the segment $[-1, 1]$, while the last coefficient is equal to ± 1. Adding to the right-hand side the relation (10) multiplied by $-\delta \sigma_{n+1}$, where $\delta > 0$ is small enough, we obtain a representation $q + \varepsilon d_1 = \mu_1 e_1 + \cdots + \mu_n e_n + \mu_{n+1} e_{n+1}$ with the coefficients strictly between -1 and 1. This means that $q + \varepsilon d_1$ (for $0 < \varepsilon < 2$) is an interior point of M_1, i.e., q (if it is vertex) is illuminated by the direction d_1.

Thus all the points (1) with a fixed combination of signs for $\sigma_1 = \pm 1, \cdots, \sigma_{n-2} = \pm 1$, (if vertices) are illuminated by the three directions d_1, d_2, d_3. Since there are 2^{n-2} combinations of signs for $\sigma_1, \cdots, \sigma_{n-2}$, the vertices of M_1 can be illuminated by $3 \cdot 2^{n-2}$ directions. But if some directions illuminate the vertices of a convex polytope, they illuminate the whole boundary of it. Consequently $c(M_1) \leq 3 \cdot 2^{n-2}$. ∎

EXAMPLE 42.2: We show that the estimate $c(M) \leq 3 \cdot 2^{n-2}$ in Theorem 42.1 is best possible. Let $M_1 \subset \mathbf{R}^2$ be the regular hexagon and M_2 be an $(n-2)$-dimensional cube. Then $M = M_1 \oplus M_2$ is an n-dimensional zonotope distinct from a parallelotope. Let a_1, a_2, a_3 be pairwise non-adjacent vertices of M_1 and $b_j, j = 1, \cdots, 2^{n-2}$, be the vertices of the cube M_2. Then all the points $a_i + b_j$ are vertices of M. The number of these vertices is equal to $3 \cdot 2^{n-2}$. Since no two of them can be illuminated by the same direction, $c(M) \geq 3 \cdot 2^{n-2}$. Consequently, according to Theorem 42.1, $c(M) = 3 \cdot 2^{n-2}$. ∎

In [B-SP 5, B-SP 6], it was proved that the same estimate (as in Theorem 42.1) holds for any zonoid which is not a parallelotope.

But we do not clarify the proof from [B-SP 5] and [B-SP 6] here, since the definition of belt bodies allows us to establish a more general result, namely (see also [Bt 14], [B-M 1])

Theorem 42.3: For any belt body $M \subset \mathbf{R}^n$, which is not a parallelotope, the inequality

$$
c(M) \leq 3 \cdot 2^{n-2}
$$

holds. □

We mention that for belt polytopes Theorem 42.3 is obtained in [Mar 2].

For proving the theorem, we establish some auxiliary statements.

<u>LEMMA 42.4:</u> *Let a be a boundary point of a compact, convex body $M \subset \mathbf{R}^n$ and e be a unit vector such that the ray l, emanating from a with the direction vector e, is a tangential ray of M. If a direction determined by a vector $p \in \mathbf{R}^n$ illuminates the point a, then for any $k > 0$ the direction determined by $p + ke$ also illuminates the point a.*   □

PROOF: Since the direction given by p illuminates a, there is some $\lambda > 0$ such that $a + \lambda p \in \operatorname{int} M$. Denote by $r > 0$ a number such that the r-neighbourhood of $a + \lambda p$ belongs to int M. Let now k be an arbitrary positive number. Since l is a tangential ray, for every $\varepsilon > 0$ a point $b \in M$ of the form $b = a + \mu(e + h)$ can be found, where $|h| < \varepsilon$ and $\mu > 0$. We look at the point $c = a + \lambda p - k\lambda h$. For a sufficiently small ε (namely for $\varepsilon < \frac{r}{k\lambda}$) this point c lies in an r-neighbourhood of $a + \lambda p$, i.e., $c \in \operatorname{int} M$, see Fig. 249. Hence the point

$$f = \frac{k\lambda}{\mu + k\lambda} b + \frac{\mu}{\mu + k\lambda} c$$

from the segment $[b, c]$ is an *interior point* of M (since $b \in M, c \in \operatorname{int} M$). But this point is also representable by

$$f = \frac{k\lambda}{\mu + k\lambda}(a + \mu(e + h)) + \frac{\mu}{\mu + k\lambda}(a + \lambda p - k\lambda h)$$

$$= a + \frac{\mu\lambda}{\mu + k\lambda}(p + ke).$$

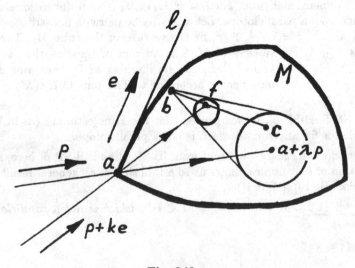

Fig. 249

Therefore, by $x = \frac{\mu\lambda}{\mu+k\lambda}$, the point $f = a + x(p + ke)$ is an *interior point* of M, and the direction $p + ke$ illuminates a. ∎

LEMMA 42.5: *Let $M \subset \mathbf{R}^n$ be a compact, convex body and e be a belt vector of M. Denote by $\pi : \mathbf{R}^n \to \mathbf{R}^{n-1}$ the projection of $\mathbf{R}^n$ parallel to e onto some subspace $\mathbf{R}^{n-1} \subset \mathbf{R}^n$. Let $b \in \mathrm{rbd}\pi(M)$ and $a \in \pi^{-1}(b) \cap M$. If the boundary point b of the body $\pi(M)$ is illuminated (in the subspace $\mathbf{R}^{n-1}$) by the direction p, then there exists a real number $k_o > 0$ such that for $k > k_o$ the boundary point $a \in \mathrm{bd}\, M$ is illuminated (in $\mathbf{R}^n$) by at least one of the directions $p \pm ke$.* □

PROOF: Let $\lambda > 0$ be a number such that $b \in \lambda p \in \mathrm{ri}\ (M)$. Let, furthermore, $c \in \mathrm{int}\ M$ be a point such that $\pi(c) = b + \lambda p$. Then the point a is illuminated by the direction $p' = c - a$. Thus,

$$\pi(p') = \pi(c) - \pi(a) = (b + \lambda p) - b = \lambda p,$$

and therefore $p' = \lambda p + \mu e$, where μ is some real number. Hence a is illuminated by the direction $\lambda p + \mu e$ or, equivalently, a is illuminated by the direction $p + \frac{\mu}{\lambda}e$.

Since e is a belt vector of M, at least one of the rays emanating from a in the directions $\pm e$ is tangential. Let this hold for e, say. Then by Lemma 42.4 the point $a \in \mathrm{bd}\, M$ is illuminated by the direction $(p + \frac{\mu}{\lambda}e) + ke$ for any $k > 0$. If even the ray with direction $-e$ is tangential, then, analogously, the point a is illuminated by the direction $(p + \frac{\mu}{\lambda}e) - ke$ for any $k > 0$. Denoting by k_o the number $|\frac{\mu}{\lambda}|$, we conclude that, for $k > k_o$, the point a is illuminated by one of the directions $p \pm ke$. ∎

LEMMA 42.6: *Let M, e and π be given as in the preceding lemma. Let, furthermore, $P \subset \mathrm{rbd}\ \pi(M)$ be a compact set which is illuminated (in $\mathbf{R}^{n-1}$) by a direction $p \in \mathbf{R}^{n-1}$. Then there exists a real number $k_o > 0$ such that for any real number $k > k_o$ the set $\pi^{-1}(P) \cap M$ (belonging to the boundary of M) is illuminated by two directions $p + ke$ and $p - ke$.* □

PROOF: Let $a \in \pi^{-1}(P) \cap M$. By the previous lemma, there exists a number $k_o(a)$ such that for $k > k_o(a)$ the point a is illuminated by at least one of the directions $p \pm ka$. Besides a, some neighbourhood $U(a)$ of a (in the boundary of M) has this property (Lemma 42.4) and, by compactness arguments, there is a finite number of points $a_1, \cdots, a_s \in \pi^{-1}(P) \cap M$ such that the corresponding neighbourhoods $U(a_1), \cdots, U(a_s)$ cover the whole set $\pi^{-1}(P) \cap M$.

We now take the largest of the numbers $k_o(a_1), \cdots, k_o(a_s)$ and denote it by k_o. Then for $k > k_o$, the set $U(a_1) \cup \cdots \cup (a_s)$ is illuminated by the directions $p + ke$ and $p - ke$, i.e., by these two directions the whole set $\pi^{-1}(P) \cap M$ is illuminated. ∎

Theorem 42.7: Let $M \subset \mathbf{R}^n$ be a compact, convex body and e be a belt vector of M. We denote by $\pi : \mathbf{R}^n \to \mathbf{R}^{n-1}$ the projection of $\mathbf{R}^n$ parallel to e onto

$\mathbf{R}^{n-1}$. *Then $c(M) \le 2c(\pi(M))$, where $c(\pi(M))$ is the illumination number of $\pi(M)$ with respect to $\mathbf{R}^{n-1}$.*   □

PROOF: Let $c(\pi(M))$ be denoted by c, i.e., there exist directions $p_1, \cdots, p_c$ in $\mathbf{R}^{n-1}$ by which the whole boundary of $\pi(M) \subset \mathbf{R}^{n-1}$ is illuminated. Let V_i be the set of points $b \in \text{rbd } \pi(M)$ which are illuminated by p_i, $i \in \{1, \cdots, c\}$. The sets $V_1, \cdots, V_s$ are open with respect to rbd $\pi(M)$, and they cover the whole boundary rbd $\pi(M)$ of the body $\pi(M) \subset \mathbf{R}^{n-1}$. Hence there are *compact* sets $F_1, \cdots, F_s$ (correspondingly contained in $V_1, \cdots, V_s$) whose union also contains the whole boundary of $\pi(M)$ in $\mathbf{R}^{n-1}$. By the previous lemma, there is a number $k_o > 0$ such that each of the sets $\pi^{-1}(F_i) \cap M$ is illuminated by two directions $p_i \pm ke$ for $k > k_o$, $i = 1, \cdots, c$. Consequently for $k > k_o$ the whole set

$$Q = \pi^{-1}(\text{rbd } \pi(M)) \cap M = (\pi^{-1}(F_1) \cap M) \cup \cdots \cup (\pi^{-1}(F_c) \cap M)$$

is illuminated by the 2c directions $p_1 \pm ke, \cdots, p_c \pm ke$.

The directions $p_1 \pm ke, \cdots, p_c \pm ke$ illuminate some *open* set $V \subset \text{bd } M$, containing the compact set Q. Therefore it remains to illuminate the compact set $(\text{bd } M) \backslash V$. The whole set $(\text{bd } M) \backslash Q$ is illuminated by the two directions $\pm e$. In particular, the compact set $(\text{bd } M) \backslash V$ is illuminated by $\pm e$. In view of compactness arguments, there is a number $k > k_o$ such that two directions $e + \frac{1}{k}p_1, -e + \frac{1}{k}p_1$ (or, what is the same, the directions $p_1 \pm ke$) illuminate the set $(\text{bd } M) \backslash V$. This means that the boundary of M is illuminated by the directions $p_1 \pm ke, \cdots, p_c \pm ke$. ∎

LEMMA 42.8: *Let $\pi : \mathbf{R}^n \to \mathbf{R}^{n-1}$ be a projection of the space $\mathbf{R}^n$ onto the $(n-1)$-subspace $\mathbf{R}^{n-1}$. If $e \in R^n$ is a belt vector of a compact, convex body $M \subset \mathbf{R}^n$ which is not parallel to the direction of that projection, then $e_1 = \text{norm } \pi(e)$ is a belt vector of the body $\pi(M)$ in $\mathbf{R}^{n-1}$.*   □

PROOF: Let $L_1 \subset \mathbf{R}^{n-1}$ be a supporting line of the body $\pi(M) \subset \mathbf{R}^{n-1}$ parallel to e_1, and $a \in \pi^{-1}(b) \cap M$, i.e., $a \in M$ and $\pi(a) = b$. Let L be the line through a which is parallel to e. Then $\pi(L) = L_1$ (since $e_1 = \text{norm } \pi(e)$). As far as e is a belt vector of the body M, there exists a sequence $x_1, x_2, \cdots \in M$ such that all x_k are distinct from a, $\lim_{k \to \infty} \text{norm } (x_k - a)$ exists and is equal to $\pm e$. Then $y_k = \pi(x_k)$ is a point of the body $\pi(M) \subset \mathbf{R}^{n-1}$ and $\lim_{k \to \infty} \text{norm } (y_k - b) = \lambda e_1$, where $\lambda \ne 0$ (since e is not parallel to the direction of the projection π). ∎

Theorem 42.9: Let $M \subset \mathbf{R}^n$ be a compact, convex body. If M has n linearly independent belt vectors, then $c(M) \le 2^n$. Moreover, if M has an additional belt vector, then $c(M) \le 3 \cdot 2^{n-2}$.   □

PROOF: We use induction over n. For $n = 2$, the theorem trivially follows from Theorem 42.1. For the step from $n-1$ to n, let $e_1, \cdots, e_n$ be linearly independent belt vectors of the body M. Denote by π the projection of the space $\mathbf{R}^n$ parallel to e_n onto the subspace $\mathbf{R}^{n-1}$, containing $e_1, \cdots, e_{n-1}$. By

Lemma 42.8, the vectors $e_1, \cdots, e_{n-1}$ are belt vectors of the body $\pi(M) \subset \mathbf{R}^{n-1}$. The induction hypothesis says that $c(\pi(M)) \leq 2^{n-1}$, and therefore $c(M) \leq 2c(\pi(M)) \leq 2^n$ (cf. Theorem 42.7). If the body M has (besides $\pm e_1, \cdots, \pm e_n$) an additional belt vector $e = \lambda_1 e_1 + \cdots + \lambda_n e_n$, then two of the coefficients $\lambda_1, \cdots, \lambda_{n-1}$ have to be nonzero ones (for getting this coefficient system, a suitable changing of indices might be necessary). Hence, the body $\pi(M)$ has, besides $\pm e_1, \cdots, \pm e_{n-1}$, an additional belt vector. Thus by the induction hypothesis we have $c(\pi(M)) \leq 3 \cdot 2^{n-3}$, and by Theorem 42.7 this yields $c(M) \leq 2c(\pi(M)) \leq 3 \cdot 2^{n-2}$. ∎

We underline that the assertions above (such as the Theorems 42.7 and 42.9) are established without the assumoption that M is a belt body.

Now we are ready to prove Theorem 42.3.

PROOF: Let $M \subset \mathbf{R}^n$ be a belt body and $B(M) \subset S^{n-1}$ denote the set of all its belt vectors. Assume that the system $B(M)$ lies in one hyperplane Γ, whose unit normals are given by $\pm p$. Then, by Definiton 41.5, the set $H(M)$ consists of only two vectors, namely $\pm p$. This contradicts the compactness of M. Hence the set $B(M)$ cannot lie in only one hyperplane, i.e., the body M has n linearly independent belt vectors $e_1, \cdots, e_n$. By Theorem 42.9, this yields $c(M) \leq 2^n$. We now assume that, besides $\pm e_1, \cdots, \pm e_n$, the body M has no further belt vectors. Denote by $\pm p_i$ the unit normals of the hyperplanes spanned by the system $\{e_1, \cdots, e_n\} \setminus \{e_i\}$, $i = 1, \cdots, n$. Then the set $H(M)$ (see Definition 41.5) only consists of the vectors $\pm p_1, \cdots, \pm p_n$. Consequently the body M has to be a parallelotope with these vectors as outer unit normals of its facets. This implies the following statement: If the belt body M is not a parallelotope, then it has, besides $\pm e_1, \cdots, \pm e_n$, at least one additional belt vector. By Theorem 42.9, this yields the inequality $c(M) \leq 3 \cdot 2^{n-2}$. ∎

Exercises:

1. Consider the omitted case in the proof of Theorem 42.1 for zonotopes when $\sigma_n = \sigma_{n-1} = \sigma_{n+1} = \pm 1$.

2. Prove that if $Z \subset \mathbf{R}^n$ is a zonoid and M is its tangential zonotope, then $c(Z) \leq c(M)$.

3. Prove that if an n-dimensional zonoid $Z \subset \mathbf{R}^n$ is the limit of a sequence $M_1, M_2, \cdots$ of its tangential zonotopes, then

$$c(Z) = \lim_{k \to \infty} c(M_k).$$

4. Let $M_1, M_2, \cdots$ be a convergent sequence of compact, convex bodies in $\mathbf{R}^n$ and M be its limit. We say that this convergence is *tangential* if $H(M_k) \subset \mathrm{cl}\, H(M)$ for all $k = 1, 2, \cdots$. Prove that in the case of tangential convergence the inequality $c(M) \leq \underline{\lim}_{k \to \infty} c(M_k)$ holds.

5. Prove that every belt body can be represented as the limit of a *tangentially convergent* sequence of belt polytopes.

6. Denote by σ the symmetry of $\mathbf{R}^n$ with respect to the origin. For every compact, convex body $M \subset \mathbf{R}^n$, the Minkowski sum $M + \sigma(M)$ is said to be the *difference body* of M. Prove that for $n = 2$ the difference body $M + \sigma(M)$ of $M \subset \mathbf{R}^2$ is a zonoid.

7. Decide whether the assertion of the previous exercise is true for $n > 2$

8. Let $\{e_1, \cdots, e_n, f_1, \cdots, f_k\}$ be a simplicial vector system (as in Exercise 6 of section 17) and M be a zonotope, whose segment summands are correspondingly parallel to the vectors $e_1, \cdots, e_n, f_1, \cdots, f_k$. Prove that $c(M) \leq 3^k \cdot 2^{n-2k}$.

9. Prove that every projection $\pi(M)$ of a belt body $M \subset \mathbf{R}^n$ is an $(n-1)$-dimensional belt body.

10. In the notation of Lemma 42.8 we say that $a \in \pi^{-1}(b) \cap M$ is a *top point* (with respect to e) if a is the *last* point that we meet moving along the line $\pi^{-1}(b)$ in the direction e (i.e., $a + \lambda e \notin M$ for $\lambda > 0$). Conducting the reasoning in Lemmas 42.5, 42.6 and in Theorem 42.7, prove that if at each top point $a \in \pi^{-1}(\text{rbd } \pi(M)) \cap M$ the ray emanating from a in the direction e is a tangential ray of M, then $c(M) = c(\pi(M)) + 1$.

11. In the previous notation, prove that if for every $b \in \text{rbd } \pi(M)$ at least one point $a \in \pi^{-1}(b) \cap M$ is a regular boundary point of M, then $\pi(M)$ is a smooth body in $\mathbf{R}^{n-1}$. Consequently $c(\pi(M)) = n$.

12. Combining the two previous exercises, prove the following theorem which is established in [De 6]: If for a direction e and a compact, convex body $M \subset \mathbf{R}^n$ each top point $a \in \pi^{-1}(\text{rbd } \pi(M)) \cap M$ is a regular boundary point of M, then $c(M) = n + 1$. Consequently $a(M) \leq n + 1$.

13. Prove that for the body M in Example 40.1 the equality $c(M) = n + 1$ holds.

§43 Solution of the Szökefalvi-Nagy problem for belt bodies

A complete solution of the Szökefalvi-Nagy problem for zonotopes is given in [Ba 1]. It is extended to the class of zonoids in [Ba 2]. In this section we will describe a further generalization. More precisely, we present a complete solution of the Szökefalvi-Nagy problem for the family of belt bodies, cf. also [B-B 2] and [B-M 1].

Theorem 43.1: Let $M \subset \mathbf{R}^n$ be a belt body. The inequality him $M \leq r, 1 \leq r \leq n$, *holds if and only if M is representable in the following form:*

$$M = M_1 \oplus \cdots \oplus M_s, \quad \dim M_i \leq r, \ i = 1, \cdots, s.$$

In other words, if a belt body M is represented as a direct sum of convex sets $M_1, \cdots, M_s$, each of which is indecomposable, then

$$\text{him } M = \max(\dim M_1, \cdots, \dim M_s). \quad \square$$

The proof is given at the end of the section, after some auxiliary statements. First of all, we prove a *key-lemma*. To formulate it, we introduce the notion of *Helly-maximal bodies*. Namely, a compact, convex body $M \subset \mathbf{R}^n$ is said to be Helly-maximal if him $M = \dim M$. For example, the complete solution of the Szökefalvi-Nagy problem in $\mathbf{R}^3$ (cf. Theorem 29.5) means that an indecomposable, compact, convex body $M \subset \mathbf{R}^3$ is Helly-maximal if and only if it is neither a stack, nor an outcut.

Lemma 43.2: A zonotope $M \subset \mathbf{R}^n$ is Helly-maximal if and only if it is indecomposable.   □

In [Ba 1], this lemma was proved in a complicated way. Here we give a simpler proof.

PROOF: The part "only if" is obvious: if md $M = n$, then the vector system $H(M)$ contains $n + 1$ minimally dependent vectors and consequently $H(M)$ is not splittable, i.e., M is not decomposable, cf. Corollary 4.3.

For proving the part "if", let $M \subset \mathbf{R}^n$ be an indecomposable zonotope with the representation $M = I_1 + \cdots + I_s$, where $I_1, \cdots, I_s$ are pairwise nonparallel segments. Passing from I_s to $I_{s-1}, I_{s-2}, \cdots$, we finally obtain a number $q \leq s$ such that the zonotope $I_1 + \cdots + I_k$ is indecomposable for $k = q, \cdots, s$, whereas $I_1 + \cdots + I_{q-1}$ is decomposable. For each $k = q-1, \cdots, s$ the zonotope $I_1 + \cdots + I_k$ is n-dimensional (indeed, if e.g. $I_1 + \cdots + I_{s-1}$ is not n-dimensional, then $I_1 + \cdots + I_s$ is decomposable, contradicting the assumption).

If $q < s$, then $I_1 + \cdots + I_s = (I_1 + \cdots + I_q) + (I_{q+1} + \cdots + I_s)$. Consequently $H(I_1 + \cdots + I_q) \subset H(I_1 + \cdots + I_s)$, i.e., md $(I_1 + \cdots + I_q) \leq$ md $(I_1 + \cdots + I_s)$. Thus it is sufficient to prove that md $(I_1 + \cdots + I_q) = n$.

Since $I_1 + \cdots + I_{q-1}$ is decomposable, we have $I_1 + \cdots + I_{q-1} = M_1 \oplus \cdots \oplus M_t$, where $M_1, \cdots, M_t$ are indecomposable zonotopes, each of which has a dimension $< n$. We are going to carry out an induction by the dimension n (for $n = 1$ the lemma is trivial). So, by the assumption of induction, for each zonotope $M_1, \cdots, M_t$ the lemma is valid, and we have to prove that the zonotope $I_1 + \cdots + I_q = M_1 \oplus \cdots \oplus M_t + I_q$ is Helly-maximal, i.e., md $(M_1 \oplus \cdots \oplus M_t + I_q) = n$.

We put $L_i = $ aff $M_i, i = 1, \cdots, t$. Hence $\mathbf{R}^n = L_1 \oplus \cdots \oplus L_t$. Denote by y the vector going along the segment I_q. Then $y = y_1 + \cdots + y_t$, where $y_i \in L_i, i = 1, \cdots, t$. It is clear that $y_1 \neq o, \cdots, y_s \neq o$ (otherwise the zonotope $M_1 \oplus \cdots \oplus M_t + I_q$ would be decomposable).

Since md is an affine invariant, without loss of generality we may assume that the decomposition $\mathbf{R}^n = L_1 \oplus \cdots \oplus L_t$ is orthogonal. Then

$$H(M_1 \oplus \cdots \oplus M_t) = H(M_1) \cup \cdots \cup H(M_t) \subset L_1 \cup \cdots \cup L_t.$$

Denote by n_i the dimension of the subspace $L_i, i = 1, \cdots, t$. According to the assumption of induction, we have md $M_i = \dim M_i = n_i$, and hence there are vectors $e_o^{(i)}, e_1^{(i)}, \cdots, e_{n_i}^{(i)} \in H(M_i) \subset L_i$ which are minimally dependent:

$$\gamma_o e_o^{(i)} + \gamma_1 e_1^{(i)} + \cdots + \gamma_{n_i} e_{n_i}^{(i)} = o, \tag{1}$$

where the coefficients are positive.

The vectors $e_1^{(i)}, \cdots, e_{n_i}^{(i)}$ form a basis in L_i. Consequently there is a vector among them which has a nonzero scalar product with y_i. We may assume (changing the enumeration of the vectors, if necessary) that $\langle e_1^{(i)}, y_i \rangle \neq 0$. Furthermore, the vectors $e_o^{(i)}, e_2^{(i)}, \cdots, e_{n_i}^{(i)}$ also form a basis in L_i. A similar reasoning shows that we may assume $\langle e_o^{(i)}, y_i \rangle \neq 0$.

Since $e_j^{(i)} \in H(M_i)$, there exists an $(n_i - 1)$-dimensional face $\Gamma_j^{(i)}$ of the zonotope M_i such that $e_j^{(i)}$ is the unit outward normal of $\Gamma_j^{(i)}$ (in the subspace L_i). We are especially interested in the faces $\Gamma_o^{(i)}$ and $\Gamma_1^{(i)}$ with the outward normals $e_o^{(i)}$ and $e_1^{(i)}$. For each $i = 1, \cdots, t-1$ we consider the zonotopes

$$\Delta^{(i)} = \sum_{j<i} M_j \oplus \Gamma_o^{(i)} \oplus \Gamma_1^{(i+1)} \oplus \sum_{j>i+1} M_j,$$

$$\Pi_o^{(i)} = \sum_{j<i} M_j \oplus \Gamma_o^{(i)} \oplus \sum_{j>i} M_j;$$

$$\Pi_1^{(i+1)} = \sum_{j\leq i} M_j \oplus \Gamma_1^{(i+1)} \oplus \sum_{j>i+1} M_j.$$

Then $\Delta^{(i)}$ is an $(n-2)$-dimensional face of the zonotope $M_1 \oplus \cdots \oplus M_t$, whereas $\Pi_o^{(i)}$ and $\Pi_1^{(i+1)}$ are its $(n-1)$-dimensional faces with the corresponding unit outward normals $e_o^{(i)}$ and $e_1^{(i+1)}$. The vector y is not parallel to the face $\Delta^{(i)}$, since $\langle e_o^{(i)}, y \rangle \neq 0$ (and $\langle e_1^{(i+1)}, y \rangle \neq 0$). Consequently $\Delta^{(i)} + I_q$ is an $(n-1)$-dimensional face of the zonotope $M_1 \oplus \cdots \oplus M_t + I_q$. Let u_i be the unit outward normal of this face (in $\mathbf{R}^n$). Each of the faces $\Pi_o^{(i)}, \Pi_1^{(i+1)}$ and $\Delta^{(i)} + I_q$ contains $\Delta^{(i)}$. Hence (up to a translation) their unit outward normals $e_o^{(i)}, e_1^{(i+1)}$ and u_i are orthogonal to the plane aff $\Delta^{(i)}$. This means that $e_o^{(i)}, e_1^{(i+1)}, u_i$ are linearly dependent, i.e. (since $e_o^{(i)}$ and $e_1^{(i+1)}$ are linearly independent)

$$u_i = \alpha_i e_o^{(i)} + \beta_i e_1^{(i+1)}. \tag{2}$$

Here $\alpha_i \neq 0, \beta_i \neq 0$, since $\langle u_i, y \rangle = 0, \langle e_o^{(i)}, y \rangle \neq 0, \langle e_1^{(i+1)}, y \rangle \neq 0$. Replacing $e_o^{(i)}, \cdots, e_{n_i}^{(i)}$ by $b_o^{(i)} = \lambda_o^{(i)} e_o^{(i)}, \cdots, b_{n_i}^{(i)} = \lambda_{n_i}^{(i)} e_{n_i}^{(i)}$ and u_i by $p_i = \mu_i u_i$ with suitable coefficients (successively for $i = 1, \cdots, t$), we can rewrite (1) and (2) in the form

$$b_o^{(i)} + \cdots + b_{n_i}^{(i)} = 0, \quad p_i = b_o^{(i)} + b_1^{(i+1)}. \tag{3}$$

From (3) we obtain

$$p_1 + \cdots + p_{t-1} + \sum_{i=1}^{t} (b_2^{(i)} + \cdots + b_{n_i}^{(i)}) + b_1^{(1)} + b_o^{(t)} = o,$$

and hence the $n + 1$ vectors

$$p_i \ (i = 1, \cdots, t - 1); \quad b_2^{(i)}, \cdots, b_{n_i}^{(i)} \ (i = 1, \cdots, t); \quad b_1^{(1)}, b_o^{(t)} \tag{4}$$

are linearly dependent. At the same time, if we remove from (4) the last vector $b_o^{(t)}$, we obtain n linearly independent vectors (indeed, the vectors $b_j^{(i)} = \lambda_j^{(i)} e_j$ for $i = 1, \cdots, t$ and $j = 1, \cdots, n_i$ are linearly independent; consequently, replacing $b_1^{(i+1)}$ by $b_1^{(i+1)} + b_o^{(i)} = p_i$ for $i = 1, \cdots, t$, we obtain again a system of linearly independent vectors). Returning to the previous vectors, we conclude that the vectors

$$u_i \ (i = 1, \cdots, t - 1), \quad e_2^{(i)}, \cdots, e_{n_i}^{(i)} \ (i = 1, \cdots, t), \quad e_1^{(1)}, e_o^{(t)} \tag{5}$$

(each of which is belonging to $H(M_1 \oplus \cdots \oplus M_t + I_q)$) are linearly dependent, whereas the vectors in (5), except for $e_o^{(t)}$, are linearly independent. Finally, changing the directions of some vectors (what is admissible, because for any zonotope Z it follows from $e \in H(Z)$ that $-e \in H(Z)$, by virtue of symmetry) we obtain a *positive* dependence between $n + 1$ vectors taken from $H(M_1 \oplus \cdots \oplus M_t + I_q)$, whereas some n of them are linearly independent. Consequently md $(M_1 \oplus \cdots \oplus M_t + I_q) = n$. ∎

Theorem 43.3: Let $M \subset \mathbf{R}^n$ *be a compact, convex body. If the set $B(M)$ of all its belt vectors is n-dimensional and not splittable, then the body M is Helly-maximal.* □

PROOF: Let $B \subset B(M)$ be a finite set of belt vectors of M which is not splittable (cf. the beginning of the proof of Theorem 5.5).

Let now Z be a zonotope whose segment summands are correspondingly parallel to the vectors of the system B. This zonotope is indecomposable (since the system B is not splittable), and therefore Lemma 43.2 shows that Z is Helly-maximal, i.e., md $Z = n$. The set $H(Z)$ is obtained in the following manner. We take any linearly independent subsystem $e_1', \cdots, e_{n-1}'$ from B and denote by $\pm p(e_1', \cdots, e_{n-1}')$ the unit normals of the hyperplane spanned by $\{e_1', \cdots, e_{n-1}'\}$. Then the set of the vectors, which are obtained in such a manner, is precisely the system $H(Z)$. Since the vectors from B are belt vectors of M, Theorem 41.2 implies that $H(Z) \subset \mathrm{cl}\, H(M)$. Furthermore, since md $Z = n$, the system $H(Z)$ contains $n + 1$ minimally dependent vectors and, by $H(Z) \subset \mathrm{cl}\, H(M)$, this is also true for the system cl $H(M)$. This shows that md $M = n$, i.e., him $M = \dim M$. ∎

In the theorem above it is not assumed that M is a belt body. If M is even a belt body, then (as the following theorem demonstrates) a stronger assertion can be derived.

Theorem 43.4: Let $M \subset \mathbf{R}^n$ *be a compact, convex belt body. Then the following properties of M are equivalent.*

(a) *The body M is indecomposable (i.e., the vector system $H(M)$ is not splittable).*

(b) *The system $B(M)$ of the belt vectors of M is not splittable.*

(c) *The body M is Helly-maximal.*   □

PROOF: We establish the implications (a) $\Rightarrow$ (b), (b) $\Rightarrow$ (c), and (c) $\Rightarrow$ (a).

(a) $\Rightarrow$ (b): Assume that the belt body M does not satisfy the condition (b), i.e., there exists a nontrivial decomposition $\mathbf{R}^n = L_1 \oplus L_2$ such that $B(M) \subset L_1 \cup L_2$. If the orthocomplements of the subspaces L_1, L_2 are denoted by N_1, N_2, then also $\mathbf{R}^n = N_1 \oplus N_2$ holds. Denote by n_1 and n_2 the dimensions of the subspaces L_1 and L_2, respectively. Since M is a belt body, the following property holds: If $q \in \mathrm{cl}\, H(M)$, then for any $\varepsilon > 0$ there exist linearly independent vectors $e_1, \cdots, e_{n-1} \in B(M)$ such that the hyperplane spanned by them has a normal vector p satisfying $\| p - q \| < \varepsilon$. The inclusion $B(M) \subset L_1 \cup L_2$ implies that either some n_1 vectors from $e_1, \cdots, e_{n-1}$ lie in L_1 and the remaining ones in L_2, or n_2 of them lie in L_2 and the remaining ones in L_1. In the first case, the vector p is orthogonal to the subspace L_1, i.e., $p \in N_1$; in the second one $p \in N_2$. Hence in any case $p \in N_1 \cup N_2$. In view of the arbitrariness of ε, we have $q \in N_1 \cup N_2$. This shows that $\mathrm{cl}\, H(M) \subset N_1 \cup N_2$, and therefore the vector systems $\mathrm{cl}\, H(M)$ and $H(M)$ are splittable. Thus the body M does not satisfy the condition (a).

(b) $\Rightarrow$ (c): We remark that for each belt body $M \subset \mathbf{R}^n$ the set $B(M)$ of all its belt vectors is n-dimensional (this follows from Definition 41.5, since any compact, convex body $K \subset \mathbf{R}^n$ has an n-dimensional set $H(K)$). Therefore Theorem 43.3 implies that if the set $B(M)$ is not splittable, then M is a Helly-maximal body, i.e., (b) implies (c).

(c) $\Rightarrow$ (a): Assume that the body M is decomposable, i.e., a representation $M = M_1 \oplus \cdots \oplus M_s$ is realizable, where each of the convex sets $M_1, \cdots, M_s$ has a dimension less than $\dim M$. Then, according to Theorem 25.4,

$$\begin{aligned} \mathrm{him}\, M &= \max(\mathrm{him}\, M_1, \cdots, \mathrm{him}\, M_s) \\ &\leq \max(\dim M_1, \cdots, \dim M_s) < \dim M, \end{aligned}$$

i.e., the body M is not Helly-maximal. Thus, if (a) is not satisfied, then also (c) is not satisfied.   ■

Now we are able to give the proof of Theorem 43.1.

PROOF: By Definition 41.5, each of the sets $M_1, \cdots, M_s$ in a decomposition $M = M_1 \oplus \cdots \oplus M_s$ is a belt body with respect to its affine hull. Furthermore, since each convex set M_i is indecomposable, it is Helly-maximal, i.e., $\mathrm{him}\, M_i = \dim M_i$ (cf. Theorem 43.4). Consequently,

$$\mathrm{him}\, M = \max(\mathrm{him}\, M_1, \cdots, \mathrm{him}\, M_s) = \max(\dim M_1, \cdots, \dim M_s). \qquad (6)$$

Exercises

1. Prove that a polytope $M \subset \mathbf{R}^n$ is a belt polytope if and only if for every $(n-2)$-face $\triangle$ of the body M^* the following holds: If for a face Φ of M^* its relative interior has a nonempty intersection with lin $\triangle$, then $\Phi \subset$ lin $\triangle$. This property of the polar polytope M^* is used in [Ba 1].

2. Prove that if $M \subset \mathbf{R}^n$ is a belt polytope, then for every face $\triangle$ of the polytope M^* (where dim $\triangle \le n-2$) the intersection (bd M^*) $\cap$ lin $\triangle$ is completely formed by faces of M^*, i.e., there is no face Φ of M^* with lin $\triangle \cap$ ri $\Phi \ne \emptyset$ and $\Phi \not\subset$ lin $\triangle$.

3. Prove that if M^* satisfies the condition indicated in the previous Exercise, then M is a belt polytope.

4. Let $M \subset \mathbf{R}^n$ be a belt polytope. Prove that for each vertex a of M there exists a positive number λ such that $-\lambda a$ is another vertex of M^* and $\{a, -\lambda a\} =$ bd $M^* \cap L_1 \cap \cdots \cap L_{n-1}$, where $L_1, \cdots, L_{n-1}$ are $n-1$ equatorial $(n-1)$-subspaces of M^*.

5. Prove that if a belt body M is represented in the form $M = M_1 \oplus \cdots \oplus M_s$, then each convex set M_i, $i = 1, \cdots, s$, is a belt body in its affine hull.

6. Let $e_1, \cdots, e_n$ be an orthonormal basis in $\mathbf{R}^n$ and $I_1, \cdots, I_{n+1}$ be segments which are correspondingly parallel to $e_1, \cdots, e_n, e_{n+1} = e_1 + \cdots + e_n$. Prove that the vector system $H(M)$ of the zonotope $M = I_1 + \cdots + I_{n+1}$ consists of the vectors

$$\pm e_i, \ i = 1, \cdots, n; \quad \pm \frac{1}{\sqrt{2}}(e_i - e_j), \ i, j = 1, \cdots, n \ (i \ne j).$$

7. In the notation of the previous exercise, prove that md $M = n$.

8. We say that a zonotope $M = I_1 + \cdots + I_s$ (where $I_1, \cdots, I_s$ are pairwise nonparallel segments) is *minimally indecomposable* if M indecomposable, but the vector sum of each $s-1$ of the segments $I_1, \cdots, I_s$ is a decomposable zonotope. Prove that every minimally indecomposable zonotope $M = I_1 + \cdots + I_s$ with $s = n+1$ is affinely equivalent to the zonotope considered in Exercise 6.

9. Prove that if $M = I_1 + \cdots + I_s \subset \mathbf{R}^n$ is a minimally indecomposable zonotope, then $n + 1 \le s \le 2n - 1$.

10. For every s, n with $n + 1 \le s \le 2n - 1$ construct a minimally indecomposable zonotope $M = I_1 + \cdots + I_s \subset \mathbf{R}^n$.

11. Prove that if a stack $M \subset \mathbf{R}^3$ is indecomposable, then it is not a belt body.

12. Prove that if an outcut $M \subset \mathbf{R}^3$ is indecomposable, then it is not a belt body.

13. Let $M \subset \mathbf{R}^n$ be a compact, convex body (not necessarily a belt body) and $M = M_1 \oplus \cdots \oplus M_s$ be its decomposition such that each of the convex sets $M_1, \cdots, M_s$ is indecomposable. Suppose that the set $B(M_i)$ of the belt vectors of M_i is not splittable in the space $L_i = $ aff M_i and lin $B(M_i) = L_i$, $i = 1, \cdots, s$. Prove that the equality (6) holds.

§44 Minimal fixing systems

L. Fejes Tóth [Fe] introduced the notion of a *primitive fixing system* for a compact, convex body $M \subset \mathbb{R}^n$. Such a system $F \subset \text{bd } M$ stabilizes M against any translation and no proper subset of F does the same. Using suitable minimally dependent systems contained in $H(M)$, we will give upper and lower bounds on the cardinalities of minimal fixing systems, see also [B-M 1], [B-MA]. For belt bodies, the exact cardinalities are deduced.

Let $M \subset \mathbb{R}^n$ be a compact, convex body and $F \subset \text{bd } M$. We say that a direction l, defined by a nonzero vector $e \in \mathbb{R}^n$, is an *outer moving direction* (with respect to F) if for every $\lambda > 0$ the relation $(-\lambda e + \text{int } M) \cap F = \emptyset$ holds. If there is no outer moving direction with respect to F, then F is said to be a *fixing system* of the body M (against translation). Visually, F is a fixing system of M if, assuming "fixing nails" at the points of F, it is impossible to translate M in any direction.

A fixing system $F \subset \text{bd } M$ is said to be *primitive*, if no proper subsystem $F' \subset F$ is a fixing system of M. We denote by $\rho(M)$ the smallest one of the integers s such that there exists a fixing system of M consisting of s points. A fixing system containing $\rho(M)$ points is said to be *minimal*. It is obvious that each minimal fixing system of M is primitive.

Theorem 44.1: A point system $F \subset \text{bd } M$ is a fixing system of a compact, convex body $M \subset \mathbb{R}^n$ if and only if each direction illuminates at least one point of F.   □

PROOF: If the direction l, defined by a vector $e \neq o$, illuminates a point $b \in F$ (i.e., $b + \lambda e \in \text{int } M$ for a positive λ), then $b \in -\lambda e + \text{int } M$, i.e., l is not an outer moving direction with respect to F. Conversely, if l is not an outer moving direction with respect to F, then l illuminates at least one point $b \in F$. ∎

Theorem 44.2: Let $F = \{b_1, \cdots, b_s\} \subset \text{bd } M$ be a fixing system for a compact, convex body $M \subset \mathbb{R}^n$. Then there exists a real number $\varepsilon > 0$ such that F remains to be fixing after an arbitrary ε-displacement. More exactly, every system $F' = \{b_1', \cdots, b_s'\} \subset \text{bd } M$ satisfying
$\| b_1' - b_1 \| < \varepsilon, \cdots, \| b_s' - b_s \| < \varepsilon$ is also a fixing system with respect to M.
□

PROOF: Since $\{b_1, \cdots, b_s\}$ is a fixing system, each direction illuminates at least one of the points $b_1, \cdots, b_s$, see Theorem 44.1. Denote by $W_i \subset S^{n-1}$ the set of all unit vectors whose directions illuminate the point b_i, $i = 1, \cdots, s$. Then $W_1, \cdots, W_s$ are open subsets of S^{n-1} which cover S^{n-1}, i.e., $W_1 \cup \cdots \cup W_s = S^{n-1}$. It follows that there exist compact sets $P_1, \cdots, P_s$ such that $P_i \subset W_i$, $i = 1, \cdots, s$, and $P_1 \cup \cdots \cup P_s = S^{n-1}$.

We fix an index $i \in \{1, \cdots, s\}$. Since $P_i \subset W_i$, for every vector $e \in P_i$ the corresponding direction illuminates the point b_i. Consequently there exist neighbourhoods $U_e(b_i) \subset \text{bd } M$, $V(e) \subset S^{n-1}$ of the points b_i, e, respectively, such that for every $e' \in V(e)$ the direction of the vector e' illuminates the points of $U_e(b_i)$. By compactness of P_i, there is a finite set $\{e_1, \cdots, e_k\} \subset P_i$ such that $P_i \subset V(e_1) \cup \cdots \cup V(e_k)$. Denote by U_i the intersection of the sets $U_e(b_i)$ taken over all $e \in \{e_1, \cdots, e_k\}$. Then for each $e \in V(e_1) \cup \cdots \cup V(e_k)$ (in particular, for every $e \in P_i$) the corresponding direction illuminates all the points belonging to U_i. This means that there exists an $\varepsilon_i > 0$ with the following property: If $b'_i \in \text{bd } M$ satisfies $\| b'_i - b_i \| < \varepsilon$, then for every $e \in P_i$ the corresponding direction illuminates the point b'_i.

Denoting by ε the minimal one of the constructed numbers ε_i, $i = 1, \cdots, s$, we obtain the following assertion: For every system $F' = \{b'_1, \cdots, b'_s\} \subset \text{bd } M$ with $\| b'_1 - b_1 \| < \varepsilon, \cdots, \| b'_s - b_s \| < \varepsilon$ and every $e \in P_1 \cup \cdots \cup P_s = S^{n-1}$, the direction of e illuminates at least one of the points $b'_1, \cdots, b'_s$. Consequently $\{b'_1, \cdots, b'_s\}$ is a fixing system for M (cf. Theorem 44.1). ∎

To formulate the following assertions, we introduce a new term: a system $F \subset \text{bd } M$ is *regular* if each point $b \in F$ is a *regular* boundary point of M.

Corollary 44.3: For each compact, convex body $M \subset \mathbf{R}^n$, there exists a regular fixing system $F \subset M$ which is minimal (and consequently primitive).   □

This assertion is an immediate consequence of Theorem 44.2. ∎

REMARK: If $F = \{b_1, \cdots, b_s\} \subset \text{bd } M$ is a primitive fixing system which is *not minimal*, then the system $\{b'_1, \cdots, b'_s\} \subset \text{bd } M$, obtained by an ε-displacement, is a fixing system for F (according to Theorem 44.2), but in general $\{b'_1, \cdots, b'_s\}$ is not primitive. For example, let $M \subset \mathbf{R}^2$ be a regular hexagon and $b_1, \cdots, b_6$ be its vertices. Then $\{b_1, \cdots, b_6\} \subset \text{bd } M$ is a primitive fixing system for M (in fact, a *maximal* one). But if we take points $b'_1, \cdots, b'_6$ which are correspondingly close to $b_1, \cdots, b_6$ and $b'_i \neq b_i$, $i = 1, \cdots, 6$, then we obtain a *nonprimitive* fixing system $\{b'_1, \cdots, b'_6\}$ with respect to M.   □

Theorem 44.4: Let $M \subset \mathbf{R}^n$ be a compact, convex body and $F = \{b_1, \cdots, b_s\} \subset \text{bd } M$ be a regular point system. Denote by $p_1, \cdots, p_s$ the unit outward normals of M at the points $b_1, \cdots, b_s$. The system $\{b_1, \cdots, b_s\}$ is a fixing one with respect to M if and only if the vector system $\{p_1, \cdots, p_s\} \subset S^{n-1}$ is not one-sided.   □

PROOF: Assume that the vector system $\{p_1, \cdots, p_s\}$ is one-sided, i.e., there exists $q \in S^{n-1}$ such that $\langle q, p_i \rangle \geq 0$ for $i = 1, \cdots, s$. This means that $M \subset \Pi_i = \{x : \langle p_i, x - b_i \rangle \leq 0\}$, whereas the ray l emanating from b_i in the direction of q does not have a common point with int Π_i. Consequently b_i is not illuminated from the direction defined by the vector q. This is true for $i = 1, \cdots, s$, i.e., the direction of q illuminates no point from $\{b_1, \cdots, b_s\}$. Thus $\{b_1, \cdots, b_s\}$ is not a fixing system (Theorem 44.1). This proves the part "only if".

Conducting this reasoning in the opposite direction, we obtain the part "if".
∎

For a compact, convex body $M \subset \mathbf{R}^n$ we denote by nos M the smallest one of the integers s such that there is a subsystem $\{p_1, \cdots, p_s\}$ of $H(M)$ which is not one-sided.

Corollary 44.5: *For every compact, convex body* $M \subset \mathbf{R}^n$, *the equality* $\overline{\rho(M) = \text{nos } M}$ *holds.*   □

This is an immediate consequence of Theorem 44.4.   ∎

Theorem 44.6: *For every compact, convex body* $M \subset \mathbf{R}^n$ *the inequalities*

$$n + 1 \leq \rho(M) \leq 2n + 1 - \text{md } M$$

hold.   □

PROOF: The inequality $\rho(M) \geq n + 1$ follows from Corollary 44.5, since each minimal, non-one-sided vector system in $\mathbf{R}^n$ contains at least $n + 1$ vectors and no more than $2n$ ones, cf. [Grü 5].

We now prove the second of the inequalities given above. Denote the number md M by m and choose $m + 1$ minimally dependent vectors $p_o, \cdots, p_m \in H(M)$. Furthermore, denote by L the m-dimensional subspace spanned by $p_o, \cdots, p_m$, by N its orthogonal complement, and by $\pi : \mathbf{R}^n \to N$ the orthogonal projection. Since M is compact, the vector system $H(M)$ is not one-sided. Consequently the vector system $\pi(H(M) \setminus L)$ is not one-sided in N. Thus, since N is $(n - m)$-dimensional, it is possible to choose no more than $2(n - m)$ vectors in $\pi(H(M) \setminus L)$ which form a system that is not one-sided in N. Assume that $r_1, \cdots, r_t \in \pi(H(M) \setminus L)$ form a non-one-sided system, where $t \leq 2(n - m)$. Let, furthermore, $q_1, \cdots, q_t$ be representatives of these vectors, i.e., $q_j \in H(M)$, $\pi(q_j) = r_j$ for $j = 1, \cdots, t$. Hence we obtain the vectors $p_o, \cdots, p_m, q_1, \cdots, q_t$ in $H(M)$, and their number is not greater than $m + 1 + 2(n - m) = 2n + 1 - \text{md } H$.

To finish the proof, it suffices to establish that the vector system $p_o, \cdots, p_m$, $q_1, \cdots, q_t$ is not one-sided in $\mathbf{R}^n$ (Corollary 44.5).

Assume, conversely, that the vector system $p_o, \cdots, p_m, q_1, \cdots, q_t$ is one-sided, i.e., there exists a nonzero vector $g \in \mathbf{R}^n$ which yields a nonpositive scalar product with each of these vectors. It follows from $\langle g, p_o \rangle \leq 0, \cdots, \langle g, p_m \rangle \leq 0$ that all these scalar products are equal to zero (since $\lambda_o p_o + \cdots + \lambda_m p_m = o$ with positive coefficients). Consequently g is orthogonal to the subspace L, i.e., $\pi(g) = g$. This means that $\langle g, q_j \rangle = \langle g, \pi(q_j) \rangle$, $j = 1, \cdots, t$, and therefore $\langle g, \pi(q_j) \rangle \leq 0$ for $j = 1, \cdots, t$. But the vectors $\pi(g_j)$, $j = 1, \cdots, t$, form a system which is not one-sided in N, and hence it follows from $\langle g, \pi(q_j) \rangle \leq 0$, $j = 1, \cdots, t$, that $g = o$, contradicting the choice of g. This contradiction shows that the system $p_o, \cdots, p_m, q_1, \cdots, q_t$ is not one-sided.   ∎

We now formulate an improved estimate for the integer $\rho(M)$ obtained in [B-MA].

Theorem 44.7: For every compact, convex body $M \subset \mathbf{R}^n$, the inequality

$$n + \frac{n}{\mathrm{md}\, M} \le \rho(M)$$

holds.   □

To prove this theorem, we first establish auxiliary propositions.

Lemma 44.8: Let H be a vector system in $\mathbf{R}^n$ which is not one-sided and $m = \mathrm{md}\, H$. Let, furthermore, $q_1, \cdots, q_{m+1}$ be minimally dependent vectors. Denote by N the orthogonal complement of the subspace $L = \mathrm{lin}\,(q_1, \cdots, q_{m+1})$ and by $\pi : \mathbf{R}^n \to N$ the orthogonal projection. Then the vector system $H' = \pi(H \setminus L)$ is not one-sided in N and $\mathrm{md}\, H' \le m$.   □

PROOF: Since the vector system H is not one-sided, the vector system $H' = \pi(H \setminus L)$ is not one-sided in N. Assume that $\mathrm{md}\, H' > m$, i.e., there exist minimally dependent vectors $b_1, \cdots, b_s \in H'$, where $s > m + 1$. Choose representatives $a_1, \cdots, a_s \in H$ of these vectors, i.e., $b_i = \pi(a_i)$, $i = 1, \cdots, s$. Denote by $\mu_1 b_1 + \cdots + \mu_s b_s = 0$ a positive dependence between $b_1, \cdots, b_s$. The coefficients $\mu_1, \cdots, \mu_s$ are defined uniquely up to a common positive multiplier. Then

$$\pi(\mu_1 a_1 + \cdots + \mu_s a_s) = \mu_1 b_1 + \cdots + \mu_s b_s = o,$$

i.e., the vector

$$c = \mu_1 a_1 + \cdots + \mu_s a_s \qquad (1)$$

belongs to L. If $c = o$, then $a_1, \cdots, a_s$ are minimally dependent and hence $\mathrm{md}\, H \ge s - 1 > m$, contradicting the notation. Consequently $c \ne o$. Since $q_1, \cdots, q_{m+1}$ are minimally dependent in the m-dimensional subspace L, there is a positive dependence between c and no more than m of the vectors $q_1, \cdots, q_{m+1}$. Without loss of generality we may assume that this positive dependence has the form

$$c + \nu_1 q_1 + \cdots + \nu_t q_t = o, \qquad (2)$$

where $1 \le t \le m$. Substituting (1) into (2), we obtain a positive dependence

$$\mu_1 a_1 + \cdots + \mu_s a_s + \nu_1 q_1 + \cdots + \nu_t q_t = o.$$

Moreover, $a_1, \cdots, a_{s-1}, q_1, \cdots, q_t$ are linearly independent (if $\alpha_1 a_1 + \cdots + \alpha_{s-1} a_{s-1} + \beta_1 q_1 + \cdots + \beta_t q_t = o$, then, applying π, we obtain $\alpha_1 b_1 + \cdots + \alpha_{s-1} b_{s-1} = o$, i.e., $\alpha_1 = \cdots = \alpha_{s-1} = 0$, since $b_1, \cdots, b_{s-1}$ are linearly independent in N; hence $\beta_1 q_1 + \cdots + \beta_t q_t = o$, i.e., $\beta_1 = \cdots = \beta_t = 0$, since $q_1, \cdots, q_t$ are linearly independent in L, by $t \le m$). Thus the vectors $a_1, \cdots, a_s, q_1, \cdots, q_t$ belonging to H are minimally dependent. But the number of these vectors is equal to $s + t \ge s + 1 > m + 2$, contradicting $\mathrm{md}\, H = m$.   ∎

Lemma 44.9: Let H be a non-one-sided vector system in $\mathbf{R}^n$ with $\operatorname{md} H = m$. Then the cardinality $|H|$ of H satisfies the inequality

$$|H| \geq n + \frac{n}{m}. \quad \square$$

PROOF: We will use induction over n. The initial step of the induction is $n = 1$ (and hence $m = 1$). In this case the assertion of the Lemma is obvious.

Assume that for any dimension smaller than n the assertion of the Lemma holds, and let $H \subset \mathbf{R}^n$ be a non-one-sided vector system with $\operatorname{md} H = m$. Choose minimally dependent vectors $q_1, \cdots, q_{m+1} \in H$ and conserve the notation L, N, π, H' introduced in the previous Lemma. Denote the integer $\operatorname{md} H'$ by m'. Then $m' \leq m$ (by Lemma 44.8). Since $\dim N = n - m < n$, we have, by the induction hypothesis,

$$|H'| \geq n - m + \frac{n-m}{m'}.$$

Consequently,

$$|H| \geq (m+1) + |H'| \geq (m+1) + \left(n - m + \frac{n-m}{m'}\right) =$$

$$n + 1 + \frac{n-m}{m'} \geq n + 1 + \frac{n-m}{m} = n + \frac{n}{m}. \quad \blacksquare$$

PROOF OF THEOREM 44.7: Let $M \subset \mathbf{R}^n$ be a compact, convex body, $\operatorname{md} M = \bar{m}$. Denote the integer $\rho(M)$ by r. Then, by Corollary 44.5, there exists a non-one-sided subsystem $H = \{q_1, \cdots, q_r\} \subset H(M)$. The integer $m = \operatorname{md} H$ is not greater than $\bar{m}$. Consequently, by Lemma 44.9,

$$\rho(M) = r = |H| \geq n + \frac{n}{m} \geq n + \frac{n}{\bar{m}} = n + \frac{n}{\operatorname{md} M}. \quad \blacksquare$$

Corollary 44.10: For every compact, convex body $M \subset \mathbf{R}^n$ the inequalities

$$n + \frac{n}{\operatorname{md} M} \leq \rho(M) \leq 2n + 1 - \operatorname{md} M$$

hold. $\square$

We remark that the lower and upper estimates in Corollary 44.10 are exact (cf. Theorem 44.14).

Theorem 44.11: Let M be a compact, convex body in $\mathbf{R}^n$ which is representable as a direct vector sum: $M = M_1 \oplus \cdots \oplus M_k$. Then

$$\rho(M) = \rho(M_1) + \cdots + \rho(M_k). \quad \square$$

PROOF: Without loss of generality we may assume that aff M_i is a subspace, i.e., $o \in \operatorname{aff} M_i$, $i = 1, \cdots, k$. Denote the number $\rho(M_i)$ by ρ_i and choose (for every $i = 1, \cdots, k$) points $b_j^{(i)} \in \text{rbd } M_i$, $j = 1, \cdots, \rho_i$, which form a regular (minimal) fixing system of the body M_i in the subspace aff M_i. Besides, we choose a point $a_i \in \text{ri } M_i$, $i = 1, \cdots, k$. Furthermore, the point

$$a_1 + \cdots + a_k + \left(b_j^{(i)} - a_i \right) \in \text{bd } M, \, i = 1, \cdots, k; \, j = 1, \cdots, \rho_i,$$

is denoted by $c_j^{(i)}$. The obtained system $\left\{ c_j^{(i)} \right\}$ contains $\rho_1 + \cdots + \rho_k$ points in bd M. We are going to show that $\left\{ c_j^{(i)} \right\}$ is a fixing system of the body M. Indeed, let $e = e_1 + \cdots + e_k$ be a nonzero vector in $\mathbf{R}^n$, where e_i belongs to aff M_i, $i = 1, \cdots, k$. Since $e \neq o$, at least one of the summands $e_1, \cdots, e_k$ is nonzero, say $e_1 \neq o$. Then the direction of the vector e_1 illuminates in aff M_1 at least one of the points $b_j^{(1)}$, $j = 1, \cdots, \rho_1$. Assume that this illuminated point is $b_1^{(1)}$. This means that $b_1^{(1)} + \lambda e_1 \in \text{ri } M_1$ for λ small enough. Since a_i is a point from the relative interior of M_i, $i = 2, \cdots, k$, we have $c_1^{(1)} + \lambda e \in \text{int } M$ for λ small enough, i.e., the point $c_1^{(1)}$ is illuminated by the direction of the vector e. Thus every nonzero vector $e \in \mathbf{R}^n$ illuminates at least one of the points $c_j^{(i)}$, i.e., $\left\{ c_j^{(i)} \right\}$ is a fixing system for F. Since this system $\left\{ c_j^{(i)} \right\}$ contains $\rho_1 + \cdots + \rho_k$ points, we obtain the inequality

$$\rho(M) \leq \rho_1 + \cdots + \rho_k.$$

Now we will prove the opposite inequality. Assume, on the contrary, that $\rho(M) < \rho_1 + \cdots + \rho_k$ and choose a fixing system $b_1, \cdots, b_\mu$ for M with $\mu < \rho_1 + \cdots + \rho_k$. We denote by $\pi_i : \mathbf{R}^n \to \text{aff } M_i$ the projection parallel to the direct vector sum of all the subspaces aff M_j except for aff M_i. Furthermore, we say that the point b_k $(k = 1, \cdots, \mu)$ belongs to the index i if $\pi_j(b_k) \in \text{ri } M_j$ for all j except for i (and consequently $\pi_i(b_k) \in \text{rbd } M_i$, since $b_k \in \text{bd } M$). We remark that it is possible that a point b_k belongs to no index, i.e., $\pi_j(b_k) \in \text{rbd } M_j$ for more than one index j. Since $\rho(M) < \rho_1 + \cdots + \rho_k$, there is an index i such that there are *less* than ρ_i points among $b_1, \cdots, b_\mu$ which belong to this index i. Let, for example, the points $b_1, \cdots, b_\nu$ belong to the index 1, where $\nu < \rho_1$, and the other points $b_{\nu+1}, \cdots, b_\mu$ do not belong to the index 1. The inequality $\nu < \rho_1$ yields the existence of a direction l in aff M_1, which illuminates (in aff M_1) none of the points $\pi_1(b_1), \cdots, \pi_1(b_\nu) \in \text{rbd } M_1$. The same direction l, considered in the space $\mathbf{R}^n$, illuminates none of the points $b_1, \cdots, b_\nu \in \text{bd } M$. Moreover, the direction l illuminates none of the points $b_{\nu+1}, \cdots, b_\mu$, since $l \parallel \text{aff } M_1$, whereas the points b_ϱ with $\nu + 1 \leq \varrho \leq \mu$ do not belong to the index 1. Indeed, for $\varrho = \nu + 1, \cdots, \mu$ there exists an index $j \geq 2$ such that $\pi_j(b_\varrho) \in \text{rbd } M_j$, and consequently the line of direction l passing through b_ϱ does not meet int M. Thus l is an outer moving direction of M and therefore $b_1, \cdots, b_\mu$ is not a fixing system, contradicting the choice of the system. This contradiction shows that

$$\rho(M) \geq \rho_1 + \cdots + \rho_k. \quad \blacksquare$$

Corollary 44.12: Let $M \subset \mathbf{R}^n$ be a compact, convex body which is represen-table as a direct vector sum: $M = M_1 \oplus \cdots \oplus M_k$. Then

$$n + \frac{\dim M_1}{\mathrm{md}\, M_1} + \cdots + \frac{\dim M_k}{\mathrm{md}\, M_k} \le \rho(M) \le 2n + k - \mathrm{md}\, M_1 - \cdots - \mathrm{md}\, M_k$$

$$= n + k + (\dim M_1 - \mathrm{md}\, M_1) + \cdots + (\dim M_k - \mathrm{md}\, M_k). \quad \square$$

This is an immediate consequence of Theorem 44.11 and Corollary 44.10, since $\rho(M_i) \le 2\dim M_i + 1 - \mathrm{md}\, M_i$ and $\dim M_1 + \cdots + \dim M_k = n$.   ∎

Corollary 44.13: Let $M \subset \mathbf{R}^n$ be a compact, convex body, which is repre-sentable as a direct vector sum: $M = M_1 \oplus \cdots \oplus M_k$. If every convex set M_i, $i = 1, \cdots, k$, is Helly-maximal (i.e., $\dim M_i = \mathrm{md}\, M_i$), then

$$\rho(M) = n + k. \quad \square$$

This is an immediate consequence of the previous Corollary.   ∎

For example, each simplex and each ball are Helly-maximal. Consequently, if a compact, convex body $M \subset \mathbf{R}^n$ is the direct vector sum of k simplices (or k balls), then $\rho(M) = n + k$. This shows that each value of $\rho(M)$ between the limits indicated in Corollary 44.10 is realizable. More detailed, the following assertion holds.

Theorem 44.14: The lower and the upper estimates contained in Corollary 44.10 are exact. Moreover, for every integer ρ satisfying

$$n + \frac{n}{m} \le \rho \le 2n + 1 - m \tag{3}$$

there exists a compact, convex body $M \subset \mathbf{R}^n$ with

$$\mathrm{md}\, M = m \quad and \quad \rho(M) = \rho. \quad \square \tag{4}$$

The simplest idea of a proof is to consider $M = B_1 \oplus \cdots \oplus B_k$, where B_i is a Helly-maximal set (for example, a ball or a simplex) that has a dimen-sion $\le m$, $i = 1, \cdots, k$, and to establish that M satisfies (4) for suitable dimensions of the sets B_i. Thus, after the realization of this idea (cf. Exer-cises 10, 11), we would possess concrete examples of bodies which satisfy (4) under the condition (3). But these bodies are *decomposable*, i.e., each of them is representable as the direct vector sum of convex sets of smaller di-mensions. In Exercises 12, 13 the way is indicated to show that, by a more complicated construction, it is possible to obtain *indecomposable* bodies, i.e., for any integer ρ satisfying (3) with $m \ge 2$ there exists a compact, convex, indecomposable body $M \subset \mathbf{R}^n$ such that $\mathrm{md}\, M = m$ and $\rho(M) = \rho$.

We now give an account of the problem for *belt bodies*. Namely, we can give a complete solution of the minimal fixing system problem for belt bodies.

Theorem 44.15: Let $M \subset \mathbf{R}^n$ be a belt body (in particular, a zonoid). Let, furthermore, $\overline{M} = M_1 \oplus \cdots \oplus M_k$, where the convex sets $M_1, \cdots, M_k$ are indecomposable. Then

$$\rho(M) = n + k. \quad \square$$

PROOF: Theorem 43.4 affirms that, if a belt body N is indecomposable, then it is Helly-maximal, i.e., dim $N = $ md N. This and Corollary 44.13 imply the statement above. ∎

We will give now a short survey on known results about fixing systems (not only minimal ones). L. Fejes Tóth [Fe] showed that fixing systems for smooth convex bodies are strongly connected with the classification of the *primitive polytopes* (i.e., compact, convex n-polytopes P whose facet system has the following property: if any facet is omitted, then the intersection of the supporting half-spaces of the remaining facets is not bounded). For a wider discussion on related polytope classes we refer to [M-S-S]. B. Grünbaum [Grü 5] showed that for every compact, convex body $M \subset \mathbf{R}^n$ the inequality $\rho(M) \le 2n$ holds, and he posed the task to show that only the parallelotopes can reach this bound. (This is covered by our Theorem 44.6.) Furthermore, we mention the following result of P. Mani [Man 1]: there are convex n-polytopes $P \subset \mathbf{R}^n$, $n \ge 3$, with $2n + 2$ vertices which are fixed by their vertex set, and no polytope with strictly less than $2n + 2$ vertices can be fixed by its vertex set. In view of containment problems for convex bodies (which are, in a sense, related to fixing system problem), a reformulation of Mani's result is given in [G-K-W]. S.Fudali [Fu 2] investigated interesting relations between fixing systems and homothetical coverings of convex bodies.

L. Fejes Tóth also mentioned that the *maximal* number of points in a primitive fixing system in the plane is 6, attained by the vertices of a hexagon with pairwise parallel sides (cf. also [To]). S. Fudali [Fu 1] showed that these plane figures are the only ones attaining this upper bound. The analogous question for $n \ge 3$ has an unexpected solution found by B. Bollobás [Bl 1]. He proves that *for any positive integer k there is a compact, convex body $M \subset \mathbf{R}^3$ with a primitive fixing system $F \subset $ bd M containing more than k points, whereas every primitive fixing system for any given compact, convex body $M \subset \mathbf{R}^n$ has a finite number of points.* The construction of the example from [Bl 1] is sketched (in an improved form) in Exercises 16–17. In [Da 2] L. Danzer formulated the conjecture that for *zonotopes* the cardinalities of primitive fixing systems are *bounded*. More detailed, the rhombic dodecahedron $M \subset \mathbf{R}^3$ has a primitive fixing system consisting of 14 points (its vertices), and L. Danzer assumed that this is the *maximal* value of the cardinality of a primitive fixing system for compact, convex bodies in $\mathbf{R}^3$. The analogous zonotope in $\mathbf{R}^n$ (cf. Exercise 6 in the previous section) has $2(2^n - 1)$ vertices, and L. Danzer assumed that this body has its vertex set as the primitive fixing system of maximal cardinality. But this conjecture is false! As it is shown in

[B-M 3], there are zonotopes in $\mathbf{R}^n$ with any cardinality of primitive fixing systems. In Exercises 19–20 we describe a construction of such a zonotope.

Exercises

1. Prove that if a compact, convex body $M \subset \mathbf{R}^n$ is smooth, then there is a fixing system $F \subset \operatorname{bd} M$ of M containing $n+1$ points.

2. Show by a counterexample that the following assertion is false: If a regular fixing system of a compact, convex body $M \subset \mathbf{R}^n$ is primitive, then it is minimal.

3. Describe all fixing systems for an n-dimensional parallelotope M. Prove that $\rho(M) = 2n$.

4. Prove that if a compact, convex body $M \subset \mathbf{R}^n$ satisfies the condition $\rho(M) = 2n$, then it is a parallelotope.

5. Prove that if for a compact, convex body $M \subset \mathbf{R}^n$ the relation $\operatorname{md} M = n$ holds, then $\rho(M) = n + 1$.

6. Prove that a compact, convex body $M \subset \mathbf{R}^3$ satisfies $\rho(M) = 5$ if and only if $\operatorname{md} M = 2$, i.e., M is either a stack, or an outcut, or a direct vector sum (but not a parallelotope).

7. Prove that every stack $M \subset \mathbf{R}^n$ satisfies $\rho(M) = 2n - 1$.

8. The results of the Exercises 6 and 7 above naturally yield the following conjecture: If $M \subset \mathbf{R}^n$ is a compact, indecomposable, convex body with $\operatorname{md} M = 2$, then $\rho(M) = 2n - 1$. Show by a counterexample that for $n \geq 4$ this conjecture is false.

9. Prove that if a compact, convex, centrally symmetric body $M \subset \mathbf{R}^n$ satisfies the condition $\operatorname{md} M = 2$, then $\rho(M) = n + q$, where q is the maximal one of the integers k, for which M is representable as a direct vector sum of k convex sets of nonzero dimensions.

10. Let $M \subset R^n$ be a smooth, compact, convex body and $F = \{a_1, \cdots, a_s\} \subset \operatorname{bd} M$ be a fixing system (not necessarily minimal). Prove that if the system F is primitive, then the outward normals of M at the points of F form a *simplicial* vector system (cf. Exercise 6 in section 17). Consequently F contains no more than $2n$ points. This gives a proof of Grünbaum's result $\rho(M) \leq 2n$ [Grü 5].

11. Let n, m, ρ be integers with $1 \leq m \leq n$ and

$$n + \frac{n}{m} \leq \rho \leq 2n + 1 - m. \tag{5}$$

Prove that there exists a decomposition

$$\mathbf{R}^n = L_1 \oplus \cdots \oplus L_{\rho-n} \tag{6}$$

such that

$$m = \dim L_1 \geq \cdots \dim L_{\rho-n} \geq 1 \tag{7}$$

(for details, see Lemma 3 from [B-MA]).

12. Let (6) be the decomposition as in the previous exercise and B_i be a Helly-maximal, convex body in L_i (for example, a ball or a simplex). Prove that the convex body $M = B_1 \oplus \cdots \oplus B_k$ satisfies the condition $\operatorname{md} M = m$ and $\rho(M) = \rho$. This gives the proof of Theorem 44.14.

13. Let n, m, ρ be integers such that $n \geq m \geq 2$ and ρ satisfies (5). Let, furthermore, $\mathbf{R}^n = L_1 \oplus \cdots \oplus L_k$ be the direct decomposition as in Exercise 11, where $k = \rho - n$. We put $\dim B_i = n_i$ and choose in B_i minimally dependent vectors $a_o^{(i)}, a_1^{(i)}, \cdots, a_{n_i}^{(i)}$, $i = 1, \cdots, k$. Let $b_j = -a_o^{(1)} - a_o^{(j)}$, $j = 2, \cdots, k$. Prove that

$$H = \{a_s^{(i)}, b_j\}, \quad i = 1, \cdots, k, \ s = 0, 1, \cdots, n_i, \ j = 2, \cdots, k,$$

is a non-one-sided vector system such that $\operatorname{md} H = m$ and every non-one-sided subsystem $H' \subset H$ contains no less than ρ vectors. (for details, see Lemma 4 from [B-MA].

14. With the help of the previous exercise, prove that it is possible to construct an *indecomposable*, compact, convex body M which satisfies the conclusion of theorem 44.14 (for details, see Theorem 3 from [B-MA].

15. Let $M \subset \mathbf{R}^n$ be a compact, convex body and $F \subset \operatorname{bd} M$ be a primitive fixing system of M. Prove that F contains only a finite number of points.

16. Let e_1, e_2, e_3 be an orthonormal basis in $\mathbf{R}^3$ and (x_1, x_2, x_3) be the corresponding coordinate system. Denote by B the ball $x_1^2 + x_2^2 + x_3^2 \leq 2$ and by M the body $\operatorname{conv}\{B \cup \{b\}\}$, where b is the point $(0, 0, -2)$. Let $C_{(\gamma)}$ be the circumference $\operatorname{bd} B \cap \Gamma_\lambda$, where Γ_λ is the plane $\{x : x_3 = \gamma\}$, $-\sqrt{2} < \gamma < \sqrt{2}$. Let $a_1^{(\varepsilon)}, \cdots, a_s^{(\varepsilon)}$ be points which divide the circumference $C_{1-\varepsilon}$ into s equal parts, $s \geq 3$. Choose an arbitrary point $q \in C_{-1}$ and denote by l_q the direction defined by the vector $q - b$. (Fig. 250). Prove that there exists a number $\varepsilon > 0$ such that (i) for any $q \in C_{-1}$ the direction l_q illuminates at least one of the points $a_1^{(\varepsilon)}, \cdots, a_s^{(\varepsilon)}$; (ii) for every $k \in \{1, \cdots, s\}$ there exists a point $q \in C_{-1}$ such that the direction l_q illuminates the point $a_k^{(\varepsilon)}$ and does not illuminate any point $a_i^{(\varepsilon)}$ with $i \neq k$.

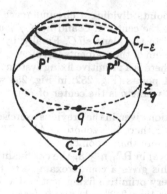

Fig. 250

17. In the notation of Exercises 16, prove that if $\varepsilon > 0$ is a suitable real number, then $F = \{b, a_1^{(\varepsilon)}, \cdots, a_s^{(\varepsilon)}\}$ is a *primitive* fixing system for the body M. Thus for any integer $s \geq 3$ there exists a primitive fixing system $F \subset \mathrm{bd}\, M$ consisting of $s + 1$ points.

It should be remarked that the above constructed body $M = \mathrm{conv}\, B \cup \{b\}$ (this example is given in [B-M 3]) has *only one* nonregular boundary point b. The initial example by B. Bollobás (Fig. 251) is a body with infinitely many nonregular boundary points.

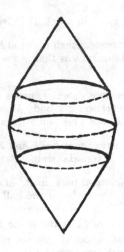

Fig. 251

18. Using the construction conducted in the previous exercises, show that for any integer $s \geq 3$ there exists a convex polytope $M \subset \mathbf{R}^3$ such that M has a primitive fixing system consisting of $s + 1$ boundary points.

19. Let $c_1, \cdots, c_m$ be points dividing a circumference $C \subset \mathbf{R}^3$ into s equal arcs, $m \geq 9$. Denote by c the center of C and by $b' \neq c$ a point such that the vector $b' - c$ is orthogonal to the plane $\mathrm{aff}\, C$. Finally, by M we denote the zonotope which is the vector sum of the segments $[c_1, p], \cdots, [c_m, p]$. Prove (for details see also [B-M 3]) that there is a primitive fixing system $F = \{b, a_1, \cdots, a_m\} \subset \mathrm{bd}\, M$ consisting of $m + 1$ points (Fig. 252; in Fig. 253 we give the view of M from the direction $b - q$, where q is the center of M).

20. Using the construction given in the previous exercise, prove that for any positive integers $n \geq 3$ and s there is a zonotope $M \subset \mathbf{R}^n$ that has a primitive fixing system containing more than s points. Thus in the class of belt bodies (even in the class of zonotopes) in $\mathbf{R}^n$, $n \geq 3$, the cardinality of primitive fixing systems is not bounded. This gives a counterexample to the conjecture of L. Danzer which says that every primitive fixing system for any zonotope $M \subset \mathbf{R}^n$, $n \geq 3$, contains no more than $2(2^n - 1)$ points (see p. 359).

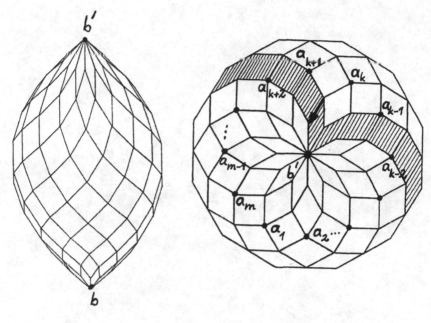

Fig. 252   Fig. 253

VIII. Some research problems

PROBLEM 1:

Let $d(x, y)$ be a metric in the n-dimensional vector space $\mathbf{R}^n$ (without any connection to the metric induced by the norm and the linear operations in $\mathbf{R}^n$). We say that the metric d is *invariant with respect to translations* if

$$d(x + a, y + a) = d(x, y)$$

for any $a, x, y \in \mathbf{R}^n$. Furthermore, we say that a metric d is *normable* if there exists a norm $\| \cdot \|$ in $\mathbf{R}^n$ such that $d(x, y) = \| x - y \|$ for any $x, y \in \mathbf{R}^n$. Finally, we say that a metric d is *bounded* if the set

$$B = \{x \in \mathbf{R}^n : d(o, x) \le 1\}$$

is bounded in $\mathbf{R}^n$. The problem is to describe a condition under which a metric d in $\mathbf{R}^n$ is normable.

Let d be a metric in $\mathbf{R}^n$ invariant with respect to translations. Then the metric d is normable if and only if each d-segment (in the sense of this metric) is linearly convex, cf. [SP-SV 1]. It is unknown whether this condition is equivalent to the following requirement: every d-convex set is linearly convex. In other words, the conjecture is the following (this is due to [SP-SV 1]): if a metric d in $\mathbf{R}^n$ is invariant with respect to translations, then each d-segment is linearly convex (i.e., d is normable) if and only if each d-convex set is linearly convex. Thus, a counterexample to this conjecture would be a metric d in $\mathbf{R}^n$ invariant with respect to translations such that every d-convex set is linearly convex, while not all d-segments are linearly convex.

We recall that not each d-segment is d-convex, cf. Example 9.3. In this connection it would be interesting to describe the metrics d for which every d-segment is a d-convex set. This problem is interesting even for the case of normable metrics in $\mathbf{R}^n$. For example, if $\Re^2$ is a *two-dimensional* Minkowski space, then every d-segment is a d-convex set. Example 10.4 gives a three-dimensional Minkowski space $\Re^3$ in which every d-segment is a d-convex set. Example 9.3 shows that, in general, this is not true for Minkowski spaces. Moreover, it is not clear whether the requirement that the unit ball $\sum \subset \Re^n$ is d-convex implies the d-convexity of any d-segment.

Finally, we indicate a connection of the considered problem to the linear convexity of d-segments. Consider the Cauchy equation

$$f(x + y) = f(x) + f(y), \qquad\qquad (*)$$

where x, y are real numbers. If a real-valued, continuous function $f(x)$ defined on $\mathbf{R}$ satisfies the equation $(*)$, then it is linear, i.e., $f(x) = kx$ (where k is a real number). Moreover, if f is a measurable function satisfying $(*)$, then it is linear as well. But with the help of Zorn's Lemma it is possible to construct a real-valued function f (not measurable) which satisfies $(*)$ and is not linear. Such an example was first constructed by G. Hamel [Ham]; cf. the more detailed account (with applications to the problem of measurement of volumes) in [Bt 16]. Moreover, it is easy to construct a Hamel function which takes the value 0 only at the point o. Then

$$d(x, y) = |f(x - y)|$$

is a metric in $\mathbf{R}$ invariant with respect to translations. Similar metrics can be indicated in $\mathbf{R}^n$, for every integer $n > 1$. And these metrics do not satisfy the requirement that each d-segment is linearly convex. Thus the linear convexity of any d-segment is not a consequence of the invariance with respect to translations. This shows that both the requirements explained above (the invariance with respect to translations and the linear convexity of d-segments) are independent.

PROBLEM 2:

Let X be a metric space with $d(x, y)$ as its metric. Let, furthermore, $M \subset X$ be a d-convex set. We say that a point $x \in M$ is *strictly d-extreme* if there are no points $x_1, x_2 \in M \setminus \{x\}$ such that the d-segment $[x_1, x_2]_d$ contains x (cf. Exercise 11 in §10). Strictly d-extreme points are introduced in [SV 4] (where they are called "d-extremal"). The set of all strictly d-extreme points of a d-convex set $M \subset X$ we denote by $stext_d M$.

Not every compact, d-convex set $M \subset X$ has strictly d-extreme points. For example, in the Minkowski space $\mathbf{R}^n$ with the norm $\| x \| = \sum\limits_{i=1}^{n} |x_i|$ (cf. Example 12.6) the cube $M = \{x : |x_i| \leq 1, \ i = 1, \cdots, n\}$ has no strictly d-extreme point. In this connection it is interesting to describe Minkowski spaces $\Re^n$ in which the following analogue of the Krein-Milman Theorem (cf. Theorem 2.2) holds:

$$M = \mathrm{conv}_d(stext_d M) \qquad\qquad (*)$$

for every compact, d-convex set $M \subset \Re^n$. For the planar case this description is given in Theorem 14.13 of the monograph [SV 4]. Certainly, this problem can be considered for an arbitrary metric space (not only for a Minkowski space $\Re^n$). For a class of metric spaces this problem is solved in [SV 4].

Namely, consider a finite, connected graph G with the following induced metric: each edge of G is considered as a linear segment of length 1, and for any vertices $x_1, x_2 \in G$ their distance is the minimal length of arcs which join x_1 and x_2 in G. The theorem proved in [SV 4] affirms that the following three conditions are equivalent:

(i)   the equality (*) holds for every compact, d-convex set $M \subset G$;

(ii) for every compact d-convex set $M \subset G$ the equality $M = \bigcup [x_1, x_2]_d$ holds, where the union is taken over all strictly d-extreme points $x_1, x_2 \in M$;

(iii) the graph G does not contain any subgraph isomorphic to the graph shown in Fig. 254, and G is *chordable* (i.e., in every simple cycle containing more than three vertices there are two nonadjacent vertices which are joined by an edge in G).

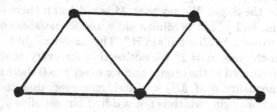

Fig. 254

It is also interesting to find another definition of strictly d-extreme points such that the equality (*) holds for every compact, d-convex set M. For example, it is possible to consider d-boundary points: a point z of a d-convex set M is said to be a *d-boundary point* [SV 4] if there exists a point $y \in M$ such that for every $x \in M \setminus \{z\}$ we have $z \notin [x, y]_d$. It is easy to show that each strictly d-extreme point of a d-convex set M is a d-boundary point of it. Moreover, if we replace $stext_d M$ by the set dM of all d-boundary points, then the relation analogous to (*) holds for any complete metric space:

$$M = \mathrm{conv}_d dM$$

for each compact, d-convex set M.

PROBLEM 3: Let $\Re^n$ be a Minkowski space. It is evident that every affine, isometric mapping $f : \Re^n \to \Re^n$ preserves the d-convexity, i.e., if $M \subset \mathbf{R}^n$ is a d-convex set, then $f(M)$ is also d-convex. There are some affine *nonisometric* mappings which preserve d-convexity, too. For example, each homothety preserves d-convexity. Moreover, if $\pi : \Re^n \to \Gamma$ is a parallel projection of $\Re^n$ onto a d-convex hyperplane Γ in the direction of a d-convex line, then also π preserves d-convexity, i.e., $\pi(M) \subset \Gamma$ is a d-convex set for every d-convex $M \subset \Re^n$. The problem is to describe all affine mappings of a Minkowski space $\Re^n$ onto itself which preserve d-convexity.

Another extension of the problem consists of the consideration of mappings $f : \Re^n \to \Re^m$ of two *different* Minkowski spaces which preserve d-convexity, too.

PROBLEM 4:

Theorem 11.14 gives a necessary and sufficient condition under which the unit ball $\sum$ of a Minkowski space $\Re^n$ satisfies the equality $\mathrm{conv}_d \sum = \sum$. But this condition is not always effective. For example, for $n \geq 3$ no polytope $\sum \subset \mathbf{R}^n$ centered at the origin (as the unit ball) satisfies the condition $\mathrm{conv}_d \sum = \sum$, cf. Theorem 12.6 in [SV 4] and Corollary 11.15 in this book. It would be interesting to find some other classes of unit balls $\sum \subset \Re^n$ for which it is possible to obtain some explicit geometrical conditions for the validity of the equality $\mathrm{conv}_d \sum = \sum$.

The Exercises 6, 9, 10 in §11 refer to another problem, also connected with the d-convex hull of the unit ball $\sum$. Namely, let $M \subset \mathbf{R}^n$ be a compact, convex body centered at the origin. We say that M is a d-hull if there exists a norm such that the unit ball $\sum$ of the Minkowski space $\Re^n$ satisfies $\mathrm{conv}_d \sum = M$. The problem is to describe all d-hulls in $\mathbf{R}^n$. The answer might be complicated enough, even in the case $n = 2$. We notice that for every compact, convex body $M \subset \mathbf{R}^n$ (centered at the origin) and for every $\varepsilon > 0$ there is a d-hull M' satisfying the inequality $\varrho(M, M') < \varepsilon$ and, moreover, that there is a body M' (centered at the origin) which is not a d-hull but satisfies $\varrho(M, M'') < \varepsilon$. In other words, the set of d-hulls and the set of bodies (centered at the origin) which are not d-hulls are both *dense* in the set of all compact, convex bodies centered at the origin.

If we limit ourselves to figures in $\mathbf{R}^2$ and, moreover, to convex polygons centered at the origin, the problem is interesting enough. For example, if at least one main diagonal of a hexagon M (centered at the origin) is parallel to a pair of its opposite sides, then M is a d-hull, and otherwise it is not a d-hull. In addition, if a hexagon M is a d-hull, it is possible that there exist *infinitely many* unit balls $\sum$ with $\mathrm{conv}_d \sum = \sum$ (cf. Fig. 255). Even for plane polygons, a complete solution would be interesting.

PROBLEM 5:

The definition of d-star-shapedness (cf. §15) uses the d-visibility in the following form: a point $x \in X$ is d-visible from a point $x_o \in X$ (where $X \subset \Re^n$ is a compact set) if the d-segment $[x_o, x]_d$ is contained in X. There are some other definitions of d-visibility. The first one requires that $[x_o, x]_d \subset \mathrm{int}\ X \cup \{x\}$. In the case of linear convexity (instead of d-convexity) this means that the whole segment $[x_o, x]$, except for the point x, is contained in $\mathrm{int}\ X$. This "internal d-visibility" is a stronger requirement than the definition used in §15. By examples it is easy to show that there are d-star-shaped sets X (as in Theorem 15.4) which do not possess the "internal d-star-shapedness", cf. Fig. 256. It is interesting to give a generalization of Theorem 15.4 for the case of "internal d-star-visibility" (even for polyhedral regions).

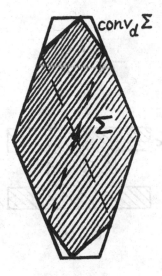

Fig. 255

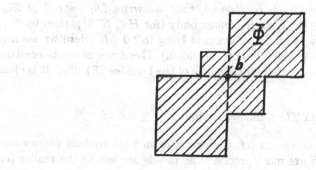

Fig. 256

One more definition of d-visibility may be obtained in the following way. We say that a point $x \in X$ is d-arc-visible from a point $x_o \in M$ if there exists a simple arc $l_x \subset (\text{int } X) \cup \{x\}$ such that its length is equal to $\| x - x_o \|$. This understanding of d-visibility is wider than that one which was considered in §15. For example, the region X in Fig. 257 is "d-arc-star-shaped", although it is not d-star-shaped in the sense of §15. The problem to generalize Krasnosel'ski's theorem in the sense of "d-arc-star-shapedness" is, perhaps, a difficult one.

Besides, in the Exercises 11 and 12 of §24 we have considered the notion of "H-star-shapedness". This circle of problems should be investigated, too.

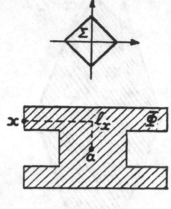

Fig. 257

PROBLEM 6:

Let M be a convex body in the normed space $\Re^n$, and let $\mathfrak{M}_d$ be the family of all d-convex sets contained in the boundary of a closed, d-convex body in $\Re^n$. If linear convexity is considered, then we write $\mathfrak{M}_l$ instead of $\mathfrak{M}_d$. Furthermore, if $M \subset \mathbf{R}^n$ is an H-convex body (for $H \subset S^{n-1}$), then by $\mathfrak{M}_H$ we denote the family of all H-convex sets lying in bd M. (Here we assume that H is a closed set which is not one-sided.) There are many interesting problems referring to the Helly dimensions of the families $\mathfrak{M}_d, \mathfrak{M}_H$. It is clear that the inequalities

$$\text{him } \mathfrak{M}_l \leq n; \quad \text{him } \mathfrak{M}_d \leq \text{him } \Re^n; \quad \text{him } \mathfrak{M}_H \leq \text{md } H$$

hold. The following theorems (cf. the exercises in §38) contain (upper and lower) bounds which are more precise. The proofs are left to the reader (cf. also [B-SP 1-4]).

Theorem 1: Let $M \subset \mathbf{R}^n$ be an arbitrary bounded, convex body with $M \neq \mathbf{R}^n$. Then the inequalities

$$\max_{F} \dim \text{aff } F \leq \text{him } \mathfrak{M}_d \leq \max_{F} \dim \text{aff } F + 1$$

hold, where the maximum is taken over the proper faces of M.   □

Corollary 2: Let $M \subset \mathbf{R}^n$ be a closed, convex body distinct from an n-simplex and from $\mathbf{R}^n$. Then him $\mathfrak{M}_l < n$. In other words, if $N_1, \cdots, N_k$ are convex sets in bd M such that each n of them have a nonempty intersection, then $N_1 \cap \cdots \cap N_k \neq \emptyset$.   □

Corollary 3: Let $M \subset \mathbf{R}^n$ be a convex n-polytope which is not a simplex. Then him $\mathfrak{M}_l = n - 1$.   □

Theorem 4: Let M be a closed, d-convex body in the Minkowski space $\mathfrak{R}^n$, with $M \neq \mathfrak{R}^n$. Then the inequalities

$$\max_F \text{him aff } F \leq \text{him } \mathfrak{M}_d \leq \max_F \text{him aff } F + 1$$

hold, where the maximum is taken over all proper faces F of M.   ☐

Theorem 5: Let $H \subset S^{n-1}$ be a set which is non-one-sided and symmetric about the origin. Furthermore, let M be a bounded, H-convex body, and R_F be a subspace which is a translate of the affine hull of some proper face F of M. Then the inequalities

$$\max_F \text{md cl } H(R_F) \leq \text{him } \mathfrak{M}_H \leq \max_F \text{md cl } H(R_F) + 1$$

hold, where the maximum is taken over all proper faces F of M.   ☐

These statements show that the combinatorial geometry on the boundary of a compact, convex body M contains sufficiently many remarkable facts (and problems). It would be interesting to develop this theory in some further directions (such as separation of faces, supporting cones etc.). In particular, it is reasonable to compare the combinatorial geometries on the boundary of a compact, convex body M and on the boundary of its polar body M^*. For example, if we consider only *faces* of the bodies M, M^*, then it is interesting to find some connections between the Helly dimensions of these two families. In the case when $M \subset \mathbf{R}^n$ is a cube (and $M^* \subset \mathbf{R}^n$ is a cross-polytope), these Helly dimensions are equal to 1 and $n - 2$, respectively.

PROBLEM 7:

In Theorems 20.4 and 20.5 the upper semicontinuity of the H-convex hull is contained. In Example 20.6 it is shown that there is no corresponding lower semicontinuity property. This example is constructed in 3-space. Is it possible to construct an analogous example in $\mathbf{R}^2$? An opposite side of this question can be formulated as follows: is it true that in the plane $\mathbf{R}^2$ (for every set $H \subset S^1$ that is not one-sided) the lower semicontinuity of the H-convex hull holds? Furthermore, in Example 20.6 the set $H \subset S^2$ is infinite. Does the lower semicontinuity of the H-convex hull hold if $H \subset S^2$ is a non-one-sided, finite set? Are there any other cases when the structure of the set $H \subset S^{n-1}$ implies the lower semicontinuity of the H-convex hull? In other words, the problem is to describe different situations in which lower semicontinuity (and, consequently, continuity) of the H-convex hull holds. Are these situations connected with conditions under which the vector sum of H-convex sets is H-convex?

PROBLEM 8:

Theorem 5.8 affirms that if a compact, convex body $M \subset \mathbf{R}^n$ is not a parallelotope, then every compact, convex body N close enough to M is not a parallelotope, too. This means that the set Q of all compact, convex bodies, which are not parallelotopes, is *open* in the set V of all compact, convex bodies in $\mathbf{R}^n$, i.e., if $M \in Q$, then there exists a real number $\varepsilon > 0$ such that the ε-neighbourhood $U_\varepsilon(M)$ (in the sense of the Hausdorff metric) is contained in Q, too. In other words, the property of "being not a parallelotope" is, in a sense, "stable".

Theorems 5.5, 5.6, and 5.7 (as well as the corresponding Exercises 8, 10, and 11) present other stability properties of convex bodies. For example, the property "to be not a simplex" is stable, i.e., the set of all compact, convex bodies in $\mathbf{R}^n$, which are not simplices, is open in V. It is interesting to describe different properties of convex bodies which are stable. We remark that the property "to be not a polytope" is *not stable*, since for every compact, convex body $M \subset \mathbf{R}^n$ there are polytopes arbitrarily close to M. But the property "to be not a polytope with less than k vertices" is stable. In other words, the set of all polytopes with less than k vertices is *close* in V.

Some stable properties are connected with H-convextiy. For example, if $H \subset S^{n-1}$ is a closed set, then the property "to be not H-convex" is stable. In other words, the set of all H-convex, compact, convex sets is closed in V (and its complement is open). Our problem is to investigate different stability properties of convex bodies (connected or not connected with H-convexity).

PROBLEM 9:

Let H be a subset of $S^{n-1} \subset \mathbf{R}^n$ and $\mathbf{R}^s$ be some s-dimensional subspace of $\mathbf{R}^n$. Let, furthermore, π denote the orthogonal projection of $\mathbf{R}^n$ onto $\mathbf{R}^s$ and $H(\mathbf{R}^s)$ the set of all vectors $\frac{\pi(f)}{\|\pi(f)\|}$, where $f \in H \setminus \mathbf{R}^n$. The set H is said to be *orthogonally closed* if, for an arbitrary subspace $\mathbf{R}^s \subset \mathbf{R}^n$, the set $H(\mathbf{R}^s)$ is closed.

We indicate some properties which are connected with orthogonally closed sets $H \subset S^{n-1}$; their proofs are left to the reader. After that, we will formulate some problems.

Theorem 1: Let the subspace $\mathbf{R}^s \subset \mathbf{R}^n$ be H-convex. A set $M \subset \mathbf{R}^s$ is H-convex if and only if M is $H(\mathbf{R}^s)$-convex in $\mathbf{R}^s$. $\square$

Theorem 2: Let the set H be orthogonally closed. Then the affine hull of each H-convex set itself is H-convex. $\square$

EXAMPLE 3: We will show that the implication in the theorem above does not hold if the set H is not orthogonally closed. Let H be the sphere S^{n-1} from which two points $e, -e$ are deleted ($e \in S^{n-1}$). Then the hyperplane $\Gamma = \{x : \langle e, x \rangle = 0\}$ is not H-convex. Nevertheless, the unit ball E of the hyperplane Γ is an H-convex set. $\square$

EXAMPLE 4: In the previous example, the set H is not closed. It is easy to construct an example such that H is even closed (but not orthogonally closed) and the implication of Theorem 2 above does not hold. Consider an even circular cone Q, which is determined by the inequalities $x_1^2 \leq 2x_2x_3$, $x_2 \geq 0$, $x_3 \geq 0$ (cf. Exercise 11 in §20), and denote by H the set of all unit vectors which are (inner or outer) normals to the surface of Q. The set H is closed and symmetric about the origin (it consists of two small circumferences from the unit sphere $S^2 \subset \mathbf{R}^3$, which are symmetric about o), but it is not orthogonally closed. To describe the set H more precisely, we remark that the point $(\sqrt{2}\sin\alpha\cos\alpha, \sin^2\alpha, \cos^2\alpha)$ belongs to bd Q, and the unit, outer normal regarding bd Q at this point is given by

$$n(\alpha) = \left(\sqrt{2}\sin\alpha\cos\alpha, -\cos^2\alpha, -\sin^2\alpha\right).$$

The vectors $\pm n(\alpha)$, $0 \le \alpha \le \pi$, even exhaust the set H.

In this example the ray l described by $x_1 = 0$, $x_2 = 0$, $x_3 \ge 0$ is an H-convex set, but its affine hull aff l (i.e., the x_3-axis) is not H-convex. $\square$

Theorem 5: Let $M_1 \subset M_2 \subset M_3 \subset \cdots$ be a countable, increasing sequence of H-convex sets, and $N = \mathrm{cl}\,(\bigcup_k M_k)$. If H is orthogonally closed, then N is H-convex. $\square$

Theorem 6: Let M be some H-convex set, $a \in M$, and K be a supporting cone of M at the point a. If H is orthogonally closed, then K is H-convex. $\square$

Theorem 7: Let H be orthogonally closed. A closed, linearly convex set $M \subset \mathbf{R}^n$ is H-convex if and only if for each $a \in M$ the supporting cone of M at the point a is H-convex. $\square$

The previous theorems show that the analogy between linear convexity and H-convexity becomes deeper if the set $H \subset S^{n-1}$ is orthogonally closed. Our problem is to develop a wider theory of H-convex sets for an orthogonally closed set $H \subset S^{n-1}$.

PROBLEM 10:

Theorems 25.2 and 29.1 describe all compact, convex bodies $M \subset \mathbf{R}^n$ with him $M = 1$ and him $M = 2$ (without the assumption of central symmetry). This yields the complete solution of the Szökefalvi-Nagy problem in $\mathbf{R}^3$ (cf. Theorem 29.5). Indeed, for the bodies $M \subset \mathbf{R}^3$ satisfying the condition of Theorem 25.2 we have him $M = 1$, for those satisfying the condition of Theorem 29.4 we obtain him $M = 2$, and for the remaining bodies him $M = 3$ holds. To obtain a complete solution of the Szökefalvi-Nagy problem in $\mathbf{R}^4$, it is necessary to distinguish the bodies $M \subset \mathbf{R}^4$ with him $M = 3$ and him $M = 4$ (without the assumption of central symmetry). This is our next problem: to describe all compact, convex bodies $M \subset \mathbf{R}^4$ with him $M = 3$, i.e., with md $M = 3$. Perhaps in the general case (of bodies in $\mathbf{R}^n$) the description of all bodies with md $M = 3$ is very complicated, but for $n = 4$ the problem might be more visible.

PROBLEM 11:

Let $e_1, \cdots, e_n$ be a basis in $\mathbf{R}^n$ and $I_o, I_1, \cdots, I_s$ be a partition of the set $\{1, \cdots, n\}$, i.e., each set $I_o, I_1, \cdots, I_s$ is nonempty, $I_o \cup I_1 \cup \cdots \cup I_s = \{1, \cdots, s\}$, and $I_i \cap I_j = \emptyset$ for $i \ne j$. We define a *book* B (cf. the end of §30) in the following way: denoting by E_i the subspace spanned by the vectors e_q with $q \in I_i\, (i = 1, \cdots, s)$, we put

$$S = E_o,\ P_q = E_o + E_q \text{ for } q = 1, \cdots, s;\ B = P_1 \cup \cdots \cup P_s.$$

So B is the book with pages $P_1, \cdots, P_s$ and the spine $S = P_1 \cap \cdots \cap P_s$.

Denote the number dim E_o by k, and the number dim E_q by d_q, for $q = 1, \cdots, s$; without loss of generality, we may suppose that $d_1 \ge \cdots \ge d_s$. In this notation, it is not difficult to find md B. First we consider the case $s > k$. We choose $k+1$ minimally dependent vectors $b_1, \cdots, b_{k+1} \in E_o$ and d_i linearly independent vectors $a_1^{(i)}, \cdots, a_{d_i}^{(i)} \in E_i$ such that for $i = 1, \cdots, s$ the vectors $b_i, a_1^{(i)}, \cdots, a_{d_i}^{(i)}$ are minimally dependent. Then the vectors $a_j^{(i)}$ (for $i = 1, \cdots, k+1$ and $j = 1, \cdots, d_i$) are minimally dependent, and hence md $B \ge d_1 + \cdots + d_{k+1} - 1$. In fact, in this case equality holds. An analogous reasoning shows that if $s \le k$, then

$$\text{md } B \ge d_1 + \cdots + d_s + (k - s).$$

These considerations can be extended to a "library L" instead of the book B (cf. the end of §30). Furthermore, a book B (or a library) is said to be *maximal* if for every nonzero vector $e \in \mathbf{R}^n \setminus B$ the inequality md $(B \cup \{e\}) >$ md B holds. Now the first problem is to describe all maximal libraries (maybe it is reasonable to extend the definition of a library considered in §30). This allows to obtain an upper estimate for md M. Indeed, if $M \subset \mathbf{R}^n$ is a compact, convex body and L is a maximal library which contains $H(M)$ (or exp M^*), then md $M \le$ md L. This upper estimate holds for any compact, convex body (which is not assumed to be centrally symmetric). And for *centrally symmetric* bodies, in the case md $M \le 4$ this approach gives the *exact* values of md M (see Theorems 30.1 and 30.2 of Kincses.)

On the other hand, the Exercises 5–8, 14–18 in §30 show that it is possible (at any rate, for him $M \le 4$) to use "library arguments" for obtaining an opposite estimate (for centrally symmetric bodies), i.e., to embed $H(M) = $ norm exp M^* into a library L with md $M = $ md L. Namely, let $H \subset \mathbf{R}^n$ be a vector system that is symmetric with repsesct to the origin, and md $H = m$. We choose minimally dependent vectors $a_o, \cdots, a_m \in H$ and denote by E the subspace spanned by these vectors. Let $b_o, b_1, \cdots, b_k \in H$ be minimally dependent vectors such that $b_o \in E$ and the subspace $E \cap \text{lin}\,(b_o, \cdots, b_k)$ is one-dimensional. Then (generalizing the method of the exercises mentioned above) we obtain that $k \le \frac{m+1}{2}$. (Otherwise, among $a_o, \cdots, a_m, b_o, \cdots, b_k$ there are more than $m + 1$ minimally dependent vectors). If, in particular, $k = k_o$, where k_o is the largest integer not exceeding $\frac{m+1}{2}$, then (up to an enumeration of $a_o, \cdots, a_m$)

$$\pm b_o = \lambda_o a_o + \cdots + \lambda_k a_k = -\lambda_{k+1} a_{k+1} - \cdots - \lambda_m a_m,$$

where $\lambda_o a_o + \cdots + \lambda_m a_m = o$ is the minimal dependence of the vector system $a_o, \cdots, a_m$. This means that the vectors $a_o, \cdots, a_m, b_o, \ldots, b_k$ are contained in the book B with the pages

$$P_1 = \text{lin } (a_o, \cdots, a_k), \, P_2 = \text{lin } (a_{k+1}, \cdots, a_m), P_3 = \text{lin } (b_o, \cdots, b_k),$$

and with the one-dimensional spine $S = \text{lin } (b_o) = P_1 \cap P_2 \cap P_3$. For $k < k_o$, similar arguments hold (cf. Exercises 17 and 18 in §30), and a suitable induction gives an embedding of $H(M)$ into a library.

The problem is to describe in this way all centrally symmetric, compact, convex bodies not only with md $M = 3$ and md $M = 4$, but for greater values of md M. In particular, this approach seems to be realizable for the description of the centrally symmetric, compact, convex bodies $M \subset \mathbf{R}^n$ with md $M = n - 1$. This problem was discussed with Janos Kincses in 1992.

PROBLEM 12:

First of all, we formulate some results by O. Hanner, J. Lindenstrauss, A. Hansen and A. Lima which are close to the Szökefalvi-Nagy Problem (cf. [Han], [H-L], [Li], [Lin], §7 of [D-G-K], and [Kl 1]).

Let $m > k \geq 2$ be integers. A compact, convex body $M \subset \mathbf{R}^n$ is said to possess the $m.k.intersection$ $property$, when the following assertion is true: For every collection $M_1, \cdots, M_m$ of translates of the body M the relation $M_1 \cap \cdots \cap M_m \neq \emptyset$ holds if each k of them have a common point. If the body M possesses the $m.k.$intersection property for a fixed k and an arbitrary $m > k$, then M is said to have the $\infty.k.$intersection property. The least integer r for which the body M possesses the $\infty.r + 1.$intersection property is evidently equal to the $Helly$ $dimension$ him M of the body M. To indicate the difference between this problem and the problem of Helly-dimensional classification, we consider an example.

The regular octahedron $M \subset \mathbf{R}^3$ possesses the 3.2.intersection property. In other words, if M_1, M_2, M_3 are three translates of M, and every two of them have a common point, then $M_1, \cap M_2 \cap M_3 \neq \emptyset$. However, the regular octahedron $M \subset \mathbf{R}^3$ does not possess the 4.2.intersection property, since him $M = 3$. This means that there exist four translates M_1, M_2, M_3, M_4 of M such that every two of them (consequently, every three of them) have a common point, while $M_1 \cap M_2 \cap M_3 \cap M_4 = \emptyset$. Moreover, except for affine images of the cube and the regular octahedron, there are no other compact, convex bodies in $\mathbf{R}^3$ with the 3.2.intersection property. These facts are particular cases of the following two general results of O. Hanner, cf. [Han].

Theorem 1: If a compact, convex body $M \subset \mathbf{R}^n$ possesses the 4.2.intersection property, then it possesses the $\infty.2$.intersection property, i.e., M is a parallelotope.   □

Theorem 2: Up to affine transformations, there exists only a finite number of compact, convex bodies in $\mathbf{R}^n$ with the 3.2.intersection property. More detailed, a compact, convex body $M \subset \mathbf{R}^n$ possesses the 3.2.intersection property *if and only if M is a centrally symmetric, convex polytope such that for every two disjoint faces F_1, F_2 of M (of any dimensions) there are parallel supporting hyperplanes Γ_1, Γ_2 of M with $F_1 \subset \Gamma_1, F_2 \subset \Gamma_2$.* □

The "Hanner polytopes", i.e., the convex polytopes with the 3.2.intersection property were described in Theorem 2 quite completely, but not effectively. The following result, recently obtained by A. Hansen and A. Lima [H-L], lists effectively all Hanner polytopes.

Theorem 3: A polytope $M \subset \mathbf{R}^n$ that is symmetric with respect to the origin is a Hanner polytope if and only if it can be obtained from n segments, which are, correspondingly, situated in the axes of a cartesian coordinate system and have a common midpoint at the origin, by the operations of the vector sum and convex hull (in any order). □

The term "intersection property" was introduced by J. Lindenstrauss [Lin] who has established the following result:

Theorem 4: Let q, k be integers such that $k \geq 2$, and

$$q > \frac{4k - 3 + \sqrt{1 + 8(k-1)^2}}{2}.$$

If a centrally symmetric, compact, convex body $M \subset \mathbf{R}^n$ possesses the q.k.intersection property, then it possesses the ∞.k.intersection property, i.e., him $M \leq k - 1$. □

For $k = 2$ the result of Lindenstrauss yields the fact that the 4.2.intersection property implies the ∞.2.intersection property (for centrally symmetric, compact, convex bodies). This result is in complete concordance with Hanner's Theorem 2. For $k = 3$, the result of Lindenstrauss affirms that the 8.3.intersection property implies ∞.3.intersection property. This is only a rough estimate. The exact estimate for $k = 3$ was obtained by A. Lima [Li]:

Theorem 5: If a centrally symmetric, compact, convex body $M \subset \mathbf{R}^n$ possesses the 4.3.intersection property, then it possesses the ∞.3.intersection property, i.e., him $M \leq 2$. □

We remark that, in fact, Lindenstrauss and Lima obtained more general results. They worked with balls in Banach spaces (not necessarily finite-dimensional). This explains, why their results only refer to *centrally symmetric* bodies. In particular, A. Lima [Li] has obtained necessary and sufficient conditions for the m.k.intersection property (in terms of Functional Analysis).

We now describe general (geometrical and algebraic) conditions for the m.k.intersection property in the *non-centrally-symmetric case*, obtained in [Bt 15].

Let $M \subset \mathbf{R}^n$ be a compact, convex body and $a_1, \cdots, a_s$ be boundary points of it. We say that the points $a_1, \cdots, a_s \in M$ are *separable*, if the corresponding supporting cones $\text{supcone}_{a_i} M$, $i = 1, \cdots, s$, are separable.

Theorem 6: Let $M \subset \mathbf{R}^n$ be a compact, convex body and $m > k \geq 2$ be integers. The body M does not possess the m.k.intersection property if and only if there exist boundary points $a_1, \cdots, a_m$ of M such that $a_1, \cdots, a_m$ are separable, while any k of them are not separable.   □

Theorem 7: Let $M \subset \mathbf{R}^n$ be a compact, convex body and $m > k \geq 2$ be integers. The body M possesses the m.k.intersection property if and only if for every m of its boundary points either these m points are not separable, or among them there are k points which are separable.   □

We now formulate an equivalent necessary and sufficient condition for the m.k.intersection property in terms of Linear Algebra. This condition generalizes the invariant md H. First we introduce some definitions.

Let $M \subset \mathbf{R}^n$ be a compact, convex body and a be a boundary point of it. We say that a unit vector v is *adjacent* to the point a if v belongs to the polar cone $(\text{supcone}_a M)^*$, i.e., the half-space $\{x : \langle v, x - a \rangle \leq 0\}$ contains the body M.

Theorem 8: Let $M \subset \mathbf{R}^n$ be a compact, convex body and $a_1, \cdots, a_s$ be boundary points of M. The points $a_1, \cdots, a_s$ are separable if and only if there exist minimally dependent vectors $e_1, \cdots, e_q \in cl\, H(M)$, each of which is adjacent to (at least) one of the points $a_1, \cdots, a_s$.   □

Combining this Theorem with the geometrical conditions for the m.k.intersection property (Theorems 6 and 7), we obtain the following algebraic conditions:

Theorem 9: Let $M \subset \mathbf{R}^n$ be a compact, convex body and $m > k \geq 2$ be integers. The body M does not possess the m.k.intersection property if and only if there exist boundary points $a_1, \cdots, a_m$ of M such that (i) there exist minimally dependent vectors $e_1, \cdots, e_q \in cl\, H(M)$, each of which is adjacent to (at least) one of the points $a_1, \cdots, a_m$ and (ii) for any k points chosen from $a_1, \cdots, a_m$ it is impossible to find a system of minimally dependent vectors in $cl\, H(M)$ each of which is adjacent to (at least) one of the chosen points.
□

Theorem 10: Let $M \subset \mathbf{R}^n$ be a compact, convex body and $m > k \geq 2$ be integers. The body M possesses the m.k.intersection property if and only if for every m boundary points of M (say $a_1, \cdots, a_m$) one of the following conditions holds: (i) there is no system of minimally dependent vectors in $cl\, H(M)$ each of which is adjacent to (at least) one of the points $a_1, \cdots, a_m$; (ii) it is possible to choose k points among $a_1, \cdots, a_m$ and a system of minimally dependent vectors in $cl\, H(M)$ each of which is adjacent to (at least one) of the chosen points.   □

Finally, we formulate some problems. The first one is to generalize Lima's Theorem 5 for the case of *non-centrally-symmetric* bodies. More detailed: is it true that for an arbitrary compact, convex body $M \subset \mathbf{R}^n$ the 4.3.intersection property implies the ∞.3.intersection property (i.e., him $M = 2$)? If "yes", this means that the complete list of bodies with the 4.3.intersection property is given by Theorem 29.1 (or 29.7). It is expected that Theorems 6-9 above contain an adequate apparatus for solving this problem.

Moreover, as we have seen, the 3.2.intersection property does not coincide with the 4.2.intersection property (or, what is the same, with the ∞.2.intersection property). Maybe this is the unique case in this sense. More exactly, is it true that for $m > k > 2$ the $m.k$.intersection property *always* implies the $\infty.k$.intersectional property?

PROBLEM 13:

The Borsuk problem refers to the maximal value of the number $a(M)$ over the set of all compact, convex bodies $M \subset \mathbf{R}^n$, see §31. A more detailed problem connected with the Borsuk numbers $a(M)$ consists of the geometrical description of the compact, convex bodies $M \subset \mathbf{R}^n$ with $a(M) = r$. In other words, we would like to solve the "equation" $a(M) = r$ for a given integer r, i.e., to list the compact, convex bodies $M \subset \mathbf{R}^n$ which satisfy the condition $a(M) = r$. We will call this the *reverse Borsuk problem*. Since counterexamples to Borsuk's partition conjecture are known (cf. §31), the following question is natural: What is the minimal one of the integers n for which such a counterexample in $\mathbf{R}^n$ exists? Moreover, one can even ask whether such counterexamples can be constructed already for $n = 4$. Moreover, the reverse Borsuk problem is interesting even in 3-space. As we have indicated, in the case $n = 2$ this reversed problem is solved, cf. Theorems 31.1 and 31.5. For the case $n = 3$, the reverse Borsuk problem is open up today. Namely, in view of Theorem 31.5 one can show that already the process of completing a regular tetrahedron $T \subset \mathbf{R}^3$ to a body of constant width is not realizable in a unique manner. Let T have the edge-length d. Consider the intersection B of the four balls of radius d having the vertices of T as their centers (Fig. 258). This body B is not a body of constant width, since its width regarding to the plane, that is orthogonal to two skew edges of T, is larger than d. To obtain a body of constant width, we remove from B three parts outside of T, each lying between two neighbouring facet planes and corresponding to one pair of skew edges of T. This can be done in two different ways, namely with respect to three edges of T having a common vertex (Fig. 259) and with respect to three edges around one facet of T (Fig. 260). These removed parts are replaced by spindle-shaped sets which are obtained by rotating the circular arcs from bd B lying in the facet planes of T around the corresponding edges of T, see Fig. 261. This yields two different types of bodies of constant width d (Fig. 262 and Fig. 263), and therefore the reverse Borsuk problem seems to be very complicated (and interesting), even in $\mathbf{R}^3$. However, we formulate a question referring to a stronger exact criterion for the convex bodies $M \subset \mathbf{R}^3$ satisfying $a(M) = 4$. Let $M \subset \mathbf{R}^3$ have the property that it cannot be completed to a body of constant width in a unique way. Is it true that $a(M) = 4$ if and only if the centers of the circumspheres of all the completions for M coincide?

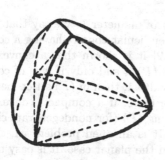

Fig. 258

Fig. 259

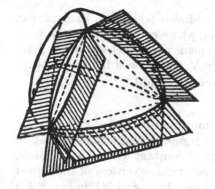

Fig. 260

Fig. 261

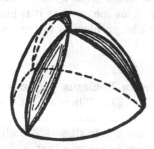

Fig. 262

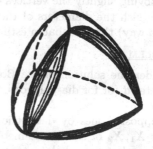

Fig. 263

PROBLEM 14:

Let $M \subset \mathbf{R}^n$ be a compact, convex body of diameter 1. We say that M has the property (P) if any n mutually perpendicular chords, having a common point, have total length ≥ 1. In [M-M 2] it is shown that a convex body $M \subset \mathbf{R}^2$ of diameter 1 has the property (P) if and only if it is of constant width 1. (Moreover, if M is of constant width 1, then in (P) strict inequality holds for nondegenerate chords.) Furthermore, if a compact, convex body $M \subset \mathbf{R}^n$, $n \geq 2$, of diameter 1 is satisfying (P) (for nondegenerate chords), then it has to be of constant width 1. It is an open problem whether for $n \geq 3$ we have the same equivalence as in the planar case. It is easy to show that the positive answer for $\mathbf{R}^{n+1}$ implies the same for the n-dimensional case. Namely, $M \subset \mathbf{R}^n$ can be enclosed in a convex body $M' \subset \mathbf{R}^{n+1}$ of constant width 1 (cf. [C-G]), and we can consider chords of M' with a common endpoint, one "parallel" to the $(n+1)$st basis vector and of length 0, cf. [M-M 2] for a respective lemma.

One can formulate analogous assertions for Minkowski spaces. Then, instead of mutual perpendicularity of the chords $[p_i, q_i]$, we can assume that for $i < j$ the direction of $[p_j, q_j]$ lies in a supporting plane of the unit ball at the point $\frac{q_i - p_i}{\|q_i - p_i\|}$. (For the latter question, posed by V. Soltan, and further problems related to the present one, we refer once more to [M-M 2].)

PROBLEM 15:

To solve the Borsuk problem in $\mathbf{R}^3$, B. Grünbaum [Grü 2] used a covering set V (cf. Fig. 138-142). It was obtained from a regular octahedron by cutting off three vertices (Fig. 140). But if a body M of constant width 1 is contained in Grünbaum's covering set V, then the other three vertices of the former octahedron can be cut off, too (since they do not belong to M). In Fig. 264, a projection of V into a diagonal plane of the octahedron is shown. So we obtain a covering set V' for all sets of diameter 1 which is *smaller* than Grünbaum's covering set. Now it is possible to get a partition of V' into four parts of diameters smaller than 1, analogous to the partition given by B. Grünbaum but moving slightly the vertices of the pieces. This means that it is possible to diminish the diameters of the four parts. The problem is to improve (in such a way) the Grünbaum estimate $0, 9887$.

PROBLEM 16:

Consider the solution of the Borsuk problem for Reuleaux polygons M of diameter d (or for dia-polygons, cf. Exercise 2 in §31). There are $2k+1$ vertices $a_o, a_1, \cdots, a_{2k}$ of M (cf. Fig. 147), and each of them is connected with two opposite ones by diametrical chords. If we try to dissect M into two pieces X_1, X_2 of smaller diameter, then we have to assume that a_o belongs to X_1, and a_k, a_{k+1} belong to X_2. Continuing this, we see that $a_o, a_1, a_2, \cdots$ belong to X_1, whereas $a_k, a_{k+1}, a_{k+2}, \cdots$ belong to X_2. But finally we come to a contradiction, since the number of vertices is odd, and we obtain two

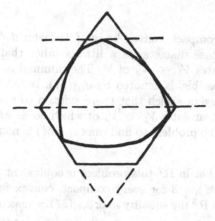

Fig. 264

diametrical vertices in X_1 or in X_2. It is possible to say that there are "sufficiently many" diametrical chrods in M, and their "topology" is such that it is "locally" possible to dissect the set of vertices into two subsets X_1, X_2, but this is "globally" impossible.

The problem is to construct higher-dimensional analogues, i.e., to find a polytope $M \subset \mathbf{R}^n$ of diameter d which has "sufficiently many" diametrical chords with a "sufficiently complicated topology". For example, if M has "sufficiently many" diametrical chords emanating from a vertex and M admits a transitive group of motions, then for every vertex $a \in M$ there are "sufficiently many" diametrical chords with endpoint a. Perhaps one can construct a polytope $M \subset \mathbf{R}^4$ in this way such that it is impossible to divide its vertex set into 5 groups each of which contains no pair of diametrical vertices. This would give a counterexample to the Borsuk hypothesis in $\mathbf{R}^4$, and the Borsuk problem would have a complete answer.

If a polytope $M \subset \mathbf{R}^4$ with sufficiently many diametrical chords and a "sufficiently complicated topology" could be constructed, then a computer-aided calculation could show whether this is a counterexample. Namely, consider the graph G whose vertices coincide with vertices of the investigated polytope M, and two vertices are connected by an edge of G if and only if the corresponding chord in M is diametrical. The problem is to brush the vertices of G by 5 colors in such a way that two vertices connected by an edge have always different colors. If this coloring problem is not solvable for G, then M is a counterexample. It might be convenient to use computers for sorting out the partitions of the vertices of G into 5 groups, and to determinate whether one of the partitions gives a solution of the coloring problem for G. If, after a complete sorting, there is no five-coloring partition of G, then the Borsuk problem in $\mathbf{R}^4$ would be solved.

PROBLEM 17:

Let $M \subset \mathbf{R}^n$ be a compact, convex body of diameter d. We take a compact, convex body V whose diameter is a little smaller than d, and we try to cover M by translates $V_1, \cdots, V_s$ of V. The minimal one of the integers s, for which this is possible, is denoted by $a_{(t)}(M)$. In other words, $a_{(t)}(M)$ is the least of the integers s such that there exists a compact, convex body V with diam $V < d$, translates $V_1 \cdots, V_s$ of which cover M. It is obvious that $a_{(t)}(M) \geq a(M)$. The problem to find $\max_M a_{(t)}(M)$ is named the "translative Borsuk problem".

It is easy to show that in $\mathbf{R}^2$ this problem is equivalent to Borsuk's original problem, i.e., $a_{(t)}(M) = 3$ for every compact, convex figure $M \subset \mathbf{R}^2$. It is possible that also in $\mathbf{R}^3$ the equality $\max_M a_{(t)}(M) = \max_M a(M) = 4$ holds. The question is whether also in $\mathbf{R}^n$ the equality $\max_M a_{(t)}(M) = \max_M a(M)$ holds, i.e., whether the Borsuk problem is, in general, equivalent to the translative Borsuk problem.

PROBLEM 18:

Let L_1, L_2 be parallel hyperplanes in $\mathbf{R}^n$ and M_1, M_2 be compact, convex bodies in L_1 and L_2, respectively. Denote by c_1, c_2 the numbers $c(M_1), c(M_2)$, i.e., c_i is the minimal number of directions which illuminate the whole boundary of the body M_i in the space L_i, $i = 1, 2$. It is easy to show that $c(\mathrm{conv}\ (M_1 \cup M_2)) \leq c_1 + c_2$, i.e., the boundary of the body conv $(M_1 \cup M_2) \subset \mathbf{R}^n$ can be illuminated by $c_1 + c_2$ directions. The converse inequality is, in general, false. But there are cases for which it would be interesting to decide whether the converse inequality holds. For example, let M_1 be an $(n - 1)$-dimensional compact, convex set with $o \notin \mathrm{aff}\ M_1$, and M_2 be the set symmetric to M_1 with respect to the origin. In this case, the mentioned converse inequality affirms that

$$c(\mathrm{conv}\ (M_1 \cup M_2)) = 2c(M_1).$$

It is unknown whether this is true. If yes, this means that the illumination problem in $\mathbf{R}^{n-1}$ is reduced to the analogous problem for *centrally symmetric* bodies in $\mathbf{R}^n$.

PROBLEM 19:

Trivially, each segment is a Helly-maximal convex body (in its own one-dimensional carrier space). Therefore, each zonotope is the vector sum of one-dimensional Helly-maximal convex sets, and we can formulate a basic statement on zonotopes (cf. Lemma 43.2) in the following manner: a non-decomposable convex body, which is the vector sum $I_1 + \cdots + I_s$ of one-dimensional Helly-maximal sets, is itself Helly-maximal. The question is, whether this remains valid for arbitrary (not necessarily one-dimensional) Helly-maximal sets. In other words, let $M_1, \cdots, M_s$ be Helly-maximal convex sets such that their vector sum is an indecomposable body in $\mathbf{R}^n$. Is this body Helly-maximal? We note that this problem has a positive solution for the class of belt bodies, since the vector sum of belt bodies is itself a belt body, and therefore under the condition of indecomposability it is Helly-maximal (cf. Theorem 43.1).

PROBLEM 20:

We will say that a compact, convex body $M \subset \mathbf{R}^n$ is a zonotope of order q if $M = M_1 + \cdots + M_s$ with dim $M_i \leq q$ for all $i \in \{1, \cdots, s\}$; here we do not assume that the sets $M_1, \cdots, M_s$ are necessarily polytopes. It is clear that the family of the zonotopes of order 1 coincides with the set of usual zonotopes, i.e., the family of the zonotopes of order q contains that of the usual zonotopes. The following problem refers to an extension of Theorem 43.1 (which holds, in particular, for zonoids and therefore also for zonotopes of order 1), namely the extension to zonotopes of order $q > 1$. Confirm (or give a counterexample to) the following hypothesis: for a zonotope $M \subset \mathbf{R}^n$ of order q and a positive integer $r \geq q$ the inequality him $M \leq r$ holds if and only if M is a direct vector sum of convex sets each of which has a dimension $\leq r$.

For $q = 1$, this statement is true (Lemma 43.2). Furthermore, for $q = r = 2$ the correctness of this hypothesis was shown in [B-B-Ch]. We remark that if the previous problem has a positive solution (at least for the case dim $M_i \leq 2$, $i = 1, \cdots, s$), then for $q = 2$ (and arbitrary $r \geq 2$) the answer to the present problem is also affirmative. In fact, each two-dimensional, compact, indecomposable, convex set M_i is Helly-maximal. Thus for $q = 2$ the present problem reduces to the previous one. The authors do not know whether there are further advances in this direction.

PROBLEM 21:

Let $M \subset \mathbf{R}^n$ be a compact, convex body with the origin in its interior. We say that closed half-spaces $P_1, \cdots, P_s$ with their bounding hyperplanes through the origin form a *face-separating system* of M, if for every face $F \subset$ bd M there is at least one index $i \in \{1, \cdots, s\}$ such that $F \cap P_i = \emptyset$. (We recall that each face $F \subset$ bd M is a *closed* set, cf. §2.) The problem is to find the minimal integer s, for which there exists a face-separating system $P_1, \cdots, P_s$ of M. Denote this minimal integer by $c^*(M)$. It is easy to show that $c^*(M) = c(M^*)$, i.e., the considered problem is the "polar version" of the illumination problem (see §34). The face-separating problem (i.e., to find $c^*(M)$ for a given body M) is particularly interesting for bodies M centered at the origin, even for zonoids or zonotopes.

Thus, an equivalent formulation of the problem (in the particular case of zonoids) is the following: Let $M \subset \mathbf{R}^n$ be a zonoid (even a zonotope) centered at the origin. We are interested in the number $c(M^*)$, i.e., in the solution of the illumination problem for the polar body M^*. Is it true that in this case the inequality $c(M^*) \leq 2n$ holds?

PROBLEM 22:

Let $M \subset \mathbf{R}^n$ be a compact, convex body and $e_1, \cdots, e_p$ be nonzero vectors whose directions illuminate the whole boundary of M (see §34). This system of directions is said to be *primitive* if bd M is not illuminated by any proper subsystem of $\{e_1, \cdots, e_p\}$. If $c(M) = c$ and $e_1, \cdots, e_c$ are vectors whose directions illuminate bd M, then this system of directions is obviously primitive. In other words, $c(M)$ is the *minimal* one of the integers p for which a primitive illuminating system $e_1, \cdots, e_p$ for the body M exists. The problem is to find the *maximal* number p for which such a primitive system exists. Denote this maximal number by $c_{\max}(M)$.

It is easy to see that if $M \subset \mathbf{R}^3$ is a pyramid with a regular n-gon as its base, then $c_{\max}(M) = n + 1$, since for every vertex of M there is a direction which illuminates that vertex and does not illuminate any other one. Furthermore, if $M \subset \mathbf{R}^3$ is a cone with a circle as its base then $c_{\max}(M) = \infty$, since for any integer $s \geq 4$ there exists a primitive illuminating system for M consisting of s directions [SV 5].

Our problem is to clarify connections between $c_{\max}(M)$ and the largest of the integers s for which a primitive fixing system $\{a_1, \cdots, a_s\}$ of the body $M \subset \mathbf{R}^n$ exists.

PROBLEM 23:

B. Grünbaum [Grü 5] has formulated the problem which is analogous to the previous one, but connected with *inner illumination* (see §37). More precisely, let $M \subset \mathbf{R}^n$ be a compact, convex body, and $a_1, \cdots, a_s \in$ bd M be points which realize an inner illuminating system with respect to the boundary of M. The system $\{a_1, \cdots, a_s\}$ is said to be *primitive* if no proper subsystem of it realizes the inner illumination of bd M. The *minimal* one of the integers s for which there exists a system $\{a_1, \cdots, a_s\}$ realizing the inner illumination of bd M is evidently equal to $p(M)$, see once more §37. The *maximal* number s for which there exists a primitive system $\{a_1, \cdots, a_s\}$ realizing the inner illumination of bd M we denote by $p_{\max}(M)$. In other words, $p_{\max}(M)$ is the minimal one of the integers s for which *every* system $\{a_1, \cdots, a_{p+1}\} \subset$ bd M illuminating bd M from inside is not primitive. The Grünbaum problem consists of finding $\max_M p_{\max}(M)$ over the set of all compact, convex bodies $M \subset R^n$. It seems to be sufficiently interesting to find $\max_M p_{\max}(M)$ if M runs only over the family of all belt bodies (or zonoids).

Problem 24:

In [She] and [K-P], the notion of *shadow-boundary* of a convex polytope $P \subset \mathbf{R}^n$ with respect to a point $x \in \mathbf{R}^n \setminus P$ was introduced, namely as intersection of P and the union of all its supporting lines through x. Analogously, in [Mar 1] this notion was modified for parallel illumination, i.e., the shadow-boundary of P with respect to a direction u in $\mathbf{R}^n$ is the intersection of P and the union of all its supporting lines having this direction u. Moreover, if each of the mentioned supporting lines has exactly one point in common with P, then the shadow-boundary is said to be *sharp*. Any sharp shadow-boundary is homeomorphic to an $(n-2)$-sphere, and it should be remarked that the converse implication does not hold. Namely, let $C \subset \mathbb{R}^3$ be a compact, right, circular cylinder, and let C_1, C_2 be two half-circles taken from the boundary of the top and bottom 2-faces of C, respectively. These half-circles are chosen in such a manner that a union of their orthogonal projections onto a 2-plane, this projection taken along one generator of C, yield a complete circle. Let, further on, B_1 and B_2 be the two generators of C connecting the respective endpoints of C_1 and C_2. Now one sees that $B_1 \cup B_2 \cup C_1 \cup C_2$ is a shadow-boundary of the convex body $K = \text{conv} (B_1 \cup B_2 \cup C_1 \cup C_2)$ along the described projection direction, and it is homeomorphic to an $(n-2)$-sphere. But it is not sharp.

It should be mentioned that there is a natural connection between shadow-boundaries of convex bodies $K \subset \mathbb{R}^n$ and the covering problem of Gohberg-Markus-Hadwiger. Namely, after suitably arranging light sources or illumination directions with respect to K, one has to guarantee that the intersection of all the corresponding shadow-boundaries is empty. If this is realized, then bd K is completely illuminated (cf. [Mar 5]). For example, if the compact, convex body $K \subset \mathbb{R}^n$ is a centrally symmetric polytope illuminated by pairwise opposite directions, then the smallest number of sharp shadow-boundaries having empty intersection is half the number of smaller homothets by which K can be covered.

In [K-P], [Mar 1], and [M-M 1] upper and lower bounds on the numbers of sharp shadow-boundaries of convex n-polytopes (with respect to exterior points and directions) for a prescribed number m of facets have been obtained, and in [M-S] these considerations were extended to the class of compact, convex bodies $K \subset \mathbb{R}^n$ and to numbers of illuminated regions in bd K. By all these results, special classes of convex polytopes have been characterized, such as parallelotopes, simplices, $(n-2)$-fold both-way infinite cylinders over planar m-gons and others.

For example, in [M-M 1] the following theorem was proved: *If P is a convex n-polytope with m facets, then the number $\delta(P)$ of realizable sharp shadow-boundaries in* bd P *satisfies*

$$2^{n-2}\left[\binom{m-n+1}{2} + \binom{m-n+1}{1} + \binom{m-n+1}{0}\right] - 1 \leq \delta(P),$$

with equality precisely if P is an $(n-2)$-fold pyramid over a planar convex $(m-n+2)$-gon.

For showing this, the arrangement $A(P)$ of the facet hyperplanes (i.e., of the m affine hulls of the $(n-1)$-faces) can be used. Namely, two inner points x_1, x_2 of the same n-cell (in the partition of $\mathbb{R}^n$ generated by $A(P)$) yield the same sharp shadow-boundary, since they are separated from P by the same facet hyperplanes. And otherwise, if x_1, x_2 are inner points of different such n-cells (say K_1 and K_2, respectively), then they correspond with the same sharp shadow-boundary if and only if K_1 and K_2 are opposite unbounded n-cells of the described partition of $\mathbf{R}^n$. This means that the cells K_1, K_2 lie in different half-spaces with respect to each facet hyperplane of P. Thus, $\delta(P)$ equals the number of all n-cells generated by $A(P)$ (without P itself) in the projective augmentation of $\mathbf{R}^n$.

In these terms, the theorem above reads as follows: *Let A be an arrangement of m hyperplanes in the real projecitve n-space, where all m hyperplanes have empty intersection and, moreover, no three of them have an $(n-2)$-dimensional intersection. Then the number of projective n-cells, generated by A, is at least*

$$2^{n-2}\left[\sum_{i=1}^{3}\binom{m-n+2}{i}\right].$$

Equality occurs if and only if A can be constructed in the following way: Consider an arrangement of $m-n+2$ lines $L_1, \cdots, L_{m-n+2}$ in a 2-flat, no three passing through one point. This 2-flat is said to be spanned by the origin and two unit vectors e_1, e_2. Consider now the third unit vector e_3, and all 2-flats spanned by the L_i's and e_3, and the one spanned by $o; e_1, e_2$. Consider the 3-flats spanned by them and the fourth unit vector e_4, and the one spanned by $o; e_1, e_2, e_3$. Repeat this until m hyperplanes are constructed.

It should be noticed that this statement generalizes results of R.W. Shannon [Sha]. A further generalization is also taken from [M-M 1]: *Let A be as above, where instead of no three even no q of the m hyperplanes have an intersection of dimension at least $n-q+1$, with $3 \leq q \leq n+1$. Then the corresponding number of projective n-cells is at least*

$$2^{n-q+1}\sum_{i=0}^{q-1}\binom{m-n+q-2}{i},$$

and equality is attained precisely if A is constructable in the following manner: Suppose $3 \leq q \leq n+1 \leq m$, and imagine in a $(q-1)$-flat a set of $m-n+q-1$ flats $F_1, \cdots, F_{m-n+q-1}$, each of dimension $q-2$ and no q of them having nonempty intersection. Say this $(q-1)$-flat is spanned by $o; e_1, \cdots, e_{q-1}$. Take the q-th unit vector e_q and consider all $(q-1)$-flats spanned by the F_i's and e_q, and the one spanned by $o; e_1, \cdots, e_{q-1}$. Repeat this as above, until m hyperplanes are obtained.

Finally, if the intersection of the m hyperplanes of A is s-dimensional $(-1 \leq s \leq n-1)$ and no q of them have an at least $(n-q+1)$-dimensional intersection with $1 \leq q \leq n+1$, then the number of projective n-cells is at least

$$2^{m-s-q} \sum_{i=0}^{q-1} \binom{m-n+s+q-1}{i}.$$

As one sees, the statements from [M-M 1] are of the following form. In the real projective n-space, let there be given an arrangement A of m hyperplanes such that any p-flat is contained in at most $a(p)$ of these hyperplanes, with $0 \le p \le n-1$. Estimate the number of the projective n-cells from below. For the sharp lower bound

$$F(n; m; a(0), \cdots, a(n-1)),$$

the inequality

$$\begin{aligned} F(n; m; a(0), \cdots, a(n-1)) \\ \ge \min_{1 \le b \le a(n-1)} [F(n; m-b; a(0), \cdots, a(n-1)) \\ + F(n-1; n-b; a(0)-b, \cdots, a(n-2)-b)] \end{aligned}$$

holds. The simplest non-settled case is the following: we have m different lines in the plane such that through any point at most $a(\ge 3)$ of these lines pass. In this case, the inequality above reduces to

$$F(2; m; a, 1) \ge F(2; m-1; a, 1) - \left[-\frac{m-1}{a-1}\right].$$

However, if $m > a$ and $m - 2$ is not a multiple of $a - 1$, this is not sharp. In fact, equality could hold only if the lines would contain exactly $-\left[-\frac{m-1}{a-1}\right]$ points of intersection with the remaining $m - 1$ lines. Hence through each point of intersection of the m lines at least three of the lines would pass, without all lines being concurrent. However, this contradicts the theorem about ordinary vertices (cf. [Sha], §3). We suspect that the result, implied in this case by the above inequality, is not even asymptotically sharp for $a \ge 3$ fixed and $m \to \infty$.

The last question coming from [M-M 1] is possibly more fascinating, since the extremal configurations can be conjectured and are perhaps geometrically nice. Namely, R.W. Shannon [Sha] derived sharp lower bounds on the numbers of k-flats and k-cells ($0 \le k \le n-1$) and $0 \le k \le n$, respectively) in a projective arrangement A of m hyperplanes with empty intersection. The minima are attained e.g. by the "near pencil" (i.e., $m - n + 1$ of the m hyperplanes of A intersect in an $(n-2)$-flat). Analogously to the third theorem from above (on arrangements where no $q, 3 \le q \le n+1$, hyperplanes have an at least $(n - q + 1)$-dimensional intersection), one can ask about the minima of these numbers if the above configurations are excluded by requiring each q of the hyperplanes to be in general position. Are these minima attained e.g. by the configurations described in this third theorem? Seemingly, methods from [M-M 1] (referring to the number of projective n-cells) and from [Sha] (referring to the "local" combinatorial structure of the arrangement) together are not sufficient to settle this problem.

Connected with the statements above, we repeat an interesting question posed in [M-S]. Namely, let P be a convex n-polytope with m facets. What is the minimal number of all shadow-boundaries with respect to point sources, and that of all illuminated regions, in the strict sense, with respect to point sources and with respect to parallel illumination?

PROBLEM 25:

Generalizing the concept of sharp shadow-boundaries from the previous problem, one might consider the number of k-dimensional subcomplexes $(0 \leq k \leq n - 2)$ of boundaries of convex n-polytopes $P \subset \mathbf{R}^n$ arising as such shadow-boundaries, i.e., the set of those points $p \in P$ for which, with respect to some (parallel) projection onto a $(k + 1)$-subspace, the image of p lies on the boundary of the image of P, cf. [B-E-S] and [Mar 5]. In particular, for $k = 0$ the projecting flats are $(n - 1)$-dimensional, and the sharp shadow-boundaries consist of two vertices of P lying in these projecting hyperplanes which are themselves different parallel supporting hyperplanes of this polytope. More precisely, let $X = \{x_1, \cdots, x_m\} \subset \mathbf{R}^n$ (with $x_i \neq x_j$ for $i \neq j$) be a point set whose convex hull is an n-polytope P. The points $x_i, x_j \in X$ are called antipodal (strictly antipodal) if there are different parallel supporting hyperplanes H', H'' of $P = \text{conv } X$ with $x_i \in H'$, $x_j \in H''$ (with $\{x_i\} = P \cap H'$, $\{x_j\} = P \cap H''$). These notions are due to B. Grünbaum and V. Klee (cf. [Grü 4] and [Grü 6], p. 420). We denote by $a(X)$ $(sa(X))$ the number of antipodal (strictly antipodal) pairs x_i, x_j from X, and we may and will suppose that each point from X is a boundary point (a vertex) of $P = \text{conv } X$.

There are several interpretations for the quantities $a(X)$ and $sa(X)$. For example, $sa(X)$ is half the vertex number of the difference body $P + (-P)$ of P. Another interpretation is the Minkowskian analogue of the Euclidean problem investigated in [Er 2], namely the question for the upper estimate on the number of Euclidean diameters of X. Let $\mathbf{R}^n$ be endowed with a Minkowski metric. Then one might ask for the number of Minkowskian diameters, i.e., for pairs x_i, x_j whose Minkowski distance is maximal. This is $\leq a(X)$, and if X is of constant Minkowski width (i.e., $P + (-P)$ is a Minkowskian ball), then equality holds.

For results on the numbers $a(X)$ and $sa(X)$ we refer to [D-G], [Grü 4], [E-F], [M-M 3,4], [N-S 1,2] and [Ng]. Moreover, P. Erdös asked already in [Er 1] for the maximum number of points in $\mathbf{R}^3$ if all angles among them are acute. The answer was given by H.T. Croft [Cr 1] and, much shorter, by K. Schütte [Schü]: this maximum is equal to 5.

L. Danzer and B. Grünbaum [D-G] explicitly stated the problem referring to *strictly antipodal sets*: what is the upper bound on the number of points in $\mathbf{R}^n$ such that each two of them are strictly antipodal? They also clarified the relation to the acute angled sets from above: an acute angled set is strictly antipodal. Later, B. Grünbaum [Grü 4] sharpened the results of Croft and Schütte by showing that the strictly antipodal sets in $\mathbf{R}^3$ have at most 5 points. Lastly, in [E-F] P. Erdös and Z. Füredi explicitely conjectured that a strictly antipodal set in $\mathbf{R}^n$ has cardinality at most c^n, where the constant c has to be smaller than 2. Moreover, they conjectured that if we have $2^n + 1$ points in $\mathbf{R}^n$, then the largest angle is at least some constant greater than $\frac{\pi}{2}$. Theorem 2.2 in [E-F] states that there exists an acute angled set in $\mathbf{R}^n$ of cardinality at least about $(1,15)^n$, and without proof it is announced that there exists such a set with cardinality $(2^{1/4} - o(1))^{n-1}$, which is about $(1,19)^n$. A similar statement is given in §3 of [M-M 3].

The planar case of the strictly antipodal problem is easy to solve in a direct way, but it also follows from [D-G], since an *antipodal set* in $\mathbf{R}^2$ has at most 4 points (with equality only for the vertex set of a parallelogram).

Summarizing all these statements, one should call this problem the Erdös-Danzer-Grünbaum-Füredi problem.

A natural generalization of (strictly) antipodal sets is the following. Let $S^k = \{S_1, \cdots, S_m\}$ be a finite set of k-dimensional simplices in $\mathbf{R}^n$, $1 \leq k \leq n-2$. The set S^k is said to be *(strictly) k-antipodal* if for any $i \neq j$ there are different parallel supporting hyperplanes H', H'' of the convex hull P of $\bigcup\limits_{i=1}^{m} S_i$ such that $S_i \subset H', S_j \subset H''$ (resp. $S_i = P \cap H', S_j = P \cap H''$). One can conjecture that, analogously to the Danzer-Grünbaum theorem for $k = 0$ (cf. [D-G]), a k-antipodal set of k-simplices has at most 2^{n-k} elements, which is attained for 2^{n-k} k-simplices on 2^{n-k} parallel k-faces of a cube. (Besides in [M-M 3], this conjecture was also formulated by I. Bárány and V. Soltan.)

In [M-M 4] it was shown that if $S^1 = \{S_1, \cdots, S_m\}$ is a strictly antipodal set of m segments (1-simplices) in $\mathbf{R}^3$, which are pairwise skew, then $m \leq 3$ (where this bound is sharp). Moreover, I. Talata (oral communication) has shown that an antipodal set of $(n-2)$-simplices in $\mathbf{R}^n$ has at most 4 elements.

A lot of further open problems and conjectures related to the numbers $a(X)$ and $sa(X)$ and more general concepts can be found in the papers mentioned above, see also chapter F in [C-F-G].

PROBLEM 26:

From our point of view, the following problem is highly geometric in nature, and it is also discussed in the problem book [C-F-G], §F 3, as well as in [G-G-L], chapter 17, §5.3.1. Moreover, D. Larman [L] has suggested to investigate this problem in connection with the Borsuk partition problem, cf. [C-F-G], section D 14.

What is the maximum cardinality of a set X in $\mathbf{R}^n$ if this set determines at most two different positive distances occuring between pairs from X? (Under this assumption, we shall say that X is a 2-*distance set*. An example is given by the $\frac{1}{2}n(n+1)$ edge-midpoints of the regular n-simplex.) In $\mathbf{R}^n$, this maximum is between $\binom{n+1}{2}$ and $\binom{n+2}{2}$, and in some dimensions the upper estimate is almost attained. In general, however, only for the lower estimate examples are known. Similarly, if there are at most s different positive distances, then the maximum number of points of X is between $\binom{n+1}{s}$ and $\binom{n+s}{s}$. However, for given n and s arbitrary, this problem is not to be awaited to be solved in the near future, since in this sense it has been investigated by excellent mathematicians in the past 50 years, and the progress was rather slow.

With respect to known results, we shall give now an account which is more extensive than that in section F3 of [C-F-G].

L.M. Kelly [Kel] showed that a 2-distance set in $\mathbf{R}^2$ has at most 5 points, and this estimate is sharp (take the vertices of a regular pentagon). H. Golomb [Gol] proved the same, but, moreover, found all sets in $\mathbf{R}^2$ all triples of which determine isosceles triangles. H.T. Croft [Cr 2] proved that 2-distance sets in $\mathbf{R}^3$ have at most 6 points, and he gave four examples: the vertices of the regular octahedron, of the semiregular prism over an equilateral triangle (with square lateral facets), and two regular pentagonal pyramids with lateral edges, either equal to the side or to the diagonal of the pentagonal base. He also investigated sets in $\mathbf{R}^3$ all triples of which determine isosceles triangles. Much more was proved by S.J. Einhorn and I.J. Schoenberg [E-S]. First, in $\mathbf{R}^n$ there are only finitely many 2-distance sets of cardinality at least $n+2$ (up to similarity); in particular, they obtained a finite upper bound for their cardinality, namely 5^n. Furthermore, they described a method that, in a sense, yields a mechanical way to obtain all 2-distance sets in $\mathbf{R}^n$. For the plane, they found all (namely 6) 2-distance sets with 4 points: the vertices of a square, of a rhombus with an angle of 60 degrees, 4 vertices of a regular pentagon, 3 vertices of an equilateral triangle and its center, the center and 3 points on the perimeter of a circle, dividing the perimeter to arcs of angles with 30, 30, 300 degrees, and with 60, 150, 150 degrees, respectively. (Of course, this can be read off from the paper [Gol], but in [E-S] this is explicit.) In addition, the only planar 2-distance set with 5 points is the vertex set of the regular pentagon. Furthermore, Einhorn and Schoenberg verified that in $\mathbf{R}^3$ there are precisely 26 non-planar 2-distance 5-point sets.

Moreover, there are exactly six 2-distance 6-point sets; those found by Croft, and two subsets of the vertices of the Platonic icosahedron without opposite vertices (the two regular pentagonal pyramids of Croft also have vertex sets which are subsets of that of the regular icosahedron). Finally, Einhorn and Schoenberg showed that there is no 2-distance 7-point set, and that in $\mathbf{R}^n$ the cardinality of an s-distance set (i.e., there are exactly s different positive distances) is bounded by a function of n and s, but this function is very large.

P. Delsarte, J.M. Goethals, and J.J. Seidel [D-G-S] obtained upper estimates with respect to s-distance sets on S^{n-1}, in particular (for $s=2$) leading to $\frac{1}{2}n(n+3)$, which is sharp for $n=2, 6$ and 22.

D.G. Larman, C.A. Rogers, and J.J. Seidel [L-R-S] derived the upper bound $\frac{1}{2}(n+1)(n+4)$ for 2-distance sets in $\mathbf{R}^n$, and finally A. Blokhuis [Blo 2] obtained the best general upper bound for 2-distance sets in $\mathbf{R}^n$ up to now, namely $\frac{1}{2}(n+1)(n+2)$ (but the exact maximum is unknown for $n \geq 4$). This bound is just 1 greater than the bound of Delsarte, Goethals and Seidel for S^{n-1}, which itself was sharp for $n = 2, 6, 22$. Thus, for these values of n the estimate from their paper is almost sharp. The corresponding proof is also reproduced in the survey paper [B-S], where, moreover, an example of an s-distance set in $\mathbf{R}^n$ with $\binom{n+1}{s}$ elements is constructed. Namely, it suffices to consider those vectors in a hyperplane of $\mathbf{R}^{n+1}$, which have s coordinates 1, and $n + 1 - s$ coordinates 0. (Actually, they give this for $s = 2$, but the general case is clear from that.)

In [B-B-S] the upper bound $\binom{n+s}{s}$ on the cardinality of any s-distance set in $\mathbf{R}^n$ was obtained, which reduces to the bound of Blokhuis above if $s = 2$. For s fixed and n tending to infinity, the example from the Blokhuis-Seidel paper asymptotically meets that bound. (A weaker estimate for the same quantity was obtained earlier, namely in [Ba-Ba].)

Several natural questions occur. For example, what is the maximum number $g_3(s)$ of points of an s-distance set in $\mathbf{R}^3$ for small s? In particular, is the only 3-distance 12-point set in $\mathbf{R}^3$ the vertex set of the Platonic icosahedron? And is the only 5-distance 20-point set in $\mathbf{R}^3$ the vertex set of the Platonic dodecahedron? And referring to the inequalities $\binom{n+1}{s} \leq g_n(s) \leq \binom{n+s}{s}$ from [B-S], it can be thought that, for $s \geq 3$ fixed and $n \to \infty$, $g_n(s)$ is closer to the lower bound. For further related investigations we refer to [Blo 3] and [H-P].

In addition it should be remarked that the cardinality of s-distance sets in hyperbolic n-space has a finite upper bound, see [Ban] and [B-B-S-D].

Finally, we mention the strongly related problem on isosceles triangles in finite point sets: how many such triangles can occur in a set X of m points in $\mathbf{R}^n$? For partial results we refer to several papers mentioned above and section F6 in [C-F-G]. In [E-P] it was shown that the maximal number of isosceles triangles among n points in the plane lies between $cn^2 \log n$ and $cn^{5/2}$ for some constant $c > 0$, and Theorem 12.2 from [P-A] says that the number of triples spanning such triangles is at most $cn^{7/3}$.

<u>PROBLEM 27:</u>

We formulate the following problem: *to describe geometrically all compact, convex bodies $M \subset \mathbf{R}^n$ for which the minimal cardinality of fixing systems is equal to r, where r is a given positive integer.* In other words, we speak about the solutions of the *equation $\rho(M) = r$* with respect to $M \subset \mathbf{R}^n$.

Theorem 44.6 shows that this problem is restricted to $n + 1 \leq r \leq 2n$. The following two theorems are proved in [B-MA]; in fact, they are consequences of the above Theorems 25.2 and 29.1.

Theorem 1: For a compact, convex body $M \subset \mathbf{R}^n$ the equality $\rho(M) = 2n$ holds if and only if M is a parallelotope.

Theorem 2: For a compact, convex body $M \subset \mathbf{R}^n$ the equality $\rho(M) = 2n - 1$ holds if and only if $M = B \oplus P$, where the following conditions are satisfied:

(i) B is an indecomposable, compact, convex set of a dimension $q \leq n$, and P is an $(n-q)$-dimensional parallelotope;

(ii) either $q \geq 3$ and B is a stack;

   or $q \geq 4$ and B is a stack-outcut with $k = 2$ (see Theorems 29.6 and 29.1 above);

   or B is a three-dimensional outcut;

   or B is two-dimensional.

We remark that the results of J. Kincses (see section 30) allow us to obtain a solution of this problem for centrally symmetric bodies in the case $r \geq 2n-3$. It would be interesting to find the solution for some other values of r (i.e., if $r < 2n - 3$ for centrally symmetric bodies and $r < 2n - 1$ in the general case).

PROBLEM 28:

Let Q be a class of figures in the plane $\mathbf{R}^2$. We say that a convex figure $W \subset \mathbf{R}^2$ is a covering set for the class Q if any figure M from the class Q can be covered by a congruent copy of W. If no convex figure $W' \subset W$ is a covering set for the class Q, then we say that W is a *minimal, convex covering set* for the class Q.

We consider the following classes of figures in the plane:

   Q_1 is the class of all subsets of diameter 1;

   Q_2 is the class of all compact, convex subsets of diameter 1;

   Q_3 is the class of all figures of constant width 1;

   Q_4 is the class of all Reuleaux polygons of width 1;

   Q_5 is the class of all Reuleaux pentagons of width 1.

It is easy to show that the classes Q_1, Q_2, Q_3, Q_4 are equivalent in the following sense: any minimal, convex covering set for one of these classes is at the same time a minimal, convex covering set for each other class. The problem is to decide whether also the class Q_5 is equivalent (in this sense) to the previous four classes.

Bibliography

[Ai]   M. AIGNER: Combinatorial Theory. Springer, Berlin et al., 1979.

[Al]   A.D. ALEKSANDROV: A theorem on convex polyhedra (in Russian). Trudy Mat. Inst. Steklova 4 (1933), 87.

[A-Z]   A.D. ALEKSANDROV, V.A. ZALGALLER: Two-dimensional manifolds of bounded curvature (in Russian). Trudy MIAN 63, Izd. AN SSSR, Moscow 1962.

[A-K]   R.D. ANDERSON, V. KLEE: Convex functions and upper semi-continuous collections. Duke Math. J. 19 (1952), 349-357.

[Ar-Pr]   T.E. ARMSTRONG, K. PRIKRY: Liapunoff's theorem for nonatomic, finitely-additive, bounded, finite-dimensional, vector-valued measures. Trans. Amer. Math. Soc. 266 (1981), 499-514.

[Ba 1]   E. BALADZE: A complete solution of the Szökefalvi-Nagy problem for zonohedra. Soviet Math. Dokl. 34 (1987), no. 3, 458-461.

[Ba 2]   E. BALAZDE: A solution of the Szökefalvi-Nagy problem for zonoids. Dokl. Akad. Nauk SSSR 310 (1990), 11-14 (Engl. transl.: Soviet Math. Dokl. 41 (1990)).

[Ba 3]   E. BALADZE: On the semicontinuity of an invariant of Boltyanski (in Russian). Soobshch. AN GSSR 134 (1989), no. 2.

[Ba 4]   E. BALADZE: Solution of the Szökefalvi-Nagy problem for a class of convex polytopes. Submitted to Geom. Dedicata.

[B-B 1]   E. BALADZE, V.G. BOLTYANSKI: On the solution of the Szökefalvi-Nagy problem (in Russian). Soobshch. AN GSSR 120 (1985), no. 3.

[B-B 2]   E. BALADZE, V.G. BOLTYANSKI: Belt bodies and the Helly dimension (in Russian). Mat. Sbornik, 186 (1995), no. 2, 3-20.

[B-B-Ch]   E. BALADZE, V.G. BOLTYANSKI, T.A. CHABUKIANI: New results on the Szökefalvi-Nagy problem (in Russian). Trudy Tbilissi Mat. Inst. 85 (1987).

[Ban]   E. BANNAI: On s-distance subsets in real hyperbolic space. Hokkaido Math. J. 11 (1982), 201-204.

[Ba-Ba]   E. BANNAI, E. BANNAI: An upper bound for the cardinality of an s-distance subset in real Euclidean space. Combinatorica 1 (1981), 99-102.

[B-B-S]   E. BANNAI, E. BANNAI, D. STANTON: An upper bound for the cardinality of s-distance sets in real Euclidean space II. Combinatorica 3 (1983), 147-152.

[B-B-S-D]   E. BANNAI, A. BLOKHUIS, J.J. SEIDEL, PH. DELSARTE: An addition formula for hyperbolic space. J. Combin. Theory Ser. A 36 (1984), no. 3, 332-341.

[Bel]   YU. F. BELOUSOV: Theorems on covering of plane figures (in Russian). Ukrain. Geom. Sb. 20 (1977), 10-17.

[B-B]   A. BEUTELSPACHER, K. BEZDEK: The Helly dimension relative to a lattice. Coll. Math. Soc. J. Bolyai 63 (Intuitive Geometry), North-Holland, Amsterdam et al., 1994, 11-16.

[Be 1]   K. BEZDEK: The problem of illumination of the boundary of a convex body by affine subspaces. Mathematika 38 (1991), 362-375.

[Be 2]   K. BEZDEK: Hadwiger's covering conjecture and its relatives. Amer. Math.. Monthly 99 (1992), 954-956.

[Be 3]   K. BEZDEK: Hadwiger-Levi's covering problem revisited. In: New Trends in Discrete and Computational Geometry, ed. by J. Pach. Springer, Berlin et al., 1993, 199-233.

[Be 4]   K. BEZDEK: A note on the illumination of convex bodies. Geom. Dedicata 45 (1993), 89-91.

[Be 5]   K. BEZDEK: On affine subspaces that illuminate a convex set. Beitr. Algebra Geom. 35 (1994), 131-139.

[Be-Bi]   K. BEZDEK, T. BISZTRICZKY: Hadwiger's covering conjecture and low dimensional dual cyclic polytopes. Geom. Dedicata 46 (1993), 279-283.

[B-K-M]   K. BEZDEK, GY. KISS, M. MOLLARD: An illumination problem for zonoids. Israel J. Math. 81 (1993), 265-272.

[B-M-S-W]   T. BISZTRICZKY, P. MCMULLEN, R. SCHNEIDER, A.I. WEISS (EDS.): Polytopes - Abstract, Convex, and Computational. Nato ASI Series, Series C, Vol.440, Kluwer Academic Publishers, Dordrecht - Boston - London, 1994.

[Bl 1]   B. BOLLOBÁS: Fixing system for convex bodies. Studia Sci. Math. Hungarica 2 (1967), 351-354.

[Bl 2]   W. BLASCHKE: Kreis und Kugel. de Gruyter, Berlin 1956.

[Blo 1]   A. BLOKHUIS: An upper bound for the cardinality of s-distance sets in $\mathbb{E}^d$ and H^d. Eindhoven Univ. Tech. Mem. 68 (1982).

[Blo 2]   A. BLOKHUIS: An upper bound for the cardinality of 2-distance sets in Euclidean space. In: Convexity and Graph Theory (Jerusalem, 1981), Eds. M. Rosenfeld and J. Zaks, Ann. Discrete Math. 20, North-Holland, Amsterdam 1984, 65-66.

[Blo 3]   A. BLOKHUIS: Few-distance sets. Thesis, CWI Tract 7 (MR 87 f: 51023).

[B-S]   A. BLOKHUIS, J.J. SEIDEL: Few-distance sets in $\mathbf{R}^{p,q}$. Symposia Math., Vol. 28 (Rome 1983), Acad. Press, London-New York 1986, Istituto Nazionale di Alta Matematica Francesco Severi.

[Bo]   N.A. BOBYLEV: On the problem to cover bodies by homothetical bodies. Mat. Issled. 3 (1968), no. 3 (9), 19-26.

[Bo-So]   E. BOHNE, P.S. SOLTAN: Outer illumination of two-sided cones (in Russian). Studia Sci. Math. Hungar. 27 (1992), 261-266.

[Bk 1]   E.D. BOLKER: A class of convex bodies. Trans. Amer. Math. Soc. 145 (1969), 323-345.

[Bk 2]   E.D. BOLKER: Centrally symmetric polytopes. In: Proc. 12th Biannual Intern. Semin. Canad. Math. Congr. on Time Series and Stochastic Processes, Convexity and Combinatorics (Vancouver 1969), ed. by R. Pyke, Canadian Math. Congr., Montreal 1970, 255-263.

[Bt 1]   V.G. BOLTYANSKI: The problem of illuminating the boundary of a convex body (in Russian). Izv. Mold. Fil. AN SSSR (1960), no. 10 (76), 77-84.

[Bt 2]   V.G. BOLTYANSKI: On the partition of plane figures into pieces of smaller diameter (in Russian). Colloq. Math. 21 (1970), 253-263.

[Bt 3]   V.G. BOLTYANSKI: On separation systems of convex cones (in Russian). Izv. AN Arm. SSR (Matem.) 7 (1972), 325-333.

[Bt 4]   V.G. BOLTYANSKI: The tent method in the theory of extremal problems (in Russian). Uspekhi Mat. Nauk 30 (1975), 3-65.

[Bt 5]   V.G. BOLTYANSKI: On certain classes of convex sets. Soviet Math. Doklady 17 (1976), no. 1, 10-13.

[Bt 6]   V.G. BOLTYANSKI: Helly's theorem for H-convex sets. Soviet Math. Doklady 17 (1976), no. 1, 78-81.

[Bt 7]   V.G. BOLTYANSKI: Generalization of a certain theorem of B. Szökefalvi-Nagy. Soviet Math. Doklady 17 (1976), no. 3, 674-677.

[Bt 8]   V.G. BOLTYANSKI: Some theorems of combinatorial geometry (in Russian). Mat. Zametki 21 (1977), 117-124.

[Bt 9]   V.G. BOLTYANSKI: Separation of a system of convex cones in a topological vector space. Soviet Math. Doklady 32 (1985), no.1, 236-239.

[Bt 10]   V.G. BOLTYANSKI: The tent method and problems from system analysis (in Russian). In: Sb. Mt. Teor. Syst. (Ed. M. A. Krasnosel'ski). Nauka, Moscow 1986, 5-24.

[Bt 11]   V.G. BOLTYANSKI: The tent method in topological vector spaces. Soviet Math. Doklady 34 (1987), no. 1, 176-179.

[Bt 12]   V.G. BOLTYANSKI: The tent method in optimal control theory. In: Optimal Control, Intern. Series of Numerical Math., Vol. 111, ed. by R. Bulirsch, J. Stoer and K.H. Well, Birkhäuser, Basel-Boston-Berlin, 1993.

[Bt 13]   V.G. BOLTYANSKI: A new step in the solution of the Szökefalvi-Nagy problem. Discrete Comput. Geom. 8 (1992), 27-49.

[Bt 14]   V.G. BOLTYANSKI: A solution of the illumination problem for belt bodies (in Russian). Mat. Zametki 58 (1996), no. 4, 505-511.

[Bt 15]   V.G. BOLTYANSKI: On Hanner numbers, to appear.

[Bt 16]   V.G. BOLTYANSKI: Hilbert's Third Problem. Wiley & Sons, New York-Toronto-London-Sydney 1978. H

[B-Ch 1]   V.G. BOLTYANSKI, T.A. CHABUKIANI: Solution of the Szökefalvi-Nagy problem for three-dimensional convex bodies. Soviet Math. Doklady 30 (1984), no. 3, 755-757.

[B-Ch 2]   V.G. BOLTYANSKI, T.A. CHABUKIANI: On the Szökefalvi-Nagy problem (in Russian. English and Georgian summaries). Soobshch. Akad. Gruzin. SSR, 117 (1985), no.3, 477-479.

[B-G 1]   V.G. BOLTYANSKI, I.C. GOHBERG: Sätze und Probleme der kombinatorischen Geometrie. Mathematische Schülerbücherei, Nr. 61. VEB Deutscher Verlag der Wissenschaften, Berlin, 1972, 127 pp.

[B-G 2]   V.G. BOLTYANSKI, I.C. GOHBERG: The decomposition of figures into smaller parts. Popular Lectures in Mathematics. University of Chicago Press, Chicago - London, 1980. vi + 74 pp.

[B-G 2]   V.G. BOLTYANSKI, I.Z. GOHBERG: Stories about covering and illuminating of convex bodies. Nieuw Archief voor Wiskunde. Vierde serie Deel, 13 (1995), no. 1, 1-26.

[B-M 1]   V.G. BOLTYANSKI, H. MARTINI: Combinatorial geometry of belt bodies. Results Math. 28 (1995), 224-249.

[B-M 2]   V.G. BOLTYANSKI, H. MARTINI: Geometry as a method. Manuscript, 32 pp., submitted.

[B-M 3]   V.G. BOLTYANSKI, H. MARTINI: On maximal primitive fixing systems. Beitr. Alg. Geom. 37 (1996), 199–207

[B-M-SP]   V.G. BOLTYANSKI, H. MARTINI, P.S. SOLTAN: Star-shaped sets in normed spaces. Discrete Comput. Geom. 15 (1996), 63-71.

[B-Ma]   V.G. BOLTYANSKI, E. MORALES AMAYA: Minimal fixing systems for convex bodies. Journal of Applied Analysis 1 (1995), no.1, 1-13.

[B-So]   V.G. BOLTYANSKI, A. SOIFER: Geometric Etudes in Combinatorial Mathematics. Center of Excellence in Math. Education, Colorado Springs, 1991.

[B-SP 1]   V.G. BOLTYANSKI, P.S. SOLTAN: Inner illumination of the boundary of a convex body (in Russian). Mat. Sbornik 87 (129), (1972), 83-90.

[B-SP 2]   V.G. BOLTYANSKI, P.S. SOLTAN: On star-shaped sets (in Russian). Izv. AN MSSR, Ser. Fiz.-Tehn. i Mat. Nauk, 1976, no. 3, 7-11.

[B-SP 3]   V.G. BOLTYANSKI, P.S. SOLTAN: Combinatorial Geometry of Various Classes of Convex Sets (in Russian). Shtiinca, Kishinev 1978.

[B-SP 4]   V.G. BOLTYANSKI, P.S. SOLTAN: Combinatorial Geometry and Convexity Classes (in Russian). Uspekhi Mat. Nauk 33 (1978), no. 1, 3-42.

[B-SP 5]   V.G. BOLTYANSKI, P.S. SOLTAN: Solution of the Hadwiger problem for a class of convex bodies. Soviet Math. Doklady 42 (1991), no. 1, 18-22.

[B-SP 6]   V.G. BOLTYANSKI, P.S. SOLTAN: A solution of Hadwiger's problem for zonoids. Combinatorica 12 (1992), 381-388.

[B-SV]   V.G. BOLTYANSKI, V. SOLTAN: Borsuk's problem (in Russian). Mat. Zametki 22 (1977), 621-631.

[Br 1]   K. BORSUK: Über die Zerlegung einer n-dimensionalen Vollkugel in n Mengen. Verh. Internat. Math.-Kongress Zürich, 1932, Bd. II, Ed. W. Saxer, Orell Füssli Verlag, Zürich 1932, 192.

[Br 2]   K. BORSUK: Drei Sätze über die n-dimensionale euklidische Sphäre. Fund. Math. 20 (1933), 177-190.

[Br 3]   K. BORSUK: Some remarks on covering of bounded subsets of the Euclidean n-space with sets of smaller diameter. Demonstratio Math. 11 (1978), 247-251.

[B-V]   K. BORSUK, R. VAINA: On covering of bounded sets by sets with twice less diameter. Coll. Math. 42 (1979), 33-47.

[B-L]   J. BOURGAIN, J. LINDENSTRAUSS: On covering a set in $\mathbf{R}^d$ by balls of the same diameter. In: Geometric Aspects of Functional Analysis (Eds.: J. Lindenstrauss and V. Milman). Lecture Notes Math. 1469, Springer, Berlin 1991, 138-144.

[Bw]   P.L.BOWERS: The Borsuk dimension of a graph and Borsuk's partition conjecture for convex sets. Graphs and Combinatorics 6 (1990), 207-222.

[B-E-S] H. BUSEMANN, G. EWALD, G.C. SHEPHARD: Convex bodies and convexity on Grassmann cones, I-IV. Math. Ann. 151 (1963), 1-41.

[C-G]   G.D. CHAKERIAN, H. GROEMER: Convex bodies of constant width. In: Convexity and its Applications, Eds. P.M. Gruber and J. M. Wills, Birkhäuser, Basel 1983, 49-96.

[Ch]   A.B. CHARAZISHVILI: On the problem of illumination (in Russian). Soobshch. AN GSSR 71 (1973), no. 1, 289-291.

[Ci 1] D. CIESLIK: The vertex-degrees of Steiner Minimal Trees in Minkowski planes. In: Topics in Combinatorics and Graph Theory, ed. by R. Bodendieck and R. Henn. Heidelberg, 1990, 201-206.

[Ci 2] D. CIESLIK: The vertex-degrees of Steiner Minimal Trees in Banach-Minkowski spaces. Geombinatorics 3 (1994), 75-82.

[Cr 1] H.T. CROFT: On 6-point configurations in 3-space. J. London Math. Soc. 36 (1961), 289-306.

[Cr 2] H.T. CROFT: 9-point and 7-point configurations in 3-space. Proc. London Math. Soc (3) 12 (1962), 400-424.

[C-F-G] H.T. CROFT, K.J. FALCONER, R.K. GUY: Unsolved Problems in Geometry. Springer, New York et al., 1991.

[Da 1] L. DANZER: Über Durchschnittseigenschaften n-dimensionaler Kugelfamilien. J. reine angew. Math. 209 (1961), 181-203.

[Da 2] L. DANZER: Review 2942. Math. Reviews 26 (1963), 569-570.

[Da 3] L. DANZER: On the k-th diameter in $\mathbb{E}^d$ and a problem of Grünbaum. Proc. Coll. Convexity (Copenhagen 1965), Kobenhavn's Univ. Mat. Inst. 1967, p.41.

[DG]   L. DANZER, B. GRÜNBAUM: Über zwei Probleme bezüglich konvexer Körper von P. Erdös und V. L. Klee. Math. Z. 79 (1962), 95-99.

[D-G-K] L. DANZER, B. GRÜNBAUM, V. KLEE: Helly's theorem and its relatives. In: Convexity, Proc. Sympos. Pure Math. 7 (Amer. Math. Soc., Providence, R.I., 1963), 101-180.

[De 1] B.V. DEKSTER: Diameters of the pieces in Borsuk's covering. Research report M/CS 87-10 (December 1987), Mount Allison Univ., Sackville, New Brunswick.

[De 2]   B.V. DEKSTER: Borsuk's covering problem for blunt bodies. Arch. Math. 51 (1988), 87-91.

[De 3]   B.V. DEKSTER: Diameters of the pieces in Borsuk's covering. Geom. Dedicata 30 (1989), 35-41.

[De 4]   B.V. DEKSTER: The Borsuk conjecture holds for convex bodies with a belt of regular points. Geom. Dedicata 45 (1993), 301-306.

[De 5]   B.V. DEKSTER: The Borsuk conjecture holds for bodies of revolution. Journal of Geometry (to appear).

[De 6]   B.V. DEKSTER: Every convex n-dimensional body with a smooth belt can be illuminated by $n + 1$ directions. Journal of Geometry 49 (1994), 90-95.

[D-G-S]   P. DELSARTE, J.M. GOETHALS, J.J. SEIDEL: Spherical codes and designs. Geom. Dedicata 6 (1977), 363-388.

[D-L]   M. DEMBINSKI, M. LASSAK: Covering plane sets with sets of three times less diameter. Demonstratio Math. 56 (1985), 249-255.

[Doi]   J.-P. DOIGNON: Convexity in crystallographical lattices. J. Geom. 3/1 (1973), 71-85.

[Dol]   V.L. DOL'NIKOV: Generalized transversals of families of sets in R^n and connections between the Helly and Borsuk theorems. Doklady Akad. Nauk SSSR 297 (1987), 519-522.

[D-M]   A. DUBOVITSKI, A. MILYUTIN: Extremum problems with constraints (in Russian). Zh. Vychisl. Mat. i Mat. Fiz., 5 (1965), no. 3, 395-453.

[Ec]   J. ECKHOFF: Radon's theorem revisited. In: Contributions to Geometry, Proc. Geom. Sympos. (Siegen 1978), Birkhäuser, Basel 1979, 164-185.

[E]   J. ECKHOFF: Helly, Radon, and Caratheodory type theorems. In: Handbook of Convex Geometry, ed. by P.M. Gruber and J.M. Wills, North-Holland, 1993, 389-448.

[Eg 1]   H.G. EGGLESTON: Covering a three-dimensional set with sets of smaller diameter. J. London Math. Soc. 30 (1955), 11-24.

[Eg 2]   H.G. EGGLESTON: Problems in Euclidean space: Applications of Convexity. Pergamon, New York 1957.

[Eg 3]   H.G. EGGLESTON: Convexity. Cambridge Univ. Press, 1958.

[Eg 4]   H.G. EGGLESTON: Sets of constant width in finite dimensional Banach spaces. Israel J. Math. 3 (1965), 163-172.

[E-S]   S.J. EINHORN, I.J. SCHOENBERG: On Euclidean sets having only two distances between points. Indag. Math. 28 (1966), 479-504.

[Er 1]   P. ERDÖS: Some unsolved problems. Michigan Math. J. 4 (1957), 291-300.

[Er 2]   P. ERDÖS: On sets of distances of n points in Euclidean space. Magyar Tud. Akad. Mat. Kutató Int. Közl. 5 (1960), 165-169; also in: P. Erdös, The art of counting. Selected writings (Ed.: J. Spencer). MIT Press, Cambridge (Mass.) and London, 1973, 676-679.

[E-F]   P. ERDÖS, Z. FÜREDI: The greatest angle among n points in the d-dimensional Euclidean space. Combinatorial Math. (Proc. Internat. Colloq. Graph Theory Combin., Marseille-Luminy, 1981), North-Holland Math. Stud., Vol. 75, Amsterdam and New York, 1983; Ann. Discrete Math. 17.

[E-P]   P. ERDÖS, G. PURDY: Some extremal problems in geometry III, Proc. VI. Southeastern Conf. Combin., Graph Theory and Computing (Boca Raton, 1975). Congressus Numerantium 14 (1975), 291-308.

[Fe]   L. FEJES TÓTH: On primitive polyhedra. Acta Math. Acad. Sci. Hungar. 13 (1962), 379-383.

[Fu 1]   S. FUDALI: Six-point primitive fixing system in a plane. Demonstratio Math. 19 (1986), 341-348.

[Fu 2]   S. FUDALI: Fixing systems and homothetic covering. Acta Math. Hungar. 50 (3-4) (1987), 203-225.

[Ga]   D. GALE: On inscribing n-dimensional sets in a regular n-simplex. Proc. Amer. Math. Soc. 4 (1953), 222-225.

[Gar]   R. GARDNER: Geometric Tomography. Cambridge Univ. Press, to appear.

[G-S-S]   L. GERMAN, P.S. SOLTAN, V. SOLTAN: Some properties of d-convex sets. Soviet Math. Dokl. 14 (1973), 1566-1570.

[G-K]   I.Z. GOHBERG, M.G. KREIN: The basic propositions on defect numbers, root numbers, and indices of linear operators (in Russian). Uspehi Mat. Nauk 12 (1957), no. 2, 43-118 (Engl. transl.: Amer. Math. Soc. Transl. (2) 13 (1960), 185-264).

[G-M]   I.Z. GOHBERG, A.S. MARKUS: One problem on covering convex figures by similar figures (in Russian). Izv. Mold. Fil. Akad. Nauk SSSR 76 (1960), no. 1, 87-90.

[Gol]   H. GOLOMB: Advanced problems and solutions: Isosceles n-points. Amer. Math. Monthly 55 (1948), 513-514.

[G-W]   P. GOODEY, W. WEIL: Zonoids and generalizations. In: Handbook of Convex Geometry, Eds. P.M. Gruber and J.M. Wills, North-Holland, 1993, 1297-1326.

[Go]   D.B. GOODNER: Projections in normed linear spaces. Trans.
           Amer. Math. Soc. 69 (1950), 89-108.

[Gra]   R.L. GRAHAM: On partitions of an equilateral triangle. Canad.
           J. Math. 19 (1967), 394-409.

[G-K-W]   P. GRITZMANN, V. KLEE, J. WESTWATER: Polytope contain-
           ment and determination by linear probes. Proc. London Math.
           Soc. (3) 70 (1995), 691-720.

[Groe]   H. GROEMER: On complete convex bodies. Geometriae Dedicata
           20 (1986), 319-334.

[G-G-L]   M. GRÖTSCHEL, R. GRAHAM, L. LOVÁSZ (EDS.): Handbook
           of Combinatorics. Elsevier, 1994.

[Gr]   J. DE GROOT: Some special metrics in general topology. Colloq.
           Math. 6 (1958), 283-286.

[G-W 1]   P.M. GRUBER, J.M. WILLS (EDS.): Convexity and its Appli-
           cations. Birkhäuser, Basel-Boston-Stuttgart 1983.

[G-W 2]   P.M. GRUBER, J.M. WILLS (EDS.): Handbook of Convex Geo-
           metry, Vol. A and B. North-Holland, Amsterdam et al., 1993.

[Grü 1]   B. GRÜNBAUM: Borsuk's partition conjecture in Minkowski pla-
           nes. Bull. Res. Council Israel, F1, N1 (1957), 25-30.

[Grü 2]   B. GRÜNBAUM: A simple proof of Borsuk's conjecture in three
           dimensions. Proc. Cambr. Philos. Soc. 53 (1957), 776-778.

[Grü 3]   B. GRÜNBAUM: Borsuk's problem and related questions. In:
           Convexity, ed. by V. Klee, Proc. Symposia Pure Math., Vol.
           7 (Amer. Math. Soc., Providence, RI), 1963, 271-284.

[Grü 4]   B. GRÜNBAUM: Strictly antipodal sets. Israel J. Math. 1 (1963),
           5-10.

[Grü 5]   B. GRÜNBAUM: Fixing systems and inner illumination. Acta
           Math. Acad. Sci. Hungar. 15 (1964), 161-168.

[Grü 6]   B. GRÜNBAUM: Convex Polytopes. Wiley, New York 1967.

[Grü 7]   B. GRÜNBAUM: Etudes on Combinatorial Geometry and the
           Theory of Convex Bodies (in Russian; translation of [Grü 3]).
           Moscow, 1971.

[Grü 8]   B. GRÜNBAUM: Lectures on Combinatorial Geometry. Mimeo-
           graphed Notes, Univ. of Washington, Seattle, 1974.

[G-S]   R.K. GUY, J.L. SELFRIDGE: Optimal coverings of the square.
           In: Infinite and Finite Sets. Coll. Math. Soc. J. Bolyai 10 (1975),
           North-Holland, Amsterdam et al., 745-799.

[Ha 1]   H. HADWIGER: Überdeckung einer Menge durch Mengen kleine-
           ren Durchmessers. Comment. Math. Helvet. 18 (1945/46), 73-75.

[Ha 2]   H. HADWIGER: Mitteilung betreffend meine Note: Überdeckung einer Menge durch Mengen kleineren Durchmessers. Comment. Math. Helvet. 19 (1946/47), 71-73.

[Ha 3]   H. HADWIGER: Über die Zerstückung eines Eikörpers. Math. Z. 51 (1949), 161-165.

[Ha 4]   H. HADWIGER: Von der Zerlegung der Kugel in kleinere Teile. Gaz. Mat. Lisboa 15 (1954), 1-3.

[Ha 5]   H. HADWIGER: Altes und Neues über konvexe Körper. Birkhäuser, Basel-Stuttgart 1955.

[Ha 6]   H. HADWIGER: Ungelöste Probleme, Nr. 20. Elem. Math. 12 (1957), 121.

[Ha 7]   H. HADWIGER: Ungelöste Probleme, Nr. 38. Elem. Math. 15 (1960), 130-131.

[Ha 8]   H. HADWIGER: Ungelöste Probleme, Nr. 55. Elem. Math. 27 (1972), 57.

[Ha 9]   H. HADWIGER: Ungelöste Probleme, Nachtrag zu Nr. 38. Elem. Math. 17 (1962), 84-85.

[H-D-K]   H. HADWIGER, H. DEBRUNNER, V.L. KLEE: Combinatorial Geometry in the Plane. Holt, Rinehart and Winston, New York 1964.

[Ham]   G. HAMEL: Eine Basis aller Zahlen und die unstetigen Lösungen der Funktionalgleichung $f(x + y) = f(x) + f(y)$. Math. Ann. 60 (1905), 459-462.

[Han]   O. HANNER: Intersections of translates of convex bodies. Math. Scand. 4 (1956), 65-87.

[H-L]   A. HANSEN, A. LIMA: The structure of finite-dimensional Banach spaces with the 2.3. property. Acta Math. 146 (1981), 1-23.

[H-P]   H. HARBORTH, L. PIEPMEYER: Point sets with small integral distances. In: Applied Geometry and Discrete Math. (The Victor Klee Festschrift), DIMACS Ser. Discr. Math. Theor. Comp. Sci., Vol. 4 (1991), 319-324.

[H-M]   E. HEIL, H. MARTINI: Special convex bodies. In: Handbook of Convex Geometry, Eds. P.M. Gruber and J.M. Wills, North-Holland, 1993, 347-385.

[Hl]   E. HELLY: Über Mengen konvexer Körper mit gemeinschaftlichen Punkten. Jber. Deutsch. Math.-Verein 32 (1923), 175-176.

[He]   A. HÉPPÉS: On the partitioning of three-dimensional point sets into sets of smaller diameter (Hungarian). Magyar Tud. Akad. Mat. Fiz. Oszt. Kösl. 7 (1957), 413-416.

[H-R]   A. HÉPPÉS, P. RÉVÉSZ: Zum Borsukschen Zerteilungsproblem.
            Acta Math. Sci. Hungar. 7 (1956), 159-162.

[J]   H.W.E. JUNG: Über die kleinste Kugel, die eine räumliche Figur
            einschließt. J. reine angew. Math. 123 (1901), 241-257.

[K-K]   J. KAHN, G. KALAI: A counterexample to Borsuk's conjecture.
            Bull. Amer. Math. Soc. (N.S.) 29 (1993), no. 1, 60-62.

[Ke]   J.L. KELLEY: Banach spaces with the extension property. Trans.
            Amer. Math. Soc. 72 (1952), 323-326.

[Kel]   L.M. KELLY: Elementary problems and solutions: Isosceles n-
            points. Amer. Math. Monthly 54 (1947), 227-229.

[Ki]   J. KINCSES: The classification of 3- and 4-Helly-dimensional con-
            vex bodies. Geom. Dedicata 22 (1987), 283-301.

[Kl 1]   V.L. KLEE: Infinite-dimensional intersection theorems. In: Proc.
            Symposia Pure Math. 7 (Convexity), Ed. V. Klee, Amer. Math.
            Soc., Providence, RI, 1963, 349-360.

[Kl 2]   V.L. KLEE (ED.): Convexity (Proc. Symposia Pure Math. 7),
            Amer. Math. Soc., Providence, RI, 1963.

[K-P]   P. KLEINSCHMIDT, U. PACHNER: Shadow-boundaries and cuts
            of convex polytopes. Mathematika 27 (1980), 58-63.

[Kn]   R. KNAST: An approximative theorem for Borsuk's conjecture.
            Proc. Cambridge Phil. Soc. (1974), no. 1, 75-76.

[Ko 1]   D. KOLODZIEJCZYK: Some remarks on the Borsuk conjecture.
            Annales Soc. Math. Polonae, Ser. I (Comment. Math.) 28 (1988),
            78-86.

[Ko 2]   K. KOLODZIEJCZYK: On starshapedness of the union of closed
            sets in $\mathbf{R}^n$. Colloq. Math. 53 (1987), 193-197.

[Ko 3]   K. KOLODZIEJCZYK: Borsuk covering and planar sets with uni-
            que completion. Discrete Math. 122 (1993), no. 1-3, 235-244.

[Ks]   M.A. KRASNOSEL'SKI: On a criterion of starshapedness (in Rus-
            sian). Mat. Sbornik 19 (1946), 309-310.

[K-K-M]   M.G. KREIN, M.A. KRASNOSEL'SKI, D.P. MILMAN: On de-
            ficiency numbers of linear operators in Banach spaces and on
            certain geometric questions (in Russian). Sb. Trudov Mat. Inst.
            Akad. Nauk Ukr. SSR 11 (1948), 97-112.

[Kr 1]   S. KROTOSZYNSKI: Covering a plane convex body with five smal-
            ler homothetical copies. Beitr. Alg. Geom. 25 (1987), 171-176.

[Kr 2]   S. KROTOSZYNSKI: Covering a disk with smaller disks. Studia
            Sci. Math. Hungar. 28 (1993), 277-283.

[L]   D.G. LARMAN: Problem 41. In: Contributions to Geometry (J. Tölke, J.M. Wills, Eds.), Birkhäuser, Basel, 1979.

[L-R-S]   D.G. LARMAN, C.A. ROGERS, J.J. SEIDEL: On two-distance sets in Euclidean space. Bull. London Math. Soc. 9 (1977), 261-267.

[L-T]   D.G. LARMAN, N.K. TAMVAKIS: The decomposition of the n-sphere and the boundaries of plane convex domains. Annals Discrete Math. 20 (1984), 209-214.

[La 1]   M. LASSAK: An estimate concerning Borsuk's partition problem. Bull. Acad. Polon. Sci. Ser. Math. 30 (1982), 449-451.

[La 2]   M. LASSAK: Partition of sets of three-dimensional Euclidean space into subsets of two times less diameters. Demonstratio Math. 17 (1984), 355-361.

[La 3]   M. LASSAK: Solution of Hadwiger's convering problem for centrally symmetric convex bodies in E^3. J. London Math. Soc. (2) 30 (1984), 501-511.

[La 4]   M. LASSAK: Covering three-dimensional convex bodies with smaller homothetical copies. Tagungsbericht 32/1984, Math. Forschungsinstitut Oberwolfach.

[La 5]   M. LASSAK: Covering a plane convex body by four homothetical copies with the smallest positive ratio. Geom. Dedicata 21 (1986), 157-161.

[La 6]   M. LASSAK: Covering plane convex bodies with smaller homothetical copies. Coll. Math. Soc. J. Bolyai 48 (Intuitive Geometry, Siofok 1985), North-Holland, 1987, 331-337.

[La 7]   M. LASSAK: Covering the boundary of a convex body by tiles. Proc. Amer. Math. Soc. 104 (1988), 269-272.

[La 8]   M. LASSAK: Some connections between B-convexity and d-convexity. Demonstratio Math. 15 (1982), 261-270.

[La 9]   M. LASSAK: Families of convex sets closed under intersections, homotheties and uniting increasing sequences of sets. Fundamenta Math. 120 (1984), 15-40.

[L-V]   M. LASSAK, E. VASARHELYI: Covering a plane convex body with negative homothetical copies. Studia Sci. Math. Hungar. 28 (1993), 375-378.

[Lei]   K. LEICHTWEISS: Konvexe Mengen. Deutscher Verlag der Wissenschaften, Berlin 1980.

[Le 1]   H. LENZ: Zur Zerlegung von Punktmengen in solche kleineren Durchmesser. Arch. Math. 6 (1955), 413-416.

[Le 2]   H. LENZ: Über die Bedeckung ebener Punktmengen durch solche kleineren Durchmesser. Arch. Math. 7 (1956), 34-40.

[Le 3]   H. LENZ: Zerlegung ebener Bereiche in konvexe Zellen möglichst kleinen Durchmessers. Jber. Deutsche Math.-Verein. 58 (1956), 87-97.

[Lev]   F.W. LEVI: Überdeckung eines Eibereiches durch Parallelverschiebungen seines offenen Kerns. Arch. Math. 6 (1955), 369-370.

[Li]   A. LIMA: Intersection properties of balls and subspaces in Banach spaces. Trans. Amer. Math. Soc. 227 (1977), 1-62.

[Lin]   J. LINDENSTRAUSS: Extension of compact operators. Mem. Amer. Math. Soc. 48 (1964).

[L-S]   L.A. LYUSTERNIK. L.G. SHNIREL'MAN: Topological Methods for Variational Problems. ONTI, Moscow 1930.

[M-M1]   E. MAKAI, H. MARTINI: A lower bound on the number of sharp shadow-boundaries of convex polytopes. Period. Math. Hungar. 20 (1989), 249-260.

[M-M 2]   E. MAKAI, H. MARTINI: A new characterization of convex plates of constant width. Geom. Dedicata 34 (1990), 199-209.

[M-M 3]   E. MAKAI, H. MARTINI: On the number of antipodal and strictly antipodal pairs of points in finite subsets of $\mathbf{R}^d$. In: DIMACS Series in Discrete Math. and Theor. Computer Sci. 4 (1991), Amer. Math. Soc. 1991 (The Victor Klee Festschrift, ed. by P. Gritzmann and B. Sturmfels), 457-470.

[M-M 4]   E. MAKAI, H. MARTINI: On the number of antipodal and strictly antipodal pairs of points in finite subsets of $\mathbf{R}^d$, part II. Period. Math.. Hungar. 27 (1993), 185-198.

[M-S]   E. MAKAI, V. SOLTAN: Lower bounds on the numbers of shadow-boundaries and illuminated regions of a convex body. In: Coll. Math. Soc. J. Bolyai 63 (Intuitive Geometry), North-Holland, Amsterdam et al., 1994, 249-268.

[Mak]   V.V. MAKEEV: Universal coverings and projections of bodies of constant width (in Russian). Ukrain. Geom. Sb. 32 (1989), 84-88.

[Mal]   J. MALKEVITCH (ED.): Geometry's Future. Conf. Proc., sponsored by COMAP, Inc., Lexington 1991.

[Man 1]   P. MANI: On polytopes fixed by their vertices. Acta Math. Acad. Sci. Hungar. 22 (1971), 269-273.

[Man 2]   P. MANI: Inner illumination of convex polytopes. Comment. Math. Helv. 49 (1974), 65-73.

[Man 3]   P. MANI: Characterizations of convex sets. In: Handbook of Convex Geometry, Eds. P.M. Gruber and J.M. Wills. North-Holland, 1993, 19-41.

[Mar 1]   H. MARTINI: Über scharfe Schattengrenzen und Schnitte konvexer Polytope. Beitr. Alg. Geom. 19 (1985), 105-112.

[Mar 2]   H. MARTINI: Some results and problems around zonotopes. Coll. Math. Soc. J. Bolyai 48 (Intuitive Geometry), North-Holland, Amsterdam et al., 1987, 383-418.

[Mar 3]   H. MARTINI: Convex polytopes whose projection bodies and difference sets are polars. Discrete Comput. Geom. 6 (1991), 83-91.

[Mar 4]   H. MARTINI: Cross-sectional measures. Coll. Math. Soc. J. Bolyai 63 (Intuitive Geometry), North-Holland, Amsterdam et al., 1994, 269-310.

[Mar 5]   H. MARTINI: Sharp shadow-boundaries of convex bodies. Discrete Math., to appear.

[M-S-S]   P. MCMULLEN, R. SCHNEIDER, G. C. SHEPHARD: Monotopic polytopes and their intersection properties. Geom. Dedicata 3 (1974), 99-129.

[Me 1]   K. MENGER: Untersuchungen über allgemeine Metrik, I, II, III. Math. Ann. 100 (1928), 75-163.

[Me 2]   K. MENGER: Metrische Untersuchungen. Ergebnisse eines mathem. Kolloquiums (Wien) 1 (1931), 20-27.

[Mes]   H. MESCHKOWSKI: Ungelöste und unlösbare Probleme der Geometrie. B.I. Wissenschaftsverlag, Mannheim-Wien-Zürich, 1975.

[Mi]   A.D. MILKA: An analogue of the Borsuk problem (in Russian). Izv. Vyssh. Uchebn. Zaved. Mat. 1992, no. 5, 58-63.

[Na]   L. NACHBIN: A theorem of the Hahn-Banach type for linear transformations. Trans. Amer. Math. Soc. 68 (1950), 28-46.

[Nat]   I.P. NATANSON: Theory of functions of a real variable (in Russian). Moscow, 1957.

[Ne]   B.H. NEUMANN: On some affine invariants of closed convex regions. J. London Math. Soc. 14 (1939), 262-272.

[Ng]   MANH HUNG NGUEN: Lower estimates on the number of exposed diameters of a convex polytope with a small number of vertices (in Russian). Manuscript, Kishinev State Univ., Kishinev 1990, 27 pp. (Deposited in Moldavian NIINTI, 22.01.90, No. 1154-M 90).

[N-S 1]   M.H. NGUEN, V.P. SOLTAN: Lower bounds for the numbers of extremal and exposed diameters of a convex body. Studia Sci. Math. Hungar. 28 (1993), 95-100.

[N-S 2]   M.H. NGUEN, V.P. SOLTAN: Lower bounds for the number of antipodal pairs and strictly antipodal pairs of vertices of a convex polytope. Discrete Comput. Geom. 11 (1994), 149-162.

[Ni]   A. NILLI: On Borsuk's problem. Preprint (1 page), 1993, to appear in: Proc. Conf. Combinatorics, Jerusalem 1993, Eds. H. Barcelo and G. Kalai, Contemporary Math. (AMS Series).

[P-A]   J. PACH, P.K. AGARWAL: Combinatorial Geometry. Wiley & Sons, to appear.

[Pa]   I. PÁL: Ein Minimumproblem für Ovale. Math. Ann 83 (1921), 311-319.

[Pe]   J. PERKAL: Sur la subdivision des ensembles en parties de diamètre intérieure. Colloq. Math. 1 (1947), 45.

[Ra]   J. RADON: Mengen konvexer Körper, die einen gemeinsamen Punkt enthalten. Math. Ann. 83 (1921), 113-115.

[Rea]   J.R. REAY: Open problems around Radon's theorem. In: Convexity and Related Combinatorial Geometry, Proc. 2nd Univ. Oklahoma Conf. (Norman, 1980), Lecture Notes Pure Appl. Math., Dekker, New York 1982, 151-172.

[Re]   F. REULEAUX: Lehrbuch der Kinematik I. Vieweg, Braunschweig 1875 (Engl. transl.: 1876; Reprint: Dover, New York 1963).

[Ri]   A.S. RIESLING: Borsuk's problem in three-dimensional spaces of constant curvature (in Russian). Ukr. Geom. Sbornik 11 (1971), 78-83.

[Ro 1]   C.A. ROGERS: Symmetrical sets of constant width and their partitions. Mathematika 18 (1971), 105-111.

[Ro 2]   C.A. ROGERS: Some problems in the geometry of convex bodies. In: The Geometric Vein (The Coxeter Festschrift), Springer, New York-Berlin 1981, 279-284.

[Ro 3]   C.A. ROGERS: Covering a sphere with spheres. Mathematika 10 (1963), 157-164.

[R-S]   C.A. ROGERS, G.C. SHEPHARD: The difference body of convex body. Arch. Math. 8 (1957), 220-233.

[Ros]   M. ROSENFELD: Inner illumination of convex polytopes. Elem. Math. 30 (1975), 27-28.

[Sak]   S. SAKS: Theory of the Integral. 2nd ed., Dover, New York 1964.

[Sal]   G.T. SALLEE: Preassigning the boundary of diametrically-complete sets. Monatshefte Mathem. 105 (1988), 217-227.

[Schm]   P. SCHMITT: Problems in discrete and combinatorial geometry.
           In: Handbook of Convex Geometry, ed. by P.M. Gruber and J.M.
           Wills, North-Holland, 1993, 449-483.

[Schn]   R. SCHNEIDER: Convex Bodies: The Brunn-Minkowski Theory.
           Cambridge Univ. Press, 1993.

[S-W]   R. SCHNEIDER, W. WEIL: Zonoids and related topics. In: Con-
           vexity and its Applications, Eds. P.M. Gruber and J.M. Wills,
           Birkhäuser, Basel 1983, 296-317.

[Schr]   O. SCHRAMM: Illuminating sets of constant width. Mathematika
           35 (1988), 180-189.

[Schü]   K. SCHÜTTE: Minimale Durchmesser endlicher Punktmengen
           mit vorgeschriebe- nem Mindestabstand. Math. Ann. 150 (1963),
           91-98.

[Sha]   R.W. SHANNON: A lower bound on the number of cells in arran-
           gements of hyperplanes. J. Combin. Theory, Ser. A, 20 (1976),
           327-335.

[She]   G.C. SHEPHARD: Sections and projections of convex polytopes.
           Mathematika 19 (1972), 144-162.

[Si]   G. SIERKSMA: Generalizations of Helly's theorem; open pro-
           blems. In: Convexity and Related Combinatorial Geometry,
           Proc. 2nd Univ. Oklahoma Conf. (Norman, 1980), Lecture Notes
           Pure Appl. Math., Dekker, New York 1982, 173-192.

[SP 1]   P.S. SOLTAN: Inner illumination of the boundary of a convex
           body (in Russian). Mat. Sbornik 57 (1962), no. 4, 443-448.

[SP 2]   P.S. SOLTAN: On problems of covering and illuminating convex
           bodies (in Russian). Izv. AN Mold. SSR (1963), no. 1, 49-57.

[SP 3]   P.S. SOLTAN: Connections between the problems of covering and
           illuminating convex bodies (in Russian). Izv. AN Mold. SSR, Ser.
           Fiz.-Tehn. i Mat. Nauk (1966), no. 4, 91-93.

[SP 4]   P.S. SOLTAN: Inner illumination of unbounded convex bodies (in
           Russian). Dokl. Akad. Nauk SSSR 194 (1970), no. 2, 273-274.

[SP 5]   P.S. SOLTAN: Covering convex bodies by larger homothetical
           copies (in Russian). Mat. Zametki 12 (1972), no. 1, 85-90.

[SP 6]   P.S. SOLTAN: Simultaneous solution of several Steiner problems
           for graphs (in Russian). Dokl. Akad. Nauk SSSR 202 (1972), no.
           2, 294-297.

[SP 7]   P.S. SOLTAN: Helly's theorem for d-convex sets (in Russian).
           Dokl. Akad. Nauk SSSR 205 (1972), no. 3, 537-539.

[SP 8]   P.S. SOLTAN: d-Convexity and its applications to Steiner's pro-
           blem on graphs (in Russian). Proc. 18th. Internat. Sci. Coll. TH
           Ilmenau, 1973, 33-35.

[SP 9]   P.S. SOLTAN: Extremal problems on convex sets (in Russian).
            Shtiinca, Kishinev 1976.

[S-P]   P.S. SOLTAN, K. PRISAKARU: The Steiner problem on graphs
            (in Russian). Dokl. Akad. Nauk SSSR 198 (1971), no. 1, 46-49.

[SP-SV 1]   P.S. SOLTAN, V. SOLTAN: A criterion for the normability in
            a class of linear metric spaces (in Russian). Shtiinca, Kishinev,
            Mat. Issled. 10 (1975), no. 2, 275-279.

[SP-SV 2]   P.S. SOLTAN, V. SOLTAN: d-convex functions (in Russian).
            Dokl. Akad. Nauk SSSR 249 (1979), no. 3, 555-558.

[SP-SV 3]   P.S. SOLTAN, V. SOLTAN: On the X-raying of convex bodies.
            Soviet Math. Dokl. 33 (1986), 42-44.

[SV 1]   V. SOLTAN: On a class of finite-dimensional normed spaces (in
            Russian). Shtiinca, Kishinev, Mat. Issled. 42 (1976), 204-215.

[SV 2]   V. SOLTAN: On Kuratowski's problem. Bull. Acad. Polon. Sci.,
            Ser. Sci. Math. 28 (1980), no. 7-8, 369-375.

[SV 3]   V. SOLTAN: On one extension of the notion of H-convexity (in
            Russian). Dokl. Akad. Nauk SSSR 258 (1981), no. 5, 1059-1062.

[SV 4]   V. SOLTAN: Introduction to the Axiomatic Theory of Convexity
            (in Russian). Shtiinca, Kishinev 1984.

[SV 5]   V. SOLTAN: On Grünbaum's problem on the inner illumination
            of convex bodies. Acta Math. Hungar. 69 (1995), 15-25.

[St]   O. STEFANI: Covering planar sets of constant width by three
            sets of smaller diameters. Geom. Dedicata 44 (1992), 1-5.

[S-Wi]   J. STOER, C. WITZGALL: Convexity and Optimization in Finite
            Dimensions I. Grundl. math. Wiss., Bd. 163, Springer, Berlin et
            al., 1970.

[Sz]   B. SZÖKEFALVI-NAGY: Ein Satz über Parallelverschiebungen
            konvexer Körper Acta Sci. Math. 15 (1954), no. 3-4, 169-177.

[T]   J.E. TAYLOR: Zonohedra and generalized zonohedra. Amer.
            Math. Monthly 99 (1992), 108-111.

[Tho]   A.C. THOMPSON: Minkowski Geometry. Monograph, Cam-
            bridge Univ. Press, 1996.

[T-W]   J. TÖLKE, J.M. WILLS (EDS.): Contributions to Geometry,
            Proc. Geom. Sympos. Siegen 1978, Birkhäuser, Basel 1979.

[V]   V.N. VIZITEI: Problems of covering and illumination for un-
            bounded sets (in Russian). Izv. AN Mold. SSR (1961), no. 10,
            3-9.

[We 1]   B. WEISSBACH: Eine Bemerkung zur Überdeckung beschränkter
            Mengen durch Mengen kleineren Durchmessers. Beitr. Alg.
            Geom. 25 (1987), 185-191.

[We 2]   B. WEISSBACH: Polyhedral covers. Coll. Math. Soc. J. Bolyai 48
          (Intuitive Geometry), North-Holland, Amsterdam et al., 1987,
          639-646.

[We 3]   B. WEISSBACH: Einiges über Körper fester Breite. Beitr. Alg.
          Geom. 31 (1990), 69-76.

[We 4]   B. WEISSBACH: Überdeckung von Körpern fester Breite durch
          gestauchte Bilder. Preprint Math. 14/92, TU Magdeburg, 12 pp.,
          1992.

[We 5]   B. WEISSBACH: Invariante Beleuchtung konvexer Körper. Beitr.
          Alg. Geom. 37 (1996), 9-15.

[Wi]   H.S. WITSENHAUSEN: A support characterization of zonotopes.
          Mathematika 25 (1978), 13-16.

[Y-B]   I.M. YAGLOM, V.G. BOLTYANSKI: Convex Figures. Holt, Rine-
          hart and Winston, New York 1961.

[Z-R]   V.A. ZALGALLER, YU.G. RESHETNYAK: On rectifiable curves,
          additive vector-functions and displacement of segments (in Rus-
          sian). Vestnik Leningrad. Univ. Ser. Mat. Fiz. Him. 9 (1954), no.
          2, 45-67.

[Zi]   G.M. ZIEGLER: Lectures on Polytopes. Springer-Verlag, New
          York, Inc, 1995.

[Ži]   R. ŽIVALEVIČ: md $\bar{H}$ = md H. Publ. Inst. Math. 26 (1979), 307-
          311.

[Zo]   M.A. ZORN: Remark on a method in transfinite algebra. Bull.
          Amer. Math. Soc. 41 (1935), 667-670.

Author Index

Agarwal, P. K.   VII, 391, 407
Aigner, M.   207, 393
Aleksandrov, A. D.   49, 393
Alon, N.   221, 224
Anderson, R. D.   217, 393
Armstrong, T. E.   393

Baladze, E.   319, 324, 333, 346, 347, 351, 383, 392
Banach, S.   41, 166, 269, 376
Bannai, E.   391, 393, 394
Barany, I.   389
Barbier, J. E.   225
Belousov, Yu. F.   272, 394
Beutelspacher, A.   170, 394
Bezdek, K.   VII, 170, 268–272, 280, 394
Bisztriczky, T.   VII, 271, 394
Blaschke, W.   VII, 394
Blokhuis, A.   391, 394, 395
Bobylev, N. A.   272, 395
Bohne, E.   315, 395
Bolker, E. D.   319, 333, 395
Bollobas, B.   359, 362
Boltyanski, V. G.   VII, 20, 41, 91, 100, 109, 149, 154, 163, 167, 173, 174, 177, 184, 189, 209, 210, 215, 217–220, 225, 245, 246, 255, 256, 258, 261, 269–272, 275, 302, 319, 324, 339, 341, 346, 352, 360–362, 370, 383, 391, 393, 395–397, 410
Borsuk, K.   VII, 209–212, 214, 217, 219, 220, 224, 239, 252, 271, 321, 378, 380–382, 389, 397
Bourgain, T.   216, 398
Bowers, P. L.   219, 398
Busemann, H.   388, 398

Cauchy, L.   366
Chabukiani, T. A.   149, 383, 396

Chakerian, G. D.   209, 218, 219, 227, 230, 380, 398
Charazishvili, A. B.   272, 280, 398
Croft, H. T.   VII, 99, 209, 212, 220, 269, 272, 389, 390, 391, 398

Danzer, L.   VII, 91, 93, 98, 209, 215, 216, 220, 359, 375, 388, 389, 398
Debrunner, H.   VII, 91, 402
Dekster, B. V.   217, 272, 282, 398, 399
Delsarte, P.   390, 391, 394, 399
Dembinski, M.   220, 399
Denjoy, M.   323, 324
Doignon, J.-P.   170, 399
Dol'nikov, V. L.   209, 219, 399

Eckhoff, J.   VII, 91, 93, 99, 399
Eggleston, H. G.   VII, 211, 220, 227, 230, 232, 233, 235, 399
Einhorn, S. T.   390, 399
Erdös, P.   VII, 272, 388, 389, 391, 400
Ewald, G.   388, 398

Falconer, K. T.   VII, 99, 209, 212, 220, 269, 272, 389, 390, 391, 398
Farkas, J.   36, 37
Federov, E. S.   333
Fejes Toth, L.   VII, 352, 359, 400
Fudali, S.   272, 359, 400
Füredi, Z.   272, 388, 389, 400

Gale, D.   211, 212, 238, 400
Gardner, R. J.   VII, 400
Gavurin, M. K.   32
German, L.   49, 311
Goethals, T. M.   390, 391, 399
Gohberg, I. Z.   VII, VIII, 209, 215, 255, 269, 270, 272, 275, 279, 385, 396, 400
Golomb, H.   390, 400

Goodey, P.   319, 400
Goodner, D. B.   166, 401
Graham, R. L.   VII, 389, 401
Gritzmann, P.   359, 401
Groemer, H.   209, 218, 219, 227, 230, 380, 398, 401
Grötschel, M.   389, 401
de Groot, J.   49, 401
Gruber, P.M.   VII, VIII, 401
Grünbaum, B.   VII, VIII, 91, 93, 98, 209, 211–215, 220, 224, 239, 252, 270, 302, 354, 359, 360, 375, 380, 384, 388, 389, 398, 401
Guy, R. K.   VII, 99, 209, 212, 220, 269, 272, 389, 390, 391, 398, 401

Hadwiger, H.   VII, 91, 217, 221, 256, 269, 271, 275, 279, 302, 385, 401, 402
Hamel, G.   366, 402
Hanner, O.   20, 375, 376, 402
Hansen, A.   20, 375, 376, 402
Harborth, H.   391, 402
Hausdorff, F.   20, 27, 68, 270, 334
Heil, E.   209, 218, 219, 230, 402
Helly, E.   VII, 91–93, 98, 99, 102, 109, 143, 146, 148, 149, 163, 169, 170, 198, 202, 204, 207, 311, 347, 350, 358, 375, 382, 402
Heppes, A.   211, 213, 402, 403

Jung, H. W. E.   214, 403

Kahn, J.   221, 403
Kalai, G.   221, 403
Kelley, T. L.   403
Kelly, L. M.   390, 403
Kincses, J.   VIII, 198–200, 202, 203, 207, 374, 375, 403
Kiss, Gy.   268, 272, 403
Klee, V. L.   VII, VIII, 91, 93, 98, 217, 375, 388, 393, 398, 402, 403
Kleinschmidt, P.   384, 385, 403
Knast, R.   214, 216, 403
Kolodziejczyk, D.   217, 403
Kolodziejczyk, K.   107, 219, 226, 403
Krasnosel'ski, M. A.   99, 101, 106, 161, 269, 403
Krein, M. G.   269, 284, 366, 400, 403
Krotoszynski, S.   272, 403

Lange, D.   VIII
Larman, D. G.   221, 389, 391, 404

Lassak, M.   49, 214, 215, 220, 268, 270–272, 399, 404
Lebesgue, H.   320, 324
Leichtweiss, K.   97, 106, 404
Lenz, H.   214, 216, 217, 404, 405
Levi, F. W.   256, 269, 271, 405
Levin, A. T.   270
Lima, A.   20, 375–377, 402, 405
Lindenstrauss, T.   216, 375, 376, 398, 405
Lipschitz, R.   324
Lovasz, L.   389, 401
Lyusternik, L. A.   210, 220, 405

Makai, E.   VIII, 380, 385–389, 405
Makeev, V. V.   219, 405
Malkevitch, J.   VII, 405
Mani, P.   49, 302, 359, 405, 406
Markus, A. S.   255, 269–271, 275, 279, 385, 400
Martini, H.   71, 100, 209, 218, 219, 230, 269, 270, 272, 319, 333, 339, 341, 342, 346, 352, 360, 362, 380, 384–389, 396, 397, 402, 406
McMullen, P.   VII, 359, 394, 406
Menger, K.   49, 406
Meschkowski, H.   212, 406
Milka, A. D.   217, 406
Milman, D. P.   269, 366, 403
Minkowski   VII, 49, 57, 65, 72, 73, 80, 81, 90, 91, 97, 99–101, 107, 108, 160, 219, 220, 230, 238, 242, 252, 253, 269, 346, 365–368, 371, 380, 388
Mollard, M.   268, 272, 403
Morales Amaya, E.   352, 361, 391, 397

Nachbin, L.   166, 406
Natanson, I. P.   323, 324, 406
Neumann, B. H.   406
Nguen, M. H.   388, 406, 407
Nilli, A.   221, 407

Pach, J.   VII, 407
Pachner, U.   384, 385, 403
Pál, I.   210, 227, 229, 407
Perkal, J.   211, 407
Petunin, J. I.   270
Piepmeyer, L.   391, 402
Post, K. A.   270, 275
Prikry, K.   393
Prisakaru, K.   49, 409
Purdy, G.   391, 400

Radon, J.   91, 98, 407
Reay, J. R.   VII, 407
Reshetnyak, Yu. G.   319, 321, 410
Rouleaux, F.   218, 225, 232, 238, 380, 407
Revesz, P.   211, 403
Riesling, A. S.   216, 407
Rogers, C. A.   214, 216, 270, 271, 391, 404, 407
Rosenfeld, M.   302, 407

Saks, S.   323, 324, 407
Sallee, G. T.   219, 407
Schmitt, P.   VII, 209, 269, 272, 408
Schoenberg, I. J.   390, 399
Schneider, R.   VII, 319, 359, 406, 408
Schnirel'man, L. G.   210, 220, 405
Schramm, O.   216, 271, 408
Schütte, K.   388, 408
Seidel, J. J.   390, 391, 394, 395, 399, 404
Selfridge, J. L.   401
Shannon, R. W.   71, 386, 387, 408
Shephard, G. C.   270, 359, 384, 388, 398, 406, 407, 408
Sierksma, G.   VII, 408
Soifer, A.   VII, 209, 397
Soltan, P. S.   VII, 3, 4, 49, 93, 100, 101, 109, 154, 255, 268–270, 272, 273, 287, 292, 302, 315, 320, 324, 339, 341, 365, 370, 397, 400, 408, 409
Soltan, V.   VII, 49, 93, 220, 235, 245, 246, 268, 270, 302, 303, 366–368, 384, 385, 388, 389, 397, 400, 405–409

Stanton, D.   391, 394
Stefani, O.   219, 409
Stepanoff, V. V.   324
Stoer, J.   36, 409
Szökefalvi-Nagy, B.   163, 164, 166, 167, 191, 199, 203, 319, 346, 347, 373, 375, 409

Tamvakis, N. K.   211, 404
Talata, I.   389
Taylor, J. E.   333, 409
Thompson, A. C.   VII, 409
Tölke, J.   VII, 409

Ulam, S.   209

Vaina, R.   220, 397
Vasarhelyi, É.   272, 404
Vizitei, V. N.   272, 409

Weil, W.   319, 400, 408
Weiss, A. I.   VII, 394
Weissbach, B.   VIII, 212, 219, 271, 272, 282, 409, 410
Wills, J. M.   VII, 401, 409
Witsenhausen, H. S.   333, 410
Witzgall, G.   36, 409

Yaglom, I. M.   210, 219, 225, 269, 410

Zalgaller, V. A.   49, 319, 321, 393, 410
Zermelo, E.   27
Ziegler, G. M.   VII, 410
Živalevič, P.   118, 410
Zorn, M. A.   27, 89, 233, 366, 410

Subject Index

affine hull   3
affinely regular pyramid   198
α-space   67
angled space   244
antipodal points   388
– sets 389
apex   27
approximate continuity   319, 323, 324
– limit 323
asymptotic hyperplane   3
– scope 4, 62, 300, 301

ball   2
beam   190
belt   327, 333
– body 270, 319, 333, 337
– polytope 270, 333, 334, 342
– vector 327, 328, 334
belt-tangential property   339
bicone   19
body of constant width   218, 219, 225, 227, 271, 378, 380
– d-width 237
– H-width 237
book   204, 374
Borsuk conjecture   211
– number 209, 224
– problem 211, 378, 380, 382
– theorem 209, 210
boundary   3
– point 3

carrying flat   3
Cauchy equation   366
cell   170
central illumination   257
chain-closed set family   27
chordable graph   367
closed set   2
closure   2

compatible point   13
complete convex set   219, 230, 233
completion process   218
conic hull   282
convex body   3
– cone 27
– hull 2
– polyhedral set 2
– polytope 2
– set 1
counterexample to Borsuk conjecture   221
covering set   212, 224, 380
cross-polytope   11, 19, 74, 76, 91, 371
cube   22, 52, 67, 120, 197, 275, 320, 326, 389
cuboctahedron   19
cycle   201
cylinder   6, 98

d-arc-starshaped   369
d-boundary point   367
d-convex body   52, 155, 160
– cone 62
– flat 65
– hull 53, 55
– hyperplane 65, 158, 367
– set 51, 157
– subspace 65
d-convexity   49, 50
d-diameter   236
d-extreme point   65, 366
d-hull   368
d-segment   51, 365, 366
d-segment joining   55
d-separability   88
d-star-shaped set   100, 368
d-visible point   100, 107, 368
decomposability   15, 16, 17, 26, 45, 188, 350, 351

density points of measurable sets   323, 330
Denjoy's integral   323
diameter   54, 76, 77, 209, 380
diametrical chord   217, 381
diametrically maximal set   65
difference body   270, 346, 388
direct sum   15, 67, 167, 199, 311, 346, 356
direct summand   16
distance set (e.g., 2-distance set)   389
dual space   12

ε-displacement   115, 352
ε-neighbourhood   1
equatorial body   71, 73, 134
− subspace 65, 338
exposed face   9, 33
− point 10
external-local d-conicity   101, 107, 108
extreme point   7, 66

face   6, 11, 65, 141, 371
facet   7, 36, 106, 285, 359
face-separating system   383
Farkas lemma   36, 37, 38
figure of constant width   219, 235, 236, 239, 242, 244, 245, 246, 252, 380
− H-width 238
fixing system   352, 391
free direction   100

Gohberg-Markus-Hadwiger hypothesis   269, 275, 279
− number 255, 269, 279
− problem 255, 269, 279, 385
Gohberg-Markus theorem   255, 269
graph technique by Kincses   200
Grünbaum problem   239, 384

half-bounded convex set   34
half-open interval   2
Hamel function   366
Hamiltonian graph   202
Hanner polytope   20, 376
H-complete set   237
H-convex body   125, 127, 135, 155
− cone 135, 139, 140
− face 141
− halfspace 125, 138
− hull 127, 134, 371
− set 109, 143, 146, 371, 372
− supporting hyperplane 138

H-diameter   236
Helly dimension   92, 93, 143, 148, 163, 167, 169, 199, 373
− theorem 91, 98, 143
− theorem for H-convex bodies 143
Helly-maximal body   202, 347, 349, 350, 358, 382, 383
homothetic covering   255, 269, 297, 339
− problem 255, 269, 279, 319
H-star-shaped set   161, 369
H-visible point   161

illuminated boundary point   256, 301
illumination   256, 257, 260, 269, 301, 339, 352, 361, 384, 385
− problem 256, 257, 272, 339
− by a flat 268, 269, 272, 275
inproper face   6
incident flats   14
indecomposability   25, 189, 199, 347, 350, 351, 359
inner illumination   301, 302, 303, 309, 384
inscribed cone   29, 62, 293, 297
integral representation of zonoids   319, 320
interior   2
− point 2
internal d-star-shapedness   368
internal-local d-conicity   100, 101, 107, 108
intersection property   375, 376, 377, 378
− of convex sets 1, 91, 92
− of d-convex sets 53, 61, 157
− of H-convex sets 127, 152
intersectionally free convex cones   142
interval   2

join of convex bodies   17, 189, 199, 200
− of normed spaces 74, 79, 80, 83, 96

Krasnosel'ski theorem   99, 101, 106, 107, 369
Krein-Milman theorem   366

larger homothetical copy   287, 301
lattice Helly dimension   169
− polytope 169
Levi number   269
library   204, 205, 374

limit point   2
Lipschitz constant   322
− function 322
lower limit   24
− semicontinuity of exp 20, 22, 24
− semicontinuity of md 121, 123

matroid   207
maximal face   6, 141
measurable function   320, 321, 324, 330
− set 323, 330
minimal face   6
− fixing system 352, 363, 391
minimally dependent vectors   107
− indecomposable zonotope 351
Minkowski plane   57, 58, 68, 100, 108, 160, 227, 239, 365, 368
− space 49, 57, 71, 73, 101, 108, 158, 160, 227, 239, 365, 366, 367 368, 370, 388
− sum 5

near pencil   71, 387
nearly conic convex body   35, 292, 294, 297, 299, 301
nonfree direction   100
normable metric   365
normalization of a vector system   121

one-sided vector system   112, 113, 118, 121, 354, 356
open interval   2
open set   2, 262
orthogonally closed set   372, 373
outcut   190, 191, 197, 198, 351
outer moving direction   352
outward normal   10

parallelogram   5, 123, 198, 210, 242, 244, 252, 256, 278, 279
parallelotope   2, 11, 26, 76, 91, 128, 149, 164, 190, 191, 196, 214, 267, 269, 270, 279, 341, 372
particular vector of a 2-system   173
planet   333
pointed convex cone   29, 31, 315
polar body   13, 14, 22, 121, 189, 207, 333, 337, 338, 371
− cone 29, 30, 36, 37, 43
− face 15
− flat 14
polarity   12, 97

polyhedral cone   36, 315
− region 106, 161
positive basis   120
positively dependent vectors   110, 111, 112
primitive fixing system   352, 359, 391
− illuminating system 268, 384
− polytope 359
prism   19, 26, 114, 120, 148
process of segment joining   53, 54, 55, 91
proper face   6

rectifiable curve   319, 321, 323
regular boundary point   10, 121, 328, 336, 337
− dodecahedron 391
− fixing system 353
− hexagon 58, 210, 229, 230
− icosahedron 202, 390, 391
− octagon 65, 95, 101
− octahedron 67, 211, 212, 232, 326, 375, 380
− tetrahedron 120, 219, 378
− supporting hyperplane 10, 12, 14
relative boundary   3
− interior 3
Reuleaux polygon   218, 225, 380, 392
− triangle 218, 219
reverse Borsuk problem   378
rhombic dodecahedron   319, 320, 359
rotational cone   40

selfadjoint space   12
semicontinuity   20, 22, 24, 121, 123, 127, 130, 371
separability of convex cones   41, 43, 45, 47, 140, 376
− of sets 4, 5, 81, 145
− of d-convex sets 81, 83, 89, 90
separable cones   41, 43, 45, 47, 140, 376
− points 376, 377
separating hyperplane   4
shadow-boundary   384, 385, 388
simple arc   326
simplex   2, 9, 11, 26, 91, 109, 110, 114, 153, 311, 315, 332, 372
simplicial vector system   119
skeleton   71
smaller homothetic copy   289
solid cone   31, 37, 46
splitability   17, 18, 24, 26, 173, 175, 184, 349

stable property 372
stack 189, 190, 191, 197, 198, 351, 392
stack-outcut 191, 192, 392
star-shaped set 99, 100, 101, 106, 107,
    108, 368
Stepanoff-Denjoy theorem 324
strict convexity 7, 10, 50, 66, 83
– separability 5
strictly antipodal points 388
– sets 388, 389
strictly convex body 7
– normed space 50
strictly one-sided vector system 112,
    113, 114
strong separability 5, 98
support 27
supporting cone 28, 60, 135, 373
– halfspace 9
– hyperplane 9, 13, 14, 61, 71, 138
suspension 74, 96, 312
Szökefalvi-Nagy problem 163, 164,
    166, 167, 191, 199, 203, 319, 346,
    347, 373, 375, 392
– theorem 164, 166, 167

tame face 15
tangential zonotope 319, 324, 345
– convergent sequence 345
– ray 327
– top point 346
translative Borsuk problem 382

translatively movable zonoid 338, 339
two-sided cone 315

upper limit 131
– semicontinuity of $c(M)$ 267
– semicontinuity of the H-convex hull
    127, 129, 130, 131
unit ball 50, 66, 70, 74, 76, 237, 242
universal cover 212, 224, 380

vertex 7
vector function 321, 323, 324
vector-parametric representation 319,
    321, 324, 330
vector sum 5
– 2-system 173, 175, 177, 184
visible point 99

width 152, 218, 230

X-raying problem 286, 287, 302
zonogon 319
zonoid 269, 270, 320, 321, 324, 330,
    333, 336, 337, 338, 339, 345, 346,
    383, 384
zonotope 170, 270, 319, 320, 321, 324,
    326, 327, 333, 337, 338, 339, 341,
    345, 346, 359, 362, 383
– of order q 383
Zorn lemma 27, 89, 233, 366

List of Symbols

$U_\varepsilon(M)$ 1
$d(x, M)$ 1
conv F 2
int M 2
cl M 2
bd M 3
aff M 3
ri M 3
rbd M 3
ext M 7
exp M 10
sup cone$_a M$ 28
$H(K)$ 36
lin 40, 110
conv$_d M$ 53
him F 92
him $\Re^n$ 93
him$^{cl}\Re^n$ 93
him$^b\Re^n$ 93
nof$_a M$ 100
md H 112
$H(M)$ 121
md M 121
norm Q 121
conv$_H M$ 127
cov M 149
md $H(\Re^n)$ 158

$\mathfrak{V}_H$ 158
$\mathfrak{V}_d$ 158
md $\Re^n$ 158
him M 163
$a(F)$ 209
$a_n(K)$ 220
$a_\Sigma(F)$ 239
diam$_H F$ 236
diam$_d F$ 236
$b(K)$ 255
$b'(K)$ 256
$c(K)$ 256
$c'(K)$ 257
con (K, x) 282
$p(K)$ 301
$p'(K)$ 302
$q(K)$ 301
$q'(K)$ 302
bim $\mathbb{F}$ 310
limap$_{x \to x_o} f(x)$ 323
$\rho(M)$ 352
nos M 354
stext$_d M$ 366
$\mathfrak{M}_l$ 370
$\mathfrak{M}_H$ 370
$\mathfrak{M}_d$ 370

Universitext

Aksoy, A., Khamsi, M. A.: Methods in Fixed Point Theory

Aupetit, B.: A Primer on Spectral Theory

Bachem, A., Kern, W.: Linear Programming Duality

Benedetti, R., Petronio, C.: Lectures on Hyperbolic Geometry

Berger, M.: Geometry I

Berger, M.: Geometry II

Bliedtner, J., Hansen, W.: Potential Theory

Booss, B., Bleecker, D. D.: Topology and Analysis

Carleson, L., Gamelin, T.: Complex Dynamics

Carmo, M. P. do: Differential Forms and Applications

Cecil, T. E.: Lie Sphere Geometry: With Applications of Submanifolds

Chandrasekharan, K.: Classical Fourier Transforms

Charlap, L. S.: Bieberbach Groups and Flat Manifolds

Chern, S.: Complex Manifolds without Potential Theory

Chorin, A. J., Marsden, J. E.: Mathematical Introduction to Fluid Mechanics

Cohn, H.: A Classical Invitation to Algebraic Numbers and Class Fields

Curtis, M. L.: Abstract Linear Algebra

Curtis, M. L.: Matrix Groups

Dalen, D. van: Logic and Structure

Das, A.: The Special Theory of Relativity: A Mathematical Exposition

Devlin, K. J.: Fundamentals of Contemporary Set Theory

DiBenedetto, E.: Degenerate Parabolic Equations

Dimca, A.: Singularities and Topology of Hypersurfaces

Edwards, R. E.: A Formal Background to Higher Mathematics I a, and I b

Edwards, R. E.: A Formal Background to Higher Mathematics II a, and II b

Emery, M.: Stochastic Calculus in Manifolds

Foulds, L. R.: Graph Theory Applications

Frauenthal, J. C.: Mathematical Modeling in Epidemiology

Fuks, D. B., Rokhlin, V. A.: Beginner's Course in Topology

Gallot, S., Hulin, D., Lafontaine, J.: Riemannian Geometry

Gardiner, C. F.: A First Course in Group Theory

Gårding, L., Tambour, T.: Algebra for Computer Science

Godbillon, C.: Dynamical Systems on Surfaces

Goldblatt, R.: Orthogonality and Spacetime Geometry

Gouvêa, F. Q.: p-Adic Numbers

Hahn, A. J.: Quadratic Algebras, Clifford Algebras, and Arithmetic Witt Groups

Hájek, P., Havránek, T.: Mechanizing Hypothesis Formation

Hlawka, E., Schoißengeier, J., Taschner, R.: Geometric and Analytic Number Theory

Holmgren, R. A.: A First Course in Discrete Dynamical Systems

Howe, R., Tan, E. Ch.: Non-Abelian Harmonic Analysis

Humi, M., Miller, W.: Second Course in Ordinary Differential Equations
  for Scientists and Engineers

Hurwitz, A., Kritikos, N.: Lectures on Number Theory

Iversen, B.: Cohomology of Sheaves

Jost, J.: Riemannian Geometry and Geometric Analysis

Kelly, P., Matthews, G.: The Non-Euclidean Hyperbolic Plane

Universitext

Kempf, G.: Complex Abelian Varieties and Theta Functions

Kloeden, P. E., Platen E., Schurz, H.: Numerical Solution of SDE
  Through Computer Experiments

Kostrikin, A. I.: Introduction to Algebra

Krasnoselskii, M. A., Pokrovskii, A. V.: Systems with Hysteresis

Luecking, D. H., Rubel, L. A.: Complex Analysis. A Functional Analysis Approach

Ma, Zhi-Ming, Roeckner, M.: Dirichlet Forms

Mac Lane, S., Moerdijk, I.: Sheaves in Geometry and Logic

Marcus, D. A.: Number Fields

McCarthy, P. J.: Introduction to Arithmetical Functions

Meyer, R. M.: Essential Mathematics for Applied Fields

Meyer-Nieberg, P.: Banach Lattices

Mines, R., Richman, F., Ruitenberg, W.: A Course in Constructive Algebra

Moise, E. E.: Introductory Problem Courses in Analysis and Topology

Montesinos, J. M.: Classical Tessellations and Three-Manifolds

Nikulin, V. V., Shafarevich, I. R.: Geometries and Groups

Oden, J. J., Reddy, J. N.: Variational Methods in Theoretical Mechanics

Øksendal, B.: Stochastic Differential Equations

Porter, J. R., Woods, R. G.: Extensions and Absolutes of Hausdorff Spaces

Rees, E. G.: Notes on Geometry

Reisel, R. B.: Elementary Theory of Metric Spaces

Rey, W. J. J.: Introduction to Robust and Quasi-Robust Statistical Methods

Rickart, C. E.: Natural Function Algebras

Rotman, J. J.: Galois Theory

Rybakowski, K. P.: The Homotopy Index and Partial Differential Equations

Sagan, H.: Space Filling Curves

Samelson, H.: Notes on Lie Algebras

Schiff, J. L.: Normal Families of Analytic and Mesomorphic Functions

Shapiro, J. H.: Composition Operators and Classical Function Theory

Smith, K. T.: Power Series from a Computational Point of View

Smoryński, C.: Logical Number Theory I: An Introduction

Smoryński, C.: Self-Reference and Modal Logic

Stanišić, M. M.: The Mathematical Theory of Turbulence

Stillwell, J.: Geometry of Surfaces

Stroock, D. W.: An Introduction to the Theory of Large Deviations

Sunada, T.: The Fundamental Group and the Laplacian (to appear)

Sunder, V. S.: An Invitation to von Neumann Algebras

Tamme, G.: Introduction to Étale Cohomology

Tondeur, P.: Foliations on Riemannian Manifolds

Verhulst, F.: Nonlinear Differential Equations and Dynamical Systems

Zaanen, A. C.: Continuity, Integration and Fourier Theory

Springer
and the
environment

Springer

Printing: Mercedesdruck, Berlin
Binding: Buchbinderei Lüderitz & Bauer, Berlin